NOBEL LECTURES IN PHYSICS

1971–1980

NOBEL LECTURES

INCLUDING PRESENTATION SPEECHES
AND LAUREATES' BIOGRAPHIES

PHYSICS

CHEMISTRY

PHYSIOLOGY OR MEDICINE

LITERATURE

PEACE

ECONOMIC SCIENCES

NOBEL LECTURES

INCLUDING PRESENTATION SPEECHES
AND LAUREATES' BIOGRAPHIES

PHYSICS

1971–1980

EDITOR

STIG LUNDQVIST

Published for the Nobel Foundation in 1992 by

World Scientific Publishing Co. Pte. Ltd.
P O Box 128, Farrer Road, Singapore 9128
USA office: Suite 1B, 1060 Main Street, River Edge, NJ 07661
UK office: 73 Lynton Mead, Totteridge, London N20 8DH

NOBEL LECTURES IN PHYSICS (1971–1980)

ISBN 981-02-0726-3
ISBN 981-02-0727-1 (pbk)

Printed in Singapore.

FOREWORD

Since 1901 the Nobel Foundation has published annually "Les Prix Nobel" with reports from the Nobel Award Ceremonies in Stockholm and Oslo as well as the biographies and Nobel lectures of the laureates. In order to make the lectures available to people with special interests in the different prize fields the Foundation gave Elsevier Publishing Company the right to publish in English the lectures for 1901–1970, which were published in 1964–1972 through the following volumes:

Physics 1901–1970	4 volumes
Chemistry 1901–1970	4 volumes
Physiology or Medicine 1901–1970	4 volumes
Literature 1901–1967	1 volume
Peace 1910–1970	3 volumes

Elsevier decided later not to continue the Nobel project. It is therefore with great satisfaction that the Nobel Foundation has given World Scientific Publishing Company the right to bring the series up to date.

The Nobel Foundation is very pleased that the intellectual and spiritual message to the world laid down in the laureates' lectures will, thanks to the efforts of World Scientific, reach new readers all over the world.

Lars Gyllensten
Chairman of the Board

Stig Ramel
Executive Director

Stockholm, June 1991

PREFACE

The Nobel Foundation publishes annually the proceedings of the year's Prize ceremonies in a volume called *Les Prix Nobel.*

It contains the speeches given at the prize ceremony, the autobiography of each laureate as well as the Nobel Lecture. These books contain much material of great interest to the scientific community. However, they are printed only in a small number of copies, very few scientists even know that they exist and there is no advertisement.

From 1992 the material is now becoming available through a deal between the Nobel Foundation and World Scientific to publish the material from the 70's and the 80's in a series of volumes. This volume contains the materials in physics for the period 1970–1979.

The contents in this volume are not identical with the original in *Les Prix Nobel.* We have written to all the physics laureates and let them modify and update the material.

The reader may be surprised about the very short speeches of presentation. The reason is that only a few minutes are allotted to these speeches which did not permit any description of the discovery.

Stig Lundqvist

CONTENTS

Foreword v

Preface vii

1971 DENIS GABOR 1
Presentation by Erik Ingelstam 3

Biography of Denis Gabor 7
Holography, 1948–1971 11

1972 JOHN BARDEEN, LEON N COOPER and J ROBERT SCHRIEFFER 45
Presentation by Stig Lundqvist 47

Biography of John Bardeen 51
Electron-Phonon Interactions and Superconductivity 54

Biography of Leon N Cooper 71
Microscopic Quantum Interference Effects in the Theory of Superconductivity 73

Biography of J Robert Schrieffer 95
Macroscopic Quantum Phenomena from Pairing in Superconductors 97

1973 LEO ESAKI, IVAR GIAEVER and BRIAN D JOSEPHSON 109
Presentation by Stig Lundqvist 111

Biography of Leo Esaki 115
Long Journey into Tunneling 116

Biography of Ivar Giaever 135
Electron Tunneling and Superconductivity 137

Biography of Brian D Josephson 155
The Discovery of Tunnelling Supercurrents 157

1974 MARTIN RYLE and ANTONY HEWISH 165
Presentation by Hans Wilhelmsson 167

Biography of Antony Hewish 171
Pulsars and High Density Physics 174

Biography of Martin Ryle 185
Radio Telescopes of Large Resolving Power 187

1975 AAGE BOHR, BEN R MOTTELSON and JAMES RAINWATER 205
Presentation by Sven Johansson 207

Biography of Aage Bohr 211
Rotational Motion in Nuclei 213

Biography of Ben R Mottelson 235
Elementary Modes of Excitation in the Nucleus 236

Biography of James Rainwater 257
Background for the Spheroidal Nuclear Model Proposal 259

1976 BURTON RICHTER and SAMUEL C C TING 271
Presentation by Gösta Ekspong 273

Biography of Burton Richter 277
From the Psi to Charm —The Experiments of 1975 and 1976 281

Biography of Samuel Chao Chung Ting 313
The Discovery of the J Particle: A Personal Collection 316

1977 PHILIP W ANDERSON, NEVILL F MOTT and JOHN H VAN VLECK 345
Presentation by Per-Olov Löwdin 347

Biography of John Hasbrouck Van Vleck 351
Quantum Mechanics : The Key to Understanding Magnetism 353

Biography of Philip W Anderson 371
Local Moments and Localized States 376

Biography of Nevill Francis Mott 401
Electrons in Glass 403

1978 PETER LEONIDOVITCH KAPITZA, ARNO A PENZIAS and ROBERT W WILSON 415
Presentation by Lamek Hulthén 417

	Biography of Peter Leonidovitch Kapitza	421
	Plasma and the Controlled Thermonuclear Reaction	424
	Biography of Arno A Penzias	439
	The Origin of Elements	444
	Biography of Robert W Wilson	459
	The Cosmic Microwave Background Radiation	463
1979	SHELDON L GLASHOW, ABDUS SALAM and STEVEN WEINBERG	485
	Presentation by Bengt Nagel	487
	Biography of Sheldon Lee Glashow	491
	Towards a Unified Theory — Threads in a Tapestry	494
	Biography of Abdus Salam	507
	Gauge Unification of Fundamental Forces	513
	Biography of Steven Weinberg	541
	Conceptual Foundations of the Unified Theory of Weak and Electromagnetic Interactions	543
1980	JAMES W CRONIN and VAL L FITCH	561
	Presentation by Gösta Ekspong	563
	Biography of James W Cronin	567
	CP Symmetry Violation — The Search for Its Origin	570
	Biography of Val Logsdon Fitch	591
	The Discovery of Charge-Conjugation Parity Symmetry	594

Physics 1971

DENNIS GABOR

for his invention and development of the holographic method

THE NOBEL PRIZE FOR PHYSICS

Speech by professor Erik Ingelstam of the Royal Academy of Sciences
Translation

Your Majesty, Your Royal Highnesses, Ladies and Gentlemen,

Our five senses give us knowledge of our surroundings, and nature herself has many available resources. The most obvious is light which gives us the possibility to see and to be pleased by colour and shape. Sound conveys the speech with which we communicate with each other and it also allows us to experience the tone-world of music.

Light and sound are wave motions which give us information not only on the sources from which they originate, but also on the bodies through which they pass, and against which they are reflected or deflected. But light and sound are only two examples of waves which carry information, and they cover only very small parts of the electromagnetic and acoustic spectra to which our eyes and ears are sensitive.

Physicists and technologists are working continuously to improve and broaden the methods and instruments which give us knowledge about waves which lie outside our direct perception capacity. The electron microscope resolves structures which are a thousand times smaller than the wavelength of visible light. The photographic plate preserves for us a picture of a fleeting moment, which perhaps we may make use of over a long time period for measurements, or it transforms a wave-field of heat rays, X rays, or electron rays to a visible image.

And yet, important information about the object is missing in a photographic image. This is a problem which has been a key one for Dennis Gabor during his work on information theory. Because the image reproduces only the effect of the intensity of the incident wave-field, not its nature. The other characteristic quantity of the waves, phase, is lost and thereby the three dimensional geometry. The phase depends upon from which direction the wave is coming and how far it has travelled from the object to be imaged.

Gabor found the solution to the problem of how one can retain a wave-field with its phase on a photographic plate. A part of the wave-field, upon which the object has not had an effect, namely a reference wave, is allowed to fall on the plate together with the wave-field from the object. These two fields are superimposed upon one another, they interfere, and give the strongest illumination where they have the same phase, the weakest where they extinguish each other by having the opposite phase. Gabor called this plate a hologram, from the Greek *holos,* which means whole or complete, since the plate contains the whole information. This information is stored in the plate in a coded form. When the hologram is irradiated only with the reference wave, this wave is deflected in the hologram structure, and the original object's field is reconstructed. The result is a three dimensional image.

Gabor originally thought of using the principle to make an electron microscope image in two steps: first to register an object's field with electron rays in a hologram, and then to reconstruct this with visible light to make a three dimensional image with high resolution. But suitable electron sources for this were not available, and also for other technical reasons the idea could not be tested. However, through successful experiments with light Gabor could show that the principle was correct. In three papers from 1948 to 1951 he attained an exact analysis of the method, and his equations, even today, contain all the necessary information.

Holography, as this area of science is called, made its break-through when the tool, which had so far been missing, became available, namely the laser as a light source. The first laser was successfully constructed in 1960, and the basic ideas were rewarded by the 1964 Nobel Prize in physics. The laser generates continuous, coherent wave-trains of such lengths that one can reconstruct the depth in the holographic image. At about the same time a solution was discovered to the problem of getting rid of disturbing double images from the field of view. A research group at Michigan University in the United States, led by Emmett Leith, initiated this development.

The fascinated observer's admiration when he experiences the three dimensional space effect in a holographic image is, however, an unsufficient acknowledgement for the inventor. More important are the scientific and technical uses to which his idea has led. The position of each object's point in space is determined to a fraction of a light wave-length, a few tenthousandths of a millimetre, thanks to the phase in the wave-field. With this, the hologram has, in an unexpected way, enriched optical measurement techniques, and particularly interferometric measurements have been made possible on many objects. The shape of the object at different times can be stored in one and the same hologram, through illumination of it several times. When they are reconstructed simultaneously, the different wave-fields interfere with each other, and the image of the object is covered with interference lines, which directly, in light wavelengths, correspond to changes of shape between the exposures. These changes can also be, for example, vibrations in a membrane or a musical instrument.

Also, very rapid sequences of events, even in plasma physics, are amenable to analysis through hologram exposures at certain times with short light flashes from modern impulse lasers.

Gabor's original thought to use different waves for both steps within holography, has been taken up in many connections. It is especially attractive to use ultra sound waves for exposures, so that, in the second step, a sound field is reconstructed in the shape of an optical image. Despite significant difficulties there is work, with a certain amount of progress, being done in this area. Such a method should be of value for medical diagnosis, since the deflected sound field gives different information from that in X ray radiography.

Professor Gabor,

You have the honour and pleasure to have founded the basic ideas of the holographic method. Through your work and assiduous contributions of ideas you continue to add to the development of this field, and this applies especially now that you have the freedom of a professor emeritus. Your activity as a writer on culture shows that you belong to the group of physicists and technologists who are concerned about the use or damage to which technical development can lead for mankind.

The Royal Swedish Academy of Sciences wishes to give you hearty congratulations, and I now ask you to receive the Nobel Prize in physics from the hand of His Majesty the King.

Dennis Gabor

DENNIS GABOR

I was born in Budapest, Hungary, on June 5, 1900, the oldest son of Bertalan Gabor, director of a mining company, and his wife Adrienne. My life-long love of physics started suddenly at the age of 15. I could not wait until I got to the university, I learned the calculus and worked through the textbook of Chwolson, the largest at that time, in the next two years. I remember how fascinated I was by Abbe's theory of the microscope and by Gabriel Lippmann's method of colour photography, which played such a great part in my work, 30 years later. Also, with my late brother George, we built up a little laboratory in our home, where we could repeat most experiments which were modern at that time, such as wireless X-rays and radioactivity. Yet, when I reached university age, I opted for engineering instead of physics. Physics was not yet a profession in Hungary, with a total of half-a-dozen university chairs—and who could have been presumptious enough to aspire to one of these?

So I acquired my degrees, (Diploma at the Technische Hochschule Berlin, 1924, Dr-Ing. in 1927), in electrical engineering, though I sneaked over from the TH as often as possible to the University of Berlin, were physics at that time was at its apogee, with Einstein, Planck, Nernst and v. Laue. Though electrical engineering remained my profession, my work was almost always in applied physics. My doctorate work was the development of one of the first high speed cathode ray oscillographs and in the course of this I made the first iron-shrouded magnetic electron lens. In 1927 I joined the Siemens & Halske AG where I made my first of my successful inventions; the high pressure quartz mercury lamp with superheated vapour and the molybdenum tape seal, since used in millions of streeet lamps. This was also my first exercise in serendipity, (the art of looking for something and finding something else), because I was not after a mercury lamp but after a cadmium lamp, and that was not a success.

In 1933, when Hitler came to power, I left Germany and after a short period in Hungary went to England. At that time, in 1934, England was still in the depths of the depression, and jobs for foreigners were very difficult. I obtained employment with the British Thomson-Houston Co., Rugby, on an inventor's agreement. The invention was a gas discharge tube with a positive characteristic, which could be operated on the mains. Unfortunately, most of its light emission was in the short ultraviolet, so that it failed to give good efficiency with the available fluorescent powders, but at least it gave me a foothold in the BTH Research Laboratory, where I remained until the end of 1948. The years after the war were the most fruitful. I wrote, among many others, my first papers on communication theory, I developed a system of stereoscopic cinematography, and in the last year, 1948 I carried out the basic experiments in holography, at that time called "wavefront reconstruction". This again was an exercise in serendipity. The original objective was an improved electron microscope, capable of resolving atomic lattices and seeing single atoms. Three year's work, 1950—53, carried out in collaboration with the AEI Research Laboratory in Aldermaston, led to some respectable results, but still far from the goal. We had started 20 years too early. Only in recent years have certain auxiliary techniques developed to the point when electron holography could become a success. On the other hand, optical holography has become a world success after the invention and introduction of the laser, and acoustical holography has now also made a promising start.

On January 1, 1949 I joined the Imperial College of Science & Technology in London, first as a Reader in Electronics, later as Professor of Applied Electron Physics, until my retirement in 1967. This was a happy time. With my young doctorands as collaborators I attacked many problems, almost always difficult ones. The first was the elucidation of Langmuirs Paradox, the inexplicably intense apparent electron interaction in low pressure mercury arcs. The explanation was that the electrons exchanged energy not with one another, by collisions, but by interaction with an oscillating boundary layer at the wall of the discharge vessel. We made also a Wilson cloud chamber, in which the velocity of particles became measurable by impressing on them a high frequency, critical field, which produced time marks on the paths, at the points of maximum ionisation. Other developments were: a holographic microscope, a new electron-velocity spectroscope an analogue computer which was a universal, non-linear "learning" predictor, recognizer and simulator of time series, a flat thin colour television tube, and a new type of thermionic converter. Theoretical work included communication theory, plasma theory, magnetron theory and I spent several years on a scheme of fusion, in which a critical high-temperature plasma would have been established by a 1000 ampere space charge-compensated ion beam, fast enough to run over the many unstable modes which arise during its formation. Fortunately the theory showed that at least one unstable mode always remained, so that no money had to be spent on its development.

After my retirement in 1967 I remained connected with the Imperial College as a Senior Research Fellow and I became Staff Scientist of CBS Laboratories, Stamford, Conn. where I have collaborated with the President, my life-long friend, Dr. Peter C. Goldmark in many new schemes of communication and display. This kept me happily occupied as an inventor, but meanwhile, ever since 1958, I have spent much time on a new interest; the future of our industrial civilisation. I became more and more convinced that a serious mismatch has developed between technology and our social institutions, and that inventive minds ought to consider social inventions as their first priority. This conviction has found expression in three books, *Inventing the Future,* 1963, *Innovations,* 1970, and *The Mature Society,* 1972. Though I still have much unfinished technological work on my hands, I consider this as my first priority in my remaining years.

Honours

Fellow of the Royal Society, 1956.

Hon. Member of the Hungarian Academy of Sciences, 1964.

D.Sc. Univ. of London, 1964, Hon. D.Sc. Univ. of Southampton, 1970, and Technological University Delft, 1971.

Thomas Young Medal of Physical Society London, 1967.

Cristoforo Colombo Prize of Int. Inst. Communications, Genoa, 1967.

Albert Michelson Medal of The Franklin Institute, Philadelphia, 1968.

Rumford Medal of the Royal Society, 1968.

Medal of Honor of the Institution of Electrical and Electronic Engineers, 1970.

Prix Holweck of the French Physical Society, 1971.

Commander of the Order of the British Empire, 1970.

Married since 1936 to Marjorie Louise, daughter of Joseph Kennard Butler and Louise Butler of Rugby.

The following details about his last years were obtained from Ms Anne Barrett, College Archivist at the Imperial College:

Professor Denis Gabor was awarded the Nobel Prize for Physics in 1971 and gave his Nobel Lecture on Holography. In the years following his Nobel award he received honours from universities and institutions internationally. He travelled widely giving lectures, many on holography or the subject of his book *The Mature Society*.

Between 1973 and 1976 Gabor and Umberto Columbo jointly chaired a working party on the possible contribution of Science to the regeneration of natural resources. The results were published in 1978 as *Beyond the Age of Waste*.

In 1974 Gabor suffered a cerebral haemorrhage so he was unable to personally present lectures he had prepared for that year, but his large group of eminent friends rallied to present them for him. Gabor lost the power to read and write himself and his speech deteriorated but his intellect and hearing were acute, so he remained involved in the scientific world. He was able to visit the new Museum of Holography in New York and the Royal Academy holography exhibition in 1977. He became Honorary Chairman of the Board of Trustees of the New York Museum of Holography in 1978 and also sat for his holographic portrait by Hart Perry.

His health deteriorated during the latter part of 1978 and he died in 1979.

HOLOGRAPHY, 1948—1971

Nobel Lecture, December 11, 1971

by

DENNIS GABOR

Imperial Colleges of Science and Technology, London

I have the advantage in this lecture, over many of my predecessors, that I need not write down a single equation or show an abstract graph. One can of course introduce almost any amount of mathematics into holography, but the essentials can be explained and understood from physical arguments.

Holography is based on the wave nature of light, and this was demonstrated convincingly for the first time in 1801 by Thomas Young, by a wonderfully simple experiment. He let a ray of sunlight into a dark room, placed a dark screen in front of it, pierced with two small pinholes, and beyond this, at some distance, a white screen. He then saw two darkish lines at both sides of a bright line, which gave him sufficient encouragement to repeat the experiment, this time with a spirit flame as light source, with a little salt in it, to produce the bright yellow sodium light. This time he saw a number of dark lines, regularly spaced; the first clear proof that light added to light can produce darkness. This phenomenon is called interference. Thomas Young had expected it because he believed in the wave theory of light. His great contribution to Christian Huygens's original idea was the intuition that mono-

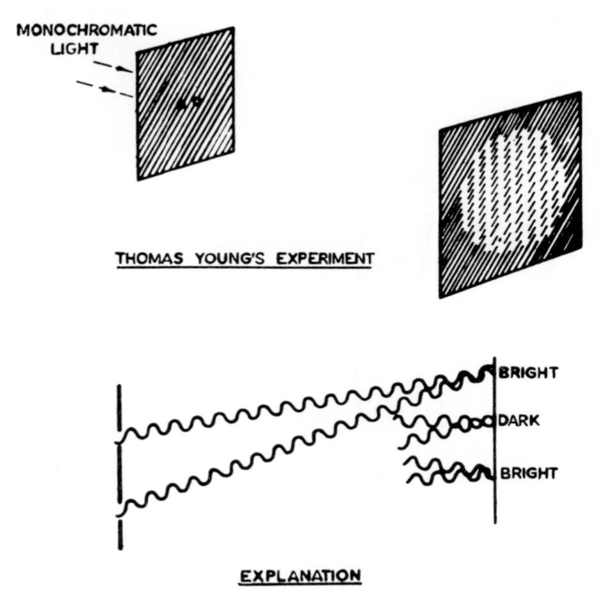

Fig. 1.
Thomas Young's Interference Experiments, 1801.

chromatic light represents regular, sinusoidal oscillations, in a medium which at that time was called "the ether". If this is so, it must be possible to produce more light by adding wavecrest to wavecrest, and darkness by adding wavecrest to wavethrough.

Light which is capable of interferences is called "coherent", and it is evident that in order to yield many interference fringes, it must be *very* monochromatic. Coherence is conveniently measured by the path difference between two rays of the same source, by which they can differ while still giving observable interference contrast. This is called the coherence length, an important quantity in the theory and practice of holography. Lord Rayleigh and Albert Michelson were the first to understand that it is a reciprocal measure of the spectroscopic line width. Michelson used it for ingenious methods of spectral analysis and for the measurement of the diameter of stars.

Let us now jump a century and a half, to 1947. At that time I was very interested in electron microscopy. This wonderful instrument had at that time produced a hundredfold improvement on the resolving power of the best light microscopes, and yet it was disappointing, because it had stopped short of resolving atomic lattices. The de Broglie wavelength of fast electrons, about 1/20 Ångström, was short enough, but the optics was imperfect. The best electron objective which one can make can be compared in optical perfection to a raindrop than to a microscope objective, and through the theoretical work of O. Scherzer it was known that it could never be perfected. The theoretical limit at that time was estimated at 4 Å, just about twice what was

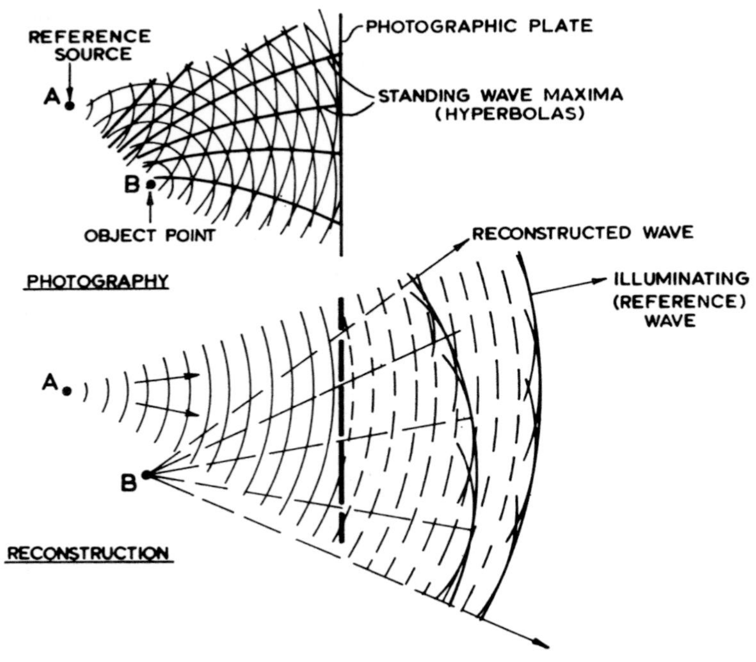

Fig. 2.
The Basic Idea of Holography, 1947.

needed to resolve atomic lattices, while the practical limit stood at about 12 Å. These limits were given by the necessity of restricting the aperture of the electron lenses to about 5/1000 radian, at which angle the spherical aberration error is about equal to the diffraction error. If one doubles this aperture so that the diffraction error is halved, the spherical aberration error is increased 8 times, and the image is hopelessly blurred.

After pondering this problem for a long time, a solution suddenly dawned on me, one fine day at Easter 1947, more or less as shown in Figure 2. Why not take a bad electron picture, but one which contains the *whole* information, and correct it by optical means? It was clear to me for some time that this could be done, if at all, only with coherent electron beams, with electron waves which have a definite phase. But an ordinary photograph loses the phase completely, it records only the intensities. No wonder we lose the phase, if there is nothing to compare it with! Let us see what happens if we add a standard to it, a "coherent background". My argument is illustrated in Figure 2, for the simple case when there is only one object point. The interference of the object wave and of the coherent background or "reference wave" will then produce interference fringes. There will be maxima wherever the phases of the two waves were identical. Let us make a hard positive record, so that it transmits only at the maxima, and illuminate it with the reference source alone. Now the phases are of course right for the reference source A, but as at the slits the phases are identical, they must be right also for B; therefore the wave of B must also appear, *reconstructed.*

A little mathematics soon showed that the principle was right, also for more than one object point, for any complicated object. Later on it turned out that in holography Nature is on the inventor's side; there is no need to take a hard positive record; one can take almost any negative. This encouraged me to complete my scheme of electron microscopy by reconstructed wavefronts, as I then called it and to propose the two-stage process shown in Figure 3. The electron microscope was to produce the interference figure between the object beam and the coherent background, that is to say the non-diffracted part of the illuminating beam. This interference pattern I called a "hologram", from the Greek word "holos"—the whole, because it contained the whole information. The hologram was then reconstructed with light, in an optical system which corrected the aberrations of the electron optics (1).

In doing this, I stood on the shoulders of two great physicists, W. L. Bragg and Frits Zernike. Bragg had shown me, a few years earlier, his "X-ray microscope" an optical Fourier-transformer device. One puts into it a small photograph of the reciprocal lattice, and obtains a projection of the electron densities, but only in certain exceptional cases, when the phases are all real, and have the same sign. I did not know at that time, and neither did Bragg, that Mieczislav Wolfke had proposed this method in 1920, but without realising it experimentally.[1] So the idea of a two-stage method was inspired by Bragg. The coherent background, on the other hand, was used with great success by Frits

[1] M. Wolfke, Phys. Zeits. *21,* 495—7, Sept. 15, 1920.

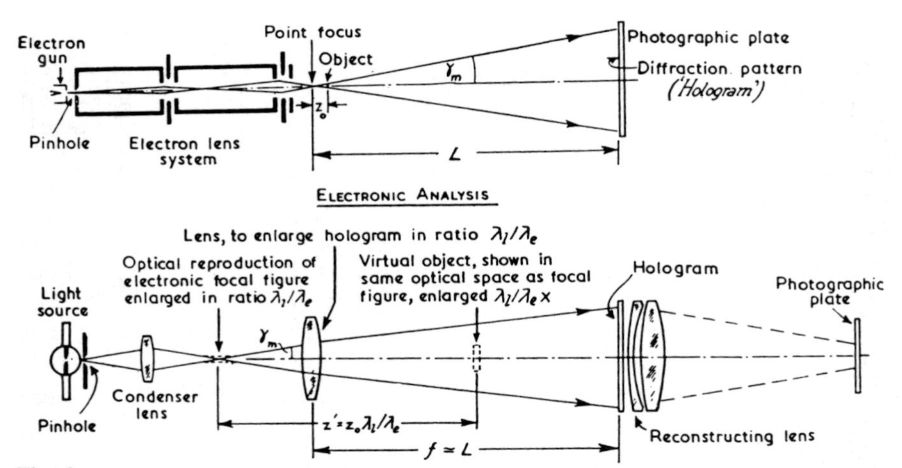

Fig. 3.
The Principle of Electron Microscopy by Reconstructed Wavefronts (Gabor, Proc. Royal Society, A, *197*, 454, 1949).

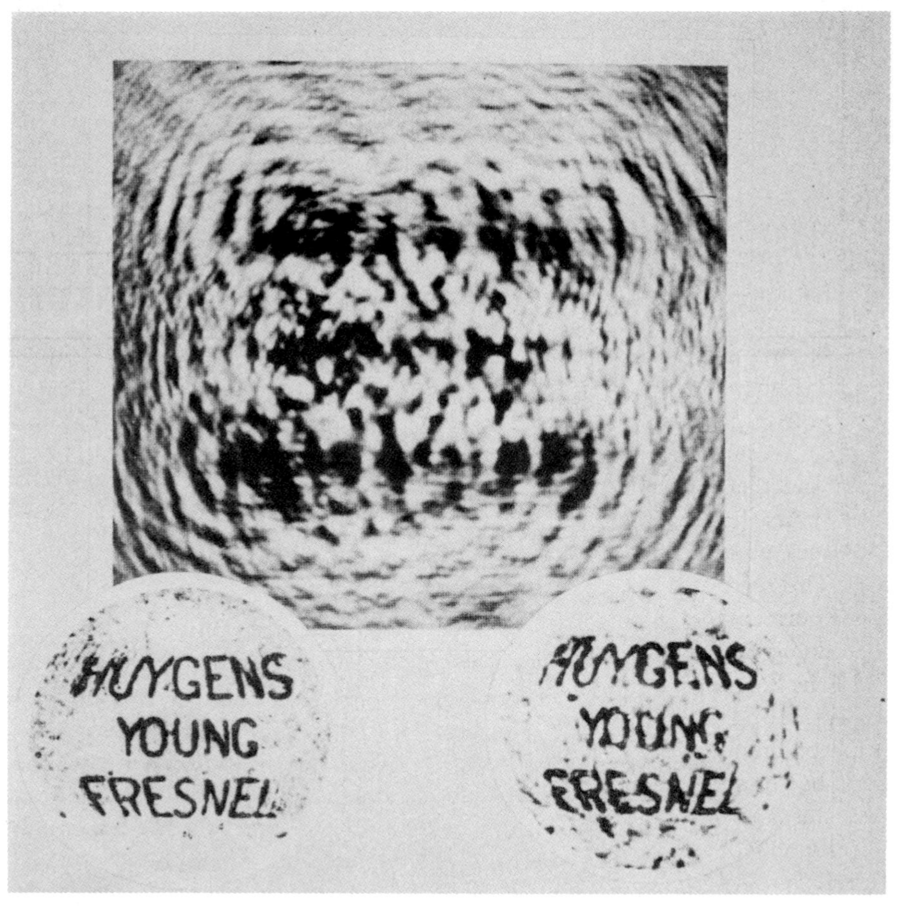

Fig 4.
First Holographic Reconstruction, 1948

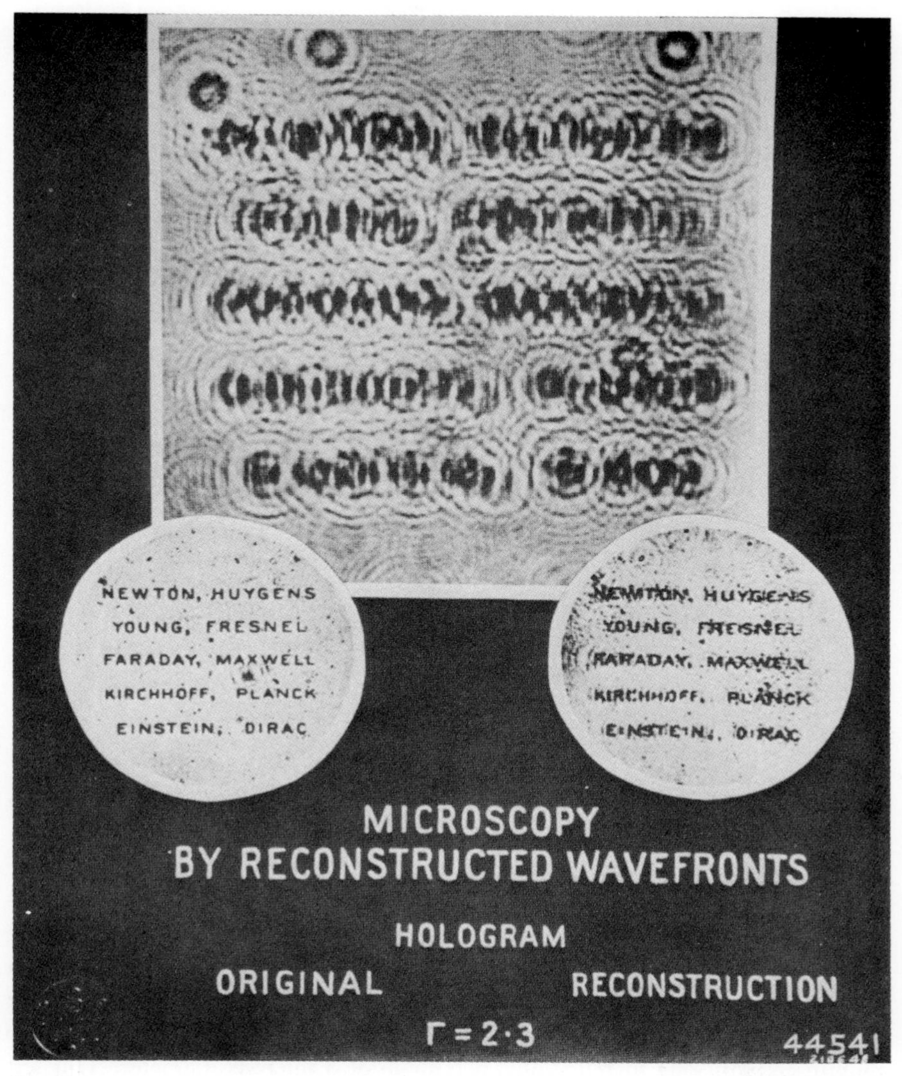

Fig. 5.
Another Example of Early Holography, 1948 (Gabor, Proc. Roy. Soc. A, *197*, 454, 1949).

Zernike in his beautiful investigations on lens aberrations, showing up their phase, and not just their intensity. It was only the reconstruction principle which had escaped them.

In 1947 I was working in the Research Laboratory of the British Thomson-Houston Company in Rugby, England. It was a lucky thing that the idea of holography came to me *via* electron microscopy, because if I had thought of optical holography only, the Director of Research, L. J. Davies, could have objected that the BTH company was an electrical engineering firm, and not in the optical field. But as our sister company, Metropolitan Vickers were makers of electron microscopes, I obtained the permission to carry out some

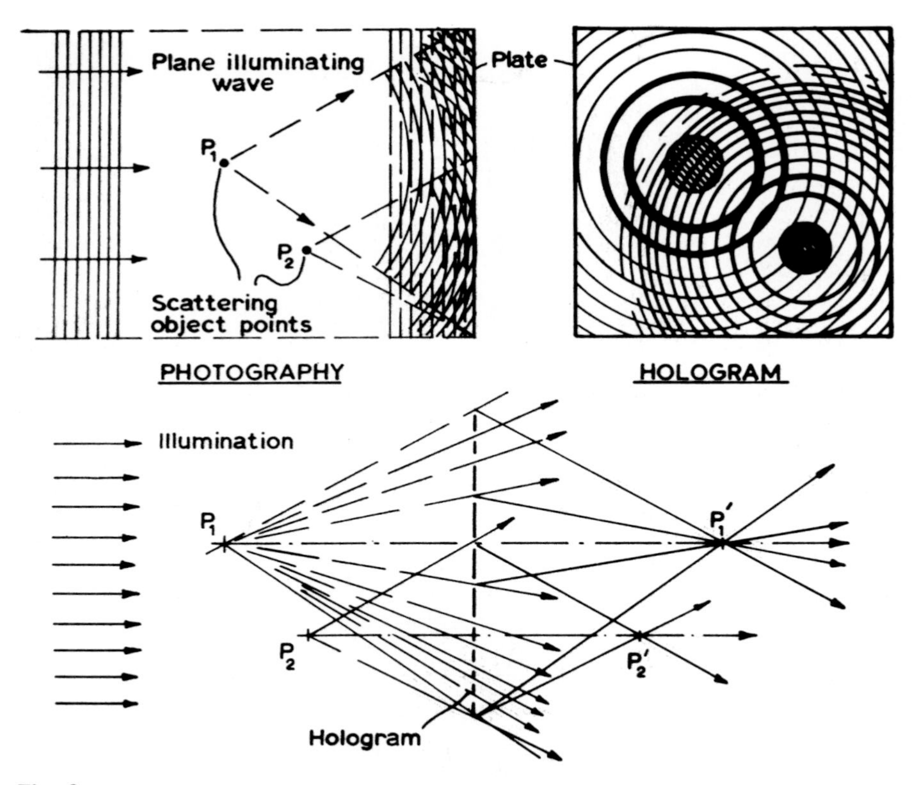

Fig. 6.
The Second Image. Explanation in Terms of Soret-lenses as Holograms of Single Object Points.

optical experiments. Figure 4 shows one of our first holographic reconstructions. The experiments were not easy. The best compromise between coherence and intensity was offered by the high pressure mercury lamp, which had a coherence length of only 0.1 mm, enough for about 200 fringes. But in order to achieve spatial coherence, we (my assistant Ivor Williams and I) had to illuminate, with one mercury line, a pinhole of 3 microns diameter. This left us with enough light to make holograms of about 1 cm diameter of objects, which were microphotographs of about 1 mm diameter, with exposures of a few minutes, on the most sensitive emulsions then available. The small coherence length forced us to arrange everything in one axis. This is now called "in line" holography, and it was the only one possible at that time. Figure 5 shows a somewhat improved experiment, the best of our series. It was far from perfect. Apart from the *schlieren,* which cause random disturbances, there was a systematic defect in the pictures, as may be seen by the distortion of the letters. The explanation is given in Figure 6. The disturbance arises from the fact that there is not one image but *two*. Each point of the object emits a spherical secondary wave, which interferes with the background and produces a system of circular Fresnel zones. Such a system is known after the optician who first produced it, a Soret lens. This is, at the same time, a positive and a

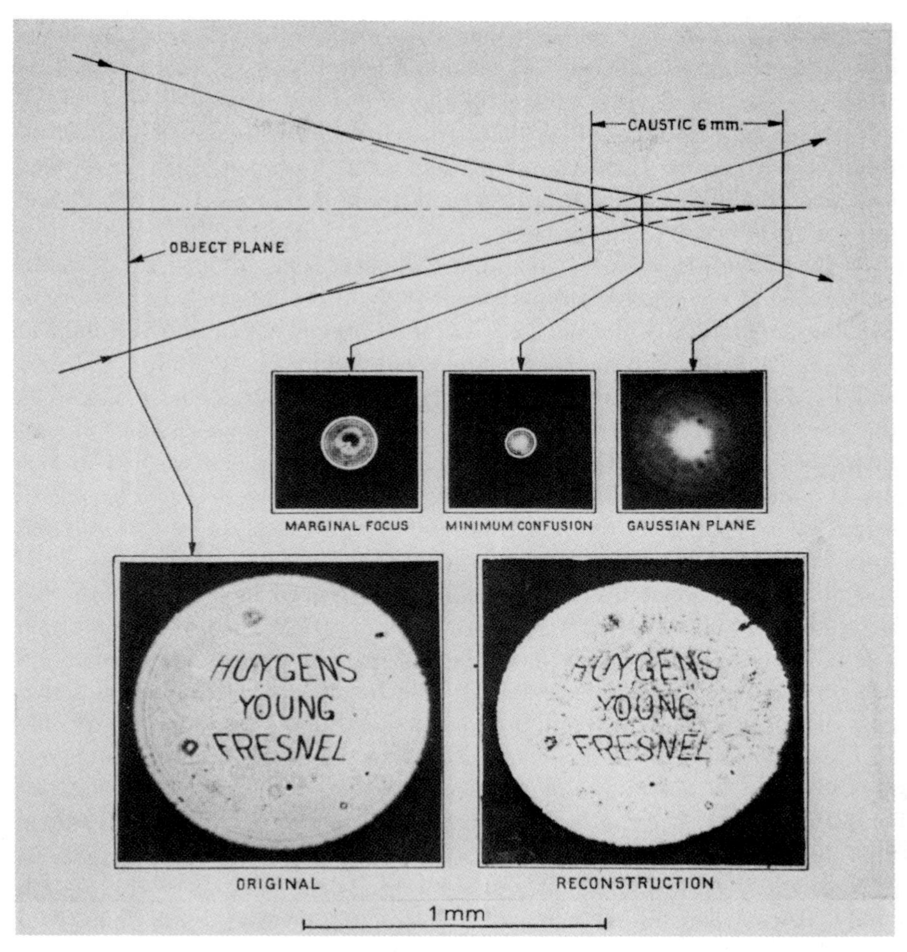

Fig. 7.
Elimination of the Second Image, by Compensation of the Spherical Aberration in the Reconstruction (Gabor, 1948; published 1951).

negative lens. One of its foci is in the original position of the object point, the other in a position conjugate to it, with respect to the illuminating wavefront. If one uses "in-line holography" both images are in line, and can be separated only by focusing. But the separation is never quite perfect, because in regular, coherent illumination every point leaves a "wake" behind it, which reaches to long distances.

I will tell later with what ease modern laser holography has got rid of this disturbance, by making use of the superior coherence of laser light which was not at my disposal in 1948. However, I was confident that I could eliminate the second image in the application which alone interested me at that time: seeing atoms with the electron microscope. This method, illustrated in Figure 7, utilized the very defect of electron lenses, the spherical aberration, in order to defeat the second image. If an electron hologram is taken with a lens with spherical aberration, one can afterwards correct *one* of the two images by suitable optics, and the other has then twice the aberration, which washes it out almost completely. Figure 7 shows that a perfectly sharp reconstruction, in which as good as nothing remains of the disturbance caused by the second image, can be obtained with a lens so bad that its definition is at least 10 times worse than the resolution which one wants to obtain. Such a very bad lens was obtained using a microscope objective the wrong way round, and using it again in the reconstruction.

So it was with some confidence that two years later, in 1950 we started a programme of holographic electron microscopy in the Research Laboratory of the Associated Electrical Industries, in Aldermaston, under the direction of Dr T. E. Allibone, with my friends and collaborators M. W. Haine, J. Dyson and T. Mulvey.[2] By that time I had joined Imperial College, and took part in the work as a consultant. In the course of three years we succeeded in considerably improving the electron microscope, but in the end we had to give up, because we had started too early. It turned out that the electron microscope was still far from the limit imposed by optical aberrations. It suffered from vibrations, stray magnetic fields, creep of the stage, contamination of the object, all made worse by the long exposures required in the weak coherent electron beam. *Now,* 20 years later, would be the right time to start on such a programme, because in the meantime the patient work of electron microscopists has overcome all these defects. The electron microscope resolution is now right up to the limit set by the sperical aberration, about 3.5 Å, and only an improvement by a factor of 2 is needed to resolve atomic lattices. Moreover, there is no need now for such very long exposures as we had to contemplate in 1951, because by the development of the field emission cathode the coherent current has increased by a factor of 3—4 orders of magnitude. So perhaps I may yet live to see the realisation of my old ideas.

My first papers on wavefront reconstruction evoked some immediate responses. G. L. Rogers (2) in Britain made important contributions to the technique, by producing among other things the first phase holograms, and also by elucidating the theory. In California Alberto Baez (3), Hussein El-Sum and P. Kirckpatrick (4) made interesting forays into X-ray holography. For my part, which my collaborator W. P. Goss, I constructed a holographic interference microscope, in which the second image was annulled in a rather complicated way by the superimposition of two holograms, "in quadrature" with one another. The response of the optical industry to this was so disappointing that we did not publish a paper on it until 11 years later, in 1966 (5). Around 1955 holography went into a long hybernation.

The revival came suddenly and explosively in 1963, with the publication of the first successful laser[3] holograms by Emmett N. Leith and Juris Upatnieks

[2] Supported by a grant of the D.S.I.R. (Direction of Scientific and Industrial Research) the first research grant ever given by that body to an industrial laboratory.

[3] I have been asked more than once why I did not invent the laser. In fact, I have thought of it. In 1950, thinking of the desirability of a strong source of coherent light, I remembered that in 1921, as a young student, in Berlin, I had heard from Einstein's own lips his wonderful derivation of Planck's law which postulated the existence of

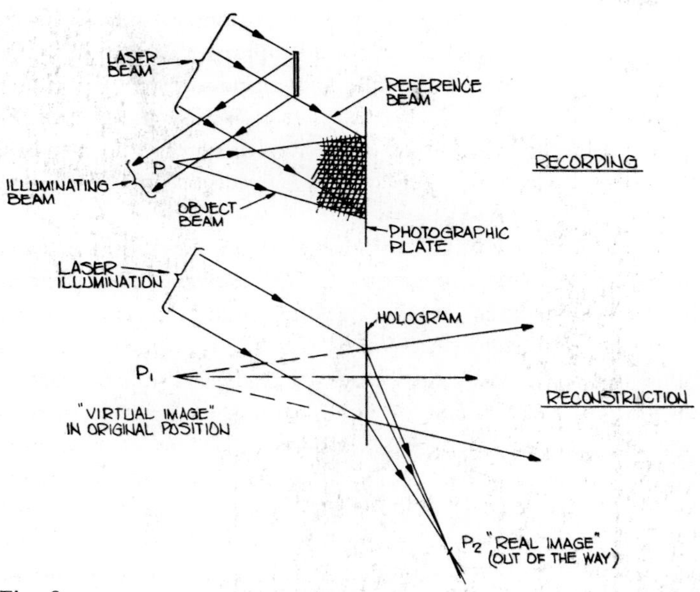

Fig. 8.
Holography with Skew Reference Beam. E. N. Leith and J. Upatnieks, 1963.

of the University of Michigan, Ann Arbor. Their success was due not only to the laser, but to the long theoretical preparation of Emmett Leith, which started in 1955. This was unknown to me and to the world, because Leith, with his collaborators Cutrona, Palermo, Porcello and Vivian applied his ideas first to the problem of the "side-looking radar" which at that time was classified (6). This was in fact two-dimensional holography with electromagnetic waves, a counterpart of electron holography. The electromagnetic waves used in radar are about 100,000 times longer than light waves, while electron waves are about 100,000 times shorter. Their results were brilliant, but to my regret I cannot discuss them for lack of time.

When the laser became available, in 1962, Leith and Upatnieks could at once produce results far superior to mine, by a new, simple and very effective method of eliminating the second image (7). This is the method of the "skew reference wave", illustrated in Figure 8. It was made possible by the great coherence length of the helium-neon laser, which even in 1962 exceeded that of the mercury lamp by a factor of about 3000. This made it possible to separate the reference beam from the illuminating beam; instead of going through the object, it could now go around it. The result was that the two reconstructed images were now separated not only in depth, but also angularly, by twice the incidence angle of the reference beam. Moreover, the intensity of

stimulated emission. I then had the idea of the pulsed laser: Take a suitable crystal, make a resonator of it by a highly reflecting coating, fill up the upper level by illuminating it through a smal hole, and discharge it explosively by a ray of its own light. I offered the idea as a Ph.D. problem to my best student, but he declined it, as too risky, and I could not gainsay it, as I could not be sure that we would find a suitable crystal.

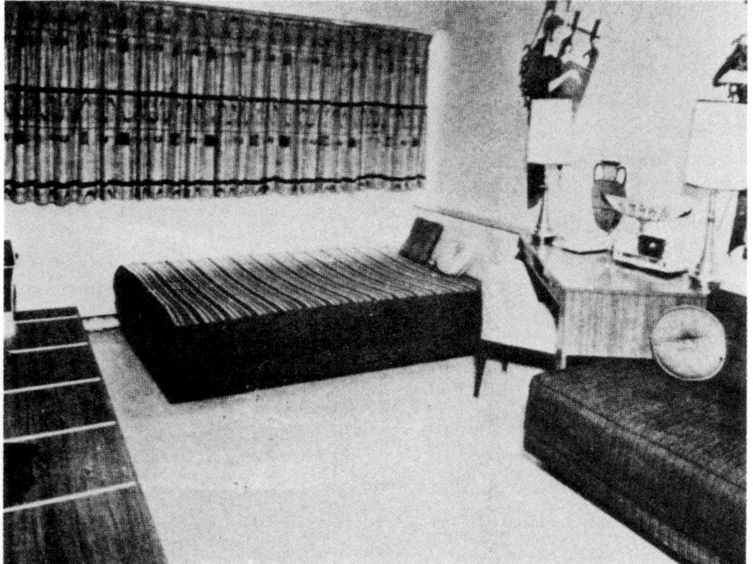

Fig. 9.
First Example of Multiple Image Storage in One Hologram. E. N. Leith and J. Upatnieks, Journal Optical Society of America, November 1964.

the coherent laser light exceeded that of mercury many millionfold. This made it possible to use very fine-grain, low speed photographic emulsions and to produce large holograms, with reasonable exposure times.

Figure 9 shows two of the earliest reconstructions made by Leith and Upatnieks, in 1963, which were already greatly superior to anything that I could produce in 1948. The special interest of these two images is, that they are reconstructions from *one* hologram, taken with different positions of the reference beam. This was the first proof of the superior storage capacity of

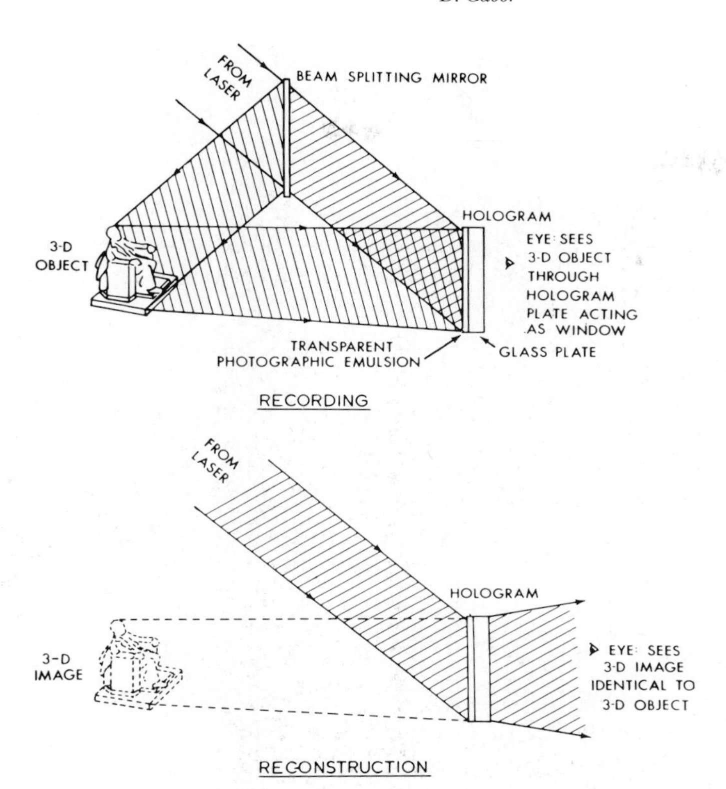

Fig. 10.
3-D Holography of a Diffusing Object with Laser Light.

Fig. 11.
Three dimensional Reconstruction of a Small Statue of Abraham Lincoln. (Courtesy of Professor G. W. Stroke, State University of New York, Stony Brook).

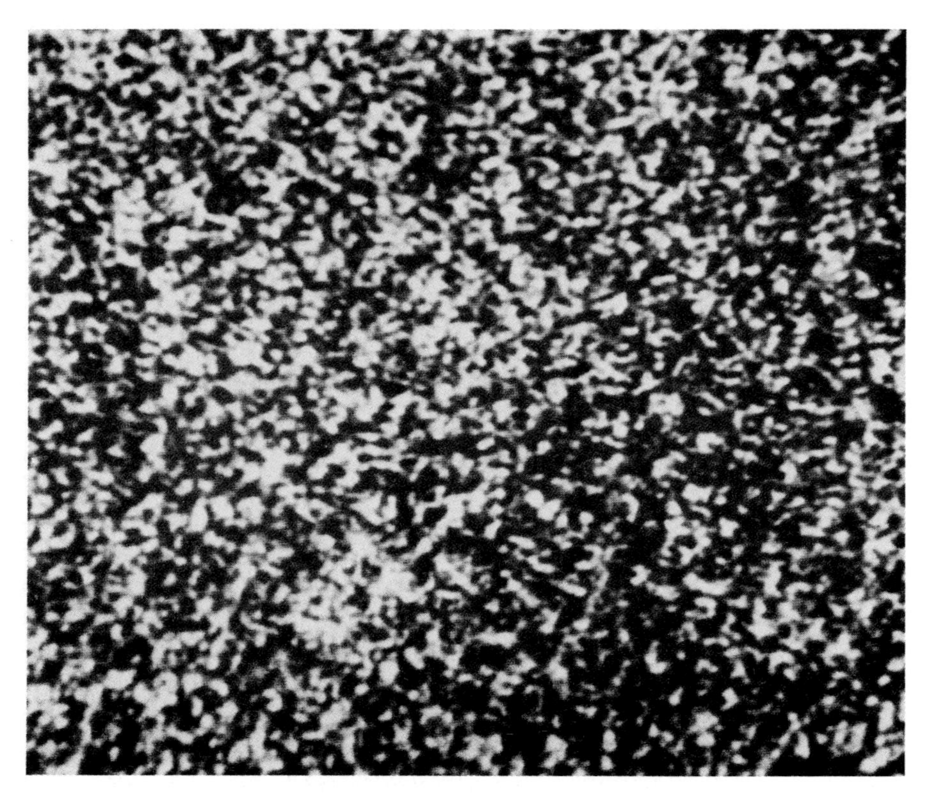

Fig. 12.
Strongly Magnified Image of a Hologram taken with Diffused Illumination. The Information is conveyed in a noiselike code. E. N. Leith and J. Upatnieks, 1964.

holograms. Leith and Upatnieks could soon store 12 different pictures in one emulsion. Nowadays one can store 100 or even 300 pages of printed matter in an area which by ordinary photography would be sufficient for one.

From then on progress became very rapid. The most spectacular result of the first year was the holography of three dimensional objects, which could be seen with two eyes. Holography was of course three dimensional from the start, but in my early, small holograms one could see this only by focusing through the field with a microscope or short-focus eyepiece. But it was not enough to make the hologram large, it was also necessary that every point of the photographic plate should see every point of the object. In the early holograms, taken with regular illumination, the information was contained in a small area, in the diffraction pattern.

In the case of rough, diffusing objects no special precautions are necessary. The small dimples and projections of the surface diffuse the light over a large cone. Figure 10 shows an example of the setup in the case of a rough object, such as a statuette of Abraham Lincoln. The reconstruction is shown in Figure 11. With a bleached hologram ("phase hologram") one has the impression of looking through a clear window at the statuette itself.

If the object is non-diffusing, for instance if it is a transparency, the information is spread over the whole hologram area by illuminating the object through a diffuser, such as a frosted glass plate. The appearance of such a "diffused" hologram is extraordinary; it looks like noise. One can call it "ideal Shannon coding", because Claude E. Shannon has shown in his Communication Theory that the most efficient coding is such that all regularities seem to have disappeared in the signal; it must be "noise-like". But where is the information in this chaos? It can be shown that it is not as irregular as it appears. It is not as if grains of sand had been scattered over the plate at random. It is rather a complicated figure, the diffraction pattern of the object, which is repeated at random intervals, but always in the same size and same orientation.

A very interesting and important property of such diffused holograms is that any small part of it, large enough to contain the diffraction pattern, contains information on the whole object, and this can be reconstructed from the fragment, only with more noise. A diffuse hologram is therefore a *distributed memory,* and this was evoked much speculation whether human memory is not perhaps, as it were, holographic, because it is well known that a good part of the brain can be destroyed without wiping out every trace of a memory. There is no time here to discuss this very exciting question. I want only to say that in my opinion the similarity with the human memory is functional only, but certainly not structural.

It is seen that in the development of holography the holograms has become always more unlike the object, but the reconstruction always more perfect. Figure 13 shows an excellent reconstruction by Leith and Upatnieks of a

Fig. 13.
Reconstruction of a Plane Transparency, Showing a Restaurant, from a Hologram taken with Diffused Illuminations (E. N. Leith and J. Upatnieks, 1964).

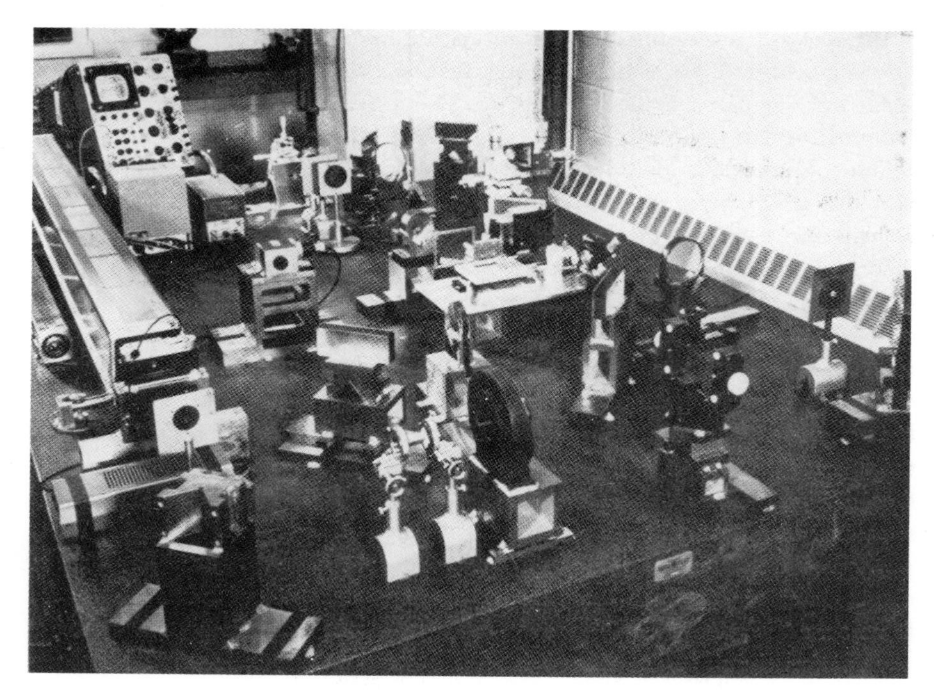

Fig. 14.
Modern Holographic Equipment.

photograph, from a diffuse hologram like the one in the previous figure.

The pioneer work carried out in the University of Michigan, Ann Arbor, led also to the stabilization of holographic techniques. Today hundreds if not thousands of laboratories possess the equipment of which an example is shown in Figure 14; the very stable granite slab or steel table, and the various optical devices for dealing with coherent light, which are now manufactured by the optical industry. The great stability is absolutely essential in all work carried out with steady-state lasers, because a movement of the order of a quarter wavelength during the exposure can completely spoil a hologram.

However, from 1965 onwards there has developed an important branch of holography where high stability is not required, because the holograms are taken in a small fraction of a microsecond, with a pulsed laser.

Imagine that you had given a physicist the problem: "Determine the size of the droplets which issue from a jet nozzle, with a velocity of 2 Mach. The sizes are probably from a few microns upwards." Certainly he would have thrown up his hands in despair! But all it takes now, is to record a simple in-line hologram of the jet, with the plate at a safe distance, with a ruby laser pulse of 20—30 nanoseconds. One then looks at the "real" image (or one reverses the illuminating beam and makes a real image of the virtual one), one dives with a microscope into the three-dimensional image of the jet and focuses the particles, one after the other. Because of the large distance, the disturbance by the second image is entirely negligible. Figure 15 shows a fine example.

As the research workers of the TRW laboratories have shown, it is possible

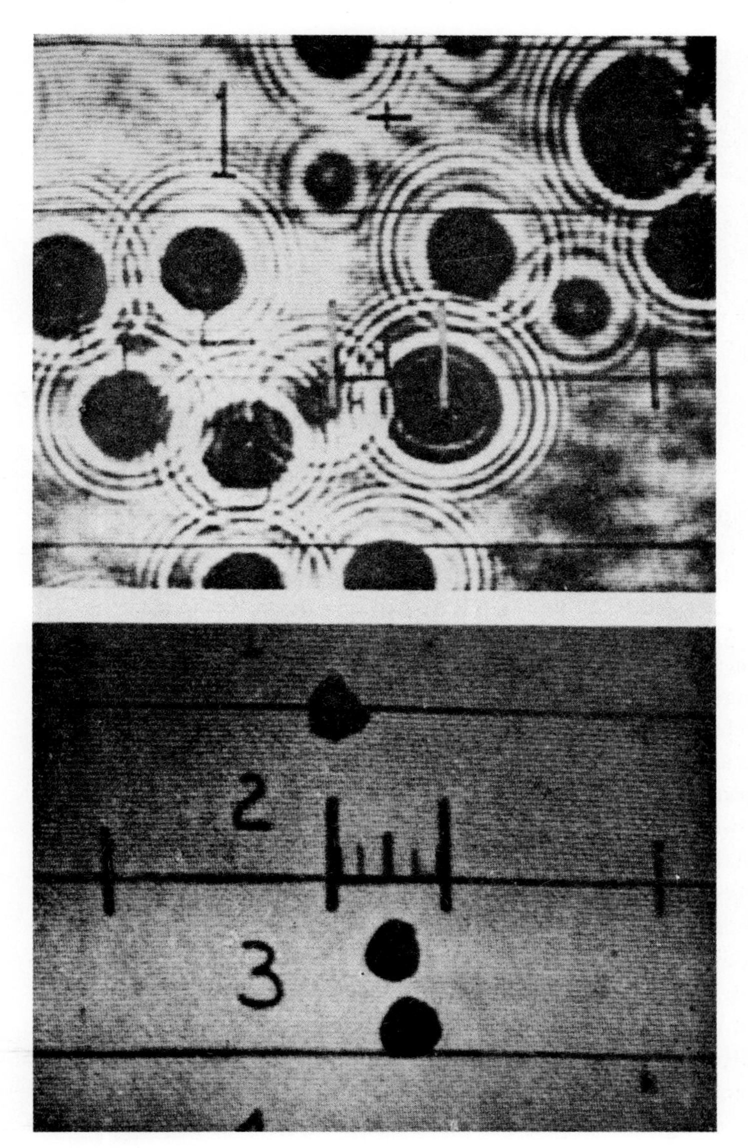

Fig. 15.
Holography of Jets. (Courtesy of Laser Holography Inc., Santa Barbara, California.)

to record in one hologram the infusoriae in several feet of dirty water, or insects in a meter of air space. Figure 16 shows two reconstructions of insects from one hologram, focusing on one after the other. The authors, C. Knox and R. E. Brooks, have also made a cinematographic record of a holographic film, in which the flight of one mosquito is followed through a considerable depth, by refocusing in every frame (9).

Another achievement of the TRW group, Ralph Wuerker and his colleagues, leads us into another branch of holography, to holographic interferometry. Figure 17 shows a reconstruction of a bullet, with its train of shockwaves, as

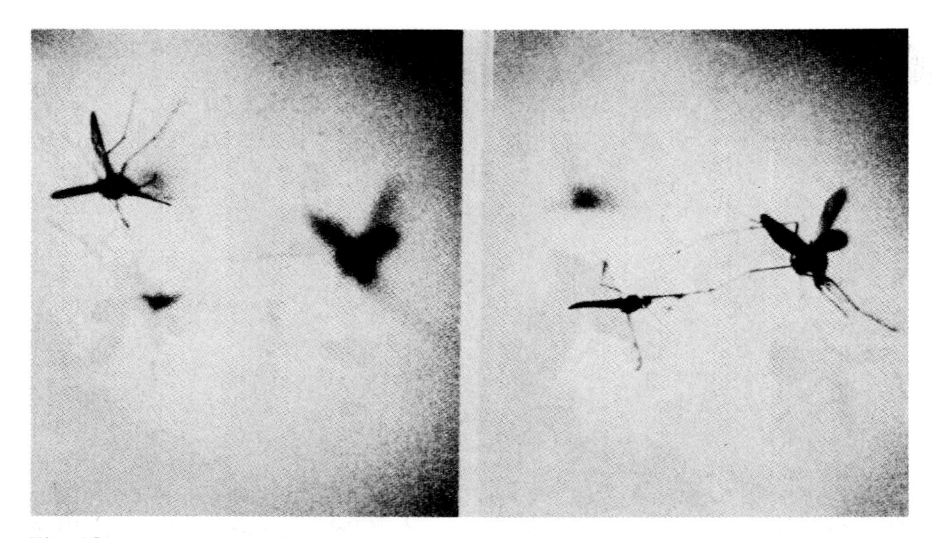

Fig. 16.
Observation of Mosquitos in Flight. Both Pictures are extracted from one Hologram. (Courtesy of C. Knox and R. E. Brooks, TRW, Redondo Beach, California.

it meets another shockwave. But it is not just an image, it is an interferometric image. The fringes show the *loci* at which the retardation of light is by integer wavelengths, relative to the quiet air, before the event. This comparison standard is obtained by a previous exposure. This is therefore a double-exposure hologram, such as will be discussed in more detail later (10).

Figure 18 shows another high achivement of pulse holography: a holographic, three-dimensional portrait, obtained by L. Siebert in the Conductron Corporation (now merged into McDonnel-Douglas Electronics Company, St Charles, Missouri). It is the result of outstanding work in the development of lasers. The ruby laser, as first realised by T. H. Maiman, was capable of short pulses, but its coherence length was of the order of a few cm only. This is no obstacle in the case of in-line holography, where the reference wave proceeds almost in step with the diffracted wavelets, but in order to take a scene of, say, one meter depth with reflecting objects one must have a coherence length of at least one meter. Nowadays single-mode pulses of 30 nanosecond duration with 10 joule in the beam and coherence lengths of 5—8 meters are available, and have been used recently for taking my holographic portrait shown in the exhibition attached to this lecture.

In 1965 R. L. Powell and K. A. Stetson in the University of Michigan, Ann Arbor, made an interesting discovery. Holographic images taken of moving objects are washed out. But if double exposure is used, first with the object at rest, then in vibration, fringes will appear, indicating the lines where the displacement amounted to multiples of a half wavelength. Figure 19 shows vibrational modes of a loudspeaker membrane, recorded in 1965 by Powell and Stetson (11), Figure 20 the same for a guitar, taken by H. A. Stetson in the laboratory of Professor Erik Ingelstam (12).

Curiously, both the interferograms of the TRW group and the vibrational

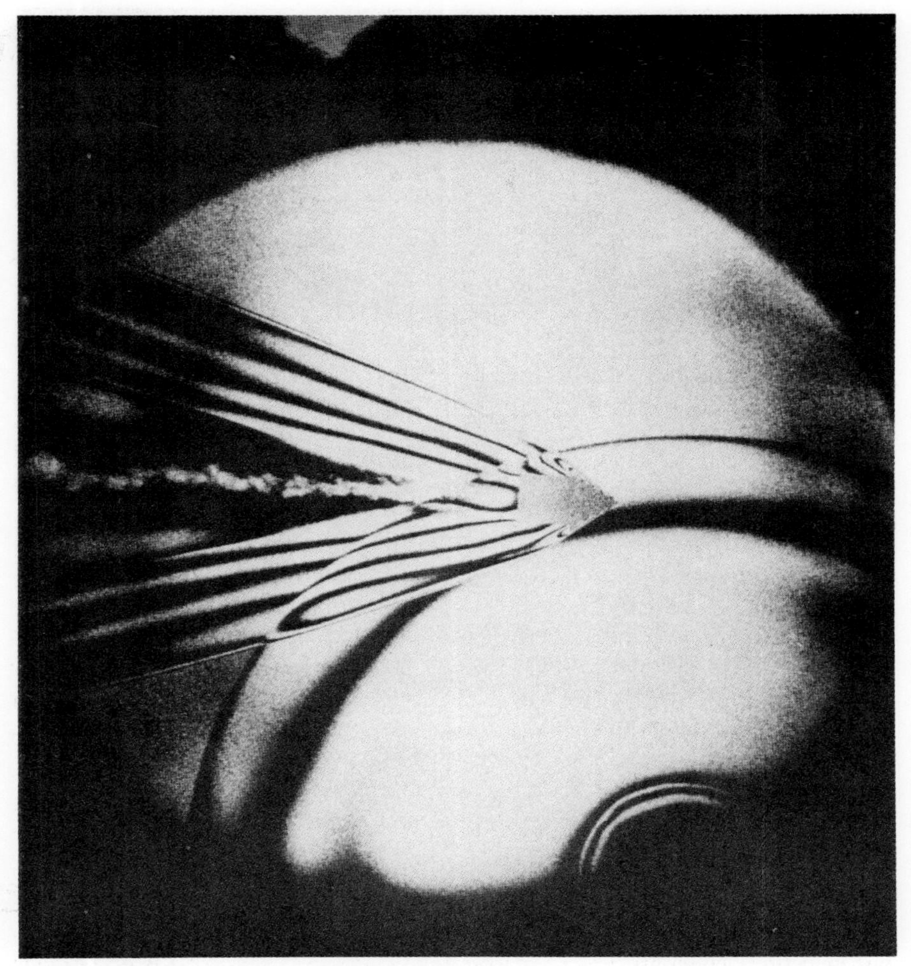

Fig. 17.
Dynamic Holographic Interferometry. This Reconstruction of a Holographic Interferogram shows the interaction of two air shock fronts and their associated flows. (Courtesy of Dr R. F. Wuerker and his associates, TRW Physical Electronics Laboratory, Redondo Beach, Calif.)

records of Powell and Stetson preceded what is really a simpler application of the interferometrical principle, and which historically ought to have come first—if the course of science would always follow the shortest line. This is the observation of small deformations of solid bodies, by double exposure holograms. A simple explanation is as follows: We take a hologram of a body in State A. This means that we freeze in the wave A by means of a reference beam. Now let us deform the body so that is assumes the State B and take a second hologram in the same emulsion with the same reference beam. We develop the hologram, and illuminate it with the reference beam. Now the two waves A and B, frozen in at different times, and which have never seen one another, will be revived simultaneously, and they interfere with one an-

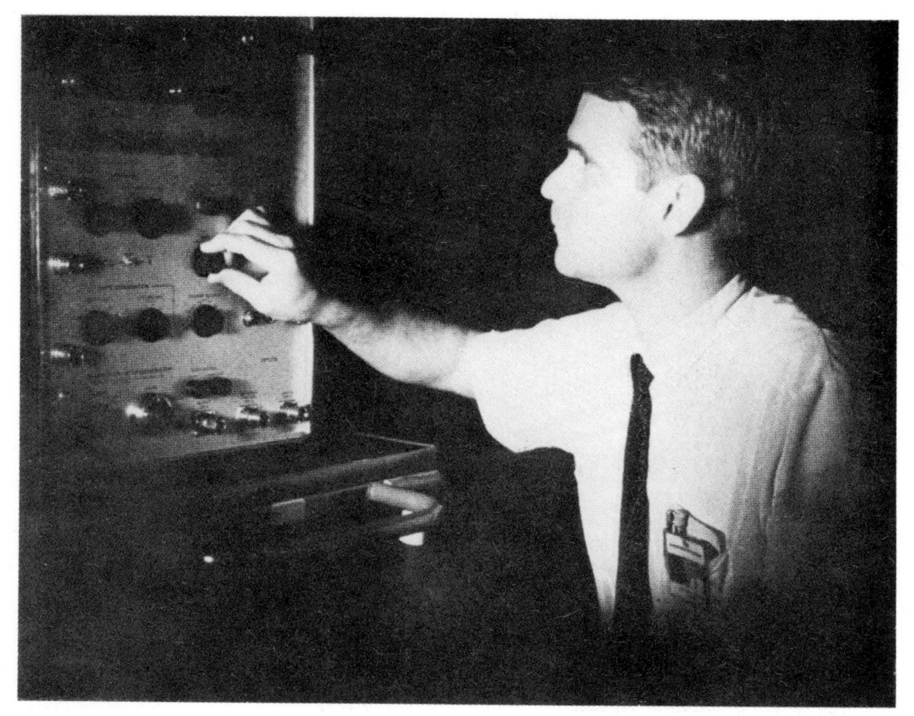

Fig. 18.
Holographic Portrait. (L. Siebert, Conductron Corporation, now merged into Mc-Donnell-Douglas Electronics Company, St Charles, Missouri.)

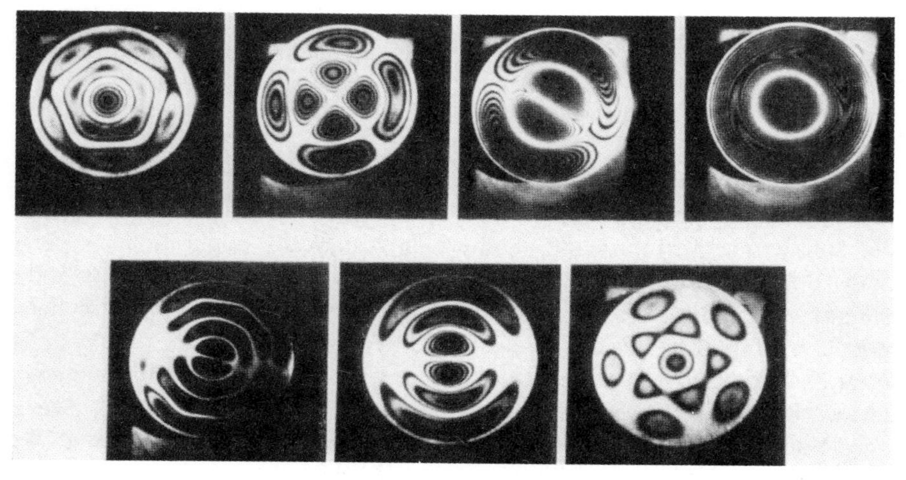

Fig. 19.
Vibrational Modes of a Loudspeaker Membrane, obtained by Holographic Interferometry. (R. L. Powell and K. A. Stetson, University of Michigan, Ann Arbor, 1965.)

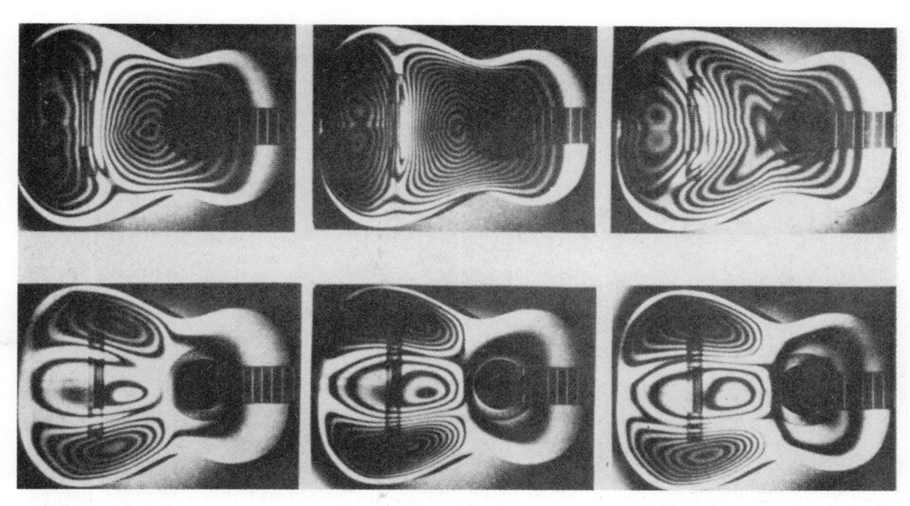

Fig. 20.
Vibrational Modes of a Guitar, Recorded by Holographic Interferometry. (Courtesy of Dr K. A. Stetson and Professor E. Ingelstam.)

other. The result is that Newton fringes will appear on the object, each fringe corresponding to a deformation of a half wavelength. Figure 21 shows a fine example of such a holographic interferogram, made in 1965 by Haines and Hillebrand. The principle was discovered simultaneously and independently also by J. M. Burch in England, and by G. W. Stroke and A. Labeyrie in Ann

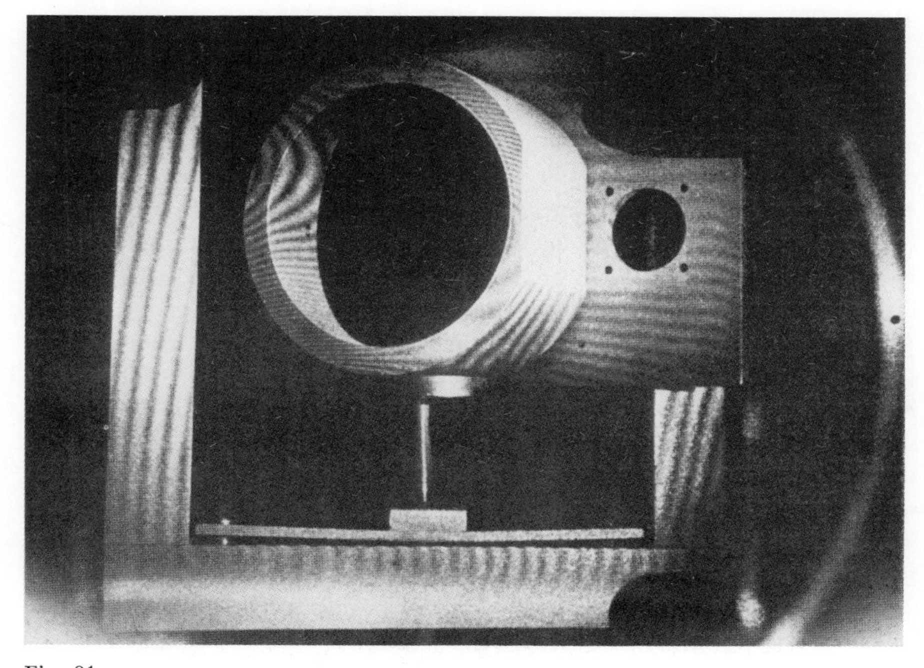

Fig. 21.
An early Example of Holographic Interferometry by Double Exposure. (Haines and Hildebrand, University of Michigan, Ann Arbor, 1965.)

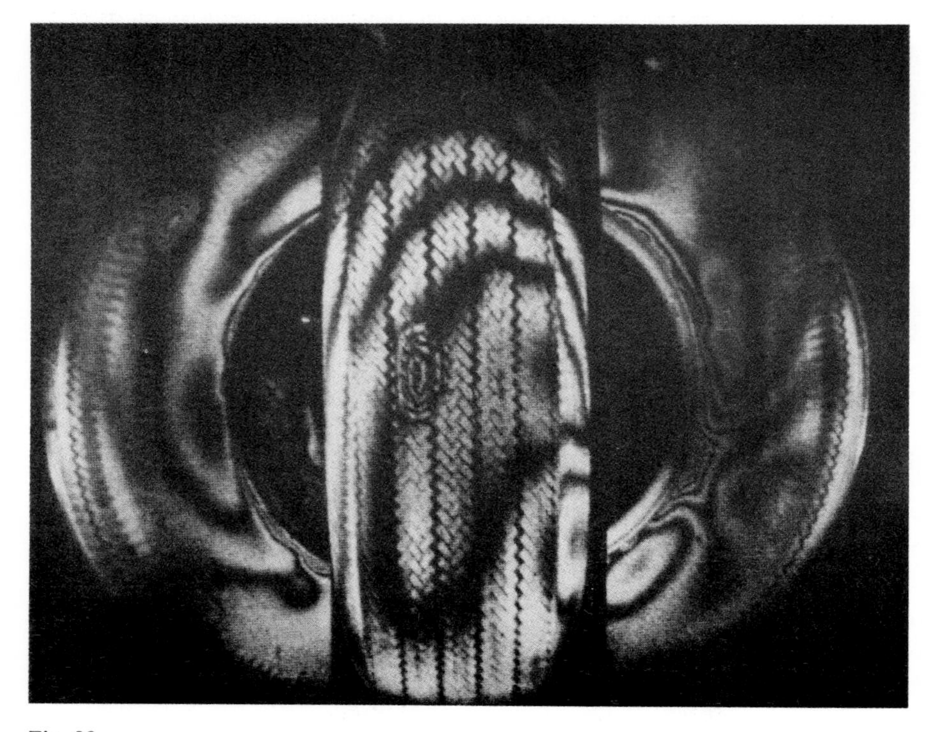

Fig. 22.
Non-destructive Testing by Holography. Double Exposure Hologram, revealing two flaws in a tyre (Courtesy of Dr Ralph Grant and GCO, Ann Arbor, Michigan).

Arbor, Michigan.

Non-destructive testing by holographic interferometry is now by far the most important industrial application of holography. It gave rise to the first industrial firm based on holography, GCO (formerly G. C. Optronics), in Ann Arbor, Michigan, and the following examples are reproduced by courtesy of GCO. Figure 22 shows the testing of a motor car tyre. The front of the tyre is holographed directly, the sides are seen in two mirrors, right and left. First a little time is needed for the tyre to settle down and a first hologram is taken. Then a little hot air is blown against it, and a second exposure is made, on the same plate. If the tyre is perfect, only a few, widely spaced fringes will appear, indicating almost uniform expansion. But where the cementing of the rubber sheets was imperfect, a little blister appears, as seen near the centre and near the top left corner, only a few thousandths of a millimeter high, but indicating a defect which could become serious. Alternatively, the first hologram is developed, replaced exactly in the original position, and the expansion of the tyre is observed "live".

Other examples of non-destructive testing are shown in Figure 23; all defects which are impossible or almost impossible to detect by other means, but which reveal themselves unmistakeably to the eye. A particularly impressive piece of equipment manufactured by GCO is shown in Figure 24. It is a holographic analyser for honeycomb sandwich structures (such as shown in the middle of

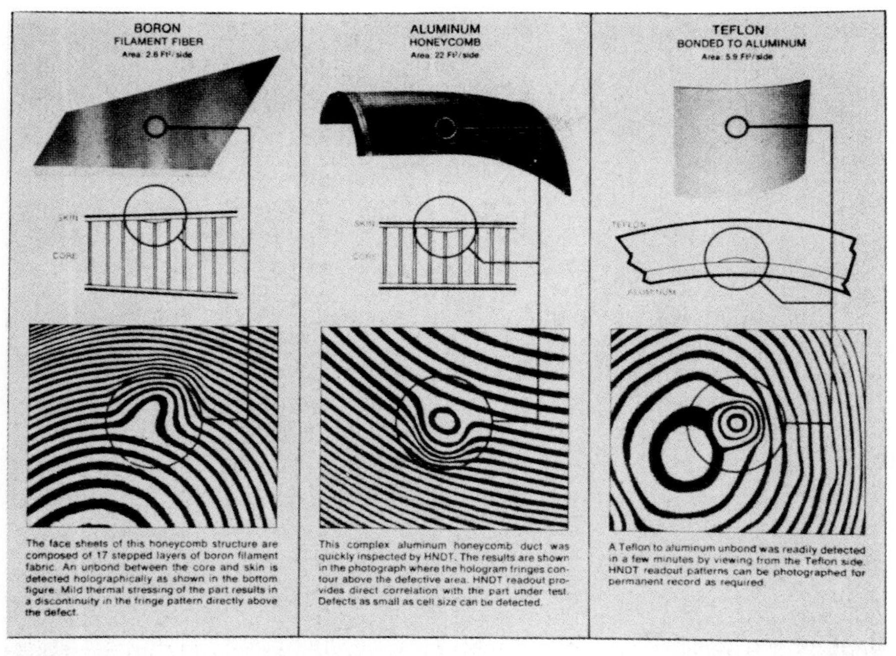

Fig. 23.
Examples of Holographic Non-destructive Testing. (Courtesy of GCO, Ann Arbor, Michigan.)

Fig. 24.
Holographic Analyzer Mark II for Sandwich Structures, GCO, Ann Arbor, Michigan.

Fig. 25.
Holographic Contour Map, made by a method initiated by B. P. Hildebrand and K. A. Haines (Journal, Optical Society of America, *57,* 155, 1967). Improved by J. Varner, University of Michigan, Ann Arbor, 1969.

Figure 23) which are used in aeroplane wings. The smallest welding defect between the aluminum sheets and the honeycomb is safely detected at one glance.

While holographic interferometry is perfectly suited for the detection of very small deformations, with its fringe unit of 1/4000 mm, it is a little too fine for the checking of the accuracy of workpieces. Here another holographic technique called "contouring" is appropriate. It was first introduced by Haines and Hildebrand, in 1965, and has been recently much improved by J. Varner, also in Ann Arbor, Michigan. Two holograms are taken of the same object, but with two wavelengths which differ by e.g. one percent. This produces *beats* between the two-fringe system, with fringe spacings corresponding to about 1/40 mm, which is just what the workshop requires (Figure 25).

From industrial applications I am now turning to another important development in holography. In 1962, just before the "holography explosion" the Soviet physicist Yu. N. Denisyuk published an important paper (13) in which he combined holography with the ingenious method of photography in natural colours, for which Gabriel Lippman received the Nobel Prize in 1908. Figure 26 a illustrates Lippmann's method and Denisyuk's idea. Lippmann produced a very fine-grain emulsion, with colloidal silver bromide, and backed the emulsion with mercury, serving as a mirror. Light falling on the emulsion was

How to view the Lippmann type reflection hologram

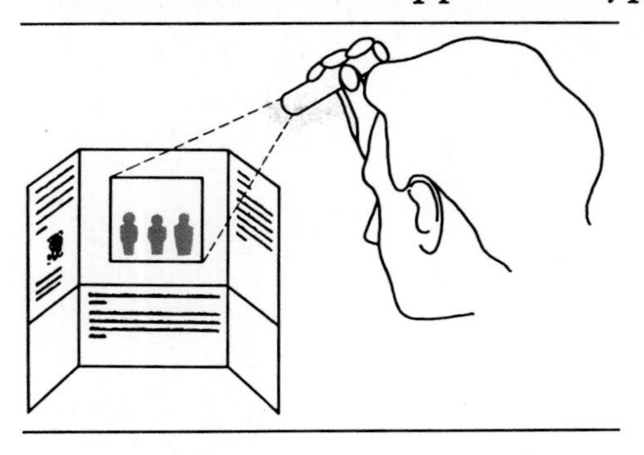

For maximum brightness (due to fulfillment of the Bragg criterion) the hologram shall be illuminated diagonally from the upper righthand corner. An ordinary penlight at a distance of about 25 cm is recommended, see figure. Other approximately point source lighting can be used, such as spotlight, slide projector, or even direct unclouded sunlight.

NB: The hologram ought to be viewed in subdued lighting, and direct overhead light be avoided. The side screens (partly book pages), as indicated in the figure, are good for screening off room light.

THE HOLOGRAM IS NOT REPRODUCED HERE DUE TO COMMERCIAL UNAVAILABILITY.

How the Lippman type reflection hologram has been constructed

The figure shows how the reference wave comes from one side of the emulsion, the signal wave from the object from the other side. The dotted line indicates how, at the reconstruction, a wave reflected from the silver layers in the emulsion is obtained, and you see in its extension backwards the object as it was at the registration. (Stroke-Labeyrie, see References.)

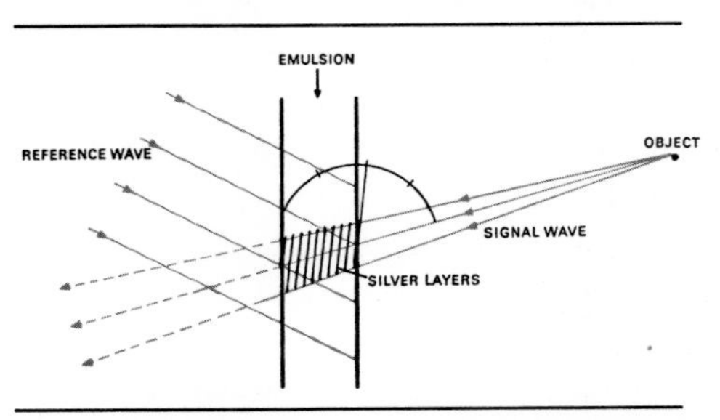

In fact, at the practical registration of a reflection hologram, the signal wave comes from the different points of the illuminated object. In order to have the reconstructed image of the object close to the hologram included, an image of the object has been transported there by means of a special lens. This gives localization of the image closely in front of and behind the hologram.

The hologram is manufactured by McDonnell Douglas Electronics Company, St. Charles, Missouri, USA.

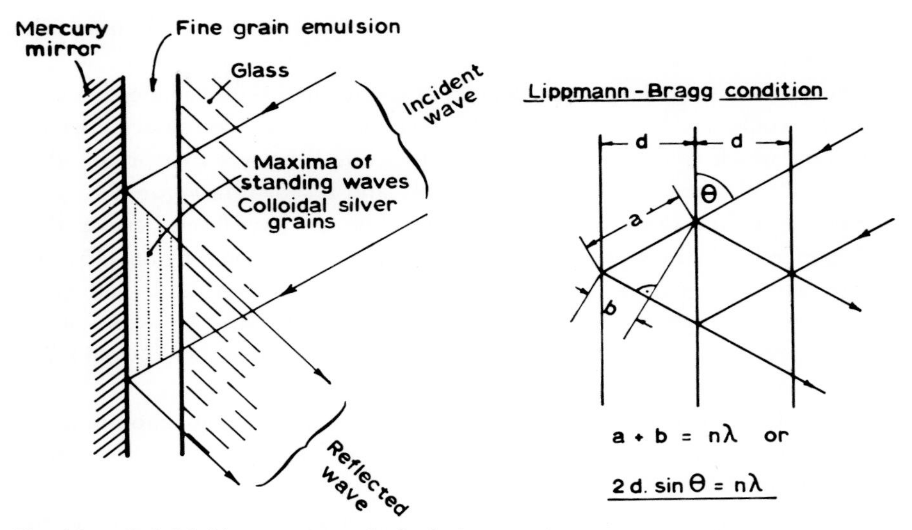

Fig. 26 a. Gabriel Lippmann's method of photography in natural colours.

reflected at the mirror, and produced a set of standing waves. Colloidal silver grains were precipitated in the maxima of the electric vector, in layers spaced by very nearly half a wavelength. After development, the complex of layers, illuminated with white light, reflected only a narrow waveband around the original colour, because only for this colour did the wavelets scattered at the Lippmann layers add up in phase.

Denisyuk's suggestion is shown in the second diagram. The object wave and the reference wave fall in from opposite sides of the emulsion. Again standing waves are produced, and Lippman layers, but these are no longer parallel to the emulsion surface, they bisect the angle between the two wavefronts. If now, and this is Denisyuk's principle, the developed emulsion is illuminated by the

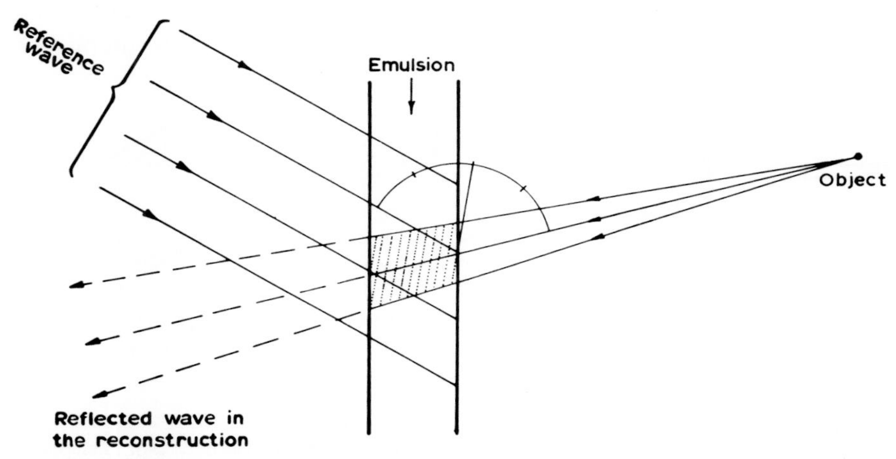

Fig. 26 b. Lippmann—Denisyuk—Stroke Reflection Hologram.

reference wave, the object will appear, in the original position and (unless the emulsion has shrunk) in the original colour.

Though Denisyuk showed considerable experimental skill, lacking a laser in 1962 he could produce only an "existence proof". A two-colour reflecting hologram which could be illuminated with white light was first produced in 1965 by G. W. Stroke and A. Labeyrie (14) and is shown in Figure 27.

Since that time single-colour reflecting holograms have been developed to high perfection by new photographic processes, by K. S. Pennington (15) and others, with reflectances approaching 100 percent, but two; and even more, three-colour holograms are still far from being satisfactory. It is one of my chief preoccupations at the present to improve this situation, but it would take too long, and it would be also rather early to enlarge on this.

An application of holography which is certain to gain high importance in the next years is information storage. I have mentioned before that holography allows storing 100—300 times more printed pages in a given emulsion than ordinary microphotography. Even without utilizing the depth dimension, the factor is better than 50. The reason is that a diffused hologram represents almost ideal coding, with full utilization of the area and of the gradation of the emulsion, while printed matter uses only about 5—10 % of the area, and the gradation not at all. A further factor arises from the utilization of the third dimension, the depth of the emulsion. This possibility was first pointed out in an ingenious paper by P. J. van Heerden (16), in 1963. Theoretically it appears possible to store one bit of information in about one wavelength cube. This is far from being practical, but the figure of 300, previously mentioned, is entirely realistic.

However, even without this enormous factor, holographic storage offers important advantages. A binary store, in the form of a checkerboard pattern on microfilm can be spoiled by a single grain of dust, by a hair or by a scratch, while a diffused hologram is almost insensitive to such defects. The holographic store, illustrated in Figure 28, is according to its author L. K. Anderson (17) (1968) only a modest beginning, yet it is capable of accessing for instance any one of 64×64 printed pages in about a microsecond. Each hologram, with a diameter of 1.2 mm can contain about 10^4 bits. Reading out this information

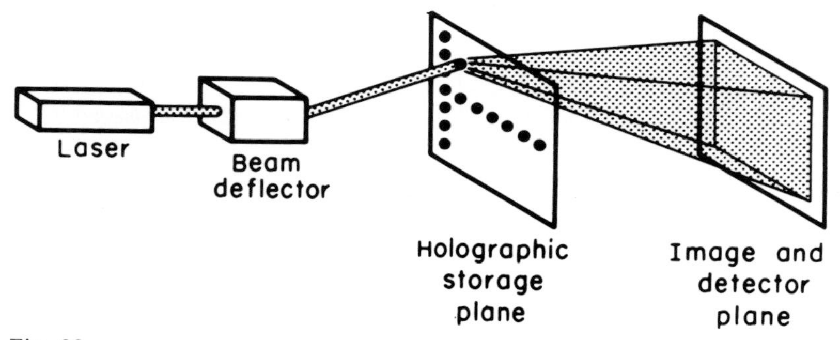

Fig. 28.
Holographic Flying Spot Store. L. K. Anderson and R. J. Collier, Bell Telephone Laboratories, 1968.

Fig. 27.
First two-colour Reflecting Hologram, Reconstructed in White Light. G. W. Stroke and A. Labeyrie, 1965.

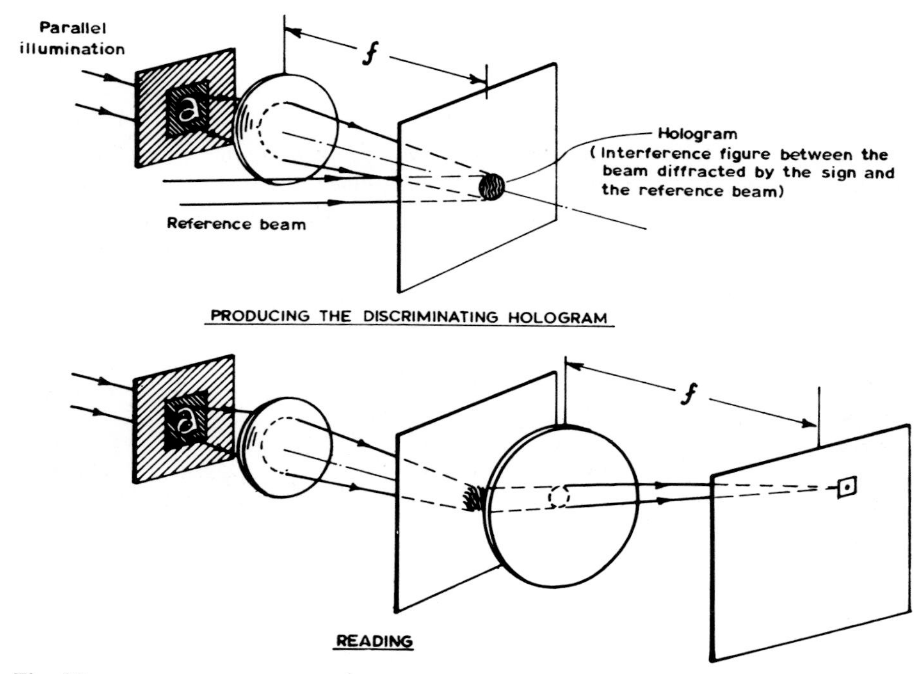

Fig. 29.
The Principle of Character Recognition by Holography.

sequentially in a microsecond would of course require an impossible waveband, but powerful parallel reading means can be provided. One can confidently expect enormous extensions of these "modest beginnings" once the project of data banks will be tackled seriously.

Another application of holography, which is probably only in an early stage, is pattern and character recognition. I can only briefly refer to the basic work which A. Vander Lugt (18) has done in the field of pattern recognition. It will be sufficient to explain the basic principle of character recognition with the aid of Figure 29.

Let us generalize a little the basic principle of holography. In all previous examples a complicated object beam was brought to interference with a simple or spherical reference beam, and the object beam was reconstructed by illuminating the hologram with the reference beam. But a little mathematics shows that this can be extended to any reference beam *which correlates sharply with itself*. The autocorrelation function is an invariant of a beam; it can be computed in any cross section. One can see at once that a spherical wave correlates sharply with itself, because it issues from a "point". But there are other beams which correlate sharply with themselves, for instance those which issue from a fingerprint, or from a Chinese ideogram, in an extreme case also those which issue from a piece of frosted glass. Hence it is quite possible for instance to translate, by means of a hologram, a Chinese ideogram into its corresponding English sentence and *vice versa*. J. N. Butters and M. Wall in Loughborough University have recently created holograms which from a portrait produce the signature of the owner, and *vice versa*.[4] In other words, a hologram can be a fairly universal translator. It can for instance translate a sign which we can read to another which a machine can read.

Figure 29 shows a fairly modest realisation of this principle. A hologram is made of a letter "a" by means of a plane reference beam. When this hologram is illuminated with the letter "a" the reference beam is reconstructed, and can activate for instance a small photocell in a certain position. This, I believe, gives an idea of the basic principle. There are of course many ways of printing letters, but it would take me too long to explain how to deal with this and other difficulties.

Fig. 30.
Laser Speckle. The appearance of e.g. a white sheet of paper, uniformly illuminated by laser light.

With character recognition devices we have already taken half a step into the future, because these are likely to become important only in the next generation of computers or robots, to whom we must transfer a little more of human intelligence. I now want to mention briefly some other problems which are half or more than half in the future.

One, which is already very actual, is the overcoming of laser speckle. Everybody who sees laser light for the first time is surprised by the rough appearance of objects which we consider as smooth. A white sheet of paper appears as if it were crawling with ants. The crawling is put into it by the restless eye, but the roughness is real. It is called "laser speckle" and Figure 30 shows a characteristic example of it. This is the appearance of a white sheet of paper in laser light, when viewed with a low-power optical system. It is not really noise; it is information which we do not want, information on the microscopic unevenness of the paper in which we are not interested. What can we do against it?

In the case of rough objects the answer is, regrettably, that all we can do is to average over larger areas, thus smoothing the deviations. This means that we must throw a great part of the information away, the wanted with the unwanted. This is regrettable but we can do nothing else, and in many cases we have enough information to throw away, as can be seen by the fully satisfactory appearance of some of the reconstructions from diffuse holograms which I have shown. However, there are important areas in which we can do much more, and where an improvement is badly needed. This is the area of microholograms, for storing and for display. They are made as diffused holograms, in order to ensure freedom from dust and scratches, but by making them diffused, we introduce speckle, and to avoid this such holograms are made nowadays much larger than would be ideally necessary. I have shown recently (19), that advantages of diffuse holograms can be almost completely retained, while the speckle can be completely eliminated by using, instead of a frosted glass, a special illuminating system. This, I hope will produce a further improvement in the information density of holographic stores.

Now let us take a more radical step into the future. I want to mention briefly two of my favourite holographic brainchilden. The first of this is Panoramic Holography, or one could also call it Holographic Art.

All the tree-dimensional holograms made so far extend to a depth of a few meters only. Would it not be possible to extend them to infinity? Could one not put a hologram on the wall, which is like a window through which one looks at a landscape, real or imaginary? I think it can be done, only it will not be a photograph but a work of art. Figure 31 illustrates the process. The artist makes a model, distorted in such a way that it appears perspectivic, and extending to any distance when viewed through a large lens, as large as the hologram. The artist can use a smaller lens, just large enough to cover both his eyes when making the model. A reflecting hologram is made of it, and illuminated with a strong, small light source. The viewer will see what the plate has seen through the lens; that is to say a scene extending to any distance, in natural colours. This scheme is under development, but considerable work

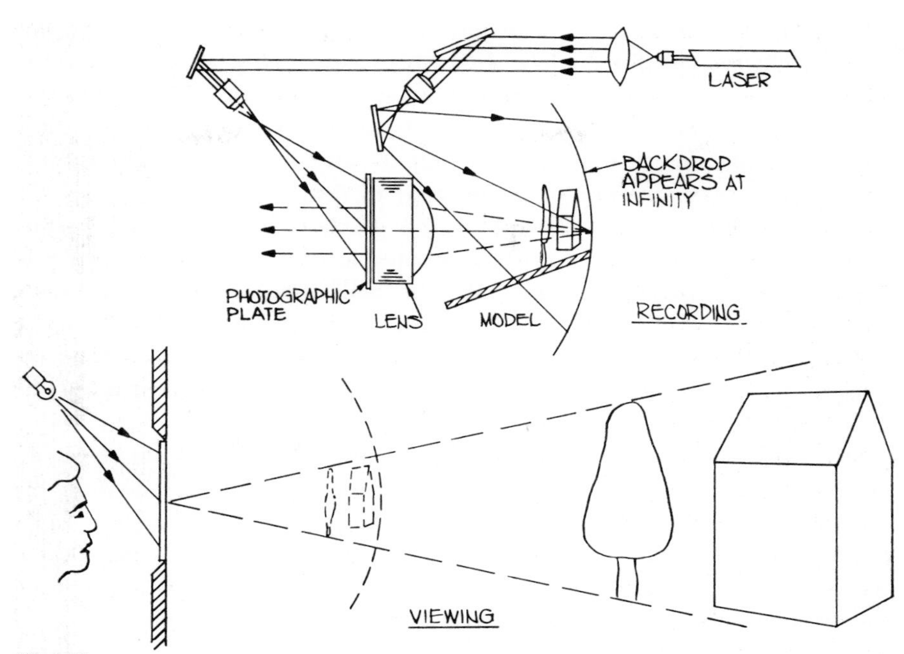

Fig. 31.
Panoramic Holography.

will be needed to make it satisfactory, because we must first greatly improve the reflectance of three-colour holograms.

An even more ambitious scheme, probably even farther in the future, is three-dimensional cinematography, without viewing aids such as Polaroids. The problem is sketched out in Figure 32. The audience (in one plane or two)

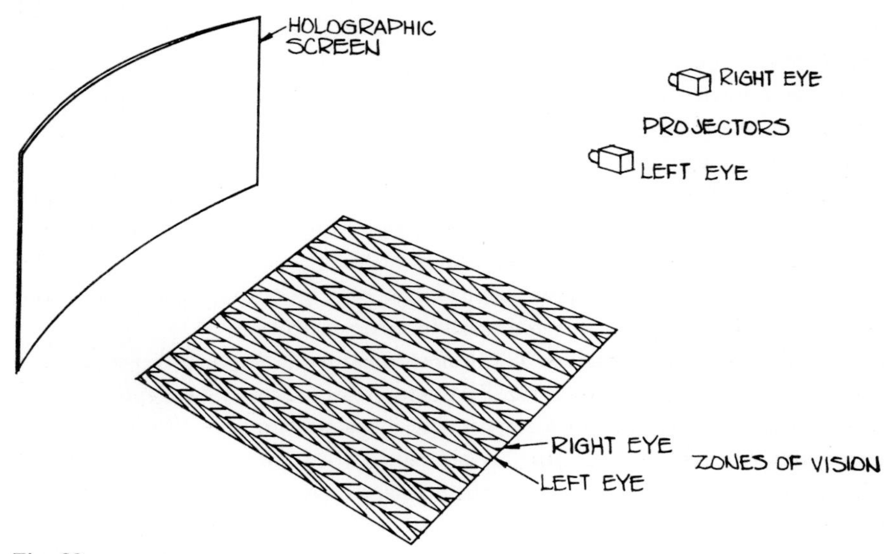

Fig. 32.
3-D Cinematography with Holographic Screen.

is covered by zones of vision, with the width of the normal eye spacing, one for the right eye, one for the left, with a blank space between two pairs. The two eyes must see two different pictures; a stereoscopic pair. The viewer can move his head somewhat to the right or left. Even when he moves one eye into the blank zone, the picture will appear dimmer but not flat, because one eye gives the impression of "stereoscopy by default".

I have spent some years of work on this problem, just before holography, until I had to realise that it is strictly unsolvable with the orthodox means of optics, lenticules, mirrors, prisms. One can make satisfactorily small screens for small theatres, but with large screens and large theatres one falls into a dilemma. If the lenticules, or the like, are large, they will be seen from the front seats; if they are small, they will not have enough definition for the back seats.

Some years ago I realised to my surprise, that holography can solve this problem too. Use a projector as the reference source, and for instance the system of left viewing zones as the object. The screen, covered with a Lippmann emulsion, will then make itself automatically into a very complicated optical system such that when a picture is projected from the projector, it will be seen only from the left viewing zones. One then repeats the process with the right projector, and the right viewing zones. Volume, (Lippmann-Denisyuk) holograms display the phenomenon of directional selectivity. If one displaces the illuminator from the original position by a certain angle, there will be no reflection. We put the two projectors at this angle (or a little more) from one another, and the effect is that the right picture will not be seen by the left eye and vice versa.

There remains of course one difficulty, and this is that one cannot practise holography on the scale of a theatre, and with a plate as large as a screen. But this too can be solved, by making up the screen from small pieces, not with the theatre but with a *model* of the theatre, seen through a lens, quite similar to the one used in panoramic holography.

I hope I have conveyed the feasibility of the scheme, but I feel sure that I have conveyed also its difficulties. I am not sure whether they will be overcome in this century, or in the next.

Ambitious schemes, for which I have a congenital inclination, take a long time for their realisation. As I said at the beginning, I shall be lucky if I shall be able to see in my lifetime the realisation of holographic electron microscopy, on which I have started 24 years ago. But I have good hope, because I have been greatly encouraged by a remarkable achievement of G. W. Stroke (20), which is illustrated in Figure 33. Professor Stroke has recently suceeded in deblurring micrographs taken by Professor Albert Crewe, Chicago, with his scanning transmission electron microscope, by a holographic filtering process, improving the resolution from 5 Angstrom to an estimated 2.5 Angstrom. This is not exactly holographic electron microscopy, because the original was not taken with coherent electrons, but the techniques used by both sides, by A. Crewe and by G. W. Stroke are so powerful, that I trust them to succeed also in the next, much greater and more important step.

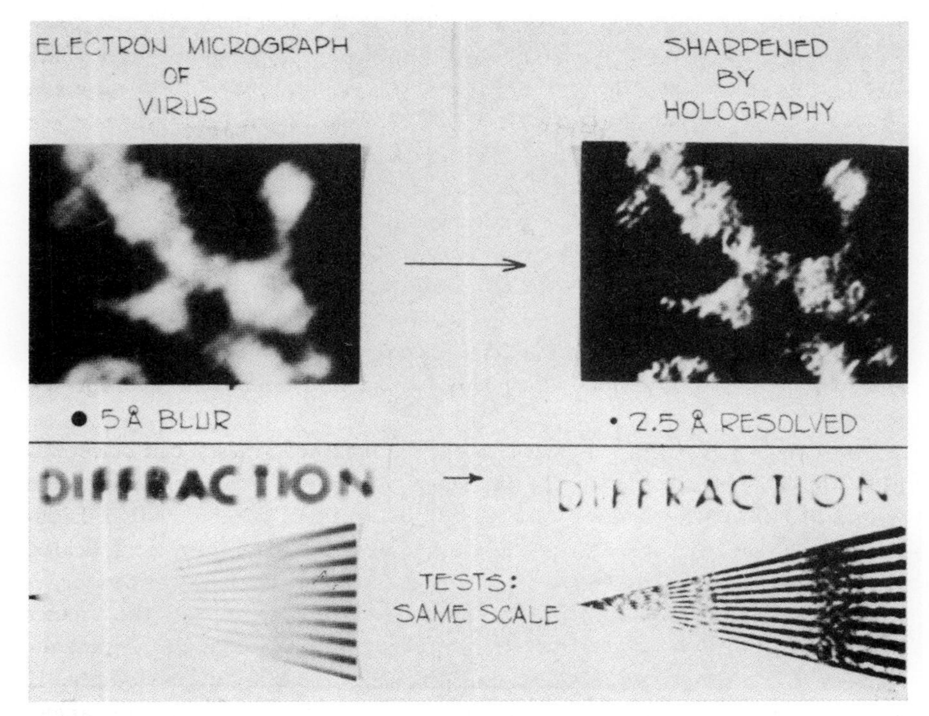

Fig. 33.
Scanning Transmission Electron Micrograph. Professor Albert Crewe, University of Chicago, holographically deblurred by Professor G. W. Stroke, 1971. The bottom photographs prove that the effect could not be obtained by hard printing, because some spatial frequencies which appear in the original with reversed phase had to be phase-corrected.

Summing up, I am one of the few lucky physicists who could see an idea of theirs grow into a sizeable chapter of physics. I am deeply aware that this has been achieved by an army of young, talented and enthusiastic researchers, of whom I could mention only a few by name. I want to express my heartfelt thanks to them, for having helped me by their work to this greatest of scientific honours.

REFERENCES

It is impossible to do justice to the hundreds of authors who have significantly contributed to the development of holography. The number of articles exceeds 2 000, and there are more than a dozen books in several languages.

An extensive bibliography may be found for instance in: T. Kallard, Editor, Holography (Optosonic Press, New York, 1969 and 1970)

BOOKS

Barrekette, E. S., Kock, W. E., Ose, T., Tsujiuchi, J. and Stroke, G. W., Eds. Applications of Holography (Plenum Press, New York, 1971).

Caulfield, H. J. and Sun Lu, The Applications of Holography (Wiley Interscience, New York, 1970).

Collier, R. J., Burckhardt, C. B. and Lin, L. H., Optical Holography (Academic Press, New York, 1971).

DeVelis, J. B. and Reynolds, G. O., Theory and Applications of Holography (Addison Wesley, Reading, Massachusetts, 1967).

Françon, M., Holographie (Masson et Cie, Paris, 1969).

Kiemle, H. und Röss, D., Einführung in die Technik der Holographie (Akademische Verlagsgesellschaft, Frankfurt am Main, 1969). *Ibid.* Introduction to Holographic Techniques (Plenum Press, 1972, in print).

Kock, W. E., Lasers and Holography (An Introduction to Coherent Optics) (Doubleday & Co., Garden City, NY. Y, 1969).

Ostrovsky, Yu. I., Holography (in Russian) Nauka, Leningrad, 1970).

Robertson E. R. and Harvey, J. M., Eds., The Engineering Uses of Holography (Cambridge University Press, 1970).

Stroke, G. W., An Introduction to Coherent Optics and Holography (Academic Press, New York: 1st Edition, 1966, Second Edition, 1969).

Viénot, J. Ch., Smigielski, P. et Royer, H., Holographie Optique (Developpements, Applications) (Dunod, Paris, 1971).

1. Gabor, D., "A New Microscopic Principles", Nature **161,** No. 4098, 777—778 (1948).
   Gabor, D., "Microscopy by Reconstructed Wavefronts", Proc. Roy. Soc. (London) **A 197,** 454—487 (1949).
   Gabor, D., "Microscopy by Reconstructed Wavefronts: II", Proc. Phys. Soc. (London) **64** (Pt. 6) No. 378 B, 449—469 (1951).
2. Rogers, G. L., "Experiments in Diffraction Microscopy", Proc. Roy. Soc. (Edinburgh) **63 A,** 193 (1952).
3. Baez, A., "Resolving Power in Diffraction Microscopy", Nature, **169,** 963—964 (1952).
4. El-Sum, H. M. A. and Kirkpatrick, P., "Microscopy by Reconstructed Wavefronts" Phys. Rev., **85,** 763 (1952).
5. Gabor, D. and Goss, W. P., "Interference Microscope with Total Wavefront Reconstruction", J. Opt. Soc. Am. **56,** 849—858 (1966).
6. Cutrona, L. J., Leith, E. N., Porcello, L. J. and Vivian, W. E., "On the Application of Coherent Optical Processing Techniques to Synthetic Aperture Radar", Proc. IEEE, **54,** 1026—1032 (1966).
7. Leith, E. N. and Upatnieks, J., "Wavefront Reconstruction with Continuous Tone Transparencies", J. Opt. Soc. Am. **53,** 522 (1963) (Abstract).
   Leith, E. N. and Upatnieks, J., "Wavefront Reconstruction with Continuous-Tone Objects", J. Opt. Soc. Am. **53,** 1377—1381 (1963).
   Leith, E. N., and Upatnieks, J., "Wavefront Reconstruction with Diffused Illumination and Three-Dimensional Objects" J. Opt. Soc. Amer., **54,** 1295 (1964.)
8. An early reference is:
   Stroke, G. W., "Theoretical and Experimental Foundations of High-Resolution Optical Holography" (Presented in Rome on 14 September 1964) in Vol. II, pp. 53—63 of Pubblicazioni IV Centenario della Nascita di Galileo Galilei (G. Barbera, Firenze, 1966).
9. Knox, C. and Brooks, R. E., "Holographic Motion Picture Microscopy" Proc. Roy. Soc. (London), **B 174,** 115—121 (1969).
10. Heflinger, L. O., Wuerker, R. F. and Brooks, R. E., J. Appl. Physics, **37,** No. 2, 642—649 (February 1966).
11. Burch, J. M., "The Application of Lasers in Production Engineering", Production Engineer, **44,** 431—442 (1965).
    Powell, R. L. and Stetson, K. A., "Interferometric Vibration Analysis by Wavefront Reconstruction", J. Opt. Soc. Am. **55,** 1593—1598 (1965).
    Stroke, G. W. and Labeyrie, A. E., "Two-Beam Interferometry by Successive Recording of Intensities in a Single Hologram", Appl. Physics Letters, **8,** No. 2, 42—44 (15 January 1966).
12. Stetson, K. A., Thesis (under direction of E. Ingelstam), Royal Institute of Technology, Stockholm (1969).
13. Denisyuk, Yu. N., "Photographic Reconstruction of the Optical Properties of an Object in its Own Scattered Radiation", Dokl. Akad. Nauk. SSR, **144,** 1275—1278 (1962).
14. Stroke, G. W. and Labeyrie, A. E., "White-Light Reconstruction of Holographic Images using the Lippmann-Bragg Diffraction Effect", Physics Letters, **20,** 368—370 (1966).
15. Pennington, K. S. and Harper, J. S., "Techniques for Producing Low-Noise, Improved-Efficiency Holograms", Appl. Optics, **9,** 1643—1650 (1970).
16. Van Heerden, P. J., "A New Method of Storing and Retrieving Information", Appl. Optics, **2,** 387—392 (1963).

17. Anderson, L. K., "Holographic Optical Memory for Bulk Data Storage", Bell Lab Rep. **46,** 318 (1968).
18. Vander Lugt, A., "Signal Detection by Complex Spatial Filtering", IEEE Trans. on Information Theory, **IT—10,** 139—145 (1964).
19. Gabor, D., "Laser Speckle and its Elimination", IBM J. of Res. and Dev. **14,** 509—514 (Sept. 1970).
20. Stroke, G. W., "Image Deblurring and Aperture Synthesis Using 'A Posteriori' Processing by Fourier-Transform Holography", Optica Acta, **16,** 401—422 (1969). Stroke, G. W. and Halioua, M., "Attainment of Diffraction-Limited Imaging in High-Resolution Electron Microscopy by 'A Posteriori' Holographic Image Sharpening", Optik (1972) (in print).

Physics 1972

JOHN BARDEEN, LEON N COOPER and J ROBERT SCHRIEFFER

for their jointly developed theory of superconductivity, usually called the BCS–theory

THE NOBEL PRIZE FOR PHYSICS

Speech by professor STIG LUNDQVIST, Chalmers University of Technology
Translation from the Swedish text

Your Royal Highnesses, Ladies and Gentlement,
The 1972 Nobel Prize for physics has been awarded to Drs John Bardeen, Leon N. Cooper and J. Robert Schrieffer for their theory of superconductivity, usually referred to as the BCS-theory.

Superconductivity is a peculiar phenomenon occurring in many metallic materials. Metals in their normal state have a certain electrical resistance, the magnitude of which varies with temperature. When a metal is cooled its resistance is reduced. In many metallic materials it happens that the electrical resistance not only decreases but also suddenly disappears when a certain critical temperature is passed which is a characteristic property of the material.

This phenomenon was discovered as early as 1911 by the Dutch physicist Kamerlingh Onnes, who was awarded the Nobel Prize for physics in 1913 for his discoveries.

The term superconductivity refers to the complete disappearance of the electrical resistance, which was later verified with an enormous accuracy. A lead ring carrying a current of several hundred ampères was kept cooled for a period of 2 1/2 years with no measurable change in the current.

An important discovery was made in the thirties, when it was shown that an external magnetic field cannot penetrate a superconductor. If you place a permanent magnet in a bowl of superconducting material, the magnet will hover in the air above the bowl, literally floating on a cushion of its own magnetic field lines. This effect may be used as an example for the construction of friction-free bearings.

Many of the properties of a metal change when it becomes superconducting and new effects appear which have no equivalent in the former's normal state. Numerous experiments have clearly shown that a fundamentally new state of the metal is involved.

The transition to the superconductive state occurs at extremely low temperatures, characteristically only a few degrees above absolute zero. For this reason practical applications of the phenomenon have been rare in the past and superconductivity has been widely considered as a scientifically interesting but exclusive curiosity confined to the low temperature physics laboratories. This state of affairs is rapidly changing and the use of superconducting devices is rapidly increasing. Superconducting magnets are often used for example in particle accelerators. Superconductivity research has in recent years resulted in substantial advances in measuring techniques and an extensive used in the computer field is also highly probable. Advanced plans for the use of superconductivity in heavy engineering are also in existence. By way of an example, it may be mentioned that the transport of electric energy to the major cities of the world with the use of superconductive lines is being planned. Looking further ahead one can see, for example, the possibility of building ultrarapid trains that run on superconducting tracks.

Superconductivity has been studied experimentally for more than sixty years. However, the central problem, the question of the physical mechanism responsible for the phenomenon remained a mystery until the late fifties. Many famous physicists tackled the problem with little success. The difficulties were related to the very special nature of the mechanism sought. In a normal metal the electrons more around individually at random, somewhat similar to the atoms in a gas, and the theory is, in principle, fairly simple. In superconductive metals the experiments suggested the existence of a collective state of the conduction electrons—a state in which the electrons are strongly coupled and their motion correlated so that there is a gigantic coherent state of macroscopic dimension containing an enormous number of electrons. The physical mechanism responsible for such a coupling remained unknown for a long time. An important step towards the solution was taken in 1950 when it was discovered simultaneously on theoretical and experimental grounds that superconductivity must be connected with the coupling of the electrons to the vibrations of the atoms in the crystal lattice. The conduction electrons are coupled to each other via these vibrations. Starting from this fundamental coupling of the electrons Bardeen, Cooper and Schrieffer developed their theory of superconductivity, published in 1957, which gave a complete theoretical explanation of the phenomenon of superconductivity.

According to their theory the coupling of the electrons to the lattice oscillations leads to the formation of bound pairs of electrons. These pairs play a fundamental role in the theory. The complete picture of the mechanism of superconductivity appeared when Bardeen, Cooper and Schrieffer showed that the motion of the different pairs is very strongly correlated and that this leads to the formation of a gigantic coherent state in which a large number of electrons participate. It is this ordered motion of the electrons in the superconductive state in contrast to the random individual motion in a normal crystal that gives superconductivity its special properties.

The theory developed by Bardeen, Cooper and Schrieffer together with extensions and refinements of the theory, which followed in the years after 1957, succeeded in explaining in considerable detail the properties of superconductors. The theory also predicted new effects and it stimulated intense activity in theoretical and experimental research which opened up new areas. These latter developments have led to new important discoveries which are being used in a number of interesting ways especially in the sphere of measuring techniques.

Developments in the field of superconductivity during the last fifteen years have been greatly inspired by the fundamental theory of superconductivity and have strikingly verified the validity and great range of the concepts and ideas developed by Bardeens, Cooper and Schrieffer.

Drs. Bardeen, Cooper and Schrieffer,

You have in your fundamental work given a complete theoretical explanation of the phenomenon of superconductivity. Your theory has also predicted new effects and stimulated an intensive activity in theoretical and experimental research. The further developments in the field of superconductivity have in a striking way confirmed the great range and validity of the concepts and ideas in your fundamental paper from 1957.

On behalf of the Royal Academy of Sciences I wish to convey to you the warmest congratulations and I now ask you to receive your prizes from the Hands of His Royal Highness the Crown Prince.

John Bardeen

JOHN BARDEEN

John Bardeen was born in Madison, Wisconsin, May 23, 1908.

He attended the University High School in Madison for several years, and graduated from Madison Central High School in 1923. This was followed by a course in electrical engineering at the University of Wisconsin, where he took extra work in mathematics and physics. After being out for a term while working in the engineering department of the Western Electric Company at Chicago, he graduated with a B.S. in electrical engineering in 1928. He continued on at Wisconsin as a graduate research assistant in electrical engineering for two years, working on mathematical problems in applied geophysics and on radiation from antennas. It was during this period that he was first introduced to quantum theory by Professor J. H. Van Vleck.

Professor Leo J. Peters, under whom his research in geophysics was done, took a position at the Gulf Research Laboratories in Pittsburgh, Pennsylvania. Dr. Bardeen followed him there and worked during the next three years (1930–33) on the development of methods for the interpretation of magnetic and gravitational surveys. This was a stimulating period in which geophysical methods were first being applied to prospecting for oil.

Because he felt his interests were in theoretical science, Dr. Bardeen resigned his position at Gulf in 1933 to take graduate work in mathematical physics at Princeton University. It was here, under the leadership of Professor E. P. Wigner, that he first became interested in solid state physics. Before completing his thesis (on the theory of the work function of metals) he was offered a position as Junior Fellow of the Society of Fellows at Harvard University. He spent the next three years there working with Professors Van Vleck and Bridgman on problems in cohesion and electrical conduction in metals and also did some work on the level density of nuclei. The Ph.D. degree at Princeton was awarded in 1936.

From 1938–41 Dr. Bardeen was an assistant professor of physics at the University of Minnesota and from 1941–45 a civilian physicist at the Naval Ordnance Laboratory in Washington, D. C. His war years were spent working on the influence fields of ships for application to underwater ordnance and minesweeping. After the war, he joined the solid-state research group at the Bell Telephone Laboratories, and remained there until 1951, when he was appointed Professor of Electrical Engineering and of Physics at the University of Illinois. Since 1959 he has also been a member of the Center for Advanced Study of the University.

Dr. Bardeen's main fields of research since 1945 have been electrical conduction in semiconductors and metals, surface properties of semiconductors, theory of superconductivity, and diffusion of atoms in solids. The Nobel Prize in Physics was awarded in 1956 to John Bardeen, Walter H. Brattain, and William Shockley for "investigations on semiconductors and the discovery of the transistor effect," carried on at the Bell Telephone Laboratories. In 1957, Bardeen and two colleagues, L. N. Cooper and J. R. Schrieffer, proposed the first successful explanation of superconductivity, which has been a puzzle since its discovery in 1908. Much of his research effort since that time has been devoted to further extensions and applications of the theory.

Dr. Bardeen died in 1991.

Born Madison, Wisconsin, May 23, 1908. Son of Dr. Charles R. and Althea Harmer Bardeen, both deceased. Dr. Bardeen was Professor of Anatomy and Dean of the Medical School at the University of Wisconsin. Stepmother, Mrs. Kenelm McCauley, Milwaukee, Wisconsin. Married Jane Maxwell, 1938. Children: James M., William A., Elizabeth A. Attended public schools and university in Madison, Washington School, Madison, 1914–17, University High School, Madison, 1917–21, Madison Central High, 1921–23.

B.S. and M.S. in E. E., University of Wisconsin, 1928 and 1929.
Geophysicist, Gulf Research and Development Corp., Pittsburgh, PA, 1930–33.
Attended Graduate College, Princeton University, 1933–35, received Ph.D. in Math. Phys., 1936.
Junior Fellow, Society of Fellows, Harvard University, 1935–38.
Assistant Professor of Physics, University of Minnesota, 1938–41.
Physicist, Naval Ordnance Laboratory, 1941–45.
Research Physicist, Bell Telephone Laboratories, 1945–51.
Professor of Electrical Engineering and of Physics, University of Illinois, and a member of the Center for Advanced Study of the University, 1951–1975.
Emeritus, 1975–1991.
Served on U.S. President's Science Advisory Committee, 1959–62.
Member of the Board of Directors, Xerox Corporation, Rochester, New York, 1961–1974. Consultant, 1952–82.
Lorentz Professor, University of Leiden, Netherlands, 1975.
Visiting Professor, Karlsruhe, 1978; Grenoble, 1981; University of California, Santa Barbara, 1981, 1984; Nihon University, Tokyo, 1982.
White House Science Council, 1982–83.

Honorary degrees: Union College (1955), Wisconsin (1960), Rose Polytechnic Inst. (1966), Western Reserve (1966), Univ. of Glasgow (1967), Princeton (1968), Rensselaer Polytechnic Inst. (1969), Notre Dame (1970), Harvard (1973), Minnesota (1973), Illinois (1974), Michigan (1974), Pennsylvania

(1976), Delhi, India (1977), Indian Inst. of Tech., Madras, India (1977), Cambridge (U.K.) (1977), Georgetown (1980), St. Andrews (1980), Clarkson (1981).

Awards: Stuart Ballentine Medal, Franklin Inst. (1952); Buckley Prize, Am. Physical Soc. (1954); John Scott Medal, Philadelphia (1955); Nobel Prize (Physics) shared with W. H. Brattain and W. Shockley (1956); Vincent Bendix Award, Amer. Soc. Eng. Educ. (1964); National Medal of Science (1965); Michelson-Morley Award, Case-Western Reserve (1968); Medal of Honor, Inst. of Electrical and Electronics Eng. (1971); Nobel Prize (Physics) shared with L. N. Cooper and J. R. Schrieffer (1972); James Madison Medal, Princeton (1973); National Inventors Hall of Fame (1974); Franklin Medal, Franklin Inst. (1975); Presidential Medal of Freedom (1977); Washington Award, Western Soc. Eng. (1983); Founders Awards, Nat. Acad. Eng. (1984); Lomonosov Prize, USSR Acad. of Sci. (1988); The Harold Pender Award, University of Pennsylvania (1988).

Academic and Professional Societies: American Physical Society (President, 1968–69); IEEE (hon. mem.); National Academy of Sciences; National Academy of Engineering; American Academy of Sciences; American Philosophical Society; Foreign Member, Royal Society of London; Foreign Member, Indian Nat. Sci. Acad.; Honorary Fellow, The Institute of Physics (London); Foreign Member, Inst. of Electronics and Telecommunications (India); Hon. Member, Japan Academy; Hon. Doctor, Venezuelan Academy; Foreign Member, USSR Academy of Sciences; Pakistan Academy of Sciences, Corr. Mem., Hungarian Acad. Sci. and Austrian Acad. Sci.

ELECTRON-PHONON INTERACTIONS AND SUPERCONDUCTIVITY

Nobel Lecture, December 11, 1972

By JOHN BARDEEN

Departments of Physics and of Electrical Engineering
University of Illinois
Urbana, Illinois

1
INTRODUCTION

Our present understanding of superconductivity has arisen from a close interplay of theory and experiment. It would have been very difficult to have arrived at the theory by purely deductive reasoning from the basic equations of quantum mechanics. Even if someone had done so, no one would have believed that such remarkable properties would really occur in nature. But, as you well know, that is not the way it happened, a great deal had been learned about the experimental properties of superconductors and phenomenological equations had been given to describe many aspects before the microscopic theory was developed. Some of these have been discussed by Schrieffer and by Cooper in their talks.

My first introduction to superconductivity came in the 1930's and I greatly profited from reading David Shoenberg's little book on superconductivity, [1] which gave an excellent summary of the experimental findings and of the phenomenological theories that had been developed. At that time it was known that superconductivity results from a phase change of the electronic structure and the Meissner effect showed that thermodynamics could be applied successfully to the superconductive equilibrium state. The two fluid Gorter—Casimir model was used to describe the thermal properties and the London brothers had given their famous phenomenological theory of the electrodynamic properties. Most impressive were Fritz London's speculations, given in 1935 at a meeting of the Royal Society in London, [2] that superconductivity is a quantum phenomenon on a macroscopic scale. He also gave what may be the first indication of an energy gap when he stated that "the electrons be coupled by some form of interaction in such a way that the lowest state may be separated by a finite interval from the excited ones." He strongly urged that, based on the Meissner effect, the diamagnetic aspects of superconductivity are the really basic property.

My first abortive attempt to construct a theory, [3] in 1940, was strongly influenced by London's ideas and the key idea was small energy gaps at the Fermi surface arising from small lattice displacements. However, this work was interrupted by several years of wartime research, and then after the war I joined the group at the Bell Telephone Laboratories where my work turned to semiconductors. It was not until 1950, as a result of the discovery of the isotope effect, that I again began to become interested in superconductivity, and shortly after moved to the University of Illinois.

The year 1950 was notable in several respects for superconductivity theory. The experimental discovery of the isotope effect [4, 5] and the independent prediction of H. Fröhlich [6] that superconductivity arises from interaction between the electrons and phonons (the quanta of the lattice vibrations) gave the first clear indication of the directions along which a microscopic theory might be sought. Also in the same year appeared the phenomenological Ginzburg—Landau equations which give an excellent description of superconductivity near T_c in terms of a complex order parameter, as mentioned by Schrieffer in his talk. Finally, it was in 1950 that Fritz London's book [7] on superconductivity appeared. This book included very perceptive comments about the nature of the microscopic theory that have turned out to be remarkably accurate. He suggested that superconductivity requires "a kind of solidification or condensation of the average momentum distribution." He also predicted the phenomenon of flux quantization, which was not observed for another dozen years.

The field of superconductivity is a vast one with many ramifications. Even in a series of three talks, it is possible to touch on only a few highlights. In this talk, I thought that it might be interesting to trace the development of the role of electron-phonon interactions in superconductivity from its beginnings in 1950 up to the present day, both before and after the development of the microscopic theory in 1957. By concentrating on this one area, I hope to give some impression of the great progress that has been made in depth of understanding of the phenomena of superconductivity. Through developments by many people, [8] electron-phonon interactions have grown from a qualitative concept to such an extent that measurements on superconductors are now used to derive detailed quantitative information about the interaction and its energy dependence. Further, for many of the simpler metals and alloys, it is possible to derive the interaction from first principles and calculate the transition temperature and other superconducting properties.

The theoretical methods used make use of the methods of quantum field theory as adopted to the many-body problem, including Green's functions, Feynman diagrams, Dyson equations and renormalization concepts. Following Matsubara, temperature plays the role of an imaginary time. Even if you are not familiar with diagrammatic methods, I hope that you will be able to follow the physical arguments involved.

In 1950, diagrammatic methods were just being introduced into quantum field theory to account for the interaction of electrons with the field of photons. It was several years before they were developed with full power for application to the quantum statistical mechanics of many interacting particles. Following Matsubara, those prominent in the development of the theoretical methods include Kubo, Martin and Schwinger, and particularly the Soviet physicists, Migdal, Galitski, Abrikosov, Dzyaloshinski, and Gor'kov. The methods were first introduced to superconductivity theory by Gor'kov [9] and a little later in a somewhat different form by Kadanoff and Martin. [10] Problems of superconductivity have provided many applications for the powerful Green's function methods of many-body theory and these applications have helped to further develop the theory.

Diagrammatic methods were first applied to discuss electron-phonon interactions in normal metals by Migdal [11] and his method was extended to superconductors by Eliashberg. [12] A similar approach was given by Nambu. [13] The theories are accurate to terms of order $(m/M)^{1/2}$, where m is the mass of the electron and M the mass of the ion, and so give quite accurate quantitative accounts of the properties of both normal metals and superconductors.

We will first give a brief discussion of the electron-phonon interactions as applied to superconductivity theory from 1950 to 1957, when the pairing theory was introduced, then discuss the Migdal theory as applied to normal metals, and finally discuss Eliashberg's extension to superconductors and subsequent developments. We will close by saying a few words about applications of the pairing theory to systems other than those involving electron-phonon interactions in metals.

2
Developments from 1950—1957

The isotope effect was discovered in the spring of 1950 by Reynolds, Serin, et al, [4] at Rutgers University and by E. Maxwell [5] at the U. S. National Bureau of Standards. Both groups measured the transition temperatures of separated mercury isotopes and found a positive result that could be interpreted as $T_c M^{1/2} \simeq$ constant, where M is the isotopic mass. If the mass of the ions is important, their motion and thus the lattice vibrations must be involved.

Independently, Fröhlich, [6] who was then spending the spring term at Purdue University, attempted to develop a theory of superconductivity based on the self-energy of the electrons in the field of phonons. He heard about the isotope effect in mid-May, shortly before he submitted his paper for publication and was delighted to find very strong experimental confirmation of his ideas. He used a Hamiltonian, now called the Fröhlich Hamiltonian, in which interactions between electrons and phonons are included but Coulomb interactions are omitted except as they can be included in the energies of the individual electrons and phonons. Fröhlich used a perturbation theory approach and found an instability of the Fermi surface if the electron-phonon interaction were sufficiently strong.

When I heard about the isotope effect in early May in a telephone call from Serin, I attempted to revive my earlier theory of energy gaps at the Fermi surface, with the gaps now arising from dynamic interactions with the phonons rather than from small static lattice displacements. [14] I used a variational method rather than a perturbation approach but the theory was also based on the electron self-energy in the field of phonons. While we were very hopeful at the time, it soon was found that both theories had grave difficulties, not easy to overcome. [15] It became evident that nearly all of the self-energy is included in the normal state and is little changed in the transition. A theory involving a true many-body interaction between the electrons seemed to be required to account for superconductivity. Schafroth [16] showed that starting with the Fröhlich Hamiltonian, one cannot derive the Meissner effect in any order of perturbation theory. Migdal's theory, [11] supposedly correct to terms of order $(m/M)^{1/2}$, gave no gap or instability at the Fermi surface and no indication of superconductivity.

Of course Coulomb interactions really are present. The effective direct Coulomb interaction between electrons is shielded by the other electrons and the electrons also shield the ions involved in the vibrational motion. Pines and I derived an effective electron-electron interaction starting from a Hamiltonian in which phonon and Coulomb terms are included from the start. [17] As is the case for the Fröhlich Hamiltonian, the matrix element for scattering of a pair of electrons near the Fermi surface from exchange of virtual phonons is negative (attractive) if the energy difference between the electron states involved is less than the phonon energy. As discussed by Schrieffer, the attractive nature of the interaction was a key factor in the development of the microscopic theory. In addition to the phonon induced interaction, there is the repulsive screened Coulomb interaction, and the criterion for superconductivity is that the attractive phonon interaction dominate the Coulomb interaction for states near the Fermi surface. [18]

During the early 1950's there was increasing evidence for an energy gap at the Fermi surface. [19] Also very important was Pippard's proposed non-local modification [20] of the London electrodynamics which introduced a new length the coherence distance, ξ_0, into the theory. In 1955 I wrote a review article [17] on the theory of superconductivity for the Handbuch der Physik, which was published in 1956. The central theme of the article was the energy gap, and it was shown that Pippard's version of the electrodynamics would likely follow from an energy gap model. Also included was a review of electron-phonon interactions. It was pointed out that the evidence suggested that all phonons are involved in the transition, not just the long wave length phonons, and that their frequencies are changed very little in the normal-superconducting transition. Thus one should be able to use the effective interaction between electrons as a basis for a true many-body theory of the superconducting state. Schrieffer and Cooper described in their talks how we were eventually able to accomplish this goal.

3
Green's Function Method for Normal Metals

By use of Green's function methods, Migdal [11] derived a solution of Fröhlich's Hamiltonian, $H = H_{el}+H_{ph}+H_{el\text{-}ph}$, for normal metals valid for abritrarily strong coupling and which involves errors only of order $(m/M)^{1/2}$. The Green's functions are defined by thermal average of time ordered operators for the electrons and phonons, respectively

$$G = -i\langle T\psi(1)\psi^{+}(2)\rangle \quad (1a)$$

$$D = -i\langle T\emptyset(1)\emptyset^{+}(2)\rangle \quad (1b)$$

Here $\psi(r,t)$ is the wave field operator for electron quasi-particles and $\emptyset(r,t)$ for the phonons, the symbols 1 and 2 represent the space-time points (r_1,t_1) and (r_2,t_2) and the brackets represent thermal averages over an ensemble.

Fourier transforms of the Green's functions for $H_0 = H_{el}+H_{ph}$ for non-interacting electrons and phonons are

$$G_0(P) = \frac{1}{\omega_n - \varepsilon_0(k) + i\delta_k} \tag{2a}$$

$$D_0(Q) = \left\{\frac{1}{\nu_n - \omega_0(q) + i\delta} - \frac{1}{\nu_n + \omega_0(q) - i\delta}\right\}, \tag{2b}$$

where $P = (k,\omega_n)$ and $Q = (q,\nu_n)$ are four vectors, $\varepsilon_0(k)$ is the bare electron quasiparticle energy referred to the Fermi surface, $\omega_0(q)$ the bare phonon frequency and ω_n and ν_n the Matsubara frequencies

$$\omega_n = (2n+1)\pi i k_B T; \quad \nu_n = 2n\pi i k_B T \tag{3}$$

for Fermi and Bose particles, respectively.

As a result of the electron-phonon interaction, $H_{el\text{-}ph}$, both electron and phonon energies are renormalized. The renormalized propagators, G and D, can be given by a sum over Feynman diagrams, each of which represents a term in the perturbation expansion. We shall use light lines to represent the bare propagators, G_0 and D_0, heavy lines for the renormalized propagators, G and D, straight lines for the electrons and curly lines for the phonons.

The electron-phonon interaction is described by the vertex

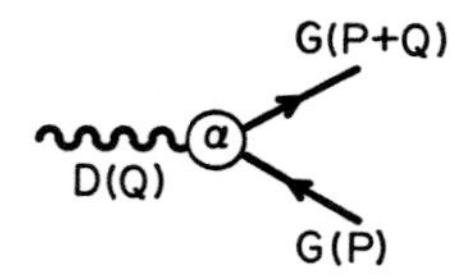

which represents scattering of an electron or hole by emission or absorption of a phonon or creation of an electron and hole by absorption of a phonon by an electron in the Fermi sea. Migdal showed that renormalization of the vertex represents only a small correction, of order $(m/M)^{1/2}$, a result in accord with the Born-Oppenheimer adiabatic-approximation. If terms of this order are neglected, the electron and phonon self-energy corrections are given by the lowest order diagrams provided that fully renormalized propagators are used in these diagrams.

The electron self-energy $\Sigma(P)$ in the Dyson equation:

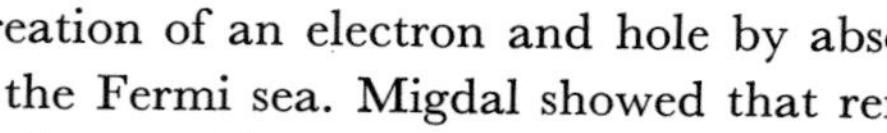

= +

$$G(P) = G_0(P) + G_0(P)\Sigma(P)G(P) \tag{4}$$

is given by the diagram

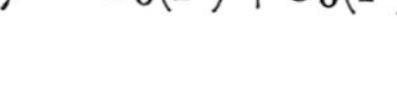

Σ= (5)

The phonon self-energy, $\pi(Q)$, defined by

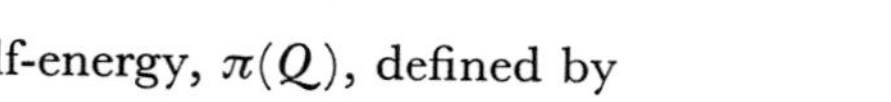

\+ = (6)

is given by

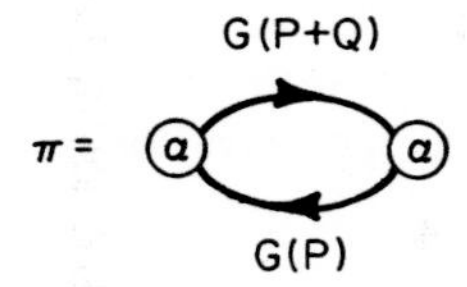

Since to order $(m/M)^{1/2}$ one can use an unrenormalized vertex function $\alpha = \alpha_0$, the Dyson equations form a closed system such that both $\Sigma(P)$ and $\pi(Q)$ can be determined. The phonon self-energy, $\pi(Q)$, gives only a small renormalization of the phonon frequencies. As to the electrons, Migdal noted that we are interested in states k very close to k_F, so that to a close approximation $\Sigma(k,\omega)$ depends only on the frequency. For an isotropic system,

$$\Sigma(k,\omega) \simeq \Sigma(k_F,\omega) \equiv \Sigma(\omega) \tag{7}$$

The renormalized electron quasi-particle energy, ω_k, is then given by a root of

$$\varepsilon(k) = \omega_k = \varepsilon_0(k) + \Sigma(\omega_k) \tag{8}$$

In the thermal Green's function formalism, one may make an analytic continuation from the imaginary frequencies, ω_n, to the real ω axis to determine $\Sigma(\omega)$.

Although $\Sigma(\omega)$ is small compared with the Fermi energy, E_F, it changes rapidly with energy and so can affect the density of states at the Fermi surface and thus the low temperature electronic specific heat. The mass renormalization factor m^*/m, at the Fermi surface may be expressed in terms of a parameter λ:

$$m^*/m = Z(k_F) = 1+\lambda = (d\varepsilon_0/dk)_F/(d\varepsilon/dk)_F \tag{9}$$

In modern notation, the experession for λ is

$$\lambda = 2\int_0^\infty d\omega \frac{\alpha^2(\omega)F(\omega)}{\omega}, \tag{10}$$

where $F(\omega)$ is the density of phonon states in energy and $\alpha^2(\omega)$ is the square of the electron-phonon coupling constant averaged over polarization directions of the phonons. Note that λ is always positive so that the Fermi surface is stable if the lattice is stable. Values of λ for various metals range from about 0.5 to 1.5. The parameter λ corresponds roughly to the $N(0)V_{\text{phonon}}$ of the BCS theory.

4 Nambu-Eliashberg Theory for Superconductors

Migdal's theory has important consequences that have been verified experimentally for normal metals, but gave no clue as to the origin of superconductivity. Following the introduction of the BCS theory, Gor'kov showed that pairing could be introduced through the anomalous Green's function

$$F(P) = i < T\psi_\uparrow\psi_\downarrow >, \tag{11}$$

Nambu showed that both types of Green's functions can be conveniently included with use of a spinor notation

$$\psi = \begin{pmatrix} \psi_\uparrow(r,t) \\ \\ \psi_\downarrow^+(r,t) \end{pmatrix} \tag{12}$$

where $\psi_\uparrow$ and $\psi_\downarrow$ are wave field operators for up and down spin electrons and a matrix Green's function with components

$$\tilde{G}_{\alpha\beta} = -\mathrm{i} < T\psi_\alpha \psi_\beta^+ > \tag{13}$$

Thus G_{11} and G_{22} are the single particle Green's functions for up and down spin particles and $G_{12} = G_{21}^* = F(P)$ is the anomalous Green's function of Gor'kov.

There are two self-energies, Σ_1 and Σ_2, defined by the matrix

$$\tilde{\Sigma} = \begin{pmatrix} \Sigma_1 & \Sigma_2 \\ \\ \Sigma_2 & \Sigma_1 \end{pmatrix} \tag{14}$$

Eliashberg noted that one can describe superconductors to the same accuracy as normal metals if one calculates the self-energies with the same diagrams that Migdal used, but with Nambu matrix propagators in place of the usual normal state Green's functions. The matrix equation for $\tilde{G}$ is

$$\tilde{G} = \tilde{G}_0 + \tilde{G}_0 \tilde{\Sigma} \tilde{G} \tag{15}$$

The matrix equation for $\tilde{\Sigma}$ yields a pair of coupled integral equations for Σ_1 and Σ_2. Again Σ_1 and Σ_2 depend mainly on the frequency and are essentially independent of the momentum variables. Following Nambu, [13] one may define a renormalization factor $Z_s(\omega)$ and a pair potential, $\Delta(\omega)$, for isotropic systems through the equations:

$$\omega Z_s(\omega) = \omega + \Sigma_1(\omega) \tag{16}$$

$$\Delta(\omega) = \Sigma_2(\omega)/Z(\omega). \tag{17}$$

Both Z_s and Δ can be complex and include quasi-particle life-time effects. Eliashberg derived coupled non-linear integral equations for $Z_s(\omega)$ and $\Delta(\omega)$ which involve the electron-phonon interaction in the function $\alpha^2(\omega)F(\omega)$.

The Eliashberg equations have been used with great success to calculate the properties of strongly coupled superconductors for which the frequency dependence of Z and Δ is important. They reduce to the BCS theory and to the nearly equivalent theory of Bogoliubov [21] based on the principle of "compensation of dangerous diagrams" when the coupling is weak. By weak coupling is meant that the significant phonon frequencies are very large compared with $k_B T_c$, so that $\Delta(\omega)$ can be regarded as a constant independent of frequency in the important range of energies extending to at most a few $k_B T_c$. In weak coupling one may also neglect the difference in quasi-particle energy renormalization and assume that $Z_s = Z_n$.

The first solutions of the Eliashberg equations were obtained by Morel and Anderson [22] for an Einstein frequency spectrum. Coulomb interactions were included, following Bogoliubov, by introducing a parameter μ^* which renormalizes the screened Coulomb interaction to the same energy range as the phonon interaction, In weak coupling, $N(0)V = \lambda - \mu^*$. They estimated λ from electronic specific heat data and μ^* from the electron density and thus the transition temperatures, T_c, for a number of metals. Order-of-magnitude agreement with experiment was found. Later work, based in large part on tunneling data, has yielded precise information on the electron-phonon interaction for both weak and strongly-coupled superconductors.

4
Analysis of Tunneling Data

From the voltage dependence of the tunneling current between a normal metal and a superconductor one can derive $\Delta(\omega)$ and thus get direct information about the Green's function for electrons in the superconductor. It is possible to go further and derive empirically from tunneling data the electron-phonon coupling, $\alpha^2(\omega)F(\omega)$, as a function of energy. That electron tunneling should provide a powerful method for investigating the energy gap in superconductors was suggested by I. Giaever, [23] and he first observed the effect in the spring of 1960.

The principle of the method is illustrated in Fig. 1. At very low temperatures, the derivative of the tunneling current with respect to voltage is proportional to the density of states in energy in the superconductor. Thus the ratio of the density of states in the metal in the superconducting phase, $\mathcal{N}_s$, to that of the same metal in the normal phase, $\mathcal{N}_n$, at an energy eV above the Fermi surface is given by

$$\frac{\mathcal{N}_s(eV)}{\mathcal{N}_n} = \frac{(dI/dV)_{ns}}{(dI/dV)_{nn}} \tag{18}$$

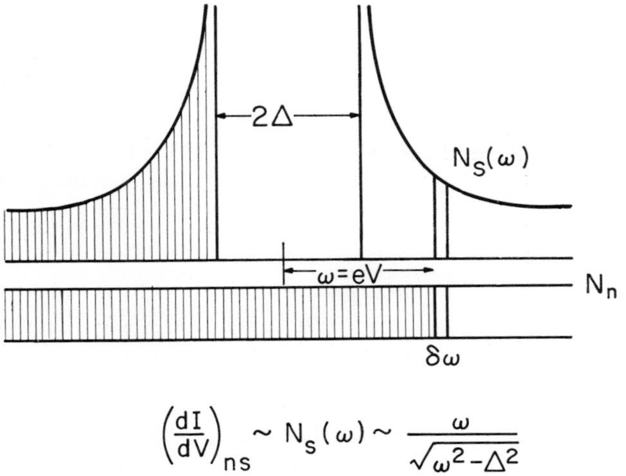

$$\left(\frac{dI}{dV}\right)_{ns} \sim N_s(\omega) \sim \frac{\omega}{\sqrt{\omega^2-\Delta^2}}$$

Tunneling from a normal metal into a superconductor

Fig. 1.
Schematic diagram illustrating tunneling from a normal metal into a superconductor near $T = 0°$K. Shown in the lower part of the diagram is the uniform density of states in energy of electrons in the normal metal, with the occupied states shifted by an energy eV from an applied voltage V across the junction. The upper part of the diagram shows the density of states in energy in the superconductor, with an energy gap 2Δ. The effect of an increment of voltage δV giving an energy change $\delta\omega$ is to allow tunneling from states in the range $\delta\omega$. Since the tunneling probability is proportional to density of states $N_s(\omega)$, the increment in current δI is proportional to $N_s(\omega)\delta V$.

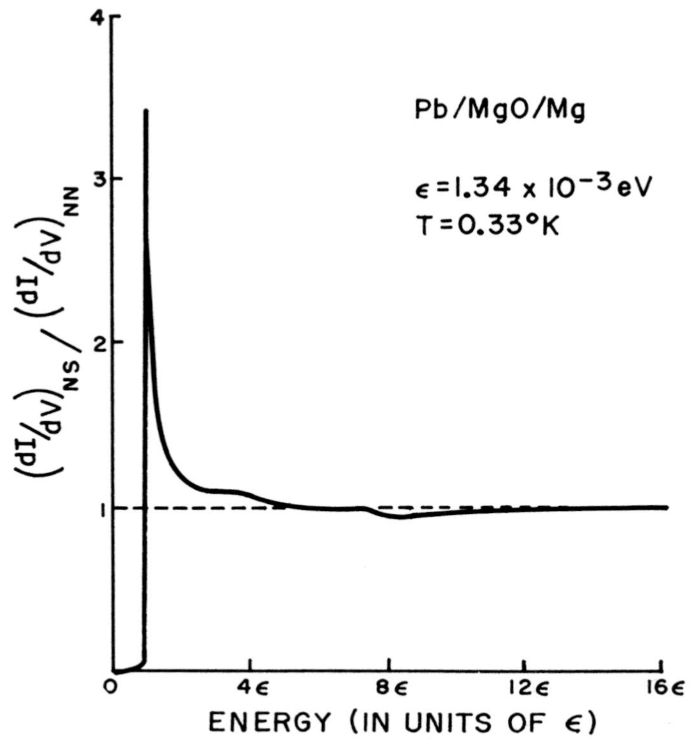

Fig. 2.
Conductance of a Pb-Mg junction as a function of applied voltage (from reference 24).

The normal density is essentially independent of energy in the range involved (a few meV). In weak coupling superconductors, for a voltage V and energy $\omega = eV$,

$$\frac{\mathcal{N}_{\mathrm{s}}(\omega)}{\mathcal{N}_{\mathrm{n}}} = \frac{\omega}{\sqrt{\omega^2 - \Delta^2}}. \tag{19}$$

As $T \to 0$ K, no current flows between the normal metal and the superconductor until the applied voltage reaches Δ/e, when there is a sharp rise in $\mathrm{d}I/\mathrm{d}V$ followed by a drop. This is illustrated in Fig. 2 for the case of Pb.

The first experiments of Giaever were on aluminum, which is a weak coupling superconductor. Good agreement was found between theory and experiment. In later measurements on tunneling into Pb, a strongly coupled superconductor, Giaever, Hart and Megerle [24] observed anomalies in the density of states that appeared to be associated with phonons, as shown in Fig. 2. These results were confirmed by more complete and accurate tunneling data on Pb by J. M. Rowell et al. [25]

In the meantime, in the summer of 1961, Schrieffer had derived numerical solutions of the Eliashberg equations working with a group engaged in developing methods for computer control using graphical display methods. [26] He and co-workers calculated the complex $\Delta(\omega)$ for a Debye frequency spectrum. Later, at the University of Pennsylvania, he together with J. W. Wilkins and D. J. Scalapino [27] continued work on the problem with a view to explaining the observed anomalies on Pb. They showed that for the general case of a complex $\Delta(\omega)$

$$\frac{(dI/dV)_{ns}}{(dI/dV)_{nn}} = \frac{N_s(\omega)}{N_n} = \text{Re}\left\{\frac{\omega}{\sqrt{\omega^2 - \Delta^2(\omega)}}\right\} \tag{20}$$

where Re represents the real part. From measurements of the ratio over the complete range of voltages, one can use Kramers-Krönig relations to obtain both the real and imaginary parts of $\Delta(\omega) = \Delta_1(\omega) + i\Delta_2(\omega)$. From analysis of the data, one can obtain the Green's functions which in turn can be used to calculate the various thermal and transport properties of superconductors. This has been done with great success, even for such strongly-coupled super conductors as lead and mercury.

For lead, Schrieffer et al, used a phonon spectrum consisting of two Lorentzian peaks, one for transverse waves and one for longitudinal and obtained a good fit to the experimental data for $T << T_c$. The calculations were extended up to T_c for Pb, Hg, and Al by Swihart, Wada and Scalapino, [28] again finding good agreement with experiment.

In analysis of tunneling data, one would like to find a phonon interaction spectrum, $\alpha^2(\omega)F(\omega)$, and a Coulomb interaction parameter, μ^*, which when inserted into the Eliashberg equations will yield a solution consistent with the tunneling data. W. L. McMillan devised a computer program such that one could work backwards and derive $\alpha^2(\omega)F(\omega)$ and μ^* directly from the tunneling data. His program has been widely used since then and has been applied to a number of superconducting metals and alloys, including, Al, Pb, Sn, the transition elements Ta and Nb, a rare earth, La, and the compound Nb_3Sn. In all cases it has been found that the phonon mechanism is dominant with reasonable values of μ^*. Peaks in the phonon spectrum agree with peaks in the phonon density of states as found from neutron scattering data, as shown in Fig. 3 for the case of Pb. In Fig. 4 is shown the real and imaginary parts of $\Delta(\omega)$ for Pb as derived from tunneling data.

One can go further and calculate the various thermodynamic and other properties. Good agreement with experiment is found for strongly coupled superconductors even when there are significant deviations from the weak coupling limits. For example, the weak-coupling BCS expression for the condensation energy at $T = 0$ K is

$$E_{BCS} = \frac{1}{2} N(0) Z_n \Delta_o^2 \tag{21}$$

where $N(0)Z_n$ is the phonon enhanced density of states and Δ_o is the gap parameter at $T = 0$ K. The theoretical expression with $Z_s(\omega)$ and $\Delta(\omega)$ derived from tunneling data, again for the case of Pb, gives [29, 30, 31]

$$E_{theor} = 0.78\, E_{BCS} \tag{22}$$

in excellent agreement with the experimental value

$$E_{exp} = (0.76 + 0.02)\, E_{BCS}. \tag{23}$$

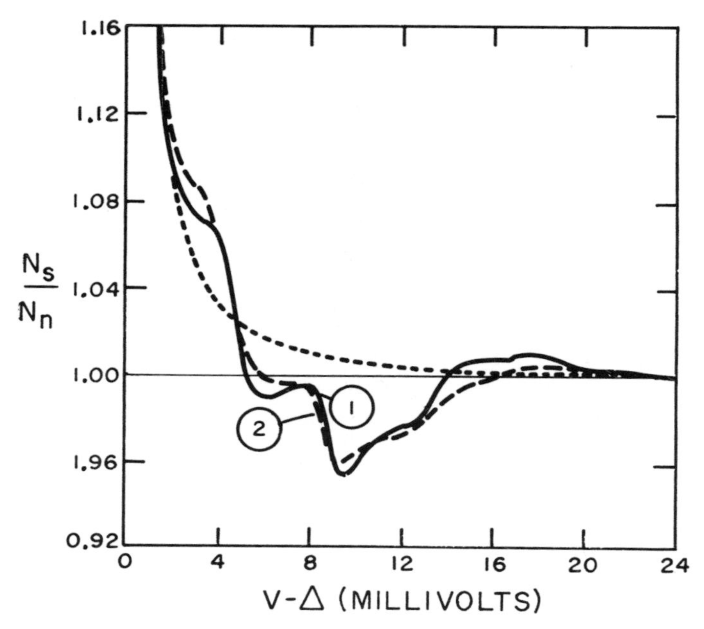

Fig. 3.
Density of states versus energy for Pb. Solid line, calculated by Schrieffer et al; long dashed ine, observed from tunneling; short dashed line, BCS weak coupling theory.

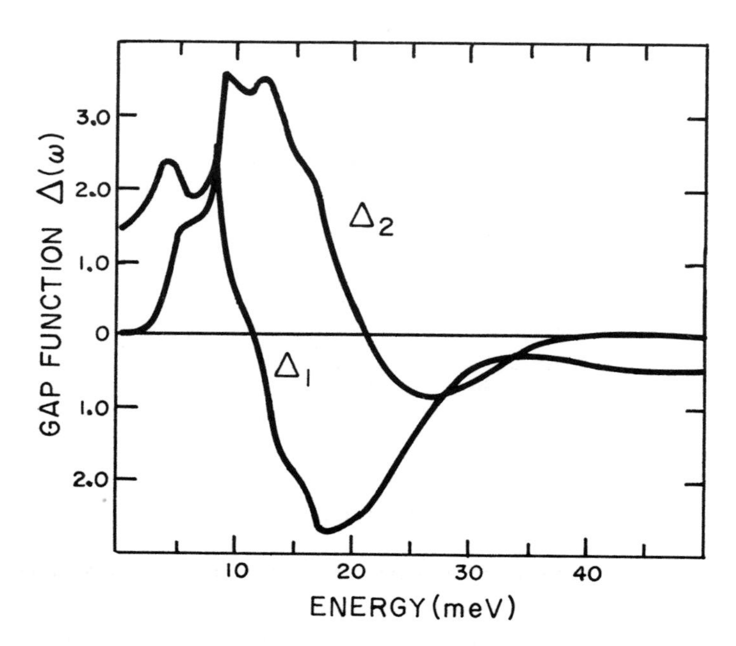

Fig. 4.
Real and imaginary parts of $\Delta(\omega) = \Delta_1(\omega)+i\Delta_2(\omega)$ versus energy for Pb. (After McMillan & Rowell).

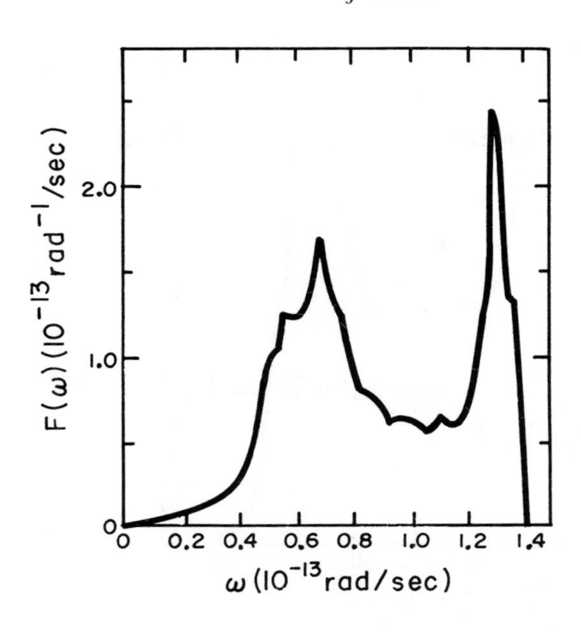

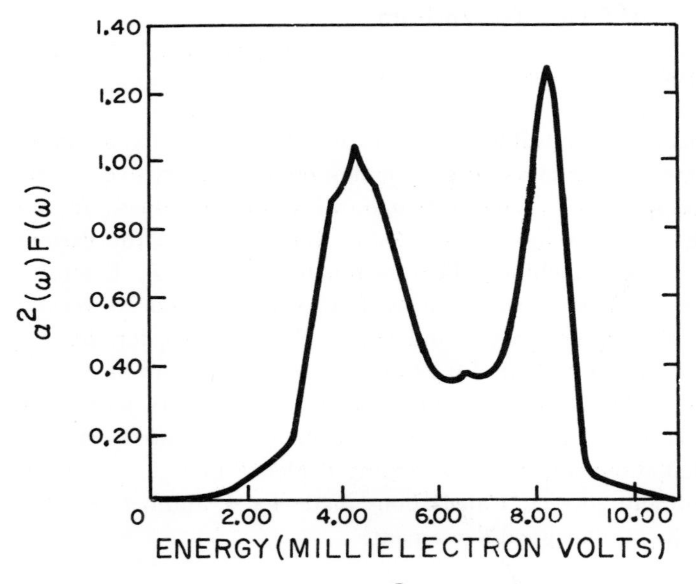

Comparison of $a^2F(a)$ and $F(\omega)$ for Pb (after McMillan and Rowell)

Fig. 5.
Comparison of α^2F for Pb derived from tunneling data with phonon density of states from neutron scattering data of Stedman et al. [8]

In Figs. 5, 6, 7, and 8 are shown other examples of $\alpha^2(\omega)F(\omega)$ derived from tunneling data for Pb, In, [31] La, [32] and Nb_3Sn. [33] In all cases the results are completely consistent with the phonon mechanism. Coulomb interactions play only a minor role, with μ^* varying only slowly from one metal to another, and generally in the range 0.1—02.

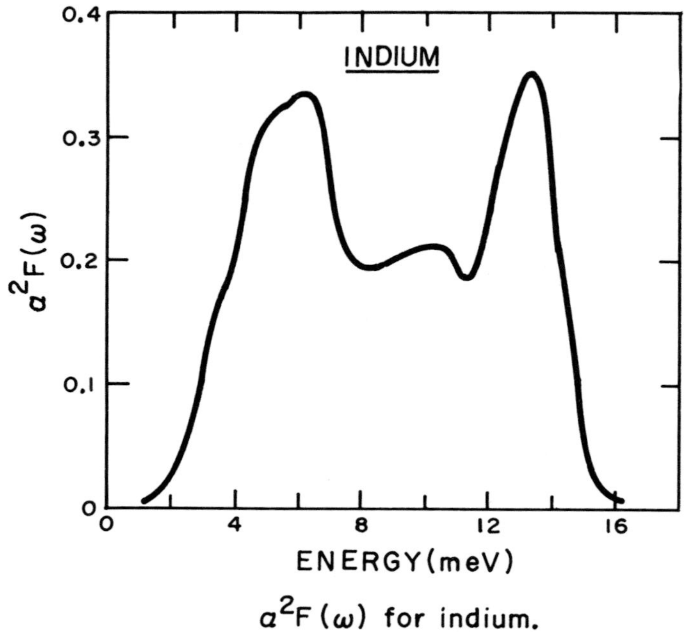

Fig. 6.
α^2F for In (after McMillan and Rowell).

As a further check, it is possible to derive the phonon density of states, $F(\omega)$ from neutron scattering data and use pseudo-potential theory to calculate the electron-phonon interaction parameter $\alpha_q(\omega)$. From these values, one can use the Eliashberg equations to calculate $Z_s(\omega)$ and $\Delta(\omega)$ and the various superconducting properties, including the transition temperature, T_c. Extensive calculations of this sort have been made by J. P. Carbotte and co-workers [34] for several of the simpler metals and alloys. For example, for the gap edge, Δ_0, in Al at $T = 0$ K they find 0.19 meV as compared with an experimental value of 0.17. The corresponding values for Pb are 1.49 meV from theory as compared with 1.35 meV from experiment. These are essentially first principles calculations and give convincing evidence that the theory as formulated is essentially correct. Calculations made for a number of other metals and alloys give similar good agreement.

Conclusions

In this talk we have traced how our understanding of the role of electron-phonon interactions in superconductivity has developed from a concept to a precise quantitative theory. The self-energy and pair potential, and thus the Green's functions, can be derived either empirically from tunneling data or directly from microscopic theory with use of the Eliashberg equations. Physicists, both experimental and theoretical, from different parts of the world have contributed importantly to these developments.

All evidence indicates that the electron-phonon interaction is the dominant mechanism in the cases studied so far, which include many simple metals,

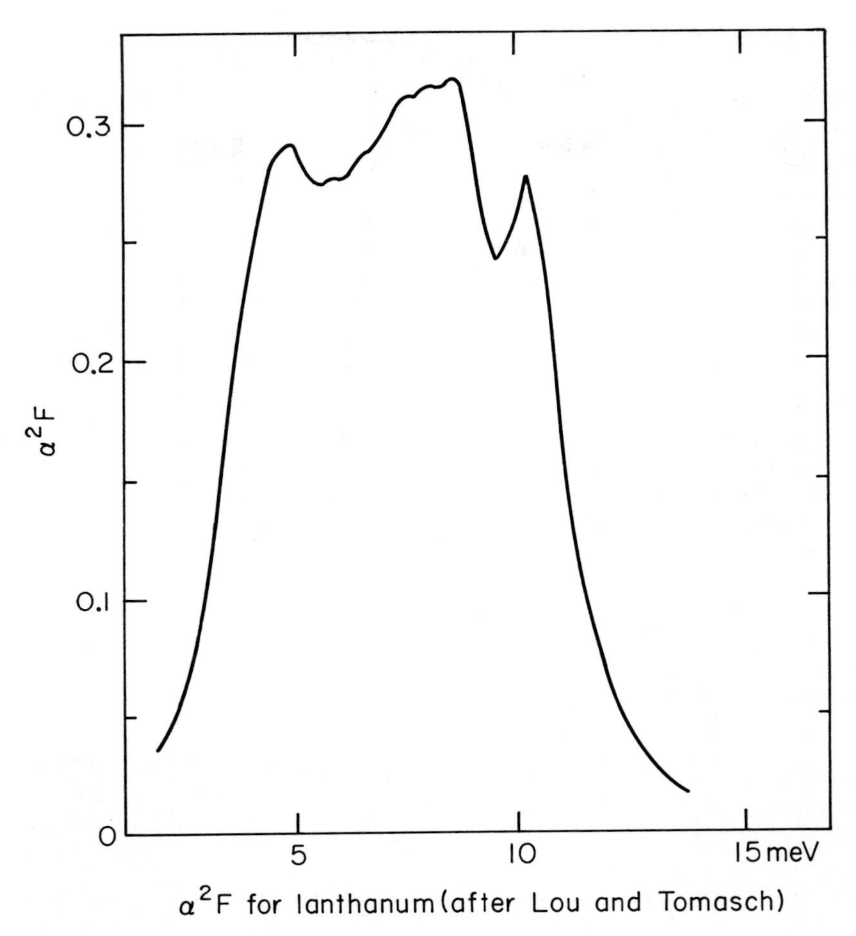

Fig. 7.
$\alpha^2 F$ for La (after Lou and Tomasch).

transition metals, a rare earth, and various alloys and compounds. Except possibly for the metallic form of hydrogen, [35] which is presumed to exist at very high pressures, it is unlikely that the phonon mechanism will yield substantially higher transition temperatures than the present maximum of about 21 K for a compound of Nb, Al and Ge.

Other mechanisms have been suggested for obtaining higher transition temperatures. One of these is to get an effective attractive interaction between electrons from exchange of virtual excitons, or electron-hole pairs. This requires a semiconductor in close proximity to the metal in a layer or sandwich structure. At present, one can not say whether or not such structures are feasible and in no case has the exciton mechanism been shown to exist. As Ginzburg has emphasized, this problem (as well as other proposed mechanisms) deserves study until a definite answer can be found. [36]

The pairing theory has had wide application to Fermi systems other than electrons in metals. For example, the theory has been used to account for

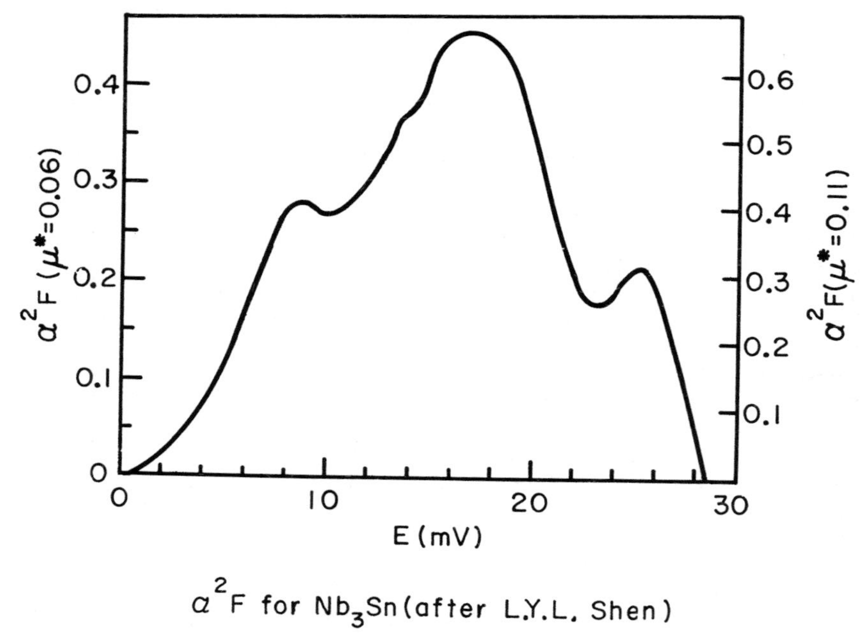

Fig. 8.
α^2F for Nb_3Sn (after Y. L. Y. Shen).

many aspects of nuclear structure. It is thought the nuclear matter in neutron stars is superfluid. Very recently, evidence has been found for a possible pairing transition in liquid He^3 at very low temperatures [37]. Some of the concepts, such as that of a degenerate vacuum, have been used in the theory of elementary particles. Thus pairing seems to be a general phenomenon in Fermi systems.

The field of superconductivity is still a very active one in both basic science and applications. I hope that these lectures have given you some feeling for the accomplishments and the methods used.

References

1. Shoenberg, D. Superconductivity, Cambridge Univ. Press, Cambridge (1938). Second edition, 1951.
2. London, F. Proc. Roy. Soc. (London) *152A*, 24 (1935).
3. Bardeen, J. Phys. Rev. *59*, 928A (1941).
4. Reynolds, C. A., Serin, B. Wright W. H. and Nesbitt, L. B. Phys. Rev. *78*, 487 (1950).
5. Maxwell, E., Phys. Rev. *78*, 477 (1950).
6. Fröhlich, H., Phys. Rev. *79*, 845 (1950); Proc. Roy. Soc. (London) Ser. A 213, 291 (1952).
7. London, F., Superfluids, New York, John Wiley and Sons, 1950.
8. For recent review articles with references, see the chapters by D. J. Scalapino and by W. L. McMillan and J. M. Rowell in Superconductivity, R. D. Parks, ed., New York, Marcel Bekker, Inc., 1969, Vol. 1. An excellent reference for the theory and earlier experimental work is J. R. Schrieffer, Superconductivity, New York, W. A. Benjamin, Inc., 1964. The present lecture is based in part on a chapter by the author in Cooperative Phenomena, H. Haken and M. Wagner, eds. to be published by Springer.

9. Gor'kov, L. P., Zh. Eksper i. teor. Fiz. *34,* 735 (1958). (English transl. Soviet Phys. — JETP *7,* 505 (1958)).
10. Kadanoff L. P. and Martin, P. C. Phys. Rev. *124,* 670 (1961).
11. Migdal, A. B., Zh. Eksper i. teor. Fiz. *34,* 1438 (1958). (English transl. Soviet Phys. — JETP *7,* 996 (1958)).
12. Eliashberg, G. M., Zh. Eksper i. teor. Fiz. *38,* 966 (1960). Soviet Phys. — JETP *11,* 696 (1960).
13. Nambu, Y., Phys. Rev. *117,* 648 (1960).
14. Bardeen, J., Phys. Rev. *79,* 167 (1950); *80,* 567 (1950); *81* 829 (1951).
15. Bardeen, J., Rev. Mod. Phys. *23,* 261 (1951).
16. Schafroth, M. R., Helv. Phys. Acta *24,* 645 (1951); Nuovo Cimento, *9,* 291 (1952).
17. For a review see Bardeen, J., Encyclopedia of Physics, S. Flugge, ed., Berlin, Springer-Verlag, (1956) Vol. XV, p. 274.
18. Bardeen, J., L. N. Cooper and J. R. Schrieffer, Phys. Rev. *108,* 1175 (1957).
19. For references, see the review article of M. A. Biondi, A. T. Forrester, M. B. Garfunkel and C. B. Satterthwaite, Rev. Mod. Phys. *30,* 1109 (1958).
20. Pippard, A. B., Proc. Roy. Soc. (London) *A216,* 547 (1954).
21. See N. N. Bogoliubov, V. V. Tolmachev and D. V. Shirkov, A New Method in the Theory of Superconductivity, New York, Consultants Bureau, Inc., 1959.
22. Morel P. and Anderson, P. W., Phys. Rev. *125,* 1263 (1962).
23. Giaever, I., Phys. Rev. Letters, *5,* 147; *5,* 464 (1960).
24. Giaever, I., Hart H. R., and Megerle K., Phys. Rev. *126,* 941 (1962).
25. Rowell, J. M., Anderson P. W. and Thomas D. E., Phys. Rev. Letters, *10,* 334 (1963).
26. Culler, G. J., Fried, B. D., Huff, R. W. and Schrieffer, J. R., Phys. Rev. Letters *8,* 339 (1962).
27. Schrieffer, J. R., Scalapino, D. J. and Wilkins, J. W., Phys. Rev. Letters *10,* 336 (1963); D. J. Scalapino, J. R. Schrieffer, and J. W. Wilkins, Phys. Rev. *148,* 263 (1966).
28. Scalapino, D. J., Wada, Y. and Swihart, J. C., Phys. Rev. Letters, *14,* 102 (1965); *14,* 106 (1965).
29. Eliashberg, G. M., Zh. Eksper i. teor. Fiz. *43,* 1005 (1962). English transl. Soviet Phys. — JETP *16,* 780 (1963).
30. Bardeen, J. and Stephen, M., Phys. Rev. *136,* A1485 (1964).
31. McMillan, W. L. and Rowell, J. M. in Reference 8.
32. Lou, L. F. and Tomasch, W. J., Phys. Rev. Lett. *29,* 858 (1972).
33. Shen, L. Y. L., Phys. Rev. Lett. *29,* 1082 (1972).
34. Carbotte, J. P., Superconductivity, P. R. Wallace, ed., New York, Gordon and Breach, 1969, Vol. 1, p. 491; J. P. Carbotte and R. C. Dynes, Phys. Rev. *172,* 476 (1968); C. R. Leavens and J. P. Carbotte, Can. Journ. Phys. *49,* 724 (1971).
35. Ashcroft N. W., Phys. Rev. Letters, *21,* 1748 (1968).
36. See V. L. Ginzburg, "The Problem of High Temperature Superconductivity," *Annual Review of Materials Science,* Vol. 2, p. 663 (1972).
37. Osheroff D. D., Gully W. J., Richardson R. C. and Lee, D. M., Phys. Rev. Lett. *29,* 1621 (1972).

LEON N COOPER

Leon Cooper was born in 1930 in New York where he attended Columbia University (A.B. 1951; A.M. 1953; Ph.D. 1954). He became a member of the Institute for Advanced Study (1954–55) after which he was a research associate of Illinois (1955–57) and later an assistant professor at the Ohio State University (1957–58). Professor Cooper joined Brown University in 1958 where he became Henry Ledyard Goddard University Professor (1966–74) and where he is presently the Thomas J. Watson, Sr. Professor of Science (1974–).

Professor Cooper is Director of Brown University's Center for Neural Science. This Center was founded in 1973 to study animal nervous systems and the human brain. Professor Cooper served as the first director with an interdisciplinary staff drawn from the Departments of Applied Mathematics, Biomedical Sciences, Linguistics and Physics. Today, Cooper, with members of the Brown Faculty, postdoctoral fellows and graduate students with interests in the neural and cognitive sciences, is working towards an understanding of memory and other brain functions, and thus formulating a scientific model of how the human mind works.

Professor Cooper has received many forms of recognition for his work in 1972, he received the Nobel Prize in Physics (with J. Bardeen and J. R. Schrieffer) for his studies on the theory of superconductivity completed while still in his 20s. In 1968, he was awarded the Comstock Prize (with J. R. Schrieffer) of the National Academy of Sciences. The Award of Excellence, Graduate Faculties Alumni of Columbia University and Descartes Medal, Academie de Paris, Université Rene Descartes were conferred on Professor Cooper in the mid 1970s. In 1985, Professor Cooper received the John Jay Award of Columbia College. He holds seven honorary doctorates.

Professor Cooper has been an NSF Postdoctoral Fellow, 1954–55, Alfred P. Sloan Foundation Research Fellow, 1959–66 and John Simon Guggenheim Memorial Foundation Fellow, 1965–66. He is a fellow of the American Physical Society and American Academy of Arts and Sciences; Sponsor, Federation of American Scientists; member of American Philosophical Society, National Academy of Sciences, Society of Neuroscience, American Association for the Advancement of Science, Phi Beta Kappa, and Sigma Xi. Professor Cooper is also on the Governing Board and Executive Committee of the International Neural Network Society and a member of the Defense Science Board.

Professor Cooper is Co-founder and Co-chairman of Nestor, Inc., an industry leader in applying neural-network systems to commercial and military applications. Nestor's adaptive pattern-recognition and risk-assessment systems simulated in small conventional computers *learn by example* to accurately classify complex patterns such as targets in sonar, radar or imaging systems, to emulate human decisions in such applications as mortgage origination and to assess risks.

Born: February 28, 1930, New York City, married, two children.
A.B. 1951; A.M. 1953; Ph.D. 1954, all at Columbia University.
Member, Institute for Advanced Study, 1954–55.
Research Associate, University of Illinois, 1955–57.
Assistant Professor, Ohio State University, 1957–58.
Associate Professor, 1958–62; Professor, 1962–66; Henry Ledyard Goddard University Professor, 1966–74; Thomas J. Watson, Sr., Professor of Science, 1974–, Brown University.
Visiting Professor, various universities and summer schools.
Consultant, various governmental agencies, industrial and educational organizations, various public lectures, international conferences and symposia.
Executive Committee of the International Neural Network Society.
Member, Defense Science Board.
NSF Postdoctoral Fellow, 1954–55.
Alfred P. Sloan Foundation Research Fellow, 1959–66.
Doctor of Sciences (honoris causa), Columbia University, 1973; University of Sussex, 1973; University of Illinois, 1974; Brown University, 1974; Gustavus Adolphus College, 1975; Ohio State University, 1976; Université Pierre et Marie Curie, Paris, 1977.
Comstock Prize (with J. R. Schrieffer), National Academy of Sciences, 1968.
Nobel Prize (with J. Bardeen and J. R. Schrieffer), 1972.
Award of Excellence, Graduate Faculties Alumni of Columbia University, 1974.
Descartes Medal, Academie de Paris, Université Rene Descartes, 1977.
John Jay Award, Columbia College, 1985.
Award for Distinguished Achievement, Columbia University, 1990.
Who's Who, Who's Who in America, Who's Who in the World, various other listings.
Fellow, American Academy of Arts and Sciences, American Physical Society; Member, American Philosophical Society, National Academy of Sciences, Society of Neuroscience, American Association for the Advancement of Science, Phi Beta Kappa, Sigma Xi; Sponsor, Federation of American Scientists.

MICROSCOPIC QUANTUM INTERFERENCE EFFECTS IN THE THEORY OF SUPERCONDUCTIVITY

Nobel Lecture, December 11, 1972

by

LEON N COOPER

Physics Department, Brown University, Providence, Rhode Island

It is an honor and a pleasure to speak to you today about the theory of superconductivity. In a short lecture one can no more than touch on the long history of experimental and theoretical work on this subject before 1957. Nor can one hope to give an adequate account of how our understanding of superconductivity has evolved since that time. The theory (1) we presented in 1957, applied to uniform materials in the weak coupling limit so defining an ideal superconductor, has been extended in almost every imaginable direction. To these developments so many authors have contributed (2) that we can make no pretense of doing them justice. I will confine myself here to an outline of some of the main features of our 1957 theory, an indication of directions taken since and a discussion of quantum interference effects due to the singlet-spin pairing in superconductors which might be considered the microscopic analogue of the effects discussed by Professor Schrieffer.

NORMAL METAL

Although attempts to construct an electron theory of electrical conductivity date from the time of Drude and Lorentz, an understanding of normal metal conduction electrons in modern terms awaited the development of the quantum theory. Soon thereafter Sommerfeld and Bloch introduced what has evolved into the present description of the electron fluid. (3) There the conduction electrons of the normal metal are described by single particle wave functions. In the periodic potential produced by the fixed lattice and the conduction electrons themselves, according to Bloch's theorem, these are modulated plane waves:

$$\emptyset_{\boldsymbol{K}}(\boldsymbol{r}) = \mathrm{u}_{\boldsymbol{K}}(\boldsymbol{r})\, \mathrm{e}^{i\boldsymbol{k}\cdot\boldsymbol{r}},$$

where $u_{\boldsymbol{K}}(\boldsymbol{r})$ is a two component spinor with the lattice periodicity. We use $\boldsymbol{K}$ to designate simultaneously the wave vector $\boldsymbol{k}$, and the spin state σ: $\boldsymbol{K} \equiv \boldsymbol{k}, \uparrow$; $-\boldsymbol{K} \equiv -\boldsymbol{k}, \downarrow$. The single particle Bloch functions satisfy a Schrödinger equation

$$\left[-\frac{\hbar^2}{2m}\nabla^2 + V_0(\boldsymbol{r})\right]\emptyset_{\boldsymbol{K}} = \mathcal{E}_{\boldsymbol{K}}\emptyset_{\boldsymbol{K}}$$

where $V_0(\boldsymbol{r})$ is the periodic potential and in general might be a linear operator to include exchange terms.

The Pauli exclusion principle requires that the many electron wave function be antisymmetric in all of its coordinates. As a result no two electrons can be

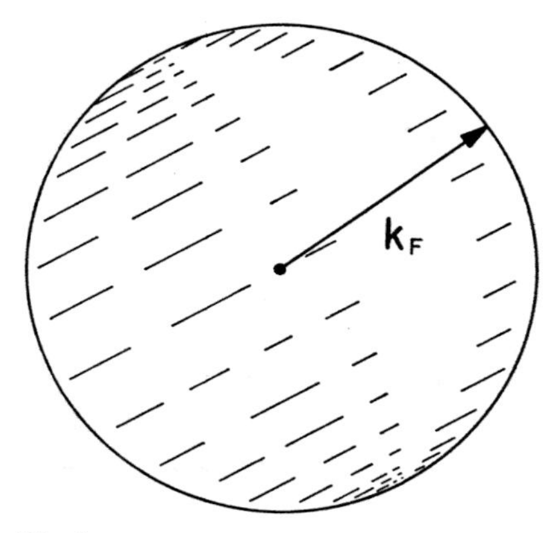

Fig. 1.
The normal ground state wavefunction, Φ_0, is a filled Fermi sphere for both spin directions.

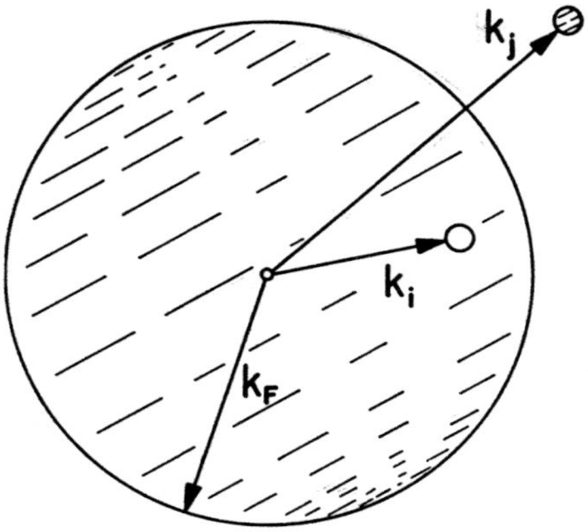

Fig. 2.
An excitation of the normal system.

in the same single particle Bloch state. The energy of the entire system is

$$W = \sum_{i=1}^{2N} \mathcal{E}_i$$

where $\mathcal{E}_i$ is the Bloch energy of the ith single electron state. The ground state of the system is obtained when the lowest N Bloch states of each spin are occupied by single electrons; this can be pictured in momentum space as the filling in of a Fermi sphere, Fig. 1. In the ground-state wave function there is no correlation between electrons of opposite spin and only a statistical correlation (through the general anti-symmetry requirement on the total wave function) of electrons of the same spin.

Single particle excitations are given by wave functions identical to the ground state except that one electron states $k_i < k_F$ are replaced by others $k_j < k_F$. This may be pictured in momentum space as opening vacancies below the Fermi surface and placing excited electrons above, Fig. 2. The energy difference between the ground state and the excited state with the particle excitation k_j and the hole excitation k_i is

$$\mathcal{E}_j - \mathcal{E}_i = \mathcal{E}_j - \mathcal{E}_F - (\mathcal{E}_i - \mathcal{E}_F) = \varepsilon_j - \varepsilon_i = |\varepsilon_j| + |\varepsilon_i|$$

where we define ε as the energy measured relative to the Fermi energy

$$\varepsilon_i = \mathcal{E}_i - \mathcal{E}_F.$$

When Coulomb, lattice-electron and other interactions, which have been omitted in constructing the independent particle Bloch model are taken into account, various modifications which have been discussed by Professor Schrieffer are introduced into both the ground state wave function and the excitations. These may be summarized as follows: The normal metal is described by a ground state Φ_0 and by an excitation spectrum which, in addition to the various collective excitations, consists of quasi-fermions which satisfy the usual anticommutation relations. It is defined by the sharpness of the Fermi surface, the finite density of excitations, and the continuous decline of the single particle excitation energy to zero as the Fermi surface is approached.

Electron Correlations that Produce Superconductivity

For a description of the superconducting phase we expect to include correlations that are not present in the normal metal. Professor Schrieffer has discussed the correlations introduced by an attractive electron-electron interaction and Professor Bardeen will discuss the role of the electron-phonon interaction in producing the electron-electron interaction which is responsible for superconductivity. It seems to be the case that any attractive interaction between the fermions in a many-fermion system can produce a superconducting-like state. This is believed at present to be the case in nuclei, in the interior of neutron stars and has possibly been observed (4) very recently in He^3. We will therefore develop the consequences of an attractive two-body interaction in a degenerate many-fermion system without enquiring further about its source.

The fundamental qualitative difference between the superconducting and normal ground state wave function is produced when the large degeneracy of the single particle electron levels in the normal state is removed. If we visualize the Hamiltonian matrix which results from an attractive two-body interaction in the basis of normal metal configurations, we find in this enormous matrix, sub-matrices in which all single-particle states except for one pair of electrons remain unchanged. These two electrons can scatter via the electron-electron interaction to all states of the same total momentum. We may envisage the pair wending its way (so to speak) over all states unoccupied by other electrons. [The electron-electron interaction in which we are interested is both weak and slowly varying over the Fermi surface. This and the fact that the energy involved in the transition into the superconducting state is small leads us to guess that only single particle excitations in a small shell near the Fermi surface play a role. It turns out, further, that due to exchange terms in the electron-electron matrix element, the effective interaction in metals between electrons of singlet spin is much stronger than that between electrons of triplet spin—thus our preoccupation with singlet spin correlations near the Fermi surface.] Since every such state is connected to every other, if the interaction is attractive and does not vary rapidly, we are presented with submatrices of the entire Hamiltonian of the form shown in Fig. 3. For purposes of illustration we have set all off diagonal matrix elements equal to the constant $-V$ and the diagonal terms equal to zero (the single particle excitation energy at the Fermi surface) as though all the initial electron levels were completely degenerate. Needless to say, these simplifications are not essential to the qualitative result.

Diagonalizing this matrix results in an energy level structure with $M—1$ levels raised in energy to $E = +V$ while one level (which is a superposition of all of the original levels and quite different in character) is lowered in energy to

$$E = -(M-1)V.$$

Since M, the number of unoccupied levels, is proportional to the volume of the container while V, the scattering matrix element, is proportional to 1/volume, the product is independent of the volume. Thus the removal of

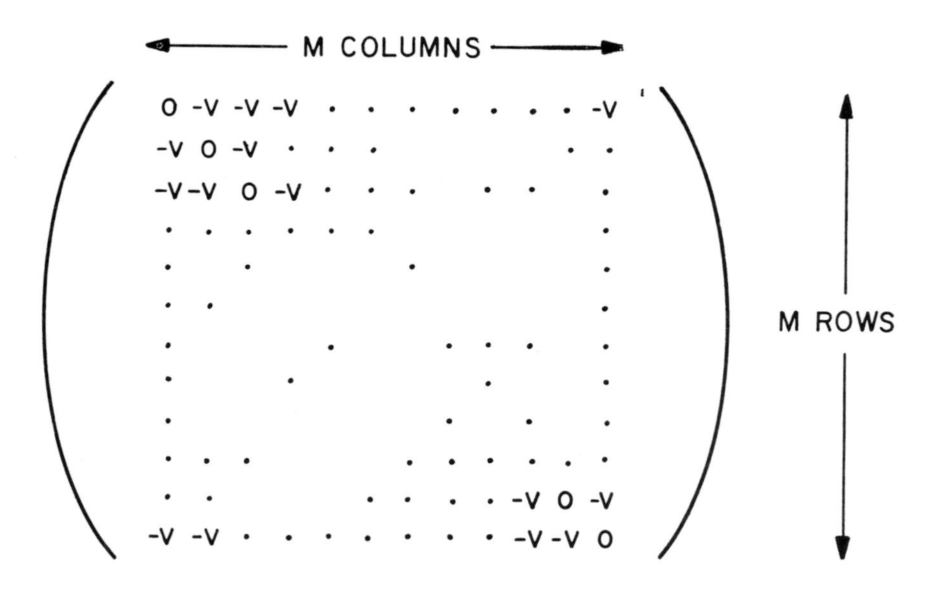

For V = 0,
M levels
at E = 0

For V > 0,M —1 levels at
E=V and one level
at E=—(M—1) V

Fig. 3.

the degeneracy produces a single level separated from the others by a volume independent energy gap.

To incorporate this into a solution of the full Hamiltonian, one must devise a technique by which all of the electrons pairs can scatter while obeying the exclusion principle. The wave function which accomplishes this has been discussed by Professor Schrieffer. Each pair gains an energy due to the removal of the degeneracy as above and one obtains the maximum correlation of the entire wave function if the pairs all have the same total momentum. This gives a coherence to the wave function in which for a combination of dynamical and statistical reasons there is a strong preference for momentum zero, singlet spin correlations, while for statistical reasons alone there is an equally strong preference that all of the correlations have the same total momentum.

In what follows I shall present an outline of our 1957 theory modified by introducing the quasi-particles of Bogoliubov and Valatin. (5) This leads to a formulation which is generally applicable to a wide range of calculations

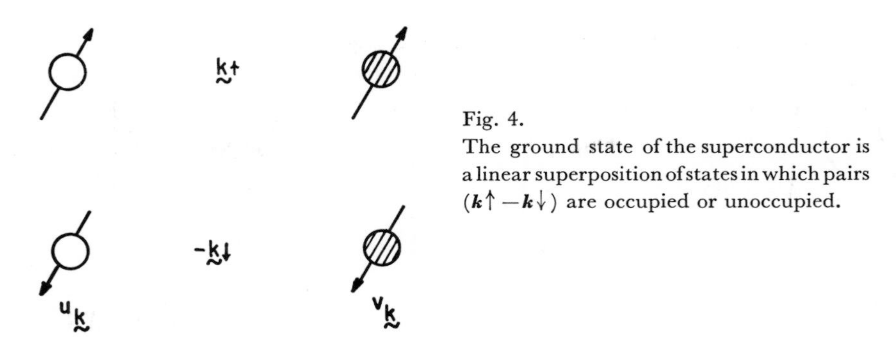

Fig. 4.
The ground state of the superconductor is a linear superposition of states in which pairs $(\boldsymbol{k}\uparrow -\boldsymbol{k}\downarrow)$ are occupied or unoccupied.

in a manner analogous to similar calculations in the theory of normal metals.

We limit the interactions to terms which scatter (and thus correlate) singlet zero-momentum pairs. To do this, it is convenient to introduce the pair operators:

$$b_{\boldsymbol{k}} = c_{-\boldsymbol{K}}c_{\boldsymbol{K}}$$
$$b^*_{\boldsymbol{k}} = c^*_{\boldsymbol{K}}\, c^*_{-\boldsymbol{K}}$$

and using these we extract from the full Hamiltonian the so-called reduced Hamiltonian

$$H_{\text{reduced}} = \sum_{k<k_f} 2|\varepsilon|\, b_{\boldsymbol{k}} b^*_{\boldsymbol{k}} + \sum_{k>k_f} 2\varepsilon b^*_{\boldsymbol{k}} b_{\boldsymbol{k}} + \sum_{kk'} V_{\boldsymbol{k}'\boldsymbol{k}} b^*_{\boldsymbol{k}'} b_{\boldsymbol{k}}$$

where $V_{\boldsymbol{k}'\boldsymbol{k}}$ is the scattering matrix element between the pair states $\boldsymbol{k}$ and $\boldsymbol{k}'$.

Ground State

As Professor Schrieffer has explained, the ground state of the superconductor is a linear superposition of pair states in which the pairs $(\boldsymbol{k}\uparrow, -\boldsymbol{k}\downarrow)$ are occupied or unoccupied as indicated in Fig. 4. It can be decomposed into two disjoint vectors—one in which the pair state $\boldsymbol{k}$ is occupied, $\emptyset_{\boldsymbol{k}}$ and one in which it is unoccupied, $\emptyset_{(\boldsymbol{k})}$:

$$\psi_0 = u_{\boldsymbol{k}}\emptyset_{(\boldsymbol{k})} + v_{\boldsymbol{k}}\emptyset_{\boldsymbol{k}}.$$

The probability amplitude that the pair state $\boldsymbol{k}$ is (is not) occupied in the ground state is then $v_{\boldsymbol{k}}(u_{\boldsymbol{k}})$. Normalization requires that $|u|^2+|v|^2 = 1$. The phase of the ground state wave function may be chosen so that with no loss o generality $u_{\boldsymbol{k}}$ is real. We can then write

$$u = (1-h)^{1/2}$$
$$v = h^{1/2}\, e^{i\varphi}$$

where

$$0 \leqslant h \leqslant 1.$$

A further decomposition of the ground state wave function of the superconductor in which the pair states $\boldsymbol{k}$ and $\boldsymbol{k}'$ are either occupied or unoccupied Fig. 5 is:

$$\psi_0 = u_{\boldsymbol{k}}u_{\boldsymbol{k}'}\emptyset_{(\boldsymbol{k}),(\boldsymbol{k}')} + u_{\boldsymbol{k}}v_{\boldsymbol{k}'}\emptyset_{(\boldsymbol{k}),\boldsymbol{k}'} + v_{\boldsymbol{k}}u_{\boldsymbol{k}'}\emptyset_{\boldsymbol{k},(\boldsymbol{k}')} + v_{\boldsymbol{k}}v_{\boldsymbol{k}'}\emptyset_{\boldsymbol{k},\boldsymbol{k}'}\ .$$

This is a Hartree-like approximation in the probability amplitudes for the occupation of pair states. It can be shown that for a fermion system the wave

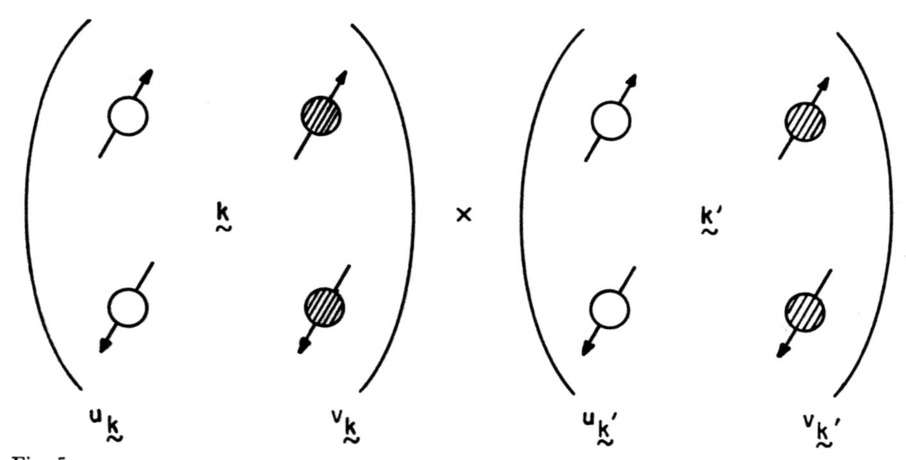

Fig. 5.
A decomposition of the ground state of the superconductor into states in which the pair states $\boldsymbol{k}$ and $\boldsymbol{k}'$ are either occupied or unoccupied.

function cannot have this property unless there are a variable number of particles. To terms of order $1/N$, however, this decomposition is possible for a fixed number of particles; the errors introduced go to zero as the number of particles become infinite. (6)

The correlation energy, W_c, is the expectation value of H_{red} for the state ψ_0

$$W_c = (\psi_0, H_{red}\psi_0) = W_c\ [h,\varphi].$$

Setting the variation of W_c with respect to h and φ equal to zero in order to minimize the energy gives

$$h = 1/2\ (1-\varepsilon/E)$$

$$E = (\varepsilon^2+|\Delta|^2)^{1/2}$$

where

$$\Delta = |\Delta|e^{i\varphi}$$

satisfies the integral equation

$$\Delta(\boldsymbol{k}) = -1/2 \sum_{\boldsymbol{k}'} V_{\boldsymbol{kk}'} \frac{\Delta(\boldsymbol{k}')}{E(\boldsymbol{k}')}.$$

If a non-zero solution of this integral equation exists, $W_c < 0$ and the "normal" Fermi sea is unstable under the formation of correlated pairs.

In the wave function that results there are strong correlations between pairs of electrons with opposite spin and zero total momentum. These correlations are built from normal excitations near the Fermi surface and extend over spatial distances typically of the order of 10^{-4} cm. They can be constructed due to the large wave numbers available because of the exclusion principle. Thus with a small additional expenditure of kinetic energy there can be a greater gain in the potential energy term. Professor Schrieffer has discussed some of the properties of this state and the condensation energy associated with it.

Single-Particle Excitations

In considering the excited states of the superconductor it is useful, as for the

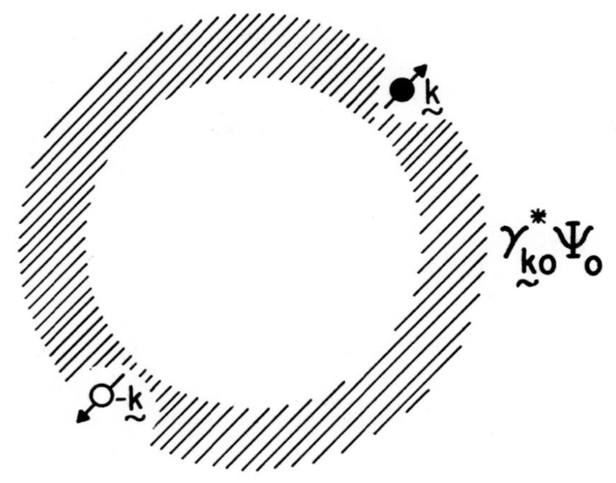

Fig. 6.
A single par icle excitation of the superconduc or in one-to-one correspondenc with an excitation of the normal fermion system.

normal metal, to make a distinction between single-particle and collective excitations; it is the single-particle excitation spectrum whose alteration is responsible for superfluid properties. For the superconductor excited (quasi-particle) states can be defined in one-to-one correspondence with the excitations of the normal metal. One finds, for example, that the expectation value of H_{red} for the excitation Fig. 6 is given by

$$E_{\boldsymbol{k}} = \sqrt{\varepsilon_{\boldsymbol{k}}^2 + |\Delta|^2}.$$

In contrast to the normal system, for the superconductor even as ε goes to zero E remains larger than zero, its lowest possible value being $E = |\Delta|$. One can therefore produce single particle excitations from the superconducting ground state only with the expenditure of a small but finite amount of energy. This is called the energy gap; its existence severely inhibits single particle processes and is in general responsible for the superfluid behavior of the electron gas. [In a gapless superconductor it is the finite value of $\Delta(\boldsymbol{r})$, the order parameter, rather than the energy gap as such that becomes responsible for the superfluid properties.] In the ideal superconductor, the energy gap appears because not a single pair can be broken nor can a single element of phase space be removed without a finite expenditure of energy. If a single pair is broken, one loses its correlation energy; if one removes an element of phase space from the system, the number of possible transitions of all the pairs is reduced resulting in both cases in an increase in the energy which does not go to zero as the volume of the system increases.

The ground state of the superconductor and the excitation spectrum described above can conveniently be treated by introducing a linear combination of c^* and c, the creation and annihilation operators of normal fermions. This is the transformation of Bogoliubov and Valatin (5):

$$\gamma^*_{\boldsymbol{k}0} = u_{\boldsymbol{k}} c^*_{\boldsymbol{K}} - v^*_{\boldsymbol{k}} c_{-\boldsymbol{K}}$$

$$\gamma^*_{\boldsymbol{k}1} = v^*_{\boldsymbol{k}} c_{\boldsymbol{K}} + u_{\boldsymbol{k}} c^*_{-\boldsymbol{K}}$$

It follows that

$$\gamma_{\boldsymbol{k}i}\,\psi_0 = 0$$

so that the γ_{ki} play the role of annihilation operators, while the γ^*_{ki} create excitations

$$\gamma^*_{ki} \cdots \gamma^*_{mj} \psi_0 = \psi_{ki}, \cdots_{mj}.$$

The γ operators satisfy Fermi anti-commutation relations so that with them we obtain a complete orthonormal set of excitations in one-to-one correspondence with the excitations of the normal metal.

We can sketch the following picture. In the ground state of the superconductor all the electrons are in singlet-pair correlated states of zero total momentum. In an m electron excited state the excited electrons are in "quasi-particle" states, very similar to the normal excitations and not strongly correlated with any of the other electrons. In the background, so to speak, the other electrons are still correlated much as they were in the ground state. The excited electrons behave in a manner similar to normal electrons; they can be easily scattered or excited further. But the background electrons—those which remain correlated—retain their special behavior; they are difficult to scatter or to excite.

Thus, one can identify two almost independent fluids. The correlated portion of the wave function shows the resistance to change and the very small specific heat characteristic of the superfluid, while the excitations behave very much like normal electrons, displaying an almost normal specific heat and resistance. When a steady electric field is applied to the metal, the superfluid electrons short out the normal ones, but with higher frequency fields the resistive properties of the excited electrons can be observed. [7]

Thermodynamic Properties, the Ideal Superconductor

We can obtain the thermodynamic properties of the superconductor using the ground state and excitation spectrum just described. The free energy of the system is given by

$$F[h, \varphi, f] = W_c(T) - TS,$$

where T is the absolute temperature and S is the entropy; f is the superconducting Fermi function which gives the probability of single-particle excitations. The entropy of the system comes entirely from the excitations as the correlated portion of the wave function is non-degenerate. The free energy becomes a function of $f(\boldsymbol{k})$ and $h(\boldsymbol{k})$, where $f(\boldsymbol{k})$ is the probability that the state $\boldsymbol{k}$ is occupied by an excitation or a quasi-particle, and $h(\boldsymbol{k})$ is the relative probability that the state $\boldsymbol{k}$ is occupied by a pair given that is not occupied by a quasi-particle. Thus some states are occupied by quasi-particles and the unoccupied phase space is available for the formation of the coherent background of the remaining electrons. Since a portion of phase space is occupied by excitations at finite temperatures, making it unavailable for the transitions of bound pairs, the correlation energy is a function of the temperature, $W_c(T)$. As T increases, $W_c(T)$ and at the same time Δ decrease until the critical temperature is reached and the system reverts to the normal phase.

Since the excitations of the superconductor are independent and in a one-to-one correspondence with those of the normal metal, the entropy of an excited configuration is given by an expression identical with that for the normal metal except that the Fermi function, $f(\boldsymbol{k})$, refers to quasi-particle excitations. The correlation energy at finite temperature is given by an expression similar to that at $T = 0$ with the available phase space modified by the occupation functions $f(\boldsymbol{k})$. Setting the variation of F with respect to h, φ, and f equal to zero gives:

$$h = 1/2\ (1-\varepsilon/E)$$

$$E = \sqrt{\varepsilon^2 + |\Delta|^2}$$

and

$$f = \frac{1}{1 + \exp(E/k_{\mathrm{B}}T)}$$

where

$$\Delta = |\Delta| \mathrm{e}^{\mathrm{i}\varphi}$$

is now temperature-dependent and satisfies the fundamental integral equation of the theory

$$\Delta_{\boldsymbol{k}}(T) = -1/2 \sum_{\boldsymbol{kk'}} V_{\boldsymbol{kk'}} \frac{\Delta_{\boldsymbol{k'}}(T)}{E_{\boldsymbol{k'}}(T)} \tanh\left(\frac{E_{\boldsymbol{k'}}(T)}{2k_{\mathrm{B}}T}\right).$$

The form of these equations is the same as that at $T = 0$ except that the energy gap varies with the temperature. The equation for the energy gap can be satisfied with non-zero values of Δ only in a restricted temperature range. The upper bound of this temperature range is defined as T_{c}, the critical temperature. For $T < T_{\mathrm{c}}$, singlet spin zero momentum electrons are strongly correlated, there is an energy gap associated with exciting electrons from the correlated part of the wave function and $E(\boldsymbol{k})$ is bounded below by $|\Delta|$. In this region the system has properties qualitatively different from the normal metal.

In the region $T > T_{\mathrm{c}}$, $\Delta = 0$ and we have in every respect the normal solution. In particular f, the distribution function for excitations, becomes just the Fermi function for excited electrons $k > k_{\mathrm{F}}$, and for holes $k < k_{\mathrm{F}}$

$$f = \frac{1}{1 + \exp(|\varepsilon|/k_{\mathrm{B}}T)}.$$

If we make our simplifications of 1957, (defining in this way an 'ideal' superconductor)

$$\begin{aligned} V_{\boldsymbol{k'k}} &= -V \qquad & |\varepsilon| < \hbar\omega_{\mathrm{av}} \\ &= 0 & \text{otherwise} \end{aligned}$$

and replace the energy dependent density of states by its value at the Fermi surface, $\mathcal{N}(0)$, the integral equation for Δ becomes

$$1 = \mathcal{N}(0)\, V \int_0^{\hbar\omega_{\mathrm{av}}} \frac{\mathrm{d}\varepsilon}{\sqrt{\varepsilon^2 + |\Delta|^2}} \tanh\left(\frac{\sqrt{\varepsilon^2 + |\Delta|^2}}{2k_{\mathrm{B}}T}\right).$$

The solution of this equation, Fig. 7, gives $\Delta(T)$ and with this f and h. We can then calculate the free energy of the superconducting state and obtain the thermodynamic properties of the system.

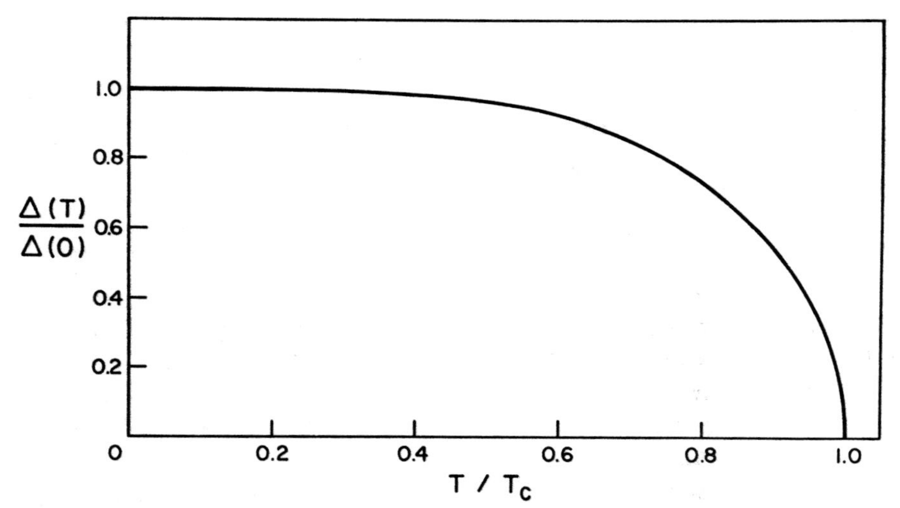

Fig. 7.
Variation of the energy gap with temperature for the ideal superconductor.

In particular one finds that at T_c (in the absence of a magnetic field) there is a second-order transition (no latent heat: $W_c = 0$ at T_c) and a discontinuity in the specific heat. At very low temperatures the specific heat goes to zero exponentially. For this ideal superconductor one also obtains a law of corresponding states in which the ratio

$$\frac{\gamma T_c^2}{H_0^2} = 0.170,$$

where

$$\gamma = 2/3\pi^2 \mathcal{N}(0) k_B^2.$$

The experimental data scatter about the number 0.170. The ratio of Δ to $k_B T_c$ is given as a universal constant

$$\Delta / k_B T_c = 1.75.$$

There are no arbitrary parameters in the idealized theory. In the region of empirical interest all thermodynamic properties are determined by the quantities γ and $\hbar w_{av}\, e^{-1/N(0)V}$. The first, γ, is found by observation of the normal specific heat, while the second is found from the critical temperature, given by

$$k_B T_c = 1.14\, \hbar\omega_{av} e^{-1/N(0)V}.$$

At the absolute zero

$$\Delta = \hbar\omega_{av} / \sinh\left(\frac{1}{\mathcal{N}(0)V}\right).$$

Further, defining a weak coupling limit $[\mathcal{N}(0)V \ll 1]$ which is one region of interest empirically, we obtain

$$\Delta \simeq 2\hbar\omega_{av} e^{-1/N(0)V}.$$

The energy difference between the normal and superconducting states becomes (again in the weak coupling limit)

$$W_s - W_n = W_c = -2\mathcal{N}(0)(\hbar\omega_{av})^2\, e^{-2/N(0)V}.$$

The dependence of the correlation energy on $(\hbar\omega_{av})^2$ gives the isotope effect, while the exponential factor reduces the correlation energy from the dimensionally expected $N(0)(\hbar\omega_{av})^2$ to the much smaller observed value. This, however, is more a demonstration that the isotope effect is consistent with our model rather than a consequence of it, as will be discussed further by Professor Bardeen.

The thermodynamic properties calculated for the ideal superconductor are in qualitative agreement with experiment for weakly coupled superconductors. Very detailed comparison between experiment and theory has been made by many authors. A summary of the recent status may be found in reference (2). When one considers that in the theory of the ideal superconductor the existence of an actual metal is no more than hinted at (We have in fact done all the calculations considering weakly interacting fermions in a container.) so that in principle (with appropriate modifications) the calculations apply to neutron stars as well as metals, we must regard detailed quantitative agreement as a gift from above. We should be content if there is a single metal for which such agreement exists. [Pure single crystals of tin or vanadium are possible candidates.]

To make comparison between theory and experiments on actual metals, a plethora of detailed considerations must be made. Professor Bardeen will discuss developments in the theory of the electron-phonon interaction and the resulting dependence of the electron-electron interaction and superconducting properties on the phonon spectrum and the range of the Coulomb repulsion. Crystal symmetry, Brillouin zone structure and the actual wave function (*S*, *P* or *D* states) of the conduction electrons all play a role in determining real metal behavior. There is a fundamental distinction between superconductors w ich always show a Meissner effect and those (type II) which allow magnetic field penetration in units of the flux quantum.

When one considers, in addition, specimens with impurities (magnetic and otherwise) superimposed films, small samples, and so on, one obtains a variety of situations, developed in the years since 1957 by many authors, whose richness and detail takes volumes to discuss. The theory of the ideal superconductor has so far allowed the addition of those extensions and modifications necessary to describe, in what must be considered remarkable detail, all of the experience actually encountered.

Microscopic Interference Effects

In its interaction with external perturbations the superconductor displays remarkable interference effects which result from the paired nature of the wave function and are not at all present in similar normal metal interactions. Neither would they be present in any ordinary two-fluid model. These "coherence effects" are in a sense manifestations of interference in spin and momentum space on a microscopic scale, analogous to the macroscopic quantum effects due to interference in ordinary space which Professor Schrieffer discussed. They depend on the behavior under time reversal of the perturbing fields. (8) It is intriguing to speculate that if one could somehow amplify them

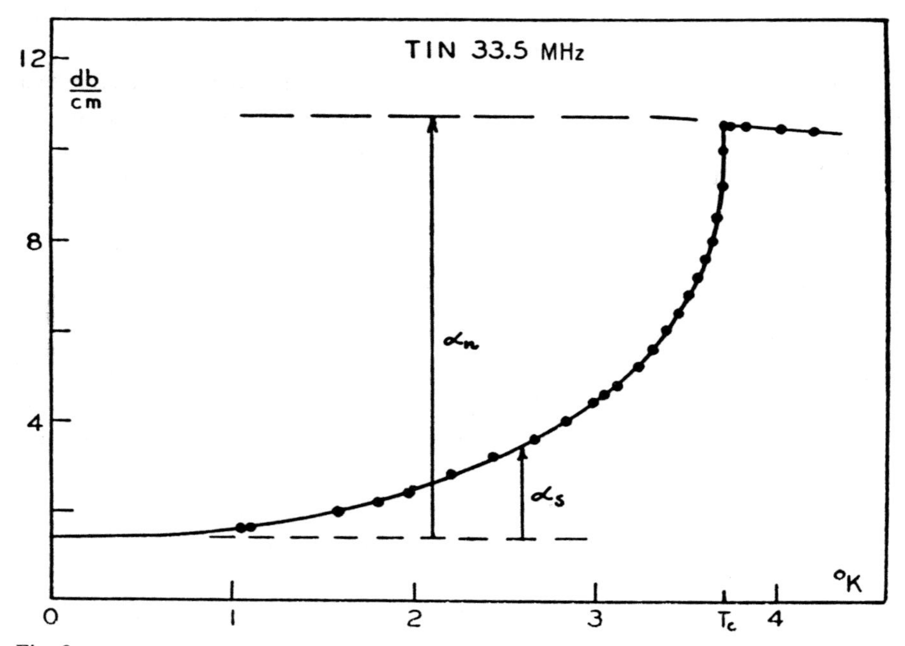

Fig. 8.
Ultrasonic attenuation as a function of temperature across the superconducting transition as measured by Morse and Bohm.

properly, the time reversal symmetry of a fundamental interaction might be tested. Further, if helium 3 does in fact display a phase transition analogous to the superconducting transition in metals as may be indicated by recent experiments (4) and this is a spin triplet state, the coherences effects would be greatly altered.

Near the transition temperature these coherence effects produce quite dramatic contrasts in the behavior of coefficients which measure interactions with the conduction electrons. Historically, the comparison with theory of the behavior of the relaxation rate of nuclear spins (9) and the attenuation of longitudinal ultrasonic waves in clean samples (10) as the temperature is decreased through T_c provided an early test of the detailed structure of the theory.

The attenuation of longitudinal acoustic waves due to their interaction with the conduction electrons in a metal undergoes a very rapid drop (10a) as the temperature drops below T_c. Since the scattering of phonons from "normal" electrons is responsible for most of the acoustic attenuation, a drop was to be expected; but the rapidity of the decrease measured by Morse and Bohm (10b) Fig. 8 was difficult to reconcile with estimates of the decrease in the normal electron component of a two-fluid model.

The rate of relaxation of nuclear spins was measured by Hebel and Slichter (9a) in zero magnetic field in superconducting aluminum from 0.94 K to 4.2 K just at the time of the development of our 1957 theory. Redfield and Anderson (9b) confirmed and extended their results. The dominant relaxation mechanism is provided by interaction with the conduction electrons so that one would expect, on the basis of a two-fluid model, that this rate should decrease below the transition temperature due to the diminishing density of "normal" electrons. The experimental results however show just the reverse. The relaxation rate does not drop but increases by a factor of more than two just below the transition temperature. Fig 13. This observed increase in the nuclear spin relaxation rate and the very sharp drop in the acoustic attenuation coefficient as the temperature is decreased through T_c impose contradictory requirements on a conventional two-fluid model.

To illustrate how such effects come about in our theory, we consider the transition probability per unit time of a process involving electronic transitions from the excited state $\boldsymbol{k}$ to the state $\boldsymbol{k}'$ with the emission to or absorption of energy from the interacting field. What is to be calculated is the rate of transition between an initial state $|i>$ and a final state $|f>$ with the absorption or emission of the energy $\hbar\omega_{|\boldsymbol{k}'-\boldsymbol{k}|}$ (a phonon for example in the interaction of sound waves with the superconductor). All of this properly summed over final states and averaged with statistical factors over initial states may be written:

$$\omega = \frac{2\pi}{\hbar}\,\frac{\sum\limits_{i,f} \exp(-W_i/k_{\mathrm{B}}T)\left|<f|H_{\mathrm{int}}|i>\right|^2 \delta(W_f - W_i)}{\sum\limits_{i} \exp(-W_i/k_{\mathrm{B}}T)}$$

We focus our attention on the matrix element $<f|H_{\mathrm{int}}|i>$. This typically contains as one of its factors matrix elements between excited states of the superconductor of the operator

$$B = \sum_{\boldsymbol{K}\,\boldsymbol{K}'} B_{\boldsymbol{K}'\boldsymbol{K}} c^*_{\boldsymbol{K}'} c_{\boldsymbol{K}}$$

where $c^*_{\boldsymbol{K}'}$ and $c_{\boldsymbol{K}}$ are the creation and annihilation operators for electrons in the states $\boldsymbol{K}'$ and $\boldsymbol{K}$, and $B_{\boldsymbol{K}'\boldsymbol{K}}$ is the matrix element between the states $\boldsymbol{K}'$ and $\boldsymbol{K}$ of the configuration space operator $B(\boldsymbol{r})$

$$B_{\boldsymbol{K}'\boldsymbol{K}} = <\boldsymbol{K}'|B(\boldsymbol{r})|\boldsymbol{K}>.$$

The operator B is the electronic part of the matrix element between the full final and initial state

$$<f|H_{\mathrm{int}}|i> = m_{fi}<f|B|i>.$$

In the normal system scattering from single-particle electron states $\boldsymbol{K}$ to $\boldsymbol{K}'$ is independent of scattering from $-\boldsymbol{K}'$ to $-\boldsymbol{K}$. But the superconducting states are linear superpositions of $(\boldsymbol{K},-\boldsymbol{K})$ occupied and unoccupied. Because of this states with excitations $\boldsymbol{k}\uparrow$ and $\boldsymbol{k}'\uparrow$ are connected not only by $c^*_{\boldsymbol{k}'\uparrow}c_{\boldsymbol{k}\uparrow}$ but also by $c^*_{-\boldsymbol{k}\downarrow}c_{-\boldsymbol{k}'\downarrow}$; if the state $|f>$ contains the single-particle excitation $\boldsymbol{k}'\uparrow$ while the state $|i>$ contains $\boldsymbol{k}\uparrow$, as a result of the superposition of occupied and unoccupied pair states in the coherent part of the wave function, these are connected not only by $B_{\boldsymbol{K}'\boldsymbol{K}}\, c^*_{\boldsymbol{K}'}c_{\boldsymbol{K}}$ but also by $B_{-\boldsymbol{K}-\boldsymbol{K}'}\, c^*_{-\boldsymbol{K}}c_{-\boldsymbol{K}'}$.

For operators which do not flip spins we therefore write:

$$B = \sum_{\boldsymbol{k}\,\boldsymbol{k}'} (B_{\boldsymbol{K}'\boldsymbol{K}}\, c^*_{\boldsymbol{K}'}c_{\boldsymbol{K}} + B_{-\boldsymbol{K}-\boldsymbol{K}'}\, c^*_{-\boldsymbol{K}}c_{-\boldsymbol{K}'}).$$

Many of the operators, B, we encounter (e.g., the electric current, or the charge density operator) have a well-defined behavior under the operation of time reversal so that

$$B_{\boldsymbol{K}'\boldsymbol{K}} = \pm B_{-\boldsymbol{K}-\boldsymbol{K}'} \equiv B_{\boldsymbol{k}'\boldsymbol{k}}.$$

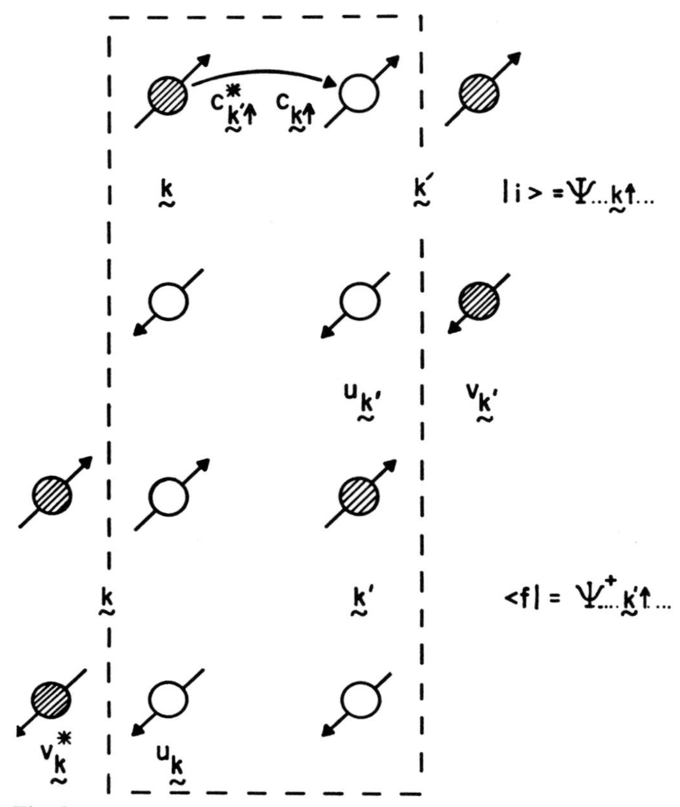

Fig. 9.
The two states $|i>$ and $<f|$ shown are connected by $c^*_{k'\uparrow}c_{k\uparrow}$ with the amplitude $u_{k'}u_k$.

Then B becomes

$$B = \sum_{k\,k'} B_{k'k} \, (c^*_{k'\uparrow}c_{k\uparrow} \pm c^*_{-k\downarrow}c_{-k'\downarrow})$$

where the upper (lower) sign results for operators even (odd) under time reversal.

The matrix element of B between the initial state, $\psi \ldots {}_{k\uparrow} \ldots$, and the final state $\psi \ldots {}_{k'\uparrow} \ldots$ contains contributions from $c^*_{k'\uparrow}c_{k\uparrow}$ Fig. 9 and unexpectedly from $c^*_{-k\downarrow}c_{-k'\downarrow}$ Fig. 10. As a result the matrix element squared $|<f|B|i>|^2$ contains terms of the form

$$|B_{k'k}|^2 \; |(u_{k'}u_k \mp v_{k'}v^*_k)|^2,$$

where the sign is determined by the behavior of B under time reversal:

upper sign $\quad$ B even under time reversal
lower sign $\quad$ B odd under time reversal.

Applied to processes involving the emission or absorption of boson quanta such as phonons or photons, the squared matrix element above is averaged with the appropriate statistical factors over initial and summed over final states; substracting emission from absorption probability per unit time, we obtain typically

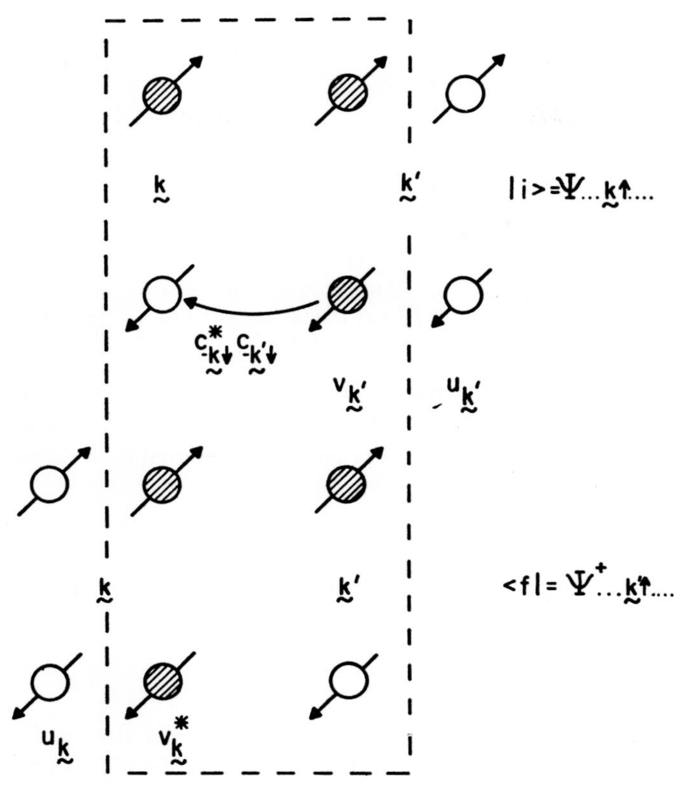

Fig. 10.
The two states $|i>$ and $<f|$ are also connected by $c^*{}_{k\downarrow}c_{-k'\downarrow}$ with the amplitude $-v_{k'}v_k^*$.

$$a = \frac{4\pi}{\hbar}|m|^2 \sum_{kk'} |(u_{k'}u_k \mp v_{k'}v_k^*)|^2 \ (f_{k'}-f_k) \ \delta(E_{k'}-E_k-\hbar\omega_{|k-k'|})$$

where f_k is the occupation probability in the superconductor for the excitation $k\uparrow$ or $k\downarrow$. [In the expression above we have considered only quasiparticle or quasi-hole scattering processes (not including processes in which a pair of excitations is created or annihilated from the coherent part of the wave function) since $\hbar\omega_{|k'-k|} < \Delta$, is the usual region of interest for the ultrasonic attenuation and nuclear spin relaxation we shall contrast.]

For the ideal superconductor, there is isotropy around the Fermi surface and symmetry between particles and holes; therefore sums of the form $\sum_k$ can be converted to integrals over the superconducting excitation energy, E:

$$\sum_k \rightarrow 2\mathcal{N}(0)\int_\Delta^\infty \frac{E}{\sqrt{E^2-\Delta^2}}\, dE$$

where $\mathcal{N}(0)\dfrac{E}{\sqrt{E^2-\Delta^2}} = \mathcal{N}(0)\dfrac{E}{\varepsilon}$ is the density of excitations in the superconductor, Fig. 11.

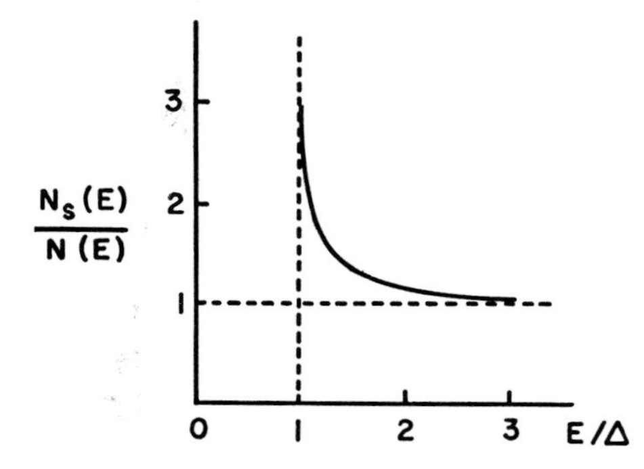

Fig. 11.
Ratio of superconducting to normal density of excitations as a function of E/Δ.

The appearance of this density of excitations is a surprise. Contrary to our intuitive expectations, the onset of superconductivity seems initially to enhance rather than diminish electronic transitions, as might be anticipated in a reasonable two-fluid model.

But the coherence factors $|(u'u \mp v'v^*)|^2$ are even more surprising; they behave in such a way as to sometimes completely negate the effect of the increased density of states. This can be seen using the expressions obtained above for u and v for the ideal superconductor to obtain

$$(u'u \mp v'v)^2 = \frac{1}{2}\left(1 + \frac{\varepsilon\varepsilon' \mp \Delta^2}{EE'}\right).$$

In the integration over $\boldsymbol{k}$ and $\boldsymbol{k}'$ the $\varepsilon\varepsilon'$ term vanishes. We thus define $(u'u \mp v'v)^2_s$; in usual limit where $\hbar\omega_{|\boldsymbol{k}'-\boldsymbol{k}|} \ll \Delta$, $\varepsilon \simeq \varepsilon'$ and $E \simeq E'$, this becomes

$$(u^2 - v^2)^2_s \rightarrow \frac{1}{2}\left(\frac{\varepsilon^2}{E^2}\right) \qquad \text{operators even under time reversal}$$

$$(u^2 + v^2)^2_s \rightarrow \frac{1}{2}\left(1 + \frac{E^2}{\Delta^2}\right) \qquad \text{operators odd under time reversal.}$$

For operators even under time reversal, therefore, the decrease of the coherence factors near $\varepsilon = 0$ just cancels the increase due to the density of states. For the operators odd under time reversal the effect of the increase of the density of states is not cancelled and should be observed as an increase in the rate of the corresponding process.

In general the interaction Hamiltonian for a field interacting with the superconductor (being basically an electromagnetic interaction) is invariant under the operation of time reversal. However, the operator B might be the electric current $\boldsymbol{j}(\boldsymbol{r})$ (for electromagnetic interactions) the electric charge density $\varrho(\boldsymbol{r})$ (for the electron-phonon interaction) or the z component of the electron spin operator, σ_z (for the nuclear spin relaxation interaction). Since under time-reversal

$\boldsymbol{j}(\boldsymbol{r}, \mathrm{t}) \rightarrow -\boldsymbol{j}(\boldsymbol{r}, -t)$ (electromagnetic interaction)
$\varrho(\boldsymbol{r}, \mathrm{t}) \rightarrow +\varrho(\boldsymbol{r}, -t)$ (electron-phonon interaction)
$\sigma_z(t) \rightarrow -\sigma_z(-t)$ (nuclear spin relaxation interaction)
these show strikingly different interference effects.

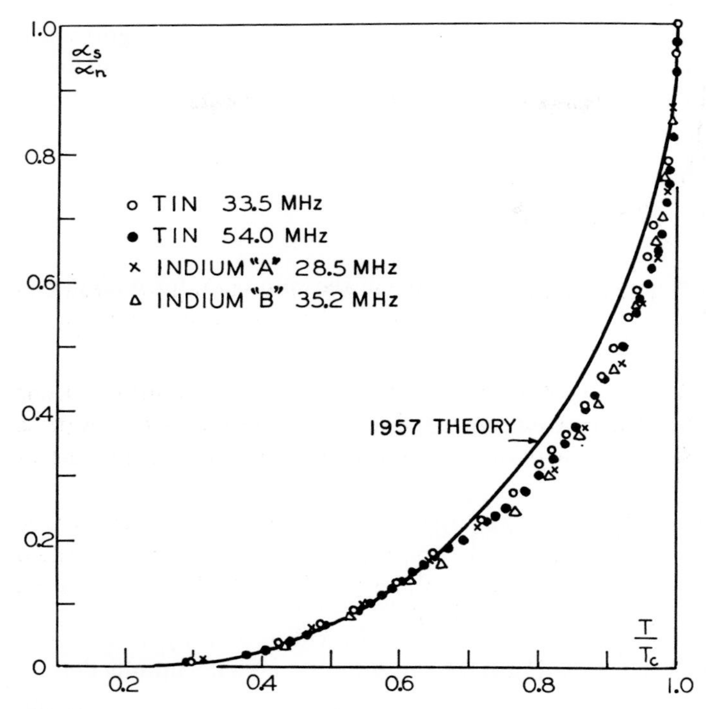

Fig. 12.
Comparison of observed ultrasonic attenuation with the ideal theory. The data are due to Morse and Bohm.

Ultrasonic attenuation in the ideal pure superconductor for $ql \gg 1$ (the product of the phonon wave number and the electron mean free path) depends in a fundamental way on the absorption and emission of phonons. Since the matrix elements have a very weak dependence on changes near the Fermi surface in occupation of states other than $\boldsymbol{k}$ or $\boldsymbol{k}'$ that occur in the normal to superconducting transition, calculations within the quasi-particle model can be compared in a very direct manner with similar calculations for the normal metal, as $B_{\boldsymbol{k}'\boldsymbol{k}}$ is the same in both states. The ratio of the attenuation in the normal and superconducting states becomes:

$$\frac{\alpha_s}{\alpha_n} = -4\int_{\Delta}^{\infty} dE(u^2-v^2)_s^2 \left(\frac{E}{\varepsilon}\right)^2 \frac{df(E)}{dE}.$$

Since $(u^2-v^2)_s^2 \to \frac{1}{2}\left(\frac{\varepsilon}{E}\right)^2$, the coherence factors cancel the density of states giving

$$\frac{\alpha_s}{\alpha_n} = 2f(\Delta(T)) = \frac{2}{1+\exp\left(\dfrac{\Delta(T)}{k_B T}\right)}.$$

Morse and Bohm (10b) used this result to obtain a direct experimental determination of the variation of Δ with T. Comparison of their attenuation data with the theoretical curve is shown in Figure 12.

In contrast the relaxation of nuclear spins which have been aligned in a magnetic field proceeds through their interaction with the magnetic moment of the conduction electrons. In an isotropic superconductor this can be shown to depend upon the z component of the electron spin operator

$$B_{K'K} = B(c^*_{k'\uparrow}c_{k\uparrow} - c^*_{-k\downarrow}c_{-k'\downarrow})$$

so that

$$B_{K'K} = -B_{-K-K'}.$$

This follows in general from the property of the spin operator under time reversal

$$\sigma_z(t) = -\sigma_z(-t).$$

The calculation of the nuclear spin relaxation rate proceeds in a manner not too different from that for ultrasonic attenuation resulting finally in a ratio of nuclear spin relaxation rates in superconducting and normal states in the same sample:

$$\frac{R_s}{R_n} = -4\int_{\Delta}^{\infty} dE(u^2+v^2)_s^2\left(\frac{E}{\varepsilon}\right)^2 \frac{df(E)}{dE}.$$

But $(u^2+v^2)_s^2$ does not go to zero at the lower limit so that the full effect of the increase in density of states at $E = \Delta$ is felt. Taken literally, in fact, this expression diverges logarithmically at the lower limit due to the infinite density of states. When the Zeeman energy difference between the spin up and spin down states is included, the integral is no longer divergent but the integrand is much too large. Hebel and Slichter, by putting in a broadening of levels phenomenologically, could produce agreement between theory and experiment. More recently Fibich (11) by including the effect of thermal phonons has obtained the agreement between theory and experiment shown in Fig. 13.

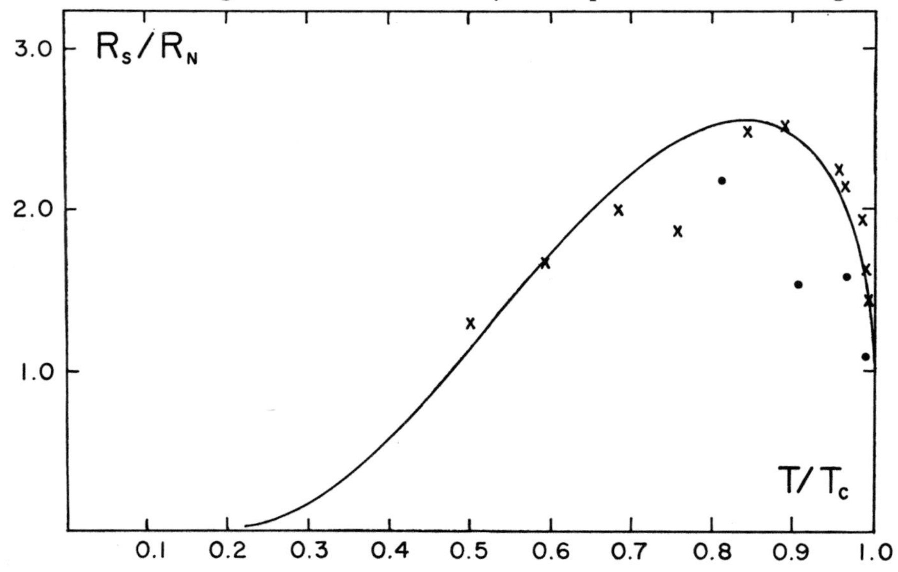

Fig. 13.
Comparison of observed nuclear spin relaxation rate with theory. The circles represent experimental data of Hebel and Slichter, the crosses data by Redfield and Anderson.

Interference effects manifest themselves in a similar manner in the interaction of electromagnetic radiation with the superconductor. Near T_c the absorption is dominated by quasi-particle scattering matrix elements of the type we have described. Near $T = 0$, the number of quasi-particle excitations goes to zero and the matrix elements that contribute are those in which quasi-particle pairs are created from ψ_0. For absorption these latter occur only when $\hbar\omega > 2\Delta$. For the linear response of the superconductor to a static magnetic field, the interference occurs in such a manner that the paramagnetic contribution goes to zero leaving the diamagnetic part which gives the Meissner effect.

The theory developed in 1957 and applied to the equilibrium properties of uniform materials in the weak coupling region has been extended in numerous directions by many authors. Professor Schrieffer has spoken of Josephson junctions and macroscopic quantum interference effects; Professor Bardeen will discuss the modifications of the theory when the electron-phonon interactions are strong. The treatment of ultrasonic attenuation, generalized to include situations in uniform superconductors in which $ql < 1$, gives a surprisingly similar result to that above. (12) There have been extensive developments using Green's function methods (13) appropriate for type II superconductors, materials with magnetic impurities and non-uniform materials or boundary regions where the order parameter is a function of the spatial coordinates. (14) With these methods formal problems of gauge invariance and/or current conservation have been resolved in a very elegant manner. (15) In addition, many calculations (16) of great complexity and detail for type II superconductors have treated ultrasonic attenuation, nuclear spin relaxation and other phenomena in the clean and dirty limits (few or large numbers of impurities). The results cited above are modified in various ways. For example, the average density of excitation levels is less sharply peaked at T_c in a type II superconductor; the coherence effects also change somewhat in these altered circumstances but nevertheless play an important role. Overall one can say that the theory has been amenable to these generalizations and that agreement with experiment is good.

It is now believed that the finite many-nucleon system that is the atomic nucleus enters a correlated state analogous to that of a superconductor. (17) Similar considerations have been applied to many-fermion systems as diverse as neutron stars, (18) liquid He^3, (19) and to elementary fermions. (20) In addition the idea of spontaneously broken symmetry of a degenerate vacuum has been applied widely in elementary particle theory and recently in the theory of weak interactions. (21) What the electron-phonon interaction has produced between electrons in metals may be produced by the van der Waals interaction between atoms in He^3, the nuclear interaction in nuclei and neutron stars, and the fundamental interactions in elementary fermions. Whatever the success of these attempts, for the theoretician the possible existence of this correlated paired state must in the future be considered for any degenerate many-fermion system where there is some kind of effective attraction between fermions for transitions near the Fermi surface.

In the past few weeks my colleagues and I have been asked many times: "What are the practical uses of your theory?" Although even a summary inspection of the proceedings of conferences on superconductivity and its applications would give an immediate sense of the experimental, theoretical and developmental work in this field as well as expectations, hopes and anticipations—from applications in heavy electrical machinery to measuring devices of extraordinary sensitivity and new elements with very rapid switching speeds for computers — I, personally, feel somewhat uneasy responding. The discovery of the phenomena and the development of the theory is a vast work to which many scientists have contributed. In addition there are numerous practical uses of the phenomena for which theory rightly should not take credit. A theory (though it may guide us in reaching them) does not produce the treasures the world holds. And the treasures themselves occasionally dazzle our attention; for we are not so wealthy that we may regard them as irrelevant.

But a theory is more. It is an ordering of experience that both makes experience meaningful and is a pleasure to regard in its own right. Henri Poincaré wrote (22):

> Le savant doit ordonner; on fait la science
> avec des faits comme une maison avec des
> pierres; mais une accumulation de faits
> n'est pas plus une science qu'un tas de
> pierres n'est une maison.

One can build from ordinary stone a humble house or the finest chateau. Either is constructed to enclose a space, to keep out the rain and the cold. They differ in the ambition and resources of their builder and the art by which he has achieved his end. A theory, built of ordinary materials, also may serve many a humble function. But when we enter and regard the relations in the space of ideas, we see columns of remarkable height and arches of daring breadth. They vault the fine structure constant, from the magnetic moment of the electron to the behavior of metallic junctions near the absolute zero; they span the distance from materials at the lowest temperatures to those in the interior of stars, from the properties of operators under time reversal to the behavior of attenuation coefficients just beyond the transition temperature.

I believe that I speak for my colleagues in theoretical science as well as myself when I say that our ultimate, our warmest pleasure in the midst of one of these incredible structures comes with the realization that what we have made is not only useful but is indeed a beautiful way to enclose a space.

References and Notes

1. Bardeen, J., Cooper, L. N. and Schrieffer, J. R., Phys. Rev. *108,* 1175 (1957).
2. An account of the situation as of 1969 may be found in the two volumes: Superconductivity, edited by R. D. Parks, Marcel Dekker, Inc., New York City (1969).
3. Sommerfeld, A., Z. Physik *47,* 1 (1928). Bloch, F., Z. Physik *52* 555 (1928).
4. Osheroff, D. D., Gully, W. J., Richardson R. C. and Lee, D. M., Phys. Rev. Letters *29,* 920 (1972); Leggett, A. J., Phys. Rev. Letters *29,* 1227 (1972).

5. Bogoliubov, N. N., Nuovo Cimento *7*, 794 (1958); Usp. Fiz. Nauk *67*, 549 (1959); Valatin, J. G., Nuovo Cimento *7*, 843 (1958).
6. Bardeen, J. and Rickayzen, G., Phys. Rev. *118*, 936 (1960); Mattis, D. C. and Lieb, E., J. Math. Phys. *2*, 602 (1961); Bogoliubov, N. N., Zubarev D. N. and Tserkovnikov, Yu. A., Zh. Eksperim. i Teor. Fiz. *39*, 120 (1960) translated: Soviet Phys. JETP *12*, 88 (1961).
7. For example, Glover, R. E. III and Tinkham, M., Phys. Rev. *108*, 243 (1957); Biondi, M. A. and Garfunkel, M. P., Phys. Rev. 116, 853 (1959).
8. The importance of the coupling of time reversed states in constructing electron pairs was emphasized by P. W. Anderson,; for example, Anderson, P. W., J. Phys. Chem. Solids *11*, 26 (1959)
9. a) Hebel, L. C. and Slichter, C. P., Phys. Rev. *113*, 1504 (1959).
   b) Redfield, A. G. and Anderson, A. G., Phys. Rev. *116*, 583 (1959).
10. a) Bommel, H. E., Phys. Rev. *96*, 220 (1954).
    b) Morse, R. W. and Bohm, H. V., Phys. Rev. *108*, 1094 (1957).
11. Fibich, M., Phys. Rev. Letters *14*, 561 (1965).
12. For example, Tsuneto T., Phys. Rev. *121*, 402 (1961).
13. Gor'kov, L. P., Zh. Eksperim. i Teor. Fiz. *34*, 735 (1958) translated: Soviet Physics JETP *7*, 505 (1958); also Martin P. C. and Schwinger J., Phys. Rev. *115*, 1342 (1959); Kadanoff L. P. and Martin P. C., Phys. Rev. *124*, 670 (1961).
14. Abrikosov, A. A. and Gor'kov, L. P., Zh. Eksperim. i Teor. Fiz. *39*, 1781 (1960) translated: Soviet Physics JETP *12*, 1243 (1961); de Gennes P. G., Superconductivity of Metals and Alloys, Benjamin, New York (1966).
15. For example, Ambegaokar V. and Kadanoff L. P., Nuovo Cimento *22*, 914 (1961).
16. For example, Caroli C. and Matricon J., Physik Kondensierten Materie *3*, 380 (1965); Maki K., Phys. Rev. *141*, 331 (1966); *156*, 437 (1967); Groupe de Supraconductivité d'Orsay, Physik Kondensierten Materie *5*, 141 (1966); Eppel D., Pesch W. and Tewordt L., Z. Physik *197*; 46 (1966); McLean F. B. and Houghton A., Annals of Physics *48*, 43 (1968).
17. Bohr, A., Mottelson, B. R. and Pines, D., Phys. Rev. *110*, 936 (1958); Migdal, A. B., Nuclear Phys. *13*, 655 (1959).
18. Ginzburg, V. L. and Kirzhnits, D. A., Zh. Eksperim. i Theor. Fiz. *47*, 2006 (1964) translated: Soviet Physics JETP *20*, 1346 (1965); Pines D., Baym G. and Pethick C., Nature *224*, 673 (1969).
19. Many authors have explored the possibility of a superconducting-like transition in He^3. Among the most recent contributions see reference 4.
20. For example. Nambu Y. and Jona-Lasinio G., Phys. Rev. *122*, 345 (1961).
21. Goldstone, J., Nuovo Cimento *19*, 154 (1961); Weinberg, S., Phys. Rev. Letters *19*, 1264 (1967).
22. Poincaré, H., La Science et l'Hypothèse, Flammarion, Paris, pg. 168 (1902). "The scientist must order; science is made with facts as a house with stones; but an accumulation of facts is no more a science than a heap of stones is a house."

J. Robert Schrieffer

J. ROBERT SCHRIEFFER

J. Robert Schrieffer was born in Oak Park, Illinois on May 31, 1931, son of John H. Schrieffer and his wife Louis (nee Anderson). In 1940, the family moved to Manhasset, New York and in 1947 to Eustis, Florida where they became active in the citrus industry.

Following his graduation from Eustis High School in 1949, Schrieffer was admitted to Massachusetts Institute of Technology, where for two years he majored in electrical engineering, then changed to physics in his junior year. He completed a bachelor's thesis on the multiple structure in heavy atoms under the direction of Professor John C. Slater. Following up on an interest in solid state physics developed while at MIT, he began graduate studies at the University of Illinois, where he immediately began research with Professor John Bardeen. After working out a problem dealing with electrical conduction on semiconductor surfaces, Schrieffer spent a year in the laboratory, applying the theory to several surface problems. In the third year of graduate studies, he joined Bardeen and Cooper in developing the theory of superconductivity, which constituted his doctoral dissertation.

He spent the academic year 1957–58 as a National Science Foundation fellow at the University of Birmingham and the Niels Bohr Institute in Copenhagen, where he continued research in superconductivity. Following a year as assistant professor at the University of Chicago, he returned to the University of Illinois in 1959 as a faculty member. In 1960 he returned to the Bohr Institute for a summer visit, during which he became engaged to Anne Grete Thomsen whom he married at Christmas of that year.

In 1962 Schrieffer joined the faculty of the University of Pennsylvania in Philadelphia, where in 1964 he was appointed Mary Amanda Wood Professor in Physics. In 1980 he was appointed Professor at the University of California, Santa Barbara and to the position of Chancellor Professor in 1984. He served as Director of the Institute for Theoretical Physics in Santa Barbara from 1984–89. In 1992 he was appointed University Professor at Florida State University and Chief Scientist of the National High Magnetic Field Laboratory.

He holds honorary degrees from the Technische Hochschule, Munich and the Universities of Geneva, Pennsylvania, Illinois, Cincinnati, Tel-Aviv, Alabama. In 1969 he was appointed by Cornell to a six-year term as a Andrew D. White Professor-at-Large.

He is a member of the American Academy of Arts and Sciences, the National Academy of Sciences of which he is a member of their council, the

American Philosophical Society, the Royal Danish Academy of Sciences and Letters and the Academy of Sciences of the USSR.

His awards include the Guggenheim Fellowship, Oliver E. Buckley Solid State Physics Prize, Comstock Prize, National Academy of Science, the Nobel Prize in Physics shared with John Bardeen and Leon N. Cooper in 1972, John Ericsson Medal, American Society of Swedish Engineers, University of Illinois Alumni Achievement Award, and in 1984 the National Medal of Science. The main thrust of his recent work has been in the area of high-temperature superconductivity, strongly correlated electrons, and the dynamics of electrons in strong magnetic fields.

The Schrieffers have three children, Bolette, Paul, and Regina.

MACROSCOPIC QUANTUM PHENOMENA FROM PAIRING IN SUPERCONDUCTORS

Nobel Lecture, December 11, 1972

by

J. R. SCHRIEFFER

University of Pennsylvania, Philadelphia, Pa.

I. INTRODUCTION

It gives me great pleasure to have the opportunity to join my colleagues John Bardeen and Leon Cooper in discussing with you the theory of superconductivity. Since the discovery of superconductivity by H. Kamerlingh Onnes in 1911, an enormous effort has been devoted by a spectrum of outstanding scientists to understanding this phenomenon. As in most developments in our branch of science, the accomplishments honored by this Nobel prize were made possible by a large number of developments preceding them. A general understanding of these developments is important as a backdrop for our own contribution.

On December 11, 1913, Kamerlingh Onnes discussed in his Nobel lecture (1) his striking discovery that on cooling mercury to near the absolute zero of temperature, the electrical resistance became vanishingly small, but this disappearance "did not take place gradually but *abruptly*." His Fig. 17 is reproduced as Fig. 1. He said, "Thus, mercury at 4.2 K has entered a new state

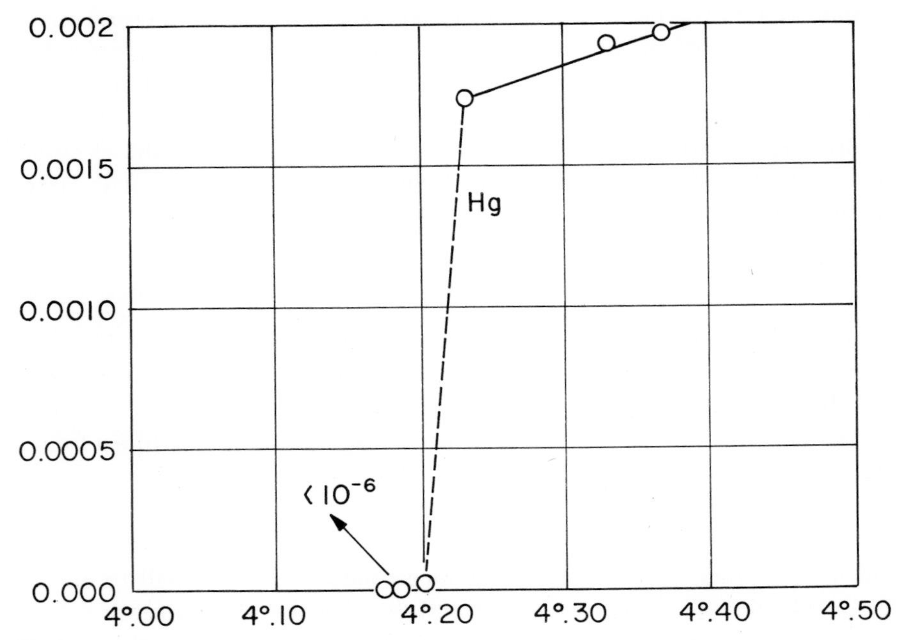

Fig. 1

which owing to its particular electrical properties can be called the state of superconductivity." He found this state could be destroyed by applying a sufficiently strong magnetic field, now called the critical field H_c. In April—June, 1914, Onnes discovered that a current, once induced in a closed loop of superconducting wire, persists for long periods without decay, as he later graphically demonstrated by carrying a loop of superconducting wire containing a persistent current from Leiden to Cambridge.

In 1933, W. Meissner and R. Ochsenfeld (2) discovered that a superconductor is a perfect diamagnet as well as a perfect conductor. The magnetic field vanishes in the interior of a bulk specimen, even when cooled down below the transition temperature in the presence of a magnetic field. The diamagnetic currents which flow in a thin penetration layer near the surface of a simply connected body to shield the interior from an externally applied field are stable rather than metastable. On the other hand, persistent currents flowing in a multiply connected body, e.g., a loop, are metastable.

An important advance in the understanding of superconductivity occurred in 1934, when C. J. Gorter and H. B. G. Casimir (3) advanced a two fluid model to account for the observed second order phase transition at T_c and other thermodynamic properties. They proposed that the total density of electrons ϱ could be divided into two components

$$\varrho = \varrho_s + \varrho_n \tag{1}$$

where a fraction ϱ_s/ϱ_n of the electrons can be viewed as being condensed into a "superfluid," which is primarily responsible for the remarkable properties of superconductors, while the remaining electrons form an interpenetrating fluid of "normal" electrons. The fraction ϱ_s/ϱ_n grows steadily from zero at T_c to unity at $T = 0$, where "all of the electrons" are in the superfluid condensate.

A second important theoretical advance came in the following year, when Fritz and Hans London set down their phenomenological theory of the electromagnetic properties of superconductors, in which the diamagnetic rather than electric aspects are assumed to be basic. They proposed that the electrical current density $\mathbf{j}_s$ carried by the superfluid is related to the magnetic vector potential $\mathbf{A}$ at each point in space by

$$\mathbf{j}_s = -\frac{1}{\Lambda c}\mathbf{A} \tag{2}$$

where Λ is a constant dependent on the material in question, which for a free electron gas model is given by $\Lambda = m/\varrho_s e^2$, m and e being the electronic mass and charge, respectively. A is to be chosen such that $\nabla \cdot \mathbf{A} = 0$ to ensure current conservation. From (2) it follows that a magnetic field is excluded from a superconductor except within a distance

$$\lambda_L = \sqrt{\Lambda c^2/4\pi}$$

which is of order 10^{-6} cm in typical superconductors for T well below T_c. Observed values of λ are generally several times the London value.

In the same year (1935) Fritz London (4) suggested how the diamagnetic property (2) might follow from quantum mechanics, if there was a "rigidity" or stiffness of the wavefunction ψ of the superconducting state such that ψ was essentially unchanged by the presence of an externally applied magnetic field. This concept is basic to much of the theoretical development since that time, in that it sets the stage for the gap in the excitation spectrum of a superconductor which separates the energy of superfluid electrons from the energy of electrons in the normal fluid. As Leon Cooper will discuss, this gap plays a central role in the properties of superconductors.

In his book published in 1950, F. London extended his theoretical conjectures by suggesting that a superconductor is a "quantum structure on a macroscopic scale [which is a] kind of solidification or condensation of the average momentum distribution" of the electrons. This momentum space condensation locks the average momentum of each electron to a common value which extends over appreciable distance in space. A specific type of condensation in momentum space is central to the work Bardeen, Cooper and I did together. It is a great tribute to the insight of the early workers in this field that many of the important general concepts were correctly conceived before the microscopic theory was developed. Their insight was of significant aid in our own work.

The phenomenological London theory was extended in 1950 by Ginzburg and Landau (5) to include a spatial variation of ϱ_s. They suggested that ϱ_s/ϱ be written in terms of a phenomenological condensate wavefunction $\psi(r)$ as $\varrho_s(r)/\varrho = |\psi(r)|^2$ and that the free energy difference ΔF between the superconducting and normal states at temperature T be given by

$$\Delta F = \int \left\{ \frac{\hbar^2}{2\bar{m}} \left| \left(\nabla + \frac{\bar{e}}{c} A(r) \right) \psi(r) \right|^2 - a(T) |\psi(r)|^2 + \frac{b(T)}{2} |\psi(r)|^4 \right\} d^3r \quad (3)$$

where $\bar{e}$, $\bar{m}$, a and b are phenomenological constants, with $a(T_c) = 0$.

They applied this approach to the calculation of boundary energies between normal and superconducting phases and to other problems.

As John Bardeen will discuss, a significant step in understanding which forces cause the condensation into the superfluid came with the experimental discovery of the isotope effect by E. Maxwell and, independently, by Reynolds, et al. (6). Their work indicated that superconductivity arises from the interaction of electrons with lattice vibrations, or phonons. Quite independently, Herbert Fröhlich (7) developed a theory based on electron-phonon interactions which yielded the isotope effect but failed to predict other superconducting properties. A somewhat similar approach by Bardeen (8) stimulated by the isotope effect experiments also ran into difficulties. N. Bohr, W. Heisenberg and other distinguished theorists had continuing interest in the general problem, but met with similar difficulties.

An important concept was introduced by A. B. Pippard (9) in 1953. On the basis of a broad range of experimental facts he concluded that a coherence length ξ is associated with the superconducting state such that a perturbation of the superconductor at a point necessarily influences the superfluid within a distance ξ of that point. For pure metals, $\xi \sim 10^{-4}$ cm. for $T \ll T_c$. He generalized the London equation (3) to a non-local form and accounted for the fact that the experimental value of the penetration depth is several times larger than the London value. Subsequently, Bardeen (10) showed that Pippard's non-local relation would likely follow from an energy gap model.

A major problem in constructing a first principles theory was the fact that the physically important condensation energy ΔF amounts typically to only 10^{-8} electron volts (e.V.) per electron, while the uncertainty in calculating the total energy of the electron-phonon system in even the normal state amounted to of order 1 e.V. per electron. Clearly, one had to isolate those correlations peculiar to the superconducting phase and treat them accurately, the remaining large effects presumably being the same in the two phases and therefore cancelling. Landau's theory of a Fermi liquid (11), developed to account for the properties of liquid He^3, formed a good starting point for such a scheme. Landau argued that as long as the interactions between the particles (He^3 atoms in his case, electrons in our case) do not lead to discontinuous changes in the microscopic properties of the system, a "quasi-particle" description of the low energy excitations is legitimate; that is, excitations of the fully interacting normal phase are in one-to-one correspondence with the excitations of a non-interacting fermi gas. The effective mass m and the Fermi velocity v_F of the quasi-particles differ from their free electron values, but aside from a weak decay rate which vanishes for states at the Fermi surface there is no essential change. It is the residual interaction between the quasi-particles which is responsible for the special correlations characterizing superconductivity. The ground state wavefunction of the superconductor ψ_0 is then represented by a particular superposition of these normal state configurations, Φ_n.

A clue to the nature of the states Φ_n entering strongly in ψ_0 is given by combining Pippard's coherence length ξ with Heisenberg's uncertainty principle

$$\Delta p \sim \hbar/\xi \sim 10^{-4} p_F \tag{4}$$

where p_F is the Fermi momentum. Thus, Ψ_0 is made up of states with quasi-particles (electrons) being excited above the normal ground state by a momentum of order Δp. Since electrons can only be excited to states which are initially empty, it is plausible that only electronic states within a momentum $10^{-4}\ p_F$ of the Fermi surface are involved significantly in the condensation, i.e., about 10^{-4} of the electrons are significantly affected. This view fits nicely with the fact that the condensation energy is observed to be of order $10^{-4}\varrho\cdot k_BT_c$. Thus, electrons within an energy $\sim v_F\Delta p \simeq kT_c$ of the Fermi surface have their energies lowered by of order kT_c in the condensation. In summary, the problem was how to account for the phase transition in which a condensation of electrons occurs in momentum space for electrons very near the Fermi surface. A proper theory should automatically account for the perfect conductivity and diamagnetism, as well as for the energy gap in the excitation spectrum.

II. The Pairing Concept

In 1955, stimulated by writing a review article on the status of the theory of superconductivity, John Bardeen decided to renew the attack on the problem.

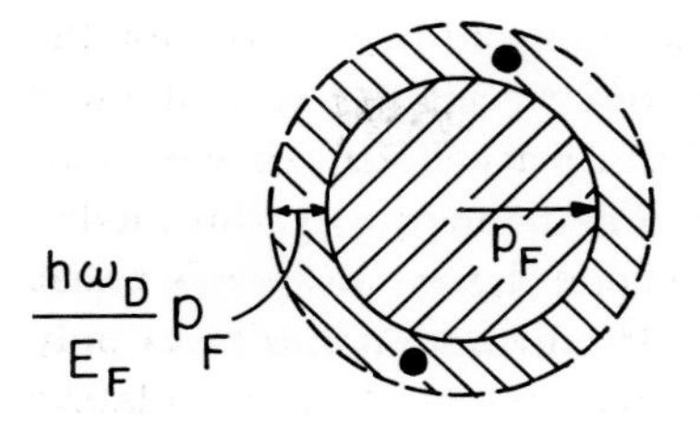

He invited Leon Cooper, whose background was in elementary particle physics and who was at that time working with C. N. Yang at the Institute for Advanced Study to join in the effort starting in the fall of 1955. I had the good fortune to be a graduate student of Bardeen at that time, and, having finished my graduate preliminary work, I was delighted to accept an invitation to join them.

We focused on trying to understand how to construct a ground state Ψ_0 formed as a coherent superposition of normal state configurations Φ_n,

$$\Psi_0 = \sum_n a_n \Phi_n \tag{5}$$

such that the energy would be as low as possible. Since the energy is given in terms of the Hamiltonian H by

$$E_0 = (\Psi_0, H\psi_0) = \sum_{n,n'} a_{n'}^* \, a_n \, (\Phi_{n'}, H\Phi_n) \tag{6}$$

we attempted to make E_0 minimum by restricting the coefficients a_n so that only states which gave negative off-diagonal matrix elements would enter (6). In this case all terms would add in phase and E_0 would be low.

By studying the eigenvalue spectrum of a class of matrices with off-diagonal elements all of one sign (negative), Cooper discovered that frequently a single eigenvalue is split off from the bottom of the spectrum. He worked out the problem of two electrons interacting via an attractive potential-V above a quiescent Fermi sea, i.e., the electrons in the sea were not influenced by V and the extra pair was restricted to states within an energy $\hbar\omega_D$ above the Fermi surface, as illustrated in Fig. 2. As a consequence of the non-zero density of quasi-particle states $\mathcal{N}(0)$ at the Fermi surface, he found the energy eigenvalue spectrum for two electrons having zero total momentum had a bound state split off from the continuum of scattering states, the binding energy being

$$E_B \cong \hbar\omega_D e^{-\frac{2}{\mathcal{N}(0)V}} \tag{7}$$

if the matrix elements of the potential are constant equal to V in the region of interaction. This important result, published in 1956 (12), showed that, regardless of how weak the residual interaction between quasi-particles is, if the interaction is attractive the system is unstable with respect to the formation of bound pairs of electrons. Further, if E_B is taken to be of order $k_B T_c$, the uncertainty principle shows the average separation between electrons in the bound state is of order 10^{-4} cm.

While Cooper's result was highly suggestive, a major problem arose. If, as we discussed above, a fraction 10^{-4} of the electrons is significantly involved in the condensation, the average spacing between these condensed electrons is roughly 10^{-6} cm. Therefore, within the volume occupied by the bound state of a given pair, the centers of approximately $(10^{-4}/10^{-6})^3 \cong 10^6$ other pairs will be found, on the average. Thus, rather than a picture of a dilute gas of strongly bound pairs, quite the opposite picture is true. The pairs overlap so strongly in space that the mechanism of condensation would appear to be destroyed due to the numerous pair-pair collisions interrupting the binding process of a given pair.

Returning to the variational approach, we noted that the matrix elements $(\Phi_{n'}, H\Phi_n)$ in (6) alternate randomly in sign as one randomly varies n and n' over the normal state configurations. Clearly this cannot be corrected to obtain a low value of E_0 by adjusting the sign of the a_n's since there are N^2 matrix elements to be corrected with only N parameters a_n. We noticed that if the sum in (6) is restricted to include only configurations in which, if any quasi-particle state, say k, s, is occupied ($s = \uparrow$ or $\downarrow$ is the spin index), its "mate" state $\bar{k}$, $\bar{s}$ is also occupied, then the matrix elements of H between such states would have a unique sign and a coherent lowering of the energy would be obtained. This correlated occupancy of pairs of states in momentum space is consonant with London's concept of a condensation in momentum.

In choosing the state $\bar{k}$, $\bar{s}$ to be paired with a given state k, s, it is important to note that in a perfect crystal lattice, the interaction between quasi-particles conserves total (crystal) momentum. Thus, as a given pair of quasi-particles interact, their center of mass momentum is conserved. To obtain the largest number of non-zero matrix elements, and hence the lowest energy, one must choose the total momentum of each pair to be the same, that is

$$k+\bar{k} = q. \tag{8}$$

States with $q \neq 0$ represent states with net current flow. The lowest energy state is for $q = 0$, that is, the pairing is such that if any state $k\uparrow$ is occupied in an admissible Φn, so is $-k\downarrow$ occupied. The choice of $\downarrow\uparrow$ spin pairing is not restrictive since it encompasses triplet and singlet paired states.

Through this reasoning, the problem was reduced to finding the ground state of the reduced Hamiltonian

$$H_{\text{red}} = \sum_{ks} \epsilon_k n_{ks} - \sum_{kk'} V_{k'k} b_{k'}^{+} b_k. \tag{9}$$

The first term in this equation gives the unperturbed energy of the quasi-particles forming the pairs, while the second term is the pairing interaction in which a pair of quasi-particles in $(k\uparrow, -k\downarrow)$ scatter to $(k'\uparrow, -k'\uparrow)$. The operators $b_k^{+} = c_{k\uparrow}^{+} c_{-k\downarrow}^{+}$, being a product of two fermion (quasi-particle) creation operators, do not satisfy Bose statistics, since $b_k^{+2} = 0$. This point is essential to the theory and leads to the energy gap being present not only for dissociating a pair but also for making a pair move with a total momentum different from the common momentum of the rest of the pairs. It is this feature which enforces long range order in the superfluid over macroscopic distances.

III. The Ground State

In constructing the ground state wavefunction, it seemed clear that the average occupancy of a pair state $(k\uparrow, -k\downarrow)$ should be unity for k far below the Fermi surface and 0 for k far above it, the fall off occurring symmetrically about k_F over a range of momenta of order

$$\Delta k \sim \frac{1}{\xi} \sim 10^4 \text{ cm}^{-1}.$$

One could not use a trial Ψ_0 as one in which each pair state is definitely occupied or definitely empty since the pairs could not scatter and lower the energy in this case. Rather there had to be an amplitude, say v_k, that $(k\uparrow, -k\downarrow)$ is occupied in Ψ_0 and consequently an amplitude $u_k = \sqrt{1-v_k^2}$ that the pair state is empty. After we had made a number of unsuccessful attempts to construct a wavefunction sufficiently simple to allow calculations to be carried out, it occurred to me that since an enormous number ($\sim 10^{19}$) of pair states $(k'\uparrow, -k'\downarrow)$ are involved in scattering into and out of a given pair state $(k\uparrow, -k\downarrow)$, the "instantaneous" occupancy of this pair state should be essentially uncorrelated with the occupancy of the other pair states at that "instant". Rather, only the *average* occupancies of these pair states are related.

On this basis, I wrote down the trial ground state as a product of operators —one for each pair state—acting on the vacuum (state of no electrons),

$$\Psi_0 = \pi_k \, (u_k + v_k b_k) \, |0>, \tag{10}$$

where $u_k = \sqrt{1-v_k^2}$. Since the pair creation operators b_k^+ commute for different k's, it is clear that Ψ_0 represents uncorrelated occupancy of the various pair states. I recall being quite concerned at the time that Ψ_0 was an admixture of states with different numbers of electrons, a wholly new concept to me, and as I later learned to others as well. Since by varying v_k the mean number of electrons varied, I used a Lagrange multiplier μ (the chemical potential) to make sure that the mean number of electrons ($\mathcal{N}_{\text{op}}$) represented by Ψ_0 was the desired number $\mathcal{N}$. Thus by minimizing

$$E_0 - \mu\mathcal{N} = (\Psi_0, [H_{\text{red}} - \mu\mathcal{N}_{\text{op}}]\Psi_0)$$

with respect to v_k, I found that v_k was given by

$$v_k^2 = \tfrac{1}{2}\left[1 - \frac{(\epsilon_k - \mu)}{E_k}\right] \tag{11}$$

where

$$E_k = \sqrt{(\epsilon_k - \mu)^2 + \Delta_k^2} \tag{12}$$

and the parameter Δ_k satisfied what is now called the energy gap equation:

$$\Delta_k = -\sum V_{k'k} \frac{\Delta_{k'}}{2E_{k'}} \tag{13}$$

From this expression, it followed that for the simple model

$$V_{k'k} = \begin{cases} V, \ |\epsilon_k - \mu| < \hbar\omega_{\text{D}} \text{ and } |\epsilon_{k'} - \mu| < \hbar\omega_{\text{D}} \\ 0, \text{ otherwise} \end{cases}$$

$$\Delta = \hbar\,\omega_{\text{D}} e - \frac{1}{\mathcal{N}(0)V} \tag{14}$$

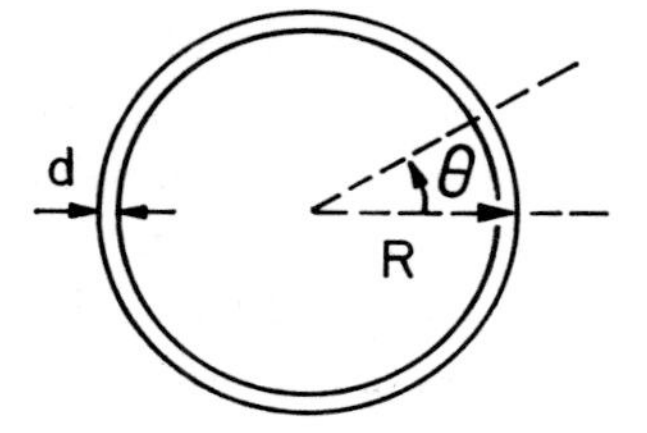

and the condensation energy at zero temperature is

$$\Delta F = \tfrac{1}{2}N(0)\Delta^2 \qquad (15)$$

The idea occurred to me while I was in New York at the end of January, 1957, and I returned to Urbana a few days later where John Bardeen quickly recognized what he believed to be the essential validity of the scheme, much to my pleasure and amazement. Leon Cooper will pick up the story from here to describe our excitement in the weeks that followed, and our pleasure in unfolding the properties of the excited states.

IV. Quantum Phenomena on a Macroscopic Scale

Superconductors are remarkable in that they exhibit quantum effects on a broad range of scales. The persistence of current flow in a loop of wire many meters in diameter illustrates that the pairing condensation makes the superfluid wavefunction coherent over macroscopic distances. On the other hand, the absorption of short wavelength sound and light by a superconductor is sharply reduced from the normal state value, as Leon Cooper will discuss. I will concentrate on the large scale quantum effects here.

The stability of persistent currents is best understood by considering a circular loop of superconducting wire as shown in Fig. 3. For an ideal small diameter wire, one would use the eigenstates $e^{im\theta}$, $(m = 0, \pm 1, \pm 2, \ldots)$, of the angular momentum L_Z about the symmetry axis to form the pairing. In the ground state no net current flows and one pairs $m\uparrow$ with $-m\downarrow$, instead of $k\uparrow$ with $-k\downarrow$ as in a bulk superconductor. In both cases, the paired states are time reversed conjugates, a general feature of the ground state. In a current carrying state, one pairs $(m+\nu)\uparrow$ with $(-m+\nu)\downarrow$, $(\nu = 0, \pm 1, \pm 2 \ldots)$, so that the total angular momentum of each pair is identical, $2\hbar\,\nu$. It is this commonality of the center of mass angular momentum of each pair which preserves the condensation energy and long range order even in states with current flow. Another set of flow states which interweave with these states is formed by pairing $(m+\nu)\uparrow$ with $(-m+\nu+1)\downarrow$, $(\nu = 0, \pm 1, \pm 2 \ldots)$, with the pair angular momentum being $(2\nu+1)\hbar$. The totality of states forms a set with all integer multiples n of $\hbar$ for allowed total angular momentum of pairs. Thus, even though the pairs greatly overlap in space, the system exhibits quantization effects as if the pairs were well defined.

There are two important consequences of the above discussion. First, the fact that the coherent condensate continues to exist in flow states shows that to scatter a pair out of the (rotating) condensate requires an increase of energy.

Crudely speaking, slowing down a given pair requires it ot give up its binding energy and hence this process will occur only as a fluctuation. These fluctuations average out to zero. The only way in which the flow can stop is if all pairs simultaneously change their pairing condition from, say, ν to $\nu-1$. In this process the system must fluctate to the normal state, at least in a section of the wire, in order to change the pairing. This requires an energy of order the condensation energy ΔF. A thermal fluctuation of this size is an exceedingly rare event and therefore the current persists.

The second striking consequence of the pair angular momentum quantization is that the magnetic flux Φ trapped within the loop is also quantized,

$$\Phi_n = n \cdot \frac{hc}{2e} \qquad (n = 0, \pm 1, \pm 2 \ldots). \tag{16}$$

This result follows from the fact that if the wire diameter d is large compared to the penetration depth λ, the electric current in the center of the wire is essentially zero, so that the canonical angular momentum of a pair is

$$L_{\text{pair}} = \frac{2e}{c} r_{\text{pair}} \times A \tag{17}$$

where r_{pair} is the center of mass coordinate of a pair and A is the magnetic vector potential. If one integrates L_{pair}, around the loop along a path in the center of the wire, the integral is nh, while the integral of the right hand side of (17) is $\frac{2e}{c} \Phi$.

A similar argument was given by F. London (4b) except that he considered only states in which the superfluid flows as a whole without a change in its internal strucutre, i.e., states analogous to the $(m+\nu)\uparrow$, $(-m+\nu)\downarrow$ set. He found $\Phi z = n \cdot hc/e$. The pairing $(m+\nu)\uparrow$, $(m+\nu+1)\downarrow$ cannot be obtained by adding ν to each state, yet this type of pairing gives an energy as low as the more conventional flow states and these states enter experimentally on the same basis as those considered by London. Experiments by Deaver and Fairbank (13), and independently by Doll and Näbauer (14) confirmed the flux quantization phenomenon and provided support for the pairing concept by showing that $2e$ rather than e enters the flux quantum. Following these experiments a clear discussion of flux quantization in the pairing scheme was given by Beyers and Yang (15).

The idea that electron pairs were somehow important in superconductivity has been considered for some time (16, 17). Since the superfluidity of liquid He^4 is qualitatively accounted for by Bose condensation, and since pairs of electrons behave in some respects as a boson, the idea is attractive. The essential point is that while a dilute gas of tightly bound pairs of electrons might behave like a Bose gas (18) this is not the case when the mean spacing between pairs is very small compared to the size of a given pair. In this case the inner structure of the pair, i.e., the fact that it is made of fermions, is essential; it is this which distinguishes the pairing condensation, with its energy gap for single pair translation as well as dissociation, from the spectrum of a Bose con-

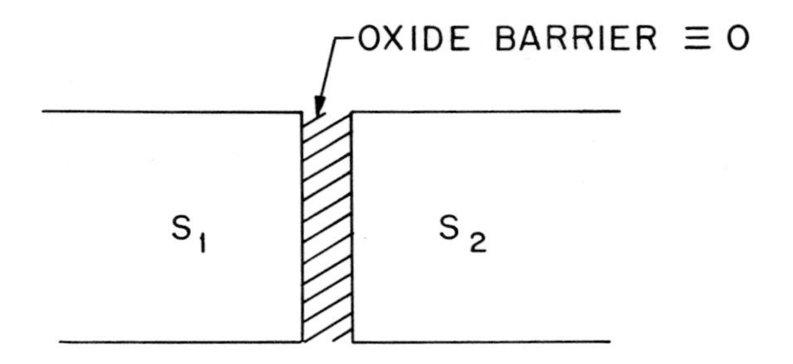

densate, in which the low energy exictations are Bose-like rather than Fermi-like as occurs in acutal superconductors. As London emphasized, the condensation is an ordering in occupying momentum space, and not a space-like condensation of clusters which then undergo Bose condensation.

In 1960, Ivar Giaever (19) carried out pioneering experiments in which electrons in one superconductor (S_1) tunneled through a thin oxide layer ($\sim$ 20—30 Å) to a second superconductor (S_2) as shown in Fig. 4. Giaever's experiments were dramatic evidence of the energy gap for quasi-particle excitations. Subsequently, Brian Josephson made a highly significant contribution by showing theoretically that a superfluid current could flow between S_1 and S_2 with zero applied bias. Thus, the superfluid wavefunction is coherent not only in S_1 and S_2 separately, but throughout the entire system, S_1—0—S_2, under suitable circumstances. While the condensate amplitude is small in the oxide, it is sufficient to lock the phases of S_1 and S_2 together, as has been discussed in detail by Josephson (20) and by P. W. Anderson (21).

To understand the meaning of phase in this context, it is useful to go back to the ground state wavefunction Ψ_0, (10). Suppose we write the parameter v_k as $|v_k|$ exp iφ and choose u_k to be real. If we expand out the k-product in Ψ_0, we note that the terms containing N pairs will have a phase factor exp (i $N\varphi$), that is, each occupied pair state contributes a phase φ to Ψ_0. Let this wavefunction, say $\Psi_0^{(1)}$ represent S_1, and have phase φ_1. Similarly, let $\Psi_0^{(2)}$ represent S_2 and have phase angle φ_2. If we write the state of the combined system as a product

$$\Psi_0^{(1,2)} = \Psi_0^{(1)}\,\Psi_0^{(2)} \tag{18}$$

then by expanding out the double product we see that the phase of that part of $\Psi_0^{(1,2)}$ which has N_1 pairs in S_1 and N_2 pairs in S_2 is $N_1\,\varphi_1+N_2\varphi_2$. For a truly isolated system, $2(N_1+N_2) = 2N$ is a fixed number of electrons; however N_1 and N_2 are not separately fixed and, as Josephson showed, the energy of the combined system is minimized when $\varphi_1 = \varphi_2$ due to tunneling of electrons between the superconductors. Furthermore, if $\varphi_1 = \varphi_2$, a current flows between S_1 and S_2

$$j = j_1 \sin(\varphi_1 - \varphi_2) \tag{19}$$

If $\varphi_1-\varphi_2 = \varphi$ is constant in time, a constant current flows with no voltage applied across the junction. If a bias voltage is V applied between S_1 and S_2, then, according to quantum mechanics, the phase changes as

$$\frac{2eV}{\hbar} = \frac{d\varphi}{dt} \tag{20}$$

Hence a constant voltage applied across such a junction produces an alternating current of frequency

$$\nu = \frac{2eV}{\hbar} = 483 \text{ THz/V}. \tag{21}$$

These effects predicted by Josephson were observed experimentally in a series of beautiful experiments (22) by many scientists, which I cannot discuss in detail here for lack of time. I would mention, as an example, the work of Langenberg and his collaborators (23) at the University of Pennsylvania on the precision determination of the fundamental constant e/h using the frequency-voltage relation obeyed by the alternating Josephson supercurrent. These experiments have decreased the uncertainty in our experimental knowledge of this constant by several orders of magnitude and provide, in combination with other experiments, the most accurate available value of the Sommerfeld fine structure constant. They have resulted in the resolution of several discrepancies between theory and experiment in quantum electrodynamics and in the development of an "atomic" voltage standard which is now being used by the United States National Bureau of Standards to maintain the U.S. legal volt.

V. Conclusion

As I have attempted to sketch, the development of the theory of superconductivity was truly a collaborative effort, involving not only John Bardeen, Leon Cooper and myself, but also a host of outstanding scientists working over a period of half a century. As my colleagues will discuss, the theory opened up the field for many exciting new developments, both scientific and technological, many of which no doubt lie in the future. I feel highly honored to have played a role in this work and I deeply appreciate the honor you have bestowed on me in awarding us the Nobel prize.

References

1. Kamerlingh Onnes, H., Nobel Lectures, Vol. 1, pp. 306–336.
2. Meissner, W. and Ochsenfeld, R., Naturwiss. *21*, 787 (1933).
3. Gorter, C. J. and Casimir, H. B. G., Phys. Z. *35*, 963 (1934); Z. Techn. Phys. *15*, 539 (1934).
4. London, F., [a] Phys. Rev. *24*, 562 (1948); [b] Superfluids, Vol. 1 (John Wiley & Sons, New York, 1950).
5. Ginzburg, V. L. and Landau, L. D., J. Exp. Theor. Phys. (U.S.S.R.) *20*, 1064 (1950).
6. Maxwell, E., Phys. Rev. *78*, 477 (1950); Reynolds, C. A., Serin, B., Wright W. H. and Nesbitt, L. B., Phys. Rev. *78*, 487 (1950).
7. Fröhlich, H., Phys. Rev. *79*, 845 (1950).

8. Bardeen, J., Rev. Mod. Phys. *23,* 261 (1951).
9. Pippard, A. B., Proc. Royal Soc. (London) A*216,* 547 (1953).
10. Bardeen, J., [a] Phys. Rev. *97,* 1724 (1955); [b] Encyclopedia of Physics, Vol. 15 (Springer-Verlag, Berlin, 1956), p. 274.
11. Landau, L. D., J. Exp. Theor. Phys. (U.S.S.R.) *30* (*3*), 1058 (920) (1956); *32* (*5*), 59 (101) (1957).
12. Cooper, L. N., Phys. Rev. *104,* 1189 (1956).
13. Deaver, B. S. Jr., and Fairbank, W. M., Phys. Rev. Letters *7,* 43 (1961).
14. Doll, R. and Näbauer, M., Phys. Rev. Letters *7,* 51 (1961).
15. Beyers, N. and Yang, C. N., Phys. Rev. Letters *7,* 46 (1961).
16. Ginzburg, V. L., Usp. Fiz. Nauk *48,* 25 (1952); transl. Fortsch. d. Phys. *1,* 101 (1953).
17. Schafroth, M. R., Phys. Rev. *96,* 1442 (1954); *100,* 463 (1955).
18. Schafroth, M. R., Blatt, J. M. and Butler, S. T., Helv. Phys. Acta *30,* 93 (1957).
19. Giaever, I., Phys. Rev. Letters *5,* 147 (1960).
20. Josephson, B. D., Phys. Letters *1,* 251 (1962); Advan. Phys. *14,* 419 (1965).
21. Anderson, P. W., in Lectures on the Many-body Problem, edited by E. R. Caianiello (Academic Press, Inc. New York, 1964), Vol. II.
22. See Superconductivity, Parks, R. D., ed. (Dekker New York, 1969).
23. See, for example, Parker, W. H. Taylor B. N. and Langenberg, D. N. Phys. Rev. Letters *18,* 287 (1967); Finnegan, T. F. Denenstein A. and Langenberg, D. N. Phys. Rev. B*4,* 1487 (1971).

Physics 1973

LEO ESAKI, IVAR GIAEVER

for their experimental discoveries regarding tunneling phenomena in semiconductors and superconductors respectively

and

BRIAN D JOSEPHSON

for his theoretical predictions of the properties of a supercurrent through a tunnel barrier, in particular those phenomena which are generally known as Josephson effects

THE NOBEL PRIZE FOR PHYSICS

Speech by professor Stig Lundqvist of the Royal Academy of Sciences
Translation from the Swedish text

Your Majesty, Your Royal Highnesses, Ladies and Gentlemen,
The 1973 Nobel Prize for physics has been awarded to Drs. Leo Esaki, Ivar Giaever and Brian Josephson for their discoveries of tunnelling phenomena in solids.

The tunnelling phenomena belong to the most direct consquences of the laws of modern physics and have no analogy in classical mechanics. Elementary particles such as electrons cannot be treated as classical particles but show both wave and particle properties. Electrons are described mathematically by the solutions of a wave equation, the Schrödinger equation. An electron and its motion can be described by a superposition of simple waves, which forms a wave packet with a finite extension in space. The waves can penetrate a thin barrier, which would be a forbidden region if we treat the electron as a classical particle. The term tunnelling refers to this wave-like property — the particle "tunnels" through the forbidden region. In order to get a notion of this kind of phenomenon let us assume that you are throwing balls against a wall. In general the ball bounces back but occasionally the ball disappears straight through the wall. In principle this could happen, but the probability for such an event is negligibly small.

On the atomic level, on the other hand, tunnelling is a rather common phenomenon. Let us instead of balls consider electrons in a metal moving with high velocities towards a forbidden region, for example a thin insulating barrier. In this case we cannot neglect the probability of tunneling. A certain fraction of the electrons will penetrate the barrier by tunnelling and we may obtain a weak tunnel current through the barrier.

The interest for tunnelling phenomena goes back to the early years of quantum mechanics, i.e. the late twenties. The best known early application of the ideas came in the model of alpha-decay of heavy atomic nuclei. Some phenomena in solids were explained by tunnelling in the early years. However, theory and experiments often gave conflicting results, no further progress was made and physicists lost interest in solid state tunnelling in the early thirties.

With the discovery of the transistor effect in 1947 came a renewed interest in the tunnelling process. Many attempts were made to observe tunnelling in semiconductors, but the results were controversial and inconclusive.

It was the young Japanese physicist Leo Esaki, who made the initial pioneering discovery that opened the field of tunnelling phenomena for research. He was at the time with the Sony Corporation, where he performed some deceptively simple experiments, which gave convincing experimental evidence for tunnelling of electrons in solids, a phenomenon which had been clouded by questions for decades. Not only was the existence of tunnelling in semiconductors established, but he also showed and explained an unforeseen aspect of tunnelling in semiconductor junctions. This new aspect led to the development of an important device, called the tunnel diode or the Esaki diode.

Esaki's discovery, published in 1958, opened a new field of research based on tunnelling in semiconductors. The method soon became of great importance in solid state physics because of its simplicity in principle and the high sensitivity of tunnelling to many finer details.

The next major advance in the field of tunnelling came in the field of superconductivity through the work of Ivar Giaever in 1960. In 1957, Bardeen, Cooper and Schrieffer had published their theory of superconductivity, which was awarded the 1972 Nobel Prize in physics. A crucial part of their theory is that an energy gap appears in the electron spectrum when a metal becomes superconducting. Giaever speculated that the energy gap should be reflected in the current-voltage relation in a tunnelling experiment. He studied tunnelling of electrons through a thin sandwich of evaporated metal films insulated by the natural oxide of the film first evaporated. The experiments showed that his conjecture was correct and his tunnelling method soon became the dominating method to study the energy gap in superconductors. Giaever also observed a characteristic fine structure in the tunnel current, which depends on the coupling of the electrons to the vibrations of the lattice. Through later work by Giaever and others the tunnelling method has developed into a new spectroscopy of high accuracy to study in detail the properties of superconductors, and the experiments have in a striking way confirmed the validity of the theory of superconductivity.

Giaver's experiments left certain theoretical questions open and this inspired the young Brian Josephson to make a penetrating theoretical analysis of tunnelling between two superconductors. In addition to the Giaever current he found a weak current due to tunelling of coupled electron pairs, called Coopers pairs. This implies that we get a supercurrent through the barrier. He predicted two remarkable effects. The first effect is that a supercurrent may flow even if no voltage is applied. The second effect is that a high frequency alternating current will pass through the barrier if a constant voltage is applied.

Josephson's theoretical discoveries showed how one can influence supercurrents by applying electric and magnetic fields and thereby control, study and exploit quantum phenomena on a macroscopic scale. His discoveries have led to the development of an entirely new method called quantum interferometry. This method has led to the development of a rich variety of instruments of extraordinary sensitivity and precision with application in wide areas of science and technology.

Esaki, Giaever and Josephson have through their discoveries opened up new fields of research in physics. They are closely related because the pioneering work by Esaki provided the foundation and direct impetus for Giaever's discovery and Giaever's work in turn provided the stimulus which led to Josephson's theoretical predictions. The close relation between the abstract concepts and sophisticated tools of modern physics and the practical applications to science and technology is strongly emphasized in these discoveries. The applications of solid state tunnelling already cover a wide range. Many devices based on tunneling are now used in electronics. The new quantum interferometry has already been used in such different applications as measurements of temperatures near the absolute zero, to detect gravitational waves, for ore prospecting, for communication through water and through mountains, to study the electromagnetic field around the heart or brain, to mention a few examples.

Drs. Esaki, Giaever and Josephson,

In a series of brilliant experiments and calculations you have explored different aspects of tunelling phenomena in solids. Your discoveries have opened up new fields of research and have given new fundamental insight about electrons in semiconductors and superconductors and about macroscopic quantum phenomena in superconductors.

On behalf of the Royal Academy of Sciences I wish to express our admiration and convey to you our warmest congratulations. I now ask you to proceed to receive your prizes from the hands of his Majesty the King.

Leo Esaki / 江崎玲於奈

LEO ESAKI

Leo Esaki was born in Osaka, Japan in 1925. Esaki completed work for a B.S. in Physics in 1947 and received his Ph.D in 1959, both from the University of Tokyo. Esaki is an IBM Fellow and has been engaged in semiconductor research at the IBM Thomas J. Watson Research Center, Yorktown Heights, New York, since 1960. Prior to joining IBM, he worked at the Sony Corp. where his research on heavily-doped Ge and Si resulted in the discovery of the Esaki tunnel diode; this device constitutes the first quantum electron device. Since 1969, Esaki has, with his colleagues, pioneered "designed semiconductor quantum structures" such as man-made superlattices, exploring a new quantum regime in the frontier of semiconductor physics.

The Nobel Prize in Physics (1973) was awarded in recognition of his pioneering work on electron tunneling in solids. Other awards include the Nishina Memorial Award (1959), the Asahi Press Award (1960), the Toyo Rayon Foundation Award for the Promotion of Science and Technology (1960), the Morris N. Liebmann Memorial Prize from IRE (1961), the Stuart Ballantine Medal from the Franklin Institute (1961), the Japan Academy Award (1965), the Order of Culture from the Japanese Government (1974), the American Physical Society 1985 International Prize for New Materials for his pioneering work in artificial semiconductor superlattices, the IEEE Medal of Honor in 1991 for contributions to and leadership in tunneling, semiconductor superlattices, and quantum wells. Dr. Esaki holds honorary degrees from Doshisha School, Japan, the Universidad Politecnica de Madrid, Spain, the University of Montpellier, France, Kwansei Gakuin University, Japan and the University of Athens, Greece. Dr. Esaki is a Director of IBM-Japan, Ltd., on the Governing Board of the IBM-Tokyo Research Laboratory, a Director of the Yamada Science Foundation and the Science and Technology Foundation of Japan. He serves on numerous international scientific advisory boards and committees, and is an Adjunct Professor of Waseda University, Japan. Currently he is a Guest Editorial writer for the Yomiuri Press. Dr. Esaki was elected a Fellow of the American Academy of Arts and Sciences in May 1974, a member of the Japan Academy on November 12, 1975, a Foreign Associate of the National Academy of Engineering (USA) on April 1, 1977, a member of the Max-Planck-Gesellschaft on March 17, 1989, and a foreign member of the American Philosophical Society in April of 1991.

LONG JOURNEY INTO TUNNELING

Nobel Lecture, December 12, 1973

by

Leo Esaki

IBM Thomas J. Watson Research Center, Yorktown Heights, N.Y., USA

I. Historical Background

In 1923, during the infancy of the quantum theory, de Broglie (1) introduced a new fundamental hypothesis that matter may be endowed with a dualistic nature—particles may also have the characteristics of waves. This hypothesis, in the hands of Schrödinger (2) found expression in the definite form now known as the Schrödinger wave equation, whereby an electron or a particle is assumed to be represented by a solution to this equation. The continuous nonzero nature of such solutions, even in classically forbidden regions of negative kinetic energy, implies an ability to penetrate such forbidden regions and a probability of tunneling from one classically allowed region to another. The concept of tunneling, indeed, arises from this quantum-mechanical result. The subsequent experimental manifestations of this concept can be regarded as one of the early triumphs of the quantum theory.

In 1928, theoretical physicists believed that tunneling could occur by the distortion, lowering or thinning, of a potential barrier under an externally applied high electric field. Oppenheimer (3) attributed the autoionization of excited states of atomic hydrogen to the tunnel effect: The coulombic potential well which binds an atomic electron could be distorted by a strong electric field so that the electron would see a finite potential barrier through which it could tunnel.

Fowler and Nordheim (4) explained, on the basis of electron tunneling, the main features of the phenomenon of electron emission from cold metals by high external electric fields, which had been unexplained since its observation by Lilienfeld (5) in 1922. They proposed a one-dimensional model. Metal electrons are confined by a potential wall whose height is determined by the work function Φ plus the fermi energy E_f, and the wall thickness is substantillay decreased with an externally applied high electric field, allowing electrons to tunnel through the potential wall, as shown in Fig. 1. They successfully derived the well-known Fowler-Nordheim formula for the current as a function of electric field F:

$$J = AF^2 \exp[-4(2m)^{1/2}\Phi^{3/2}/3\hbar F].$$

An application of these ideas which followed almost immediately came in the model for α decay as a tunneling process put forth by Gamow (6) and Gurney and Condon. (7) Subsequently, Rice (8) extended this theory to the description of molecular dissociation.

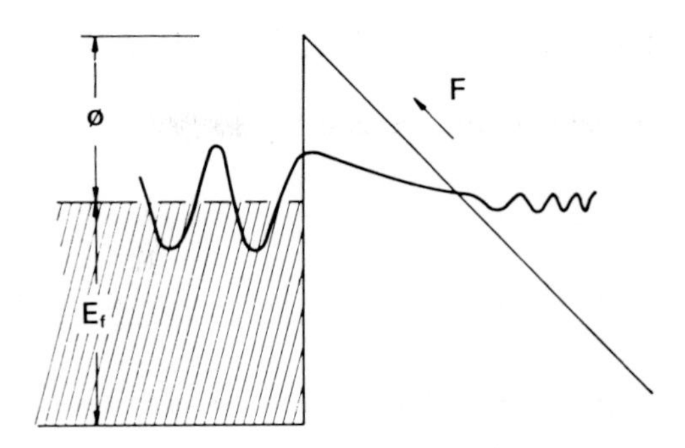

Fig. 1. Fowler-Nordheim tunneling.

$$J = AF^2 \exp(-4(2m)^{1/2}\phi^{3/2}/3hF)$$

The next important development was an attempt to invoke tunneling in order to understand transport properties of electrical contacts between two solid conductors. The problems of metal-to-metal and semiconductor-to-metal contacts are important technically, because they are directly related to electrical switches and rectifiers or detectors.

In 1930, Frenkel (9) proposed that the anomalous temperature independence of contact resistance between metals could be explained in terms of tunneling across a narrow vacuum separation. Holm and Meissner (10) then did careful measurements of contact resistances and showed that the magnitude and temperature independence of the resistance of insulating surface layers were in agreement with an explanation based on tunneling through a vacuum-like space. These measurements probably constitute the first correctly interpreted observations of tunneling currents in solids, (11) since the vacuum-like space was a solid insulating oxide layer.

In 1932, Wilson, (12) Frenkel and Joffe, (13) and Nordheim (14) applied quantum mechanical tunneling to the interpretation of metal-semiconductor contacts—rectifiers such as those made from selenium or cuprous oxide. From a most simplified energy diagram, shown in Fig. 2, the following well-known current-voltage relationship was derived:

$$J = J_s[\exp(eV/kT) - 1]$$

Apparently, this theory was accepted for a number of years until it was finally discarded after it was realized that it predicted rectification in the wrong direction for the ordinary practical diodes. It is now clear that, in the usual circumstance, the surface barriers found by the semiconductors in contact with metals, as illustrated in Fig. 2, are much too thick to observe tunneling current. There existed a general tendency in those early days of quantum mechanics to try to explain any unusual effects in terms of tunneling. In many cases, however, conclusive experimental evidence of tunneling was lacking, primarily because of the rudimentary stage of material science.

In 1934, the development of the energy-band theory of solids prompted Zener (15) to propose interband tunneling, or internal field emission, as an explanation for dielectric breakdown. He calculated the rate of transitions

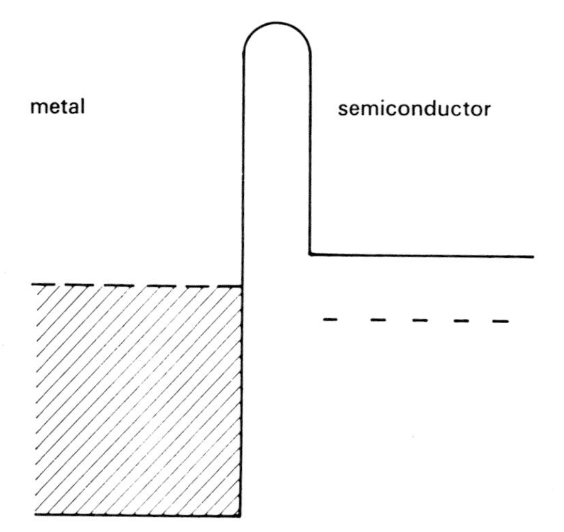

Fig. 2. Early model of metal-semiconductor rectifiers.

from a filled band to a next-higher unfilled band by the application of an electric field. In effect, he showed that an energy gap could be treated in the manner of a potential barrier. This approach was refined by Houston (16) in 1940. The Zener mechanism in dielectric breakdown, however, has never been proved to be important in reality. If a high electric field is applied to the bulk crystal of a dielectric or a semiconductor, avalanche breakdown (electron-hole pair generation) generally precedes tunneling, and thus the field never reaches a critical value for tunneling.

II. Tunnel Diode

Around 1950, the technology of Ge p-n junction diodes, being basic to transistors, was developed, and efforts were made to understand the junction properties. In explaining the reverse-bias characteristic, McAfee et al. (17) applied a modified Zener theory and asserted that low-voltage breakdown in Ge diodes (specifically, they showed a 10-V breakdown) resulted from interband tunneling from the valence band in the p-type region to the empty conduction band in the n-type region. The work of McAfee et al. inspired a number of other investigations of breakdown in p-n junctions. Results of those later studies (18) indicated that most Ge junctions broke down by avalanche, but by that time the name "Zener diodes" had already been given to the low-breakdown Si diodes. Actually, these diodes are almost always avalanche diodes. In 1957, Chynoweth and McKay (19) examined Si junctions of low-voltage breakdown and claimed that they had finally observed tunneling. In this circumstance, in 1956, I initiated the investigation of interband tunneling or internal field emission in semiconductor diodes primarily to scrutinize the elctronic structure of narrow (width) p-n junctions. This information, at the time, was also important from a technological point of view.

The built-in field distribution in p-n junctions is determined by the profile of impurities—donors and acceptors. If both the impurity distributions are

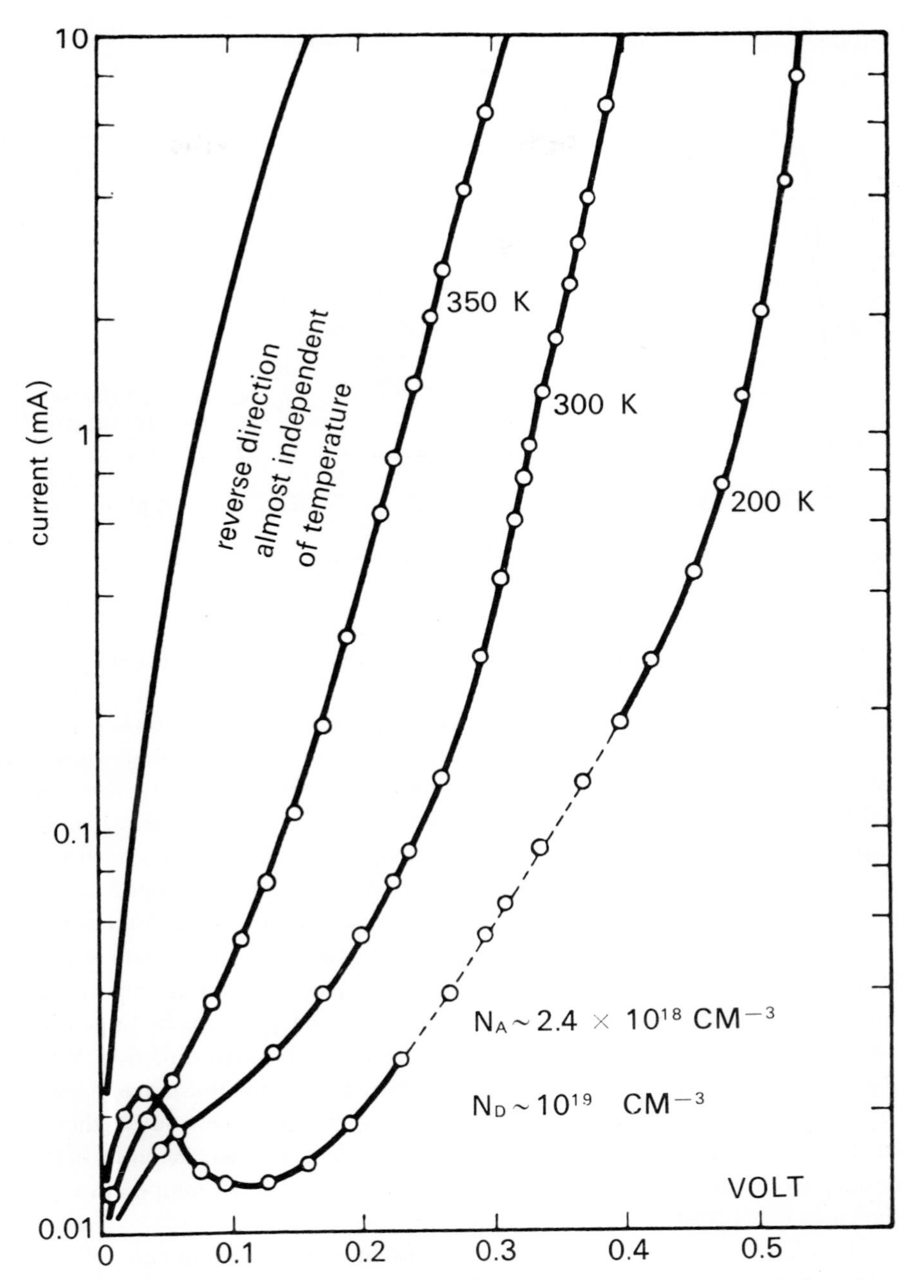

Fig. 3. Semilog plots of current-voltage characteristics in a tunnel diode, where $N_A \sim 2.4 \times 10^{18}cm^{-3}$ and $N_D \sim 10^{19}cm^{-3}$.

assumed to be step functions, the junction width W is given by W = const $[(N_A+N_D)/N_A{\cdot}N_D]^{1/2} \sim 1/N^{1/2}$, where N_A and N_D are the acceptor and donor concentrations and $N < N_A$ or N_D. Thus, first of all, we attempted to prepare heavily-doped Ge p-n junctions. Both the donor and acceptor concentrations are sufficiently high so that the respective sides of the junctions

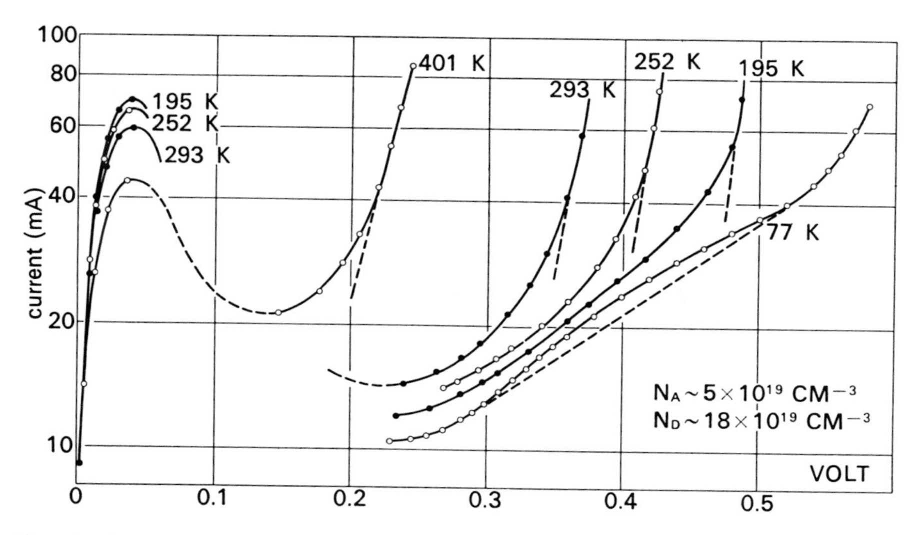

Fig. 4. Semilog plots of current-voltage characteristics in a tunnel diode, where $N_A \sim 5 \times 10^{19}\text{cm}^{-3}$ and $N_D \sim 1.8 \times 10^{19}\text{cm}^{-3}$.

are degenerate, that is, the fermi energies are located well inside the conduction or valence band.

In this study, we first obtained a backward diode which was more conductive in the reverse direction than in the forward direction. In this respect it agreed with the rectification direction predicted by the previously-mentioned old tunneling rectifier theory. The calculated junction width at zero bias was approximately 200Å, which was confirmed by capacitance measurements. In this junction, the possiblity of an avalanche was completely excluded because the breakdown occurs at much less than the threshold voltage for electron-hole pair production. The current-voltage characteristic at room temperature indicated not only that the major current-flow mechanism was convincingly tunneling in the reverse direction but also that tunneling might be responsible for current flow even in the low-voltage range of the forward direction. When the unit was cooled, we saw, for the first time, a negative resistance, appearing, as shown in Fig. 3. By further narrowing the junction width (thereby further decreasing the tunneling path), through a further increase in the doping level, the negative resistance was clearly seen at all temperatures, as shown in Fig. 4. (20)

The characteristic was analyzed in terms of interband tunneling. In the tunneling process, if it is elastic, the electron energy will be conserved. Figures 5 (a), (b), (c), and (d) show the energy diagrams of the tunnel diode at zero bias and with applied voltages, V_1, V_2, and V_3, respectively. As the bias is increased up to the voltage V_1, the interband tunnel current continues to increase, as shown by an arrow in Fig. 5 (b). However, as the conduction band in the n-type side becomes uncrossed with the valence band in the p-type side, with further increase in applied voltages, as shown in Fig. 5 (c), the current decreases because of the lack of allowed states of corresponding ener-

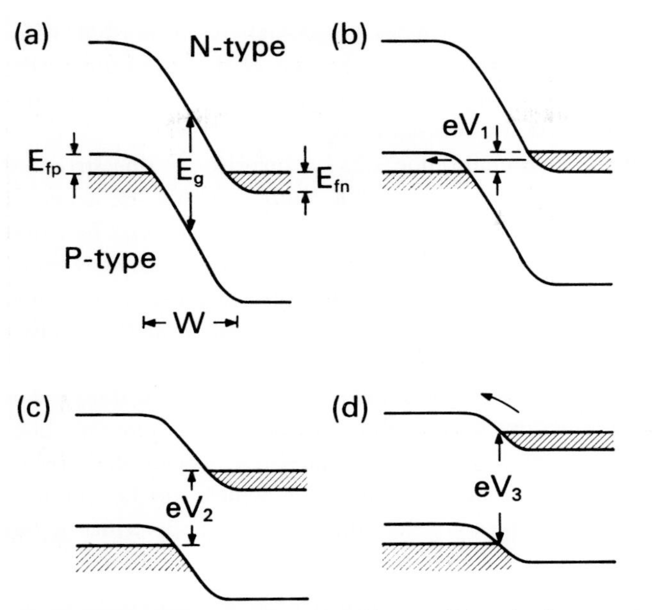

Fig. 5. Energy diagrams at varying bias-conditions in the tunnel diode.

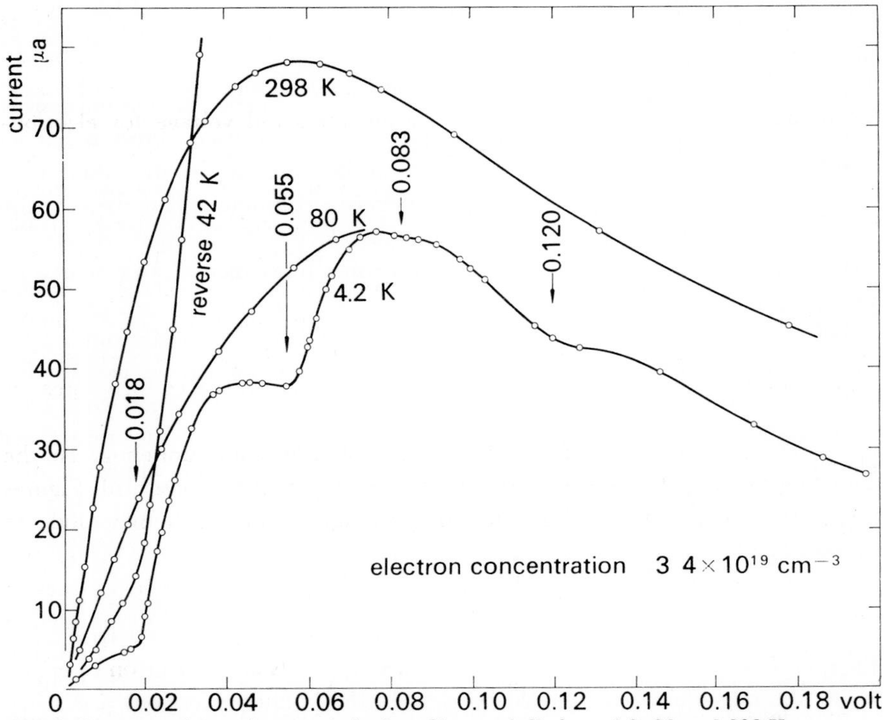

Fig. 6. Current-voltage characteristics in a Si tunnel diode at 4.2, 80 and 298 K.

gies for tunneling. When the voltage reaches V_2 or higher, the normal diffusion (or thermal) current will dominate as in the case of the usual p-n diode. Semiconductor materials other than Ge were quickly explored to obtain tunnel diodes: Si, InSb, GaAs, InAs, PbTe, GaSb, SiC, etc.

In our early study of the Si tunnel diode, (21) a surprisingly fine structure was found in the current-voltage curve at 4.2 K, indicating the existence of inelastic tunneling, as shown in Fig. 6. We were impressed with the fact that four voltages at the singularities shown in the figure agreed almost exactly with four characteristic energies due to acoustic and optical phonons, obtained from the optical absorption spectra (22) and also derived from the analysis of intrinsic recombination radiation (23) in pure silicon. The analysis of tunneling current in detail reveals not only the electronic states in the systems involved, but also the interactions of tunneling electrons with phonons, photons, plasmons, or even vibrational modes of molecular species in barriers. (24) As a result of the rich amount of information which can be obtained from a study of tunneling processes, a field called tunneling spectroscopy has emerged.

III. Negative Resistance in Metal-Oxide-Semiconductor Junctions

This talk, however, is not intended as a comprehensive review of the many theoretical and experimental investigations of tunneling, which is available elsewhere. (25) Instead, I would like to focus on only one aspect for the rest of the talk: negative resistance phenomena in semiconductors which can be observed in novel tunnel structures.

Differential negative resistance occurs only in particular circumstances, where the total number of tunneling electrons transmitted across a barrier structure per second decreases, rather than increases as in the usual case, with an increase in applied voltage. The negative resistance phenomena themselves are not only important in solid-state electronics because of possible signal amplification, but also shed light on some fundamental aspects of tunneling.

Before proceeding to the main subject, I would like to briefly outline the independent-electron theory of tunneling. (26) In tunneling, we usually deal with a *one-dimensional potential barrier* $V(x)$. The transmission coefficient D for such a barrier is defined as the ratio of the intensity of the transmitted electron wave to that of the incident wave. The most common approximation for D is the use of the semiclassical WKB form

$$D(E_x)=\exp\left[-2/\hbar \int_{x_1}^{x_2} (2m(V-E_x))^{1/2}\, dx\right] \tag{1}$$

where E_x is the kinetic energy in the direction normal to the barrier, and the quantities x_1 and x_2 are the classical turning points of an electron of energy E_x at the edges of the potential barrier. If the boundary regions are sharp, we first construct wave functions by matching values of functions as well as

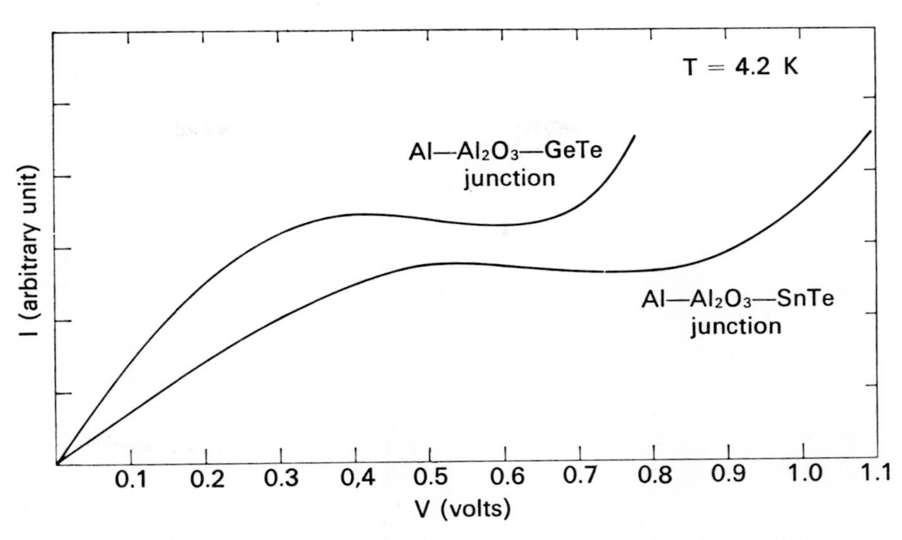

Fig. 7. Current-voltage characteristics in SnTe and GeTe tunnel junctions at 4.2 K.

their derivatives at each boundary, then calculate the transmission coefficient D.

The tunneling expression should include two basic conservation laws: 1) Conservation of the total electron energy; and 2) Conservation of the component of the electron wave vector parallel to the plane of the junction. The velocity of an incident electron associated with a state of wave number k_x is given by $1/h\ \partial E/\partial k_x$ in a one-particle approximation. Then, the tunneling current per unit area is written by

$$J = 2e/(2\pi)^3 \iiint D(E_x)(f(E) - f(E'))\, 1/\hbar\ \partial E/\partial k_x\, dk_x\, dk_y\, dk_z \qquad (2)$$

where f is the fermi distribution function or occupation probability, and E and E' are the energy of the incident electron and that of the transmitted one, respectively. The front factor $2/(2\pi)^3$ comes from the fact that the volume of a state occupied by two electrons of the opposite spin is $(2\pi)^3$ in the wave-vector space for a unit volume crystal.

The previously-mentioned tunnel diode is probably the first structure in which the negative resistance effect was observed. But, now, I will demonstrate that a similar characteristic can be obtained in a metal-oxide-semiconductor tunnel junction, (27) where the origin of the negative resistance is quite different from that in the tunnel diode. The semiconductors involved here (SnTe and GeTe) are rather unusual—more metallic than semiconducting; both of them are nonstoichiometric and higly p-type owing to high concentrations of Sn or Ge vacancies with typical carrier concentrations about 8×10^{20} and $2 \times 10^{20} \mathrm{cm}^{-3}$, respectively. The tunnel junctions were prepared by evaporating SnTe or GeTe onto an oxidized evaporated stripe of Al on quartz or sapphire substrates. In contrast to the p-n junction diodes, all ma-

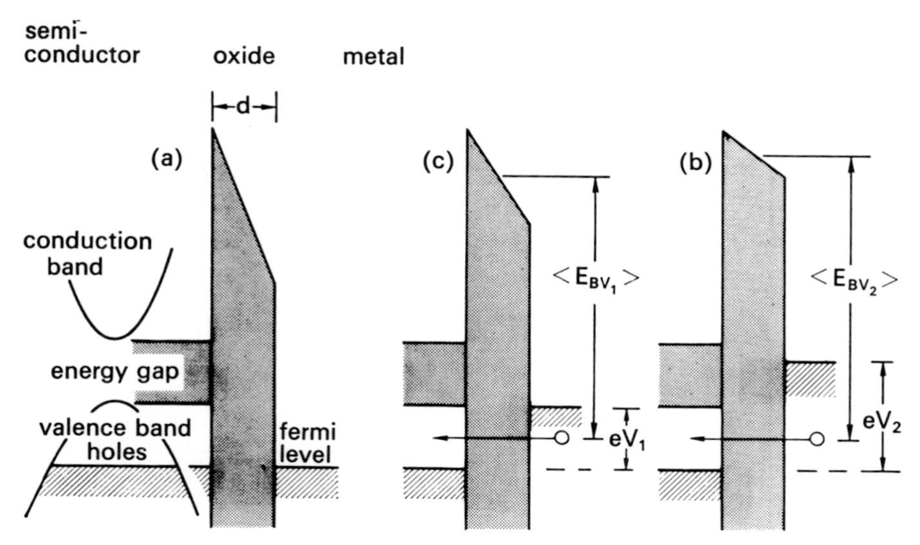

Fig, 8. Energy diagrams at varying bias-conditions in Al-Al_2O_3)SnTe or—GeTe tunnel junctions.

terials involved in these junctions are polycrystalline, although the Al oxide is possibly amorphous.

Figure 7 illustrates the current-voltage curves at 4.2 K of typical SnTe and GeTe junctions and Fig. 8 shows their energy diagrams at zero bias, and at applied voltages V_1 and V_2 from the left to the right. As is the case in the tunnel diode, until the bias voltage is increased such that the fermi level in the metal side coincides with the top of the valence band in the semiconductor side (Fig. 8 (b)), the tunnel current continues to increase. When the bias voltage is further increased (Fig. 8 (c)), however, the total number of empty allowed states or holes in the degenerate p-type semiconductor is unchanged, whereas the tunneling barrier height is raised, for instance from E_{BV_1} to E_{BV_2}, resulting in a decrease in tunneling probability determined by the exponential term, $e^{-\lambda}$, where $\lambda \sim 2d(2mE_{BV})^{1/2}/\hbar$, and E_{BV} and d are the barrier height and width, respectively. Thus a negative resistance is exhibited in the current-voltage curve. When the bias voltage becomes higher than the level corresponding to the bottom of the conduction band in the semiconductor, a new tunneling path from the metal to the conduction band is opened and one sees the current again increasing with the voltage. The rectification direction in this junction is again backward as is the case in the tunnel diode.

We might add that, in this treatment, the tunneling exponent is assumed to be determined only by the energy difference between the bottom of the conduction band in the oxide and the metal fermi energy. This assumption should be valid because this energy difference is probably much smaller than that between the top of the valence band in the oxide and the metal fermi energy.

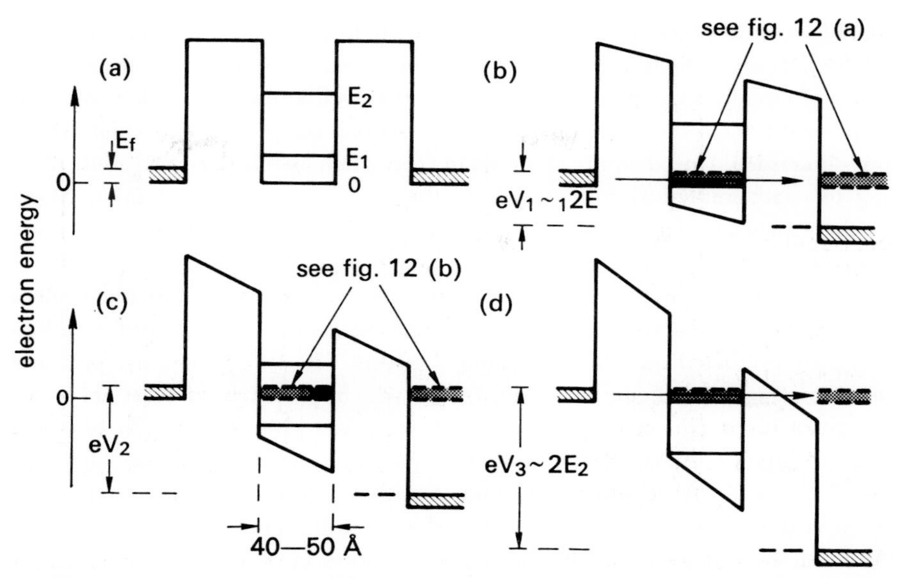

Fig. 9. Energy diagrams at varying bias-conditions in a double-barrier tunnel junction, indicating the resonant transmission in (b) and (d).

IV. Negative resistance due to Resonant Transmission

It has been known that there is a phenomenon called the resonant transmission. Historically, resonant transmission was first demonstrated in the scattering of electrons by atoms of noble gases and is known as the Ramsauer effect. In many textbooks (28) on quantum mechanics, the resonant transmission in tunneling or scattering is one of the more favored topics. In a one-dimensional double potential barrier, (29) the narrow central potential well has weakly-quantized (or quasi-stationary) bound states, of which the energies are denoted by E_1 and E_2 in Fig. 9 (a). If the energy of inicident electrons coincides with these energies, the electrons may tunnel through both barriers without any attenuation. As seen in Fig. 10 (two curves at $V = 0$), the transmission coefficient reaches unity at the electron energy $E = E_1$ or E_2. Since E_1 is a more strongly quantized state than E_2, the resonance peak at E_1 is much sharper than that at E_2. Although this sharpness depends upon the barrier thickness, one can achieve at some energy a resonance condition of 100 % transmission, whatever thickness is given to the two barriers.

This effect is quite intriguing because the transmission coefficient (or the attenuation factor) for two barriers is usually thought of as the product of two transmission coefficients, one for each barrier, resulting in a very small value for overall transmission. The situation, however, is somewhat analogous to the Fabry-Perot type interference filter in optics. The high transmissivity arises because, for certain wavelengths, the reflected waves from inside interfere destructively with the incident waves, so that only a transmitted wave remains.

This resonating condition can be extended to a periodic barrier structure. In the Kronig-Penney model of a one-dimensional crystal which consists of a series of equally-spaced potential barriers, it is well known that allowed bands of perfect transmission are separated by forbidden bands of attenuation. These one-dimensional mathematical problems can often be elegantly treated, leading to exact analytical solutions in textbooks of quantum mechanics. Many of these problems, however, are considered to be pure mathematical fantasy, far from reality.

We, recently, initiated an experimental project to materialize one-dimensional potential barriers in monocrystalline semiconductors in order to observe the predicted quantum-mechanical effects. (30) We choose n-type GaAs as a host semiconductor or a matrix in which potential barriers with the height of a fraction of one electron volt are made by inserting thin layers of $Ga_{1-x}Al_xAs$ or AlAs. Because of the similar properties of the chemical bond of Ga and Al together with their almost equal ion size, the introduction of AlAs into GaAs makes the least disturbance to the quality of single crystals. And yet the difference in the electronic structure between the two materials makes a sharp potential barrier inside the host semiconductor. We prepare the multi-layer structure with the technique of molecular beam epitaxy in ultra-high vacuum environment. Precise control of thickness and composition has been achieved by using a process control computer. (31)

With this facility, a double potential barrier structure has been prepared, (32) in which the barrier height and width are about 0.4 eV and a few tens of angstroms, respectively, and the width of the central well is as narrow as 40—50Å. From these data, the first two energies, E_1 and E_2, of the weakly-quantized states in the well are estimated to be 0.08 and 0.30 eV.

We have measured the current-voltage characteristic as well as the conductance dI/dV as a function of applied voltages in this double tunnel junction. The results at 77 K are shown in Fig. 11, and they clearly indicate two singularities in each polarity and even show a negative resistance around +0.8 volt or —0.55 volt. The applied voltages at the singularities, averaged in both polarities, are roughly twice as much as the calculated bound-state energies. This general feature is not much different a 4.2 K, although no structure is seen at room temperature.

The energy diagrams at zero bias and at applied voltages V_1, V_2 and V_3 are shown in Fig. 9. The electron densities on both the left and right GaAs sides are about $10^{18}cm^{-3}$ which gives a fermi energy of 0.04 eV at zero temperature. These electrons are considered to be classical free carriers with the effective mass, m^*, of which the kinetic energy E is given by

$$E = \frac{\hbar^2}{2m^*} (k_x^2 + k_y^2 + k_z^2).$$

On the other hand, the electrons in the central well have the weakly-quantized levels, E_1, E_2, . . ., for motion in the x direction perpendicular to the walls with a continuum for motion in the y-z plane parallel to the walls. These

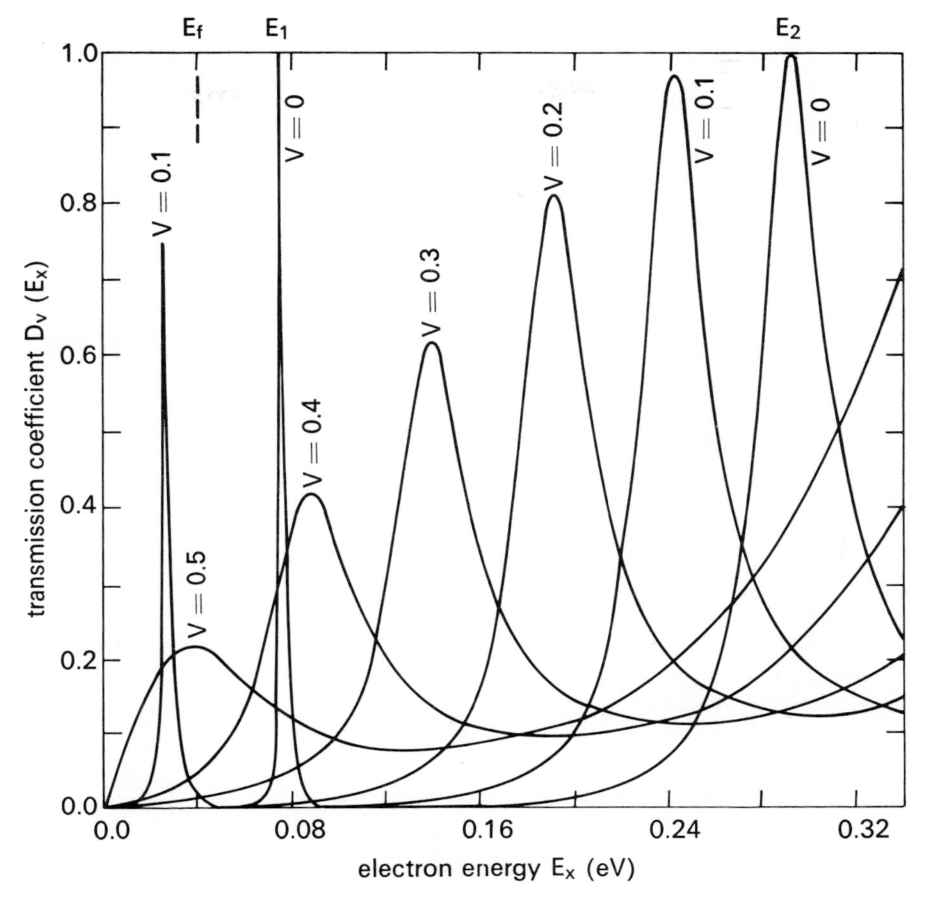

Fig. 10. Transmission coefficient versus electron energy, indicating the resonant transmission.

electrons are nearly two-dimensional, which is to say the kinetic energy E is given by

$$E = E_i + \frac{\hbar^2}{2m^*}\,(k_y{}^2 + k_z{}^2)$$

An approximation is made that the same electron effective mass, m^*, exists throughout the structure. Then an experssion for the tunneling current in this structure (33) can be derived in the framework of the previously-described tunneling formalism in Eq. 2. Using $\partial E/\partial k_x = \partial E_x/\partial k_x$, $2\pi k_t \mathrm{d}k_t = \mathrm{d}k_y \mathrm{d}k_z$ and T (temperature) $= 0$, the current is given by

$$\mathcal{J} = e/2\pi^2\hbar \int_0^{E_f} D(E_x) \int_0^{(2m^*(E_f - E_x))^{1/2}\hbar} k_t dk_t dE_x$$

$$- e/2\pi^2\hbar \int_0^{E_f - eV} D(E_x) \int_0^{(2\mathrm{m}^*(E_f - E_x - eV))^{1/2}/\hbar} k_t\, dk_t\, dE_x, \qquad (3)$$

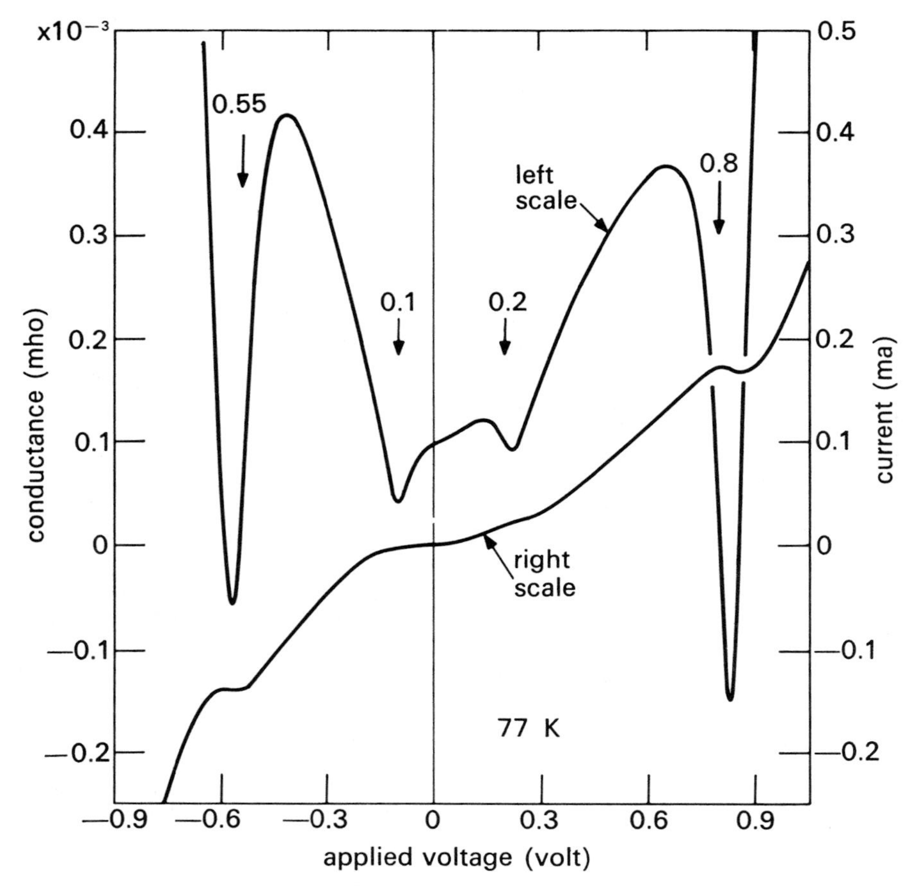

Fig. 11. Current, I, and conductance, dI/dV, versus voltage curves in a double barrier tunnel junction.

where V is the applied voltage, on which the transmission coefficient $D(E_x)$ depends. The above expression can be integrated over the transverse wave number k_t, giving

$$\mathcal{J} = em^*/2\pi^2\hbar^3 \int_0^{E_f} D_V(E_x)(E_f - E_x)\, dE_x$$

$$- em^*/2\pi^2\hbar^3 \int_0^{E_f - eV} D_V(E_x)(E_f - E_x - eV)\, dE_x \qquad (4)$$

In both Eqs. 3 and 4, the second term is nonzero only for $eV < E_f = 0.04$ eV.

Now, the transmission coefficient D_V (E_x) can be derived for each applied voltage from wave functions which are constructed by matching their values as well as derivatives at each boundary. Figure 10 shows one example of calculated D as a function of the electron energy for applied voltages

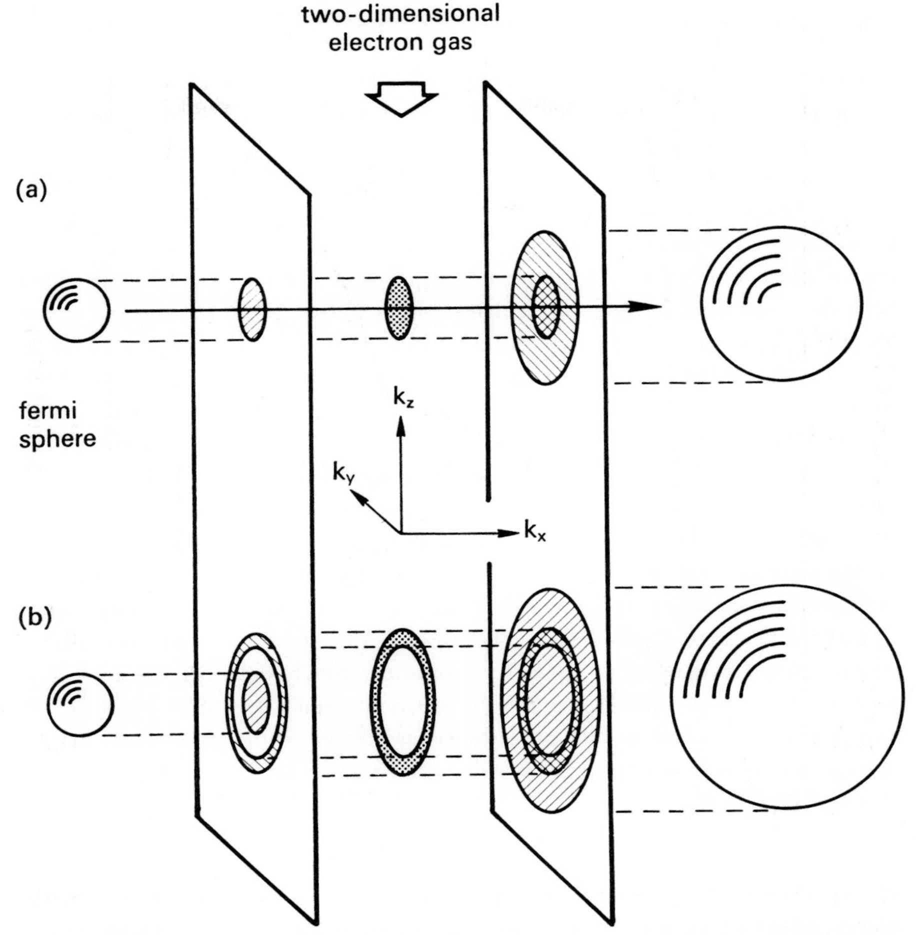

Fig. 12. Construction of shadows of energy surfaces on two k_y—k_z planes corresponding to two barriers.

between zero and 0.5 volt. The energy zero is taken at the bottom of the conduction band on the left as shown in Fig. 9. In this example, the well width is taken to be 45Å and the barrier height 0.4 eV at zero bias. The square shape for barriers and well is assumed for simplicity of calculation, although they are actually trapezoidal at any applied voltage.

Referring to Figs. 9 and 10, both the absolute values and the positions in energy for the maxima of the transmission coefficient decrease with increasing applied voltages, the origin of energy being the conduction band edge for the left outer GaAs layer. The current maxima occur at applied voltages such that the electron energies on the left coincide with the bound-state energies, as illustrated in Figs. 9 (b) and (d). This resonant transmission has been experimentally verified as shown in Fig. 11. The transmission coefficient itself at this resonance, however, is appreciably less than unity as indicated in

Fig. 10, primarily because of the asymmetric nature of the potential profile at applied voltages.

To gain an insight into this tunneling problem, particularly in view of the transverse wave-vector conservation (specular tunneling), a representation in the wave-vector space is useful and is shown in Fig. 12. Two k_y—k_z planes are shown parallel to the junction plane, corresponding to the two barriers. Figures 12 (a) and (b) show two different bias-voltage conditions. First, the Fermi sphere on the left is projected on the first screen, making a circle. A similar projection, of the two-dimensional electrons in the central well which have the same total energies as electrons in the Fermi sphere on the left at the particular applied voltage, will form a circle (Fig. 12 (a)), or a ring (Fig. 12 (b)), depending upon the value of applied voltage. If the two projected patterns have no overlap, there will be no specular tunneling current. The situation on the right screen is slightly different, since an energy sphere on the right, in which electrons have the same total energies as electrons in the Fermi sphere on the left, is rather large; mereover, its size will be increased as the applied voltage increases. Thus in this case the two projected patterns always overlap. Figures 12 (a) and (b) correspond to the bias conditions in Figs. 9 (b) and (c), respectively. With an increase in applied voltage from V_1 to V_2, the current will decrease because of a disappearance of overlapping regions, thereby causing a negative resistance. Since the current density is dependent upon the half-width of the resonant peaks shown in Fig. 10, we have observed a clear negative resistance associated with the second bound-state which is not swamped by possible excess currents arising for a variety of reasons.

V. Periodic Structure-Superlattice

The natural extension of double barriers will be to construct a series of tunnel junctions by a periodic variation of alloy composition. (30) By using the same facilities for computer-controlled molecular beam epitaxy, we tried to prepare a Kronig-Penney type one-dimensional periodic structure—a man-made superlattice with a period of 100Å. (31) The materials used here are again GaAs and AlAs or $Ga_{1-x}Al_xAs$.

The composition profile of such a structure (34) has been verified by the simultaneous use of ion sputter-etching of the specimen surface and Auger electron spectroscopy and is shown in Fig. 13. The amplitudes of the Al Auger signals serve as a measure of Al concentration near the surface within a sampling depth of only 10Å or so. The damping of the oscillatory behavior evident in the experimental data is not due to thermal diffusion or other reasons but due to a surface-roughening effect or non-uniformity in the sputter-etching process. The actual profile, therefore, is believed to be one which is illustrated by the solid line in Fig. 13. This is certainly one of the highest resolution structures ever built in monocrystalline semiconductors.

It should be recognized that the period of this superlattice is ~ 100Å—still large in comparison with the crystal lattice constant. If this period ι, how-

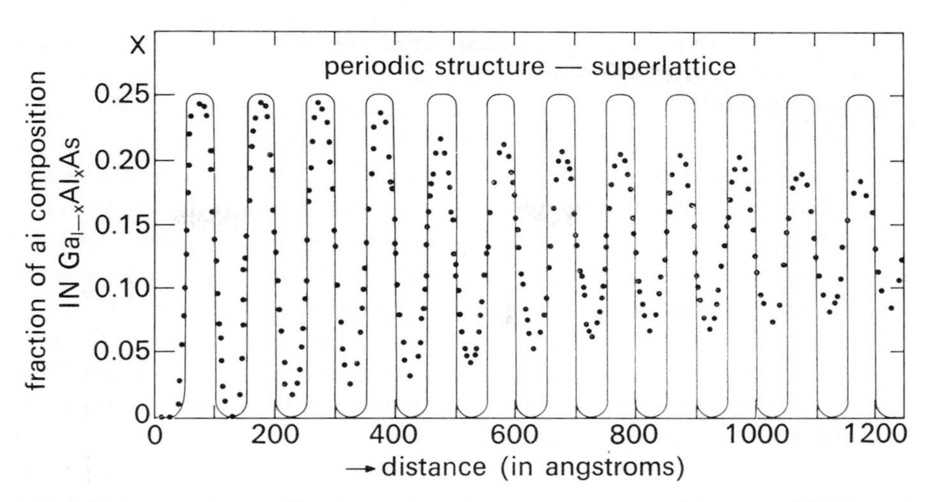

Fig. 13. Composition profile of a superlattice structure measured by a combination of ion sputter-etching and Auger electron spectroscopy.

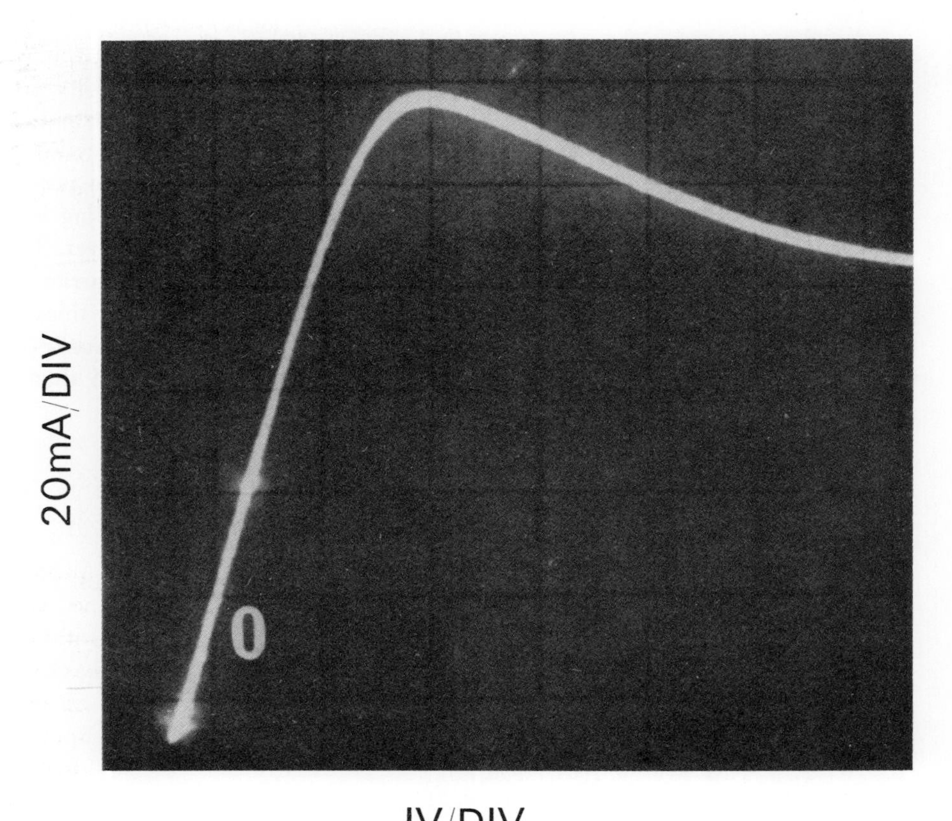

Fig. 14. Current-voltage characteristic at room temperature of a 70Å-period, GaAs-$Ga_{0.5}Al_{0.5}As$ superlattice.

ever, is still shorter than the electron mean free path, a series of narrow allowed and forbidden bands is expected, due to the subdivision of the original Brillouin zone into a series of minizones. If the electron scattering time τ, and an applied electric field F, meet a threshold condition: $eF\tau\iota/\hbar > 1$, the combined effect of the narrow energy band and the narrow wave-vector zone makes it possible for electrons to be excited beyond an inflection point in the energy-wave vector relation. This would result in a negative resistance for electrical transport in the direction of the superlattice. This can be seen in another way. The de Broglie wavelength of conduction electrons having an energy of, for instance, 0.03 eV in n-type GaAs (the effective mass $\sim 0.1\ m$). is of the order of 200Å. Therefore, an interaction of these electron waves with the Kronig-Penney type potential with a period of 100Å can be expected, and will give rise to a nonlinear transport property.

We have begun to observe such current-voltage characteristic as shown in Fig. 14. The observed negative resistance may be interesting not only from the scientific aspect but also from a practical viewpoint because one can expect, at least theoretically, that the upper limit of operating frequencies would be higher than that for any known semiconductor devices.

VI. Conclusion

I am , of course, deeply aware of important contributions made by many colleagues and my friends throughout this long journey. The subject of Section II was carried out when I was in Japan and all the rest (35) has been performed in the United States of America. Since my journey into tunneling is still continuing, I do not come to any conslusions in this talk. However, I would like to point out that many high barriers exist in this world: Barriers between nations, races and creeds. Unfortunately, some barriers are thick and strong. But I hope, with determination, we will find a way to tunnel through these barriers easily and freely, to bring the world together so that everyone can share in the legacy of Alfred Nobel.

References

1. de Broglie, L., Nature *112,* 540 (1923); Ann. de Physique (10 Ser.) *3,* 22 (1925).
2. Schrödinger, E., Ann. d. Physik (4. Folge) *79,* 361, 489 (1926).
3. Oppenheimer, J. R., Phys. Rev. *31,* 66 (1928); Proc. Nat'l. Acad. Sci. U.S. *14,* 363 (1928).
4. Fowler, R. H. and Nordheim, L., Proc. Roy. Soc. (London) *A 119,* 173 (1928).
5. Lilienfeld, J. E. Physik. Z. *23,* 506 (1922).
6. Gamow, G., Physik, Z., *51,* 204 (1928).
7. Gurney, R. W. and Condon, E. U., Nature *122,* 439 (1928).
8. Rice, O. K. Phys. Rev. *34,* 1451 (1929).
9. Frenkel, J., Phys. Rev. *36,* 1604 (1930).
10. Holm, R. and Meissner, W., Z. Physik *74,* 715 (1932), *86,* 787 (1933).
11. Holm, R., Electric Contacts, Springer-Verlag, New York, 1967, p. 118.
12. Wilson, A. H., Proc. Roy. Soc. (London) *A 136,* 487 (1932).
13. Frenkel, J. and Joffe, A., Physik. Z. Sowjetunion *1,* 60 (1932).
14. Nordheim, L., Z. Physik *75,* 434 (1932).

15. Zener, C., Proc. Roy. Soc. (London) *145,* 523 (1934).
16. Houston, W. V., Phys. Rev. *57,* 184 (1940).
17. McAfee, K. B., Ryder, E. J., Shockley, W. and Sparks, M., Phys. Rev. *83,* 650 (1951).
18. McKay, K. G. and McAfee, K. B. Phys. Rev. *91,* 1079 (1953);
McKay, K. G., Phys. Rev. *94,* 877 (1954);
Miller, S. L., Phys. *99,* 1234 (1955).
19. Chynoweth, A. G. and McKay., K. G. Phys., Rev. *106,* 418 (1957).
20. Esaki, L., Phys. Rev. *109,* 603 (1958);
Esaki, L., Solid State Physics in Electronics and Telecommunications, Proc. of Int. Conf. Brussels, 1958 (Desirant, M. and Michels, J. L., ed.), Vol. 1, Semiconductors, Part I, Academic Press, New York, 1960, p. 514.
21. Esaki, L. and Miyahara, Y., unpublished;
Esaki, L. and Miyahara, Y., Solid-State Electron. *1,* 13 (1960);
Holonyak, N., Lesk, I. A., Hall, R. N., Tiemann, J. J. and Ehrenreich, H., Phys. Rev. Letters *3,* 167 (1959).
22. Macfarlane, G. G., McLean, T. P., Quarrington, J. E. and Roberts, V., J. Phys. Chem. Solids *8,* 388 (1959).
23. Haynes, J, R., Lax, M. and Flood, W. F., J. Phys. Chem. Solids *8,* 392 (1959).
24. See, for instance, Tunneling Phenomena in Solids edited by Burstein, E. and Lundqvist, S., Plenum Press, New York, 1969.
25. Duke, C. B., Tunneling in Solids, Academic Press, New York, 1969.
26. Harrison, W. A., Phys, Rev. *123,* 85 (1961).
27. Esaki, L. and Stiles, P. J., Phys. Rev. Letters *16,* 1108 (1966); Chang, L. L., Stiles, P. J. and Esaki, L., J. Appl. Phys. *38,* 4440 (1967); Esaki, L., J. Phys. Soc. Japan, *Suppl. 21,* 589 (1966).
28. See, for instance, Bohm, D., Quantum Theory, Prentice-Hall, New Jersey, 1951.
29. Kane, E. O., page 9—11 in reference 24.
30. Esaki, L. and Tsu, R., IBM J. Res. Develop. *14,* 61 (1970); Esaki, L., Chang, L. L., Howard, W. E. and Rideout, V. L., Proc. 11th Int. Conf. Phys. Semicond., Warsaw, Poland, 1972, p. 431.
31. Chang, L. L., Esaki, L., Howard, W. E. and Ludeke, R., J. Vac. Sci. Technol. *10,* 11 (1973);
Chang, L. L., Esaki, L., Howard, W. E., Ludeke, R. and Schul, G., J. Vac. Sci. Technol. *10,* 655 (1973).
32. Chang, L. L., Esaki, L. and Tsu, R., to be published.
33. Tsu, R. and Esaki, L., Appl. Phys. Letters *22,* 562 (1973).
34. Ludeke, R., Esaki, L. and Chang, L. L., to be published.
35. Partly supported by Army Research Office, Durham, N. C.

Ivar Giaever

IVAR GIAEVER

Ivar Giaever was born in Bergen, Norway, April 5, 1929, the second of three children. He grew up in Toten where his father, John A. Giaever, was a pharmacist. He attended elementary school in Toten but received his secondary education in the city of Hamar. Next he worked one year at the Raufoss Munition Factories before entering the Norwegian Institute of Technology in 1948. He graduated in 1952 with a degree in mechanical engineering.

In 1953, Giaever completed his military duty as a corporal in the Norwegian Army, and thereafter he was employed for a year as a patent examiner for the Norwegian Government.

Giaever emigrated to Canada in 1954 and after a short period as an architect's aide he joined Canadian General Electric's Advanced Engineering Program. In 1956, he emigrated to the USA where he completed the General Electric Company's A, B and C engineering courses. In these he worked in various assignments as an applied mathematician. He joined the General Electric Research and Development Center in 1958 and concurrently started to study physics at Rensselaer Polytechnical Institute where he obtained a Ph.D. degree in 1964.

From 1958 to 1969 Dr. Giaever worked in the fields of thin films, tunneling and superconductivity. In 1965 he was awarded the Oliver E. Buckley Prize for some pioneering work combining tunneling and superconductivity. In 1969 he received a Guggenheim Fellowship and thereupon spent one year in Cambridge, England studying biophysics. Since returning to the Research and Development Center in 1970, Dr. Giaever has spent most of his effort studying the behavior of protein molecules at solid surfaces. In recognition of his work he was elected a Coolidge fellow at General Electric in May, 1973.

Dr Giaever is a member of the Institute of Electrical and Electronic Engineers, and the Biophysical Society, and he is a Fellow of the American Physical Society. Dr. Giaever has served on committees for several international conferences and presently he is a member of the Executive Committee of the Solid State division in the American Physical Society.

Ivar Giaever married Inger Skramstad in 1952 and they have four children. He became a naturalized US citizen in 1964.

Notes added

Linus Pauling is reported to have said that the Nobel Prize did not change his life — he was already famous! That was not true for me. The Nobel Prize opened a lot of doors, but also provided me with many distractions. I have, however, continued to work in biophysics, attempting to use physical methods and thoughts to solve biological problems. At the present time, I am studying the motion of mammalian cells in tissue culture by growing both normal and cancerous cells on small electrodes.

I left General Electric in 1988 to become an Institute Professor at Rensselaer (RPI) in Troy, New York 12180–3590, and concurrently I am also a Professor at the University of Oslo, Norway, sponsored by STATOIL.

On a personal note my wife and I are now the proud grandparents of almost four grandchildren.

ELECTRON TUNNELING AND SUPERCONDUCTIVITY

Nobel Lecture, December 12, 1973

by

IVAR GIÆVER

General Electric Research and Development Center, Schenectady, N.Y., USA.

In my laboratory notebook dated May 2, 1960 is the entry: "Friday, April 22, I performed the following experiment aimed at measuring the forbidden gap in a superconductor." This was obviously an extraordinary event not only because I rarely write in my notebook, but because the success of that experiment is the reason I have the great honor and pleasure of addressing you today. I shall try in this lecture, as best I can, to recollect some of the events and thoughts that led to this notebook entry, though it is difficult to describe what now appears to me as fortuitous. I hope that this personal and subjective recollection will be more interesting to you than a strictly technical lecture, particularly since there are now so many good review articles dealing with superconductive tunneling. [1, 2]

A recent headline in an Oslo paper read approximately as follows: "Master in billiards and bridge, almost flunked physics—gets Nobel Prize." The paper refers to my student days in Trondheim. I have to admit that the reporting is reasonably accurate, therefore I shall not attempt a "cover up", but confess that I almost flunked in mathematics as well. In those days I was not very interested in mechanical engineering and school in general, but I did manage to graduate with an average degree in 1952. Mainly because of the housing shortage which existed in Norway, my wife and I finally decided to emigrate to Canada where I soon found employment with Canadian General Electric. A three year Company course in engineering and applied mathematics known as the A, B and C course was offered to me. I realized this time that school was for real, and since it probably would be my last chance, I really studied hard for a few years.

When I was 28 years old I found myself in Schenectady, New York where I discovered that it was possible for some people to make a good living as physicists. I had worked on various Company assignments in applied mathematics, and had developed the feeling that the mathematics was much more advanced than the actual knowledge of the physical systems that we applied it to. Thus, I thought perhaps I should learn some physics and, even though I was still an engineer, I was given the opportunity to try it at the General Electric Research Laboratory.

The assignment I was given was to work with thin films and to me films meant photography. However I was fortunate to be associated with John Fisher who obviously had other things in mind. Fisher had started out as a mechanical engineer as well, but had lately turned his atten-

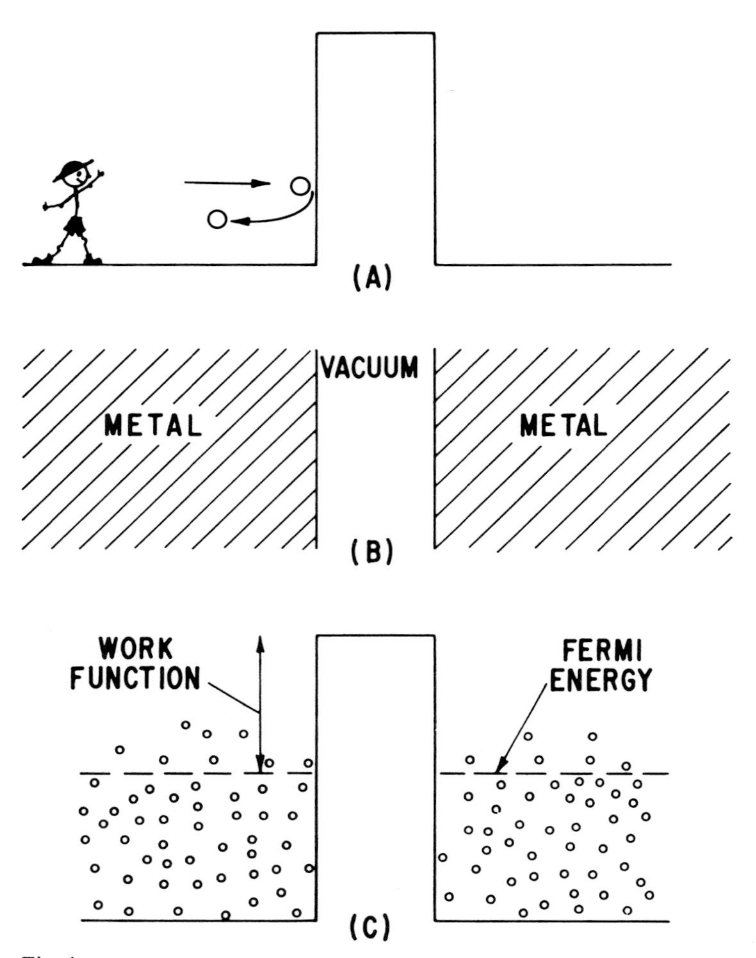

Fig. 1.
A. If a man throws a ball against a wall the ball bounces back. The laws of physics allow the ball to penetrate or tunnel through the wall but the chance is infinitesimally small because the ball is a macroscopic object. B. Two metals separated by a vacuum will approximate the above situation. The electrons in the metals are the "balls", the vacuum represents the wall. C. A pictorial energy diagram of the two metals. The electrons do not have enough energy to escape into the vacuum. The two metals can, however, exchange electrons by tunneling. If the metals are spaced close together the probability for tunneling is large because the electron is a microscopic particle.

tion towards theoretical physics. He had the notion that useful electronic devices could be made using thin film technology and before long I was working with metal films separated by thin insulating layers trying to do tunneling experiments. I have no doubt that Fisher knew about Leo Esaki's tunneling experiments at that time, but I certainly did not. The concept that a particle can go through a barrier seemed sort of strange to me, just struggling with quantum mechanics at Rensselaer Polytechnic Institute in Troy, where I took formal courses in Physics. For an engineer it sounds rather strange that if you

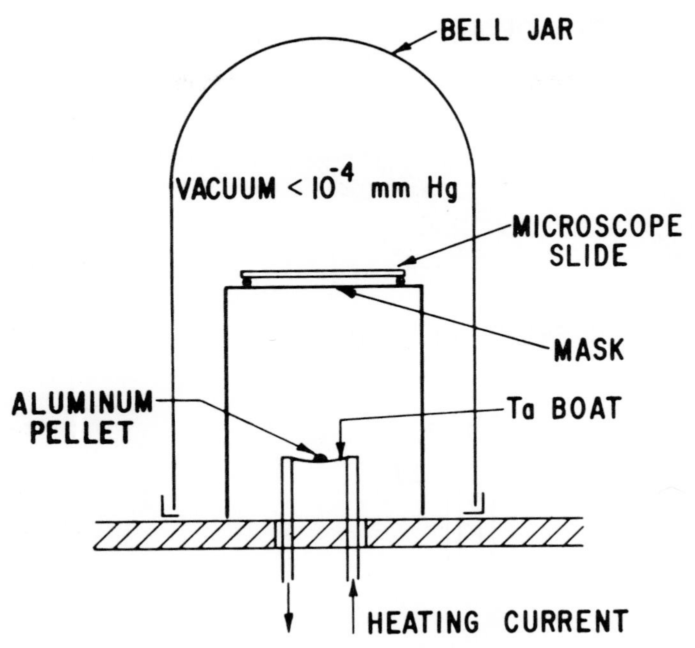

Fig. 2.
A schematic drawing of a vacuum system for depositing metal films. For example, if aluminum is heated resistively in a tantalum boat, the aluminum first melts, then boils and evaporates. The aluminum vapor will solidify on any cold substrate placed in the vapor stream. The most common substrates are ordinary microscope glass slides. Patterns can be formed on the slides by suitably shielding them with a metal mask.

throw a tennis ball against a wall enough times it will eventually go through without damaging either the wall or itself. That must be the hard way to a Nobel Prize! The trick, of course, is to use very tiny balls, and lots of them. Thus if we could place two metals very close together without making a short, the electrons in the metals can be considered as the balls and the wall is represented by the spacing between the metals. These concepts are shown in Figure 1. While classical mechanics correctly predicts the behavior of large objects such as tennis balls, to predict the behavior of small objects such as electrons we must use quantum mechanics. Physical insight relates to everyday experiences with large objects, thus we should not be too surprised that electrons sometimes behave in strange and unexpected ways.

Neither Fisher nor I had much background in experimental physics, none to be exact, and we made several false starts. To be able to measure a tunneling current the two metals must be spaced no more than about 100 Å apart, and we decided early in the game not to attempt to use air or vacuum between the two metals because of problems with vibration. After all, we both had training in mechanical engineering! We tried instead to keep the two metals apart by using a variety of thin insulators made from Langmuir films and from Formvar. Invariably, these films had pinholes and the mercury

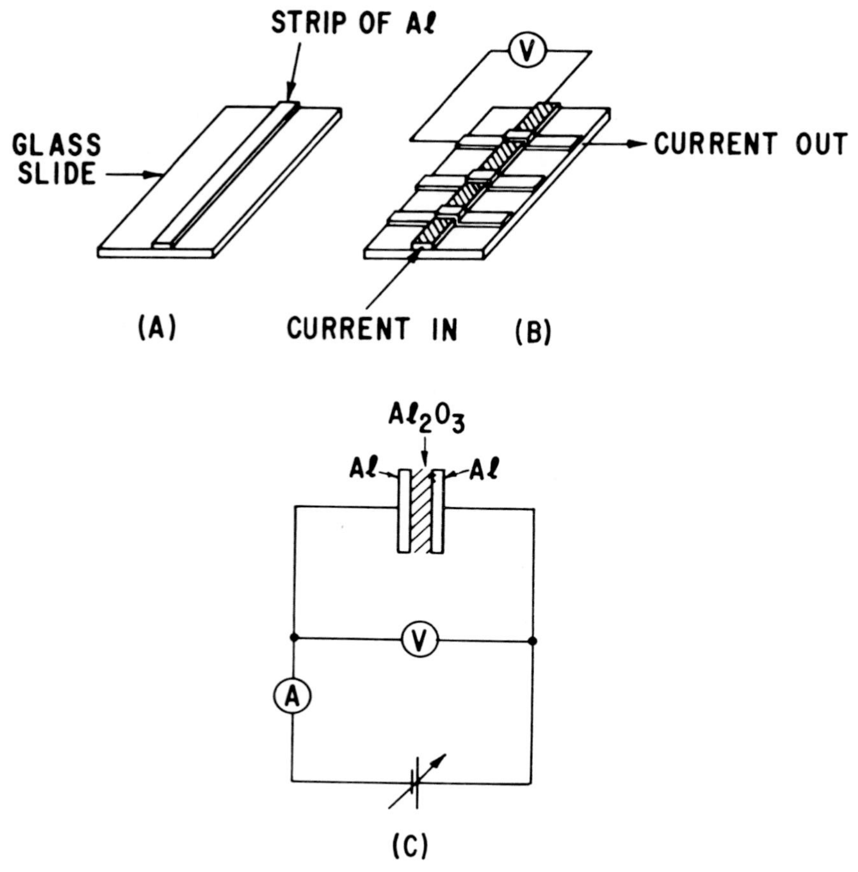

Fig. 3.
A. A microscope glass slide with a vapor deposited aluminum strip down the middle. As soon as the aluminum film is exposed to air, a protective insulating oxide forms on the surface. The thickness of the oxide depends upon such factors as time, temperature and humidity. B. After a suitable oxide has formed, cross strips of aluminum are evaporated over the first film, sandwiching the oxide between the two metal films. Current is passed along one aluminum film up through the oxide and out through the other film, while the voltage drop is monitored across the oxide. C. A schematic circuit diagram. We are measuring the current-voltage characteristics of the capacitor-like arrangement formed by the two aluminum films and the oxide. When the oxide thickness is less than 50Å or so, an appreciable dc current will flow through the oxide.

counter electrode which we used would short the films. Thus we spent some time measuring very interesting but always non-reproducible current-voltage characteristics which we referred to as miracles since each occurred only once. After a few months we hit on the correct idea: to use evaporated metal films and to separate them by a naturally grown oxide layer.

To carry out our ideas we needed an evaporator, thus I purchased my first piece of experimental equipment. While waiting for the evaporator to arrive I worried a lot—I was afraid I would get stuck in experimental physics tied

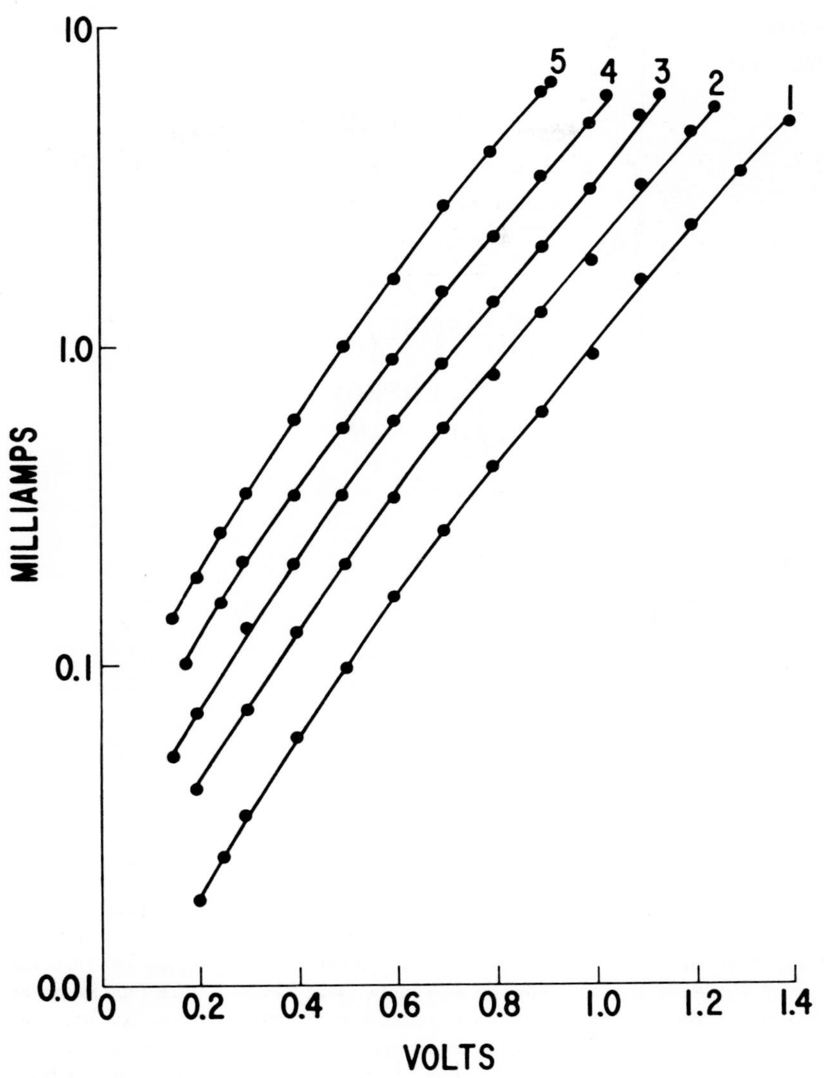

Fig. 4.
Current-voltage characteristics of five different tunnel junctions all with the same thickness, but with five different areas. The current is proportional to the area of the junction. This was one of the first clues that we were dealing with tunneling rather than shorts. In the early experiments we used a relatively thick oxide, thus very little current would flow at low voltages.

down to this expensive machine. My plans at the time were to switch into theory as soon as I had acquired enough knowledge. The premonition was correct; I did get stuck with the evaporator, not because it was expensive but because it fascinated me. Figure 2 shows a schematic diagram of an evaporator. To prepare a tunnel junction we first evaporated a strip of aluminum onto a glass slide. This film was removed from the vacuum system and heated

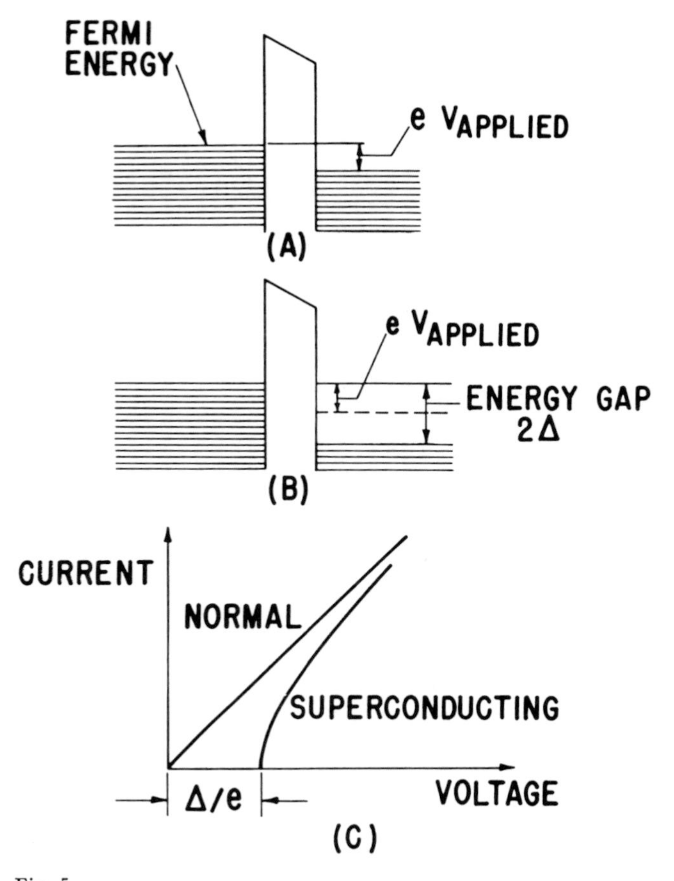

Fig. 5.
A. An energy diagram of two metals separated by a barrier. The Fermi energies in the two metals are at different levels because of the voltage difference applied between the metals. Only the left metal electrons in the energy range $e \cdot V_{App}$ can make a transition to the metal on the right, because only these electrons face empty energy states. The Pauli Principle allows only one electron in each quantum state. B. The right-hand metal is now superconducting, and an energy gap 2Δ has opened up in the electron spectrum. No single electron in a superconductor can have an energy such that it will appear inside the gap. The electrons from the metal on the left can still tunnel through the barrier, but they cannot enter into the metal on the right as long as the applied voltage is less than Δ/e, because the electrons either face a filled state or a forbidden energy range. When the applied voltage exceeds Δ/e, current will begin to flow. C. A schematic current-voltage characteristic. When both metals are in the normal state the current is simply proportional to the voltage. When one metal is superconducting the current-voltage characteristic is drastically altered. The exact shape of the curve depends on the electronic energy spectrum in the superconductor.

to oxidize the surface rapidly. Several cross strips of aluminum were then deposited over the first film making several junctions at the same time. The steps in the sample preparation are illustrated in Figure 3. This procedure solved two problems, first there were no pinholes in the oxide because it is

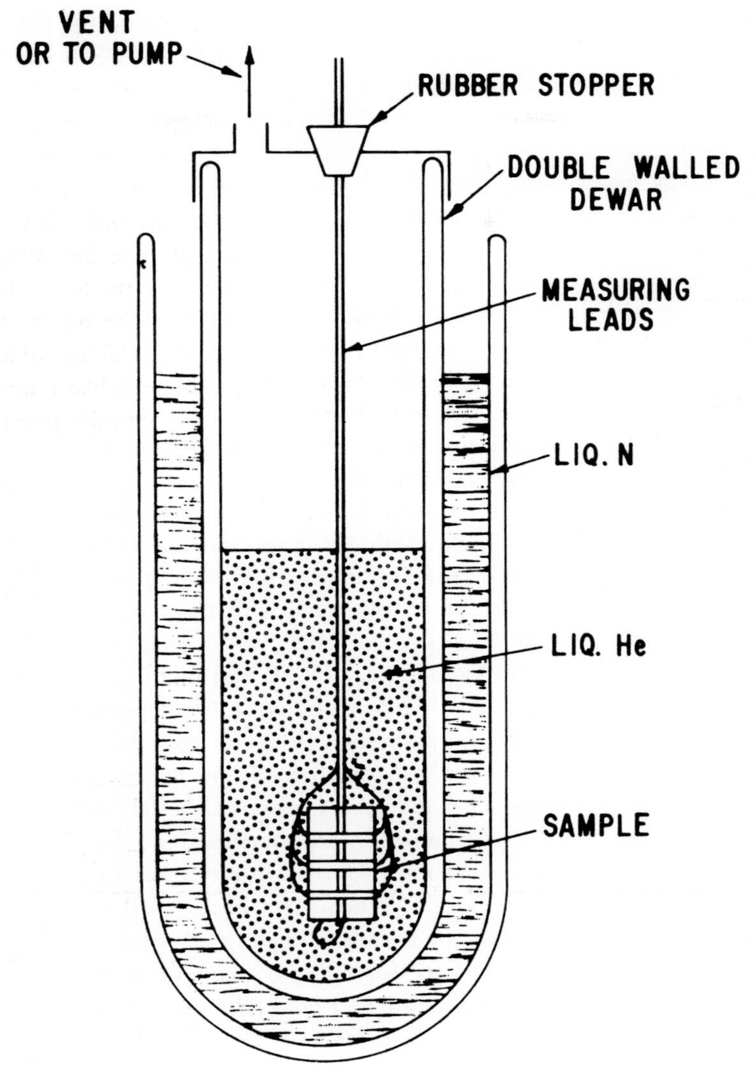

Fig. 6.
A standard experimental arrangement used for low temperature experiments. It consists of two dewars, the outer one contains liquid nitrogen, the inner one, liquid helium. Helium boils at 4.2° K at atmospheric pressure. The temperature can be lowered to about 1° K by reducing the pressure. The sample simply hangs into the liquid helium supported by the measuring leads.

self-healing, and second we got rid of mechanical problems that arose with the mercury counter electrode.

By about April, 1959, we had performed several successful tunneling experiments. The current-voltage characteristics of our samples were reasonably reproducible, and conformed well to theory. A typical result is shown in Figure 4. Several checks were done, such as varying the area and the oxide thickness of the junction as well as changing the temperature. Everything looked OK, and I even gave a seminar at the Laboratory. By this time, I had solved Schrodinger's equation enough times to believe that electrons sometimes behave as waves, and I did not worry much about that part anymore.

However, there were many real physicists at the Laboratory and they properly questioned my experiment. How did I know I did not have metallic shorts? Ionic current? Semiconduction rather than tunneling? Of course, I did not know, and even though theory and experiments agreed well, doubts about the validity were always in my mind. I spent a lot of time inventing impossible schemes such as a tunnel triode or a cold cathode, both to try to prove conclusively that I dealt with tunneling and to perhaps make my work useful. It was rather strange for me at that time to get paid for doing what I considered having fun, and my conscience bothered me. But just like quantum mechanics, you get used to it, and now I often argue the opposite point; we should pay more people to do pure research.

I continued to try out my ideas on John Fisher who was now looking into the problems of fundamental particles with his characteristic optimism and enthusiasm; in addition, I received more and more advice and guidance from Charles Bean and Walter Harrison, both physicists with the uncanny ability of making things clear as long as a piece of chalk and a blackboard were available. I continued to take formal courses at RPI, and one day in a solid state physics course taught by Professor Huntington we got to superconductivity. Well, I didn't believe that the resistance drops to exactly zero—but what really caught my attention was the mention of the energy gap in a superconductor, central to the new Bardeen-Cooper-Schrieffer theory. If the theory was any good and if my tunneling experiments were any good, it was obvious to me that by combining the two, some pretty interesting things should happen, as illustrated in Figure 5. When I got back to the GE Laboratory I tried this simple idea out on my friends, and as I remember, it did not look as good to them. The energy gap was really a many body effect and could not be interpreted literally the way I had done. But even though there was considerable skepticism, everyone urged me to go ahead and make a try. Then I realized that I did not know what the size of the gap was in units I understood—electron volts. This was easily solved by my usual method: first asking Bean and then Harrison, and, when they agreed on a few millielectron volts, I was happy because that is in a easily measured voltage range.

I had never done an experiment requiring low temperatures and liquid helium—that seemed like complicated business. However one great advantage of being associated with a large laboratory like General Electric is that there are always people around who are knowledgeable in almost any field, and better still they are willing to lend you a hand. In my case, all I had to do was go to the end of the hall where Warren DeSorbo was already doing experiments with superconductors. I no longer remember how long it took me to set up the helium dewars I borrowed, but probably no longer than a day or two. People unfamiliar with low temperature work believe that the whole field of low temperature is pretty esoteric, but all it really requires is access

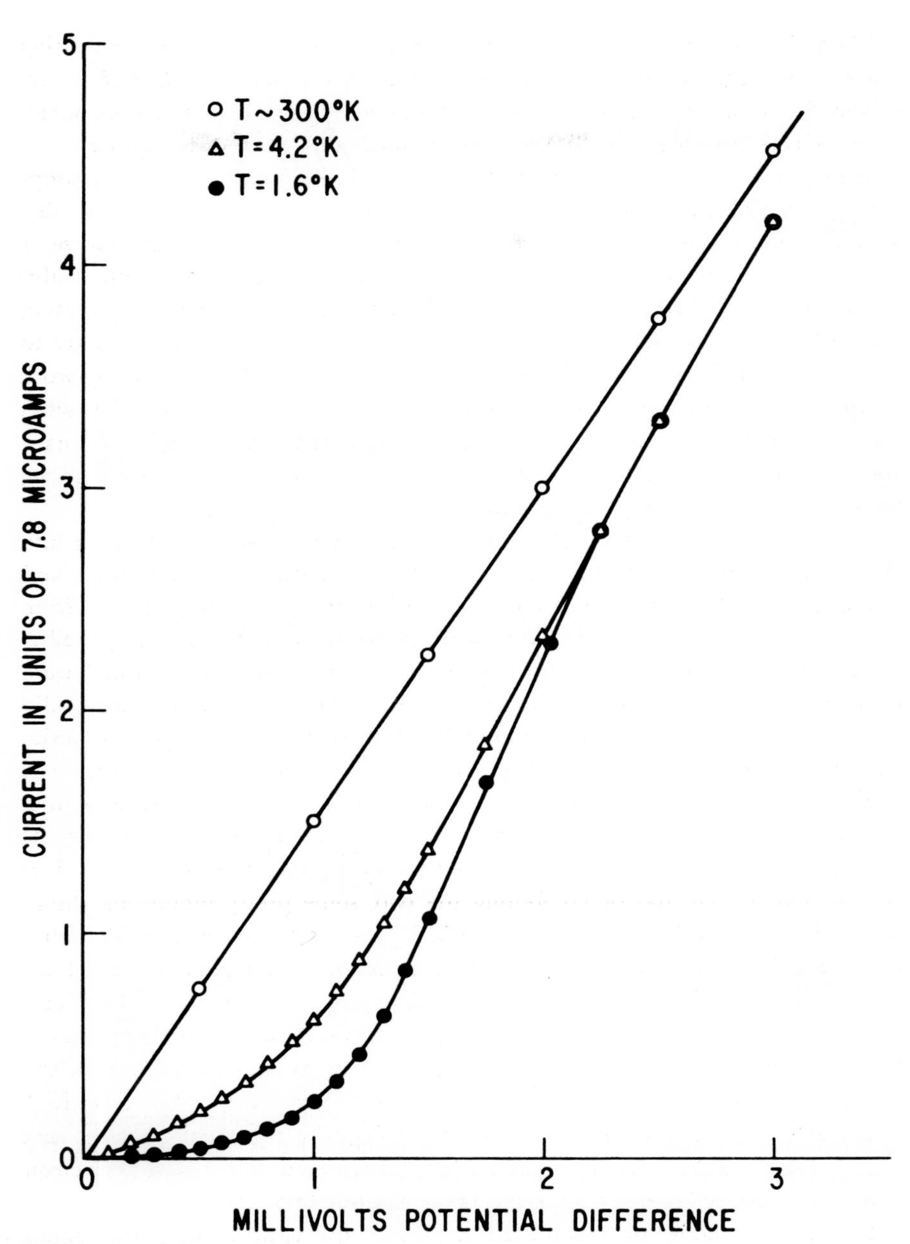

Fig. 7.
The current-voltage characteristic of an aluminum-aluminum oxide-lead sample. As soon as the lead becomes superconducting the current ceases to be proportional to the voltage. The large change between 4.2° K and 1.6° K is due to the change in the energy gap with temperature. Some current also flows at voltages less than Δ / e because of thermally excited electrons in the conductors.

to liquid helium, which was readily available at the Laboratory. The experimental setup is shown in Figure 6. Then I made my samples using the familiar aluminum-aluminum oxide, but I put lead strips on top. Both lead and alu-

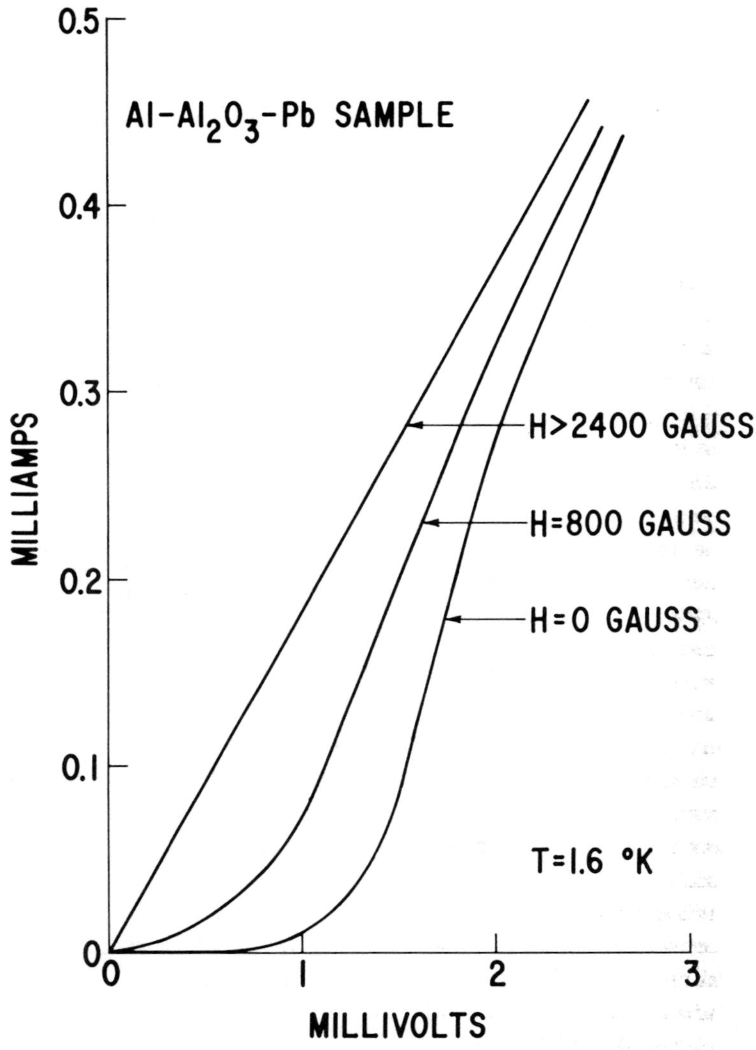

Fig. 8.
The current-voltage characteristic at 1.6° K as a function of the applied magnetic field. At 2 400 gauss the films are normal, at 0 gauss the lead film is superconducting. The reason for the change in the characteristics between 800 gauss and 0 gauss is that thin films have an energy gap that is a function of the magnetic field.

minum are superconductors, lead is superconducting at 7.2° K and thus all you need to make it superconducting is liquid helium which boils at 4.2° K. Aluminum becomes superconducting only below 1.2° K, and to reach this temperature a more complicated experimental setup is required.

The first two experiments I tried were failures because I used oxide layers which were too thick. I did not get enough current through the thick oxide to measure it reliably with the instruments I used, which were simply a standard voltmeter and a standard ammeter. It is strange to think about that

Fig. 9.
Informal discussion over a cup of coffee. From left: Ivar Giaever, Walter Harrison, Charles Bean, and John Fisher.

now, only 13 years later, when the Laboratory is full of sophisticated x-y recorders. Of course, we had plenty of oscilloscopes at that time but I was not very familiar with their use. In the third attempt rather than deliberately oxidizing the first aluminum strip, I simply exposed it to air for only a few minutes, and put it back in the evaporator to deposit the cross strips of lead. This way the oxide was no more than about 30Å thick, and I could readily measure the current-voltage characteristic with the available equipment. To me the greatest moment in an experiment is always just before I learn whether the particular idea is a good or a bad one. Thus even a failure is exciting, and most of my ideas have of course been wrong. But this time it worked! The current-voltage characteristic changed markedly when the lead changed from the normal state to the superconducting state as shown in Figure 7. That was exciting! I immediately repeated the experiment using a different sample —everything looked good! But how to make certain? It was well-known that superconductivity is destroyed by a magnetic field, but my simple setup of dewars made that experiment impossible. This time I had to go all the way across the hall where Israel Jacobs studied magnetism at low temperatures. Again I was lucky enough to go right into an experimental rig where both the temperature and the magnetic field could be controlled and I could quickly do all the proper experiments. The basic result is shown in Figure 8. Every-

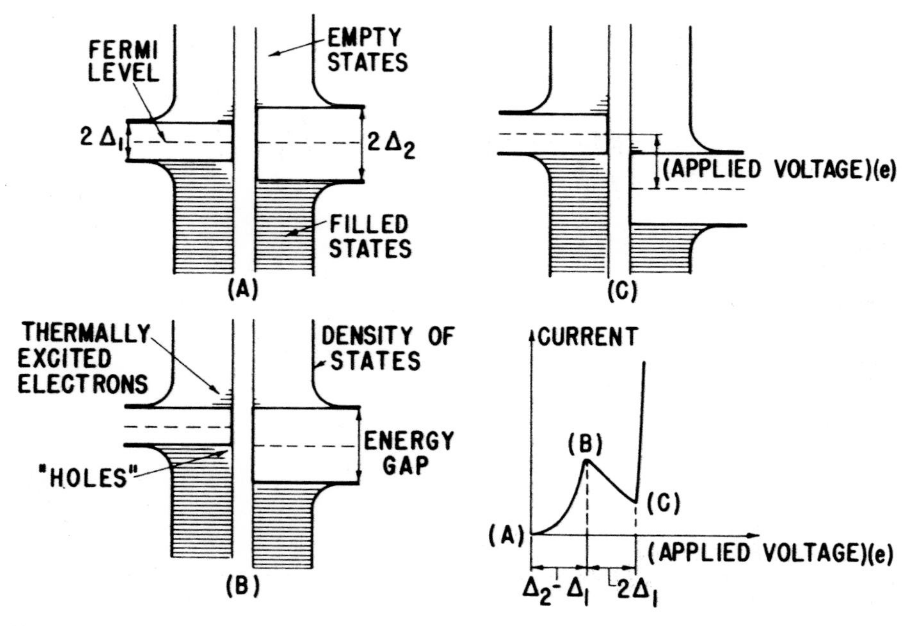

Fig. 10.
Tunneling between two superconductors with different energy gaps at a temperature larger than 0° K. A. No voltage is applied between the two conductors. B. As a voltage is applied it becomes energetically possible for more and more of the thermally excited electrons to flow from the superconductor with the smaller gap into the superconductor with the larger gap. At the voltage shown all the excited electrons can find empty states on the right. C. As the voltage is further increased, no more electrons come into play, and since the number of states the electrons can tunnel into decreases, the current will decrease as the voltage is increased. When the voltage is increased sufficiently the electrons below the gap in the superconductor on the left face empty states on the right, and a rapid increase in current will occur. D. A schematic picture of the expected current-voltage characteristic.

thing held together and the whole group, as I remember it, was very excited. In particular, I can remember Bean enthusiastically spreading the news up and down the halls in our Laboratory, and also patiently explaining to me the significance of the experiment.

I was, of course, not the first person to measure the energy gap in a superconductor, and I soon became aware of the nice experiments done by M. Tinkham and his students using infrared transmission. I can remember that I was worried that the size of the gap that I measured did not quite agree with those previous measurements. Bean set me straight with words to the effect that from then on other people would have to agree with me; my experiment would set the standard, and I felt pleased and like a physicist for the first time.

That was a very exciting time in my life; we had several great ideas to improve and extend the experiment to all sorts of materials like normal metals, magnetic materials and semiconductors. I remember many informal dis-

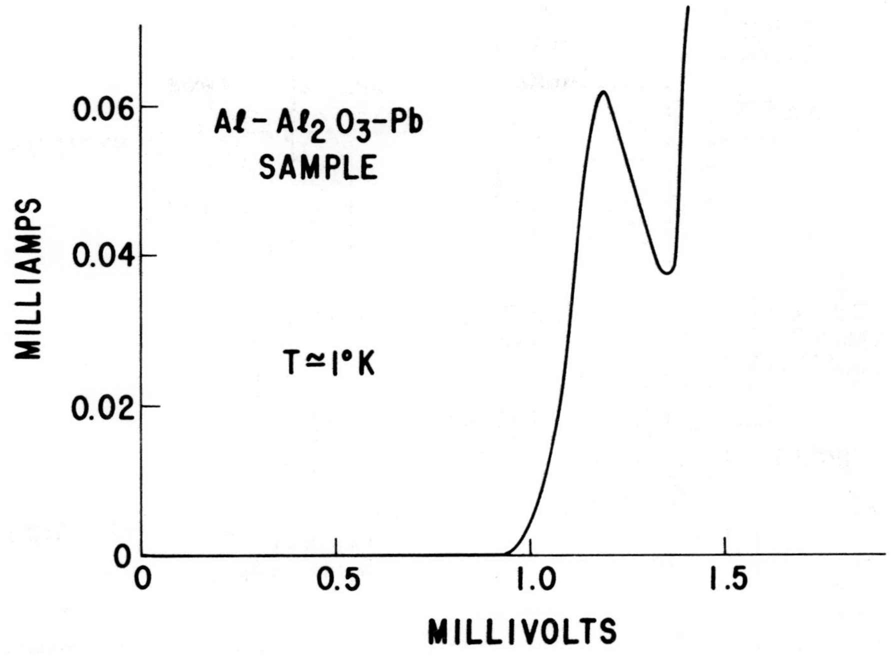

Fig. 11.
A negative resistance characteristic obtained experimentally in tunneling between two different superconductors.

cussions over coffee about what to try next and one of these sessions is in a photograph taken in 1960 which is shown in Figure 9. To be honest the picture was staged, we weren't normally so dressed up, and rarely did I find myself in charge at the blackboard! Most of the ideas we had did not work very well and Harrison soon published a theory showing that life is really complicated after all. But the superconducting experiment was charmed and always worked. It looked like the tunneling probability was directly proportional to the density of states in a superconductor. Now if this were strictly true, it did not take much imagination to realize that tunneling between two superconductors should display a negative resistance characteristic as illustrated in Figure 10. A negative resistance characteristic meant, of course, amplifiers, oscillators and other devices. But nobody around me had facilities to pump on the helium sufficiently to make aluminum become superconducting. This time I had to leave the building and reactivate an old low temperature setup in an adjacent building. Sure enough, as soon as the aluminum went superconducting a negative resistance appeared, and, indeed, the notion that the tunneling probability was directly proportional to the density of states was experimentally correct. A typical characteristic is shown in Figure 11.

Now things looked very good because all sorts of electronic devices could be made using this effect, but, of course, they would only be operative at low temperatures. We should remember that the semiconducting devices were not so advanced in 1960 and we thought that the superconducting junction

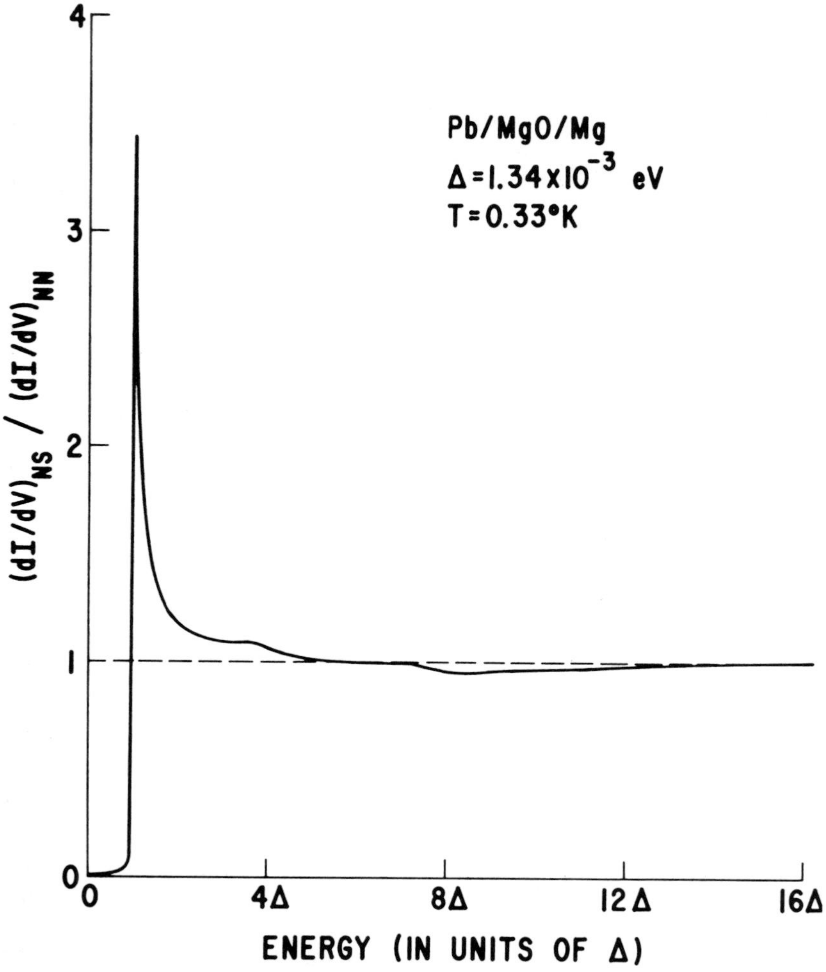

Fig. 12.
A normalized derivative of the current with respect to voltage of a lead junction at low temperature. The simple BCS-theory predicts that the derivative should approach unity asymptotically as the energy increases. Instead several wiggles are observed in the range between 4Δ and 8Δ. These wiggles are related to the phonon spectrum in lead.

would have a good chance of competing with, for example, the Esaki diode. The basic question I faced was which way to go: engineering or science? I decided that I should do the science first, and received full support from my immediate manager, Roland Schmitt.

In retrospect I realize how tempting it must have been for Schmitt to encourage other people to work in the new area, and for the much more experienced physicists around me to do so as well. Instead, at the right time, Schmitt provided me with a co-worker, Karl Megerle, who joined our Laboratory as a Research Training Fellow. Megerle and I worked well together and before long we published a paper dealing with most of the basic effects.

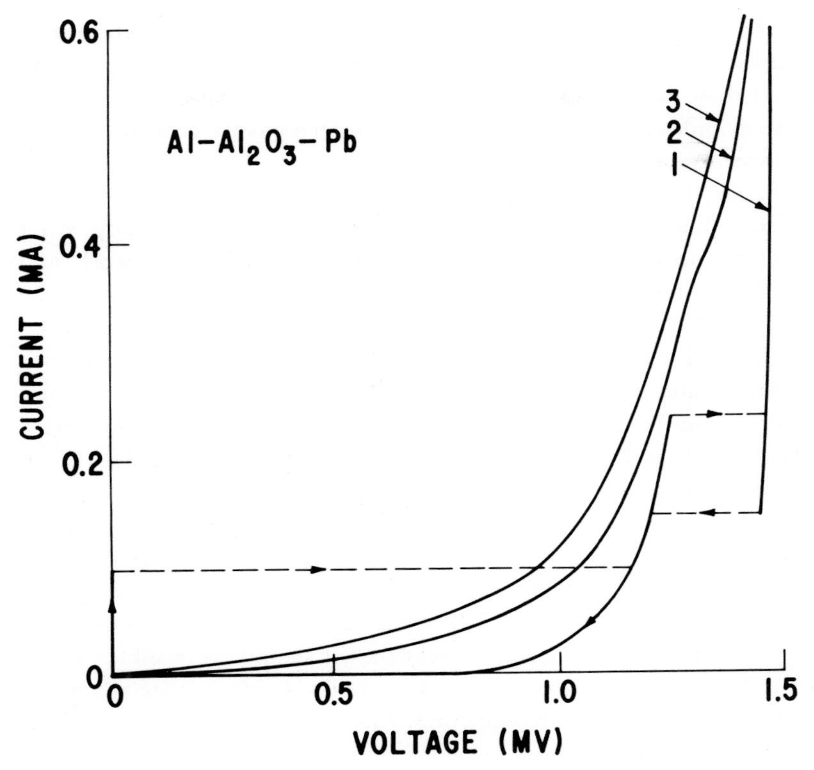

Fig. 13.
Effect of trapped magnetic field on a tunneling characteristic. Curve 1 is a virgin curve, while curve 3 is in a moderate magnetic field, and in curve 2 the magnetic field has been removed. In curve 1 we also have a small resistance-less current which we interpreted as caused by metallic shorts. In retrospect, it was actually due to the Josephson effect.

As always in physics, it is important to extend experiments to a higher energy, a greater magnetic field, or, in our case, to a lower temperature. Therefore, we joined forces with Howard Hart, who had just completed a helium 3 refrigerator that was capable of getting down to about 0.3° K. At the same time, Megerle finished a lock-in amplifier which we could use to measure directly the derivative of the current with respect to the voltage. That was really a nice looking machine with a magnet rotating past a pickup coil at eight cycles per second, but, of course, vastly inferior to the modern lock-in amplifier. We had known for some time that there were anomalies in the current-voltage characteristics of lead, and now we finally pinned them down by finding some extra wiggles in the derivative curve. This is shown in Figure 12. That made us happy because all that the tunneling experiments had done up till now was to confirm the BCS theory, and that is not what an experimentalist would really like to do. The dream is to show that a famous theory is incorrect, and now we had finally poked a hole in the theory. We speculated at the time that these wiggles were somehow associated with the phonons thought to be the cause of the attractive electron-electron interaction in a superconductor. As often happens, the theorists turned the tables on us and cleverly used these wiggles to properly extend the theory and to prove that the BCS theory indeed was correct. Professor Bardeen gave a detailed account of this in his most recent Nobel Prize lecture.

I have, so far, talked mainly about what went on at General Electric at that time; sometimes it is difficult for me to realize that Schenectady is not the center of the world. Several other people began to do tunneling work, and to mention just a few: J. M. Rowell and W. L. McMillan were really the ones who unraveled the phonon structure in a superconductor; W. J. Tomasch, of course, insisted on discovering his own effect; S. Shapiro and colleagues did tunneling between two superconductors at the same time we did; and J. Bardeen, and later M. H. Cohen *et al.*, took care of most of the theory.

Meanwhile, back at RPI, I had finished my course work and decided to do a theoretical thesis on ordered-disordered alloys with Professor Huntington because tunneling in superconductors was mainly understood. Then someone made me aware of a short paper by Brian Josephson in *Physics Letters*—what did I think? Well, I did not understand the paper, but shortly after I had the chance to meet Josephson at Cambridge and I came away impressed. One of the effects Josephson predicted was that it should be possible to pass a supercurrent with zero voltage drop through the oxide barrier when the metals on both sides were superconducting; this is now called the dc Josephson effect. We had observed this behavior many times; matter-of-fact, it is difficult not to see this current when junctions are made of tin-tin oxide-tin or lead-lead oxide-lead. The early tunnel junctions were usually made with aluminum oxide which generally is thicker and therefore thermal fluctuations suppress the dc current. In our first paper Megerle and I published a curve, which is shown in Figure 13, demonstrating such a supercurrent and also that it depended strongly on a magnetic field. However, I had a ready-made explanation for this supercurrent—it came from a metallic short or bridge. I was puzzled at the time because of the sensitivity to the magnetic field which is unexpected for a small bridge, but no one knew how a 20Å long and 20Å wide bridge would behave anyway. If I have learned anything as a scientist it is that one should not make things complicated when a simple explanation will do. Thus all the samples we made showing the Josephson effect were discarded as having shorts. This time I was too simple-minded! Later I have been asked many times if I feel bad for missing the effect? The answer is clearly no, because to make an experimental discovery it is not enough to observe something, one must also realize the significance of the observation, and in this instance I was not even close. Even after I learned about the dc Josephson effect, I felt that it could not be distinguished from real shorts, therefore I erroneously believed that only the observation of the so-called ac effect would prove or disprove Josephson's theory.

In conclusion I hope that this rather personal account may provide some slight insight into the nature of scientific discovery. My own beliefs are that the road to a scientific discovery is seldom direct, and that it does not necessarily require great expertise. In fact, I am convinced that often a newcomer to a field has a great advantage because he is ignorant and does not know all the complicated reasons why a particular experiment should not be attempted. However, it is essential to be able to get advice and help from experts in the various sciences when you need it. For me the most important ingredients were that I was at the right place at the right time and that I found so many friends both inside and outside General Electric who unselfishly supported me.

References

1. *Tunneling Phenomena in Solids* edited by Burstein, E. and Lundquist, S. Plenum Press, New York, 1969.
2. *Superconductivity* edited by Parks, R. D. Marcel Dekker, Inc., New York. 1969.

Brian Josephson

BRIAN D. JOSEPHSON

Date of birth:	4 January 1940	
Place of birth:	Cardiff High School	
Education:	Cardiff High School	
	University of Cambridge, B.A.	1960
	University of Cambridge, M.A., Ph.D	1964

Academic Career

Fellow of Trinity College, Cambridge	1962
Research Assistant Professor, University of Illinois	1964–65
Assistant Director of Research, University of Cambridge	1967–72
NSF Senior Foreign Scientist Fellow, Cornell University	1971
Reader in Physics, University of Cambridge	1972–74
Professor of Physics, University of Cambridge	1974–
Visiting Professor — Computer Science Department, Wayne State University, Detroit	1983
Visiting Professor, Indian Institute of Science, Bangalore	1984
Visiting Professor, University of Missouri-Rolla	1987

Awards

New Scientist	1969
Research Corporation	1969
Fritz London	1970

Medals

Guthrie (Institute of Physics)	1972
van der Pol	1972
Elliott Cresson (Franklin Institute)	1972
Hughes (Royal Society)	1972
Holweck (Institute of Physics and French Institute of Physics)	1972
Faraday (Institution of Electrical Engineers)	1982
Sir George Thomson (Institute of Measurement and Control)	1984

Other Information

Fellow of the Institute of Physics	
Honorary D.Sc., University of Wales	1974
Honorary Member, American Academy of Arts and Sciences	1974
Honorary Member, Institute of Electrical and Electronic Engineers	1982
Honorary D.Sc., University of Exeter	1983
Invited presentation on subject of 'Higher States of Consciousness', to US Congressional Committee	1983

THE DISCOVERY OF TUNNELLING SUPERCURRENTS

Nobel Lecture, December 12, 1973

by

BRIAN D. JOSEPHSON

Cavendish Laboratory, Cambridge, England

The events leading to the discovery of tunnelling supercurrents took place while I was working as a research student at the Royal Society Mond Laboratory, Cambridge, under the supervision of Professor Brian Pippard. During my second year as a research student, in 1961—2, we were fortunate to have as a visitor to the laboratory Professor Phil Anderson, who has made numerous contributions to the subject of tunnelling supercurrents, including a number of unpublished results derived independently of myself. His lecture course in Cambridge introduced the new concept of 'broken symmetry' in superconductors, (1) which was already inherent in his 1958 pseudospin formulation of superconductivity theory, (2) which I shall now describe.

As discussed by Cooper in his Nobel lecture last year (3), according to the Bardeen-Cooper-Schrieffer theory there is a strong positive correlation in a superconductor between the occupation of two electron states of equal and opposite momentum and spin. Anderson showed that in the idealized case where the correlation is perfect the system can be represented by a set of interacting 'pseudospins', with one pseudospin for each pair of electron states. The situation in which both states are unoccupied is represented by a pseudospin in the positive z direction, while occupation of both states is represented by a pseudospin in the negative z direction; other pseudospin orientations correspond to a superposition of the two possibilities.

The effective Hamiltonian for the system is given by

$$H = -2\sum_{k} (\varepsilon_k - \mu)\, s_{kz} - \sum_{k\neq k'} V_{kk'}\,(s_{kx}\, s_{k'x} + s_{ky}\, s_{k'y}) \qquad (1)$$

the first term being the kinetic energy and the second term the interaction energy. In this equation s_{kx}, s_{ky} and s_{kz} are the three components of the kth pseudospin, ε_k is the single-particle kinetic energy, μ the chemical potential and $V_{kk'}$ the matrix element for the scattering of a pair of electrons of equal and opposite momentum and spin. The kth pseudospin sees an effective field

$$\boldsymbol{H}_k = 2(\varepsilon_k - \mu)\,\hat{z} + 2\sum_{k\neq k'} V_{kk'}\, \boldsymbol{s}_{k'\perp} \qquad (2)$$

where $\hat{z}$ is a unit vector in the z direction and $\perp$ indicates the component of the pseudospin in the xy plane.

One possible configuration of pseudospins consistent with (2) is shown in Fig. 1 (a). All the pseudospins lie in the positive or negative z direction, and

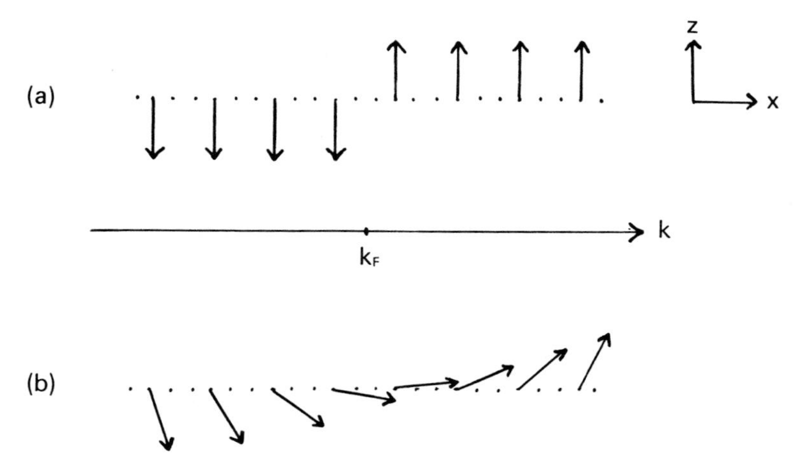

Fig. 1. Pseudospin configurations in (a) a normal metal (b) a superconductor. k_F is the Fermi momentum.

the direction reverses as one goes through the Fermi surface, since $\varepsilon_k - \mu$ changes sign there. If the interaction is attractive, however (corresponding to negative $V_{kk'}$), a configuration of lower energy exists, in which the pseudospins are tilted out of the z direction into a plane containing the z axis, and the pseudospin direction changes continuously as one goes through the Fermi surface, as in Fig. 1 (b).

The ground state of Fig. 1 (b) breaks the symmetry of the pseudospin Hamiltonian (1) with respect to rotation about the z axis, which is itself a consequence of conservation of number of electrons in the original Hamiltonian. Because of this symmetry a degenerate set of ground states exists, in which the pseudospins can lie in any plane through the z axis. The angle Φ which this plane makes with the Oxz plane will play an important role in what follows. Anderson made the observation that with a suitable interpretation of the Gor'kov theory, (4) Φ is also the phase of the complex quantity F which occurs in that theory.

I was fascinated by the idea of broken symmetry, and wondered whether there could be any way of observing it experimentally. The existence of the orginal symmetry implies that the absolute phase angle Φ would be unobservable, but the possibility of observing phase differences between the F functions in two separate superconductors was not ruled out. However, consideration of the number-phase uncertainty relation suggested that the phase difference $\Delta\Phi$ could be observed only if the two superconductors were able to exchange electrons. When I learnt of observations suggesting that a supercurrent could flow through a sufficiently thin normal region between two superconductors (5, 6), I realized that such a supercurrent should be a function of $\Delta\Phi$. I could see in principle how to calculate the supercurrent, but considered the calculation to be too difficult to be worth attempting.

I then learnt of the tunnelling experiments of Giaever, (7) described in the preceding lecture (8). Pippard (9) had considered the possibility that a Cooper pair could tunnel through an insulating barrier such as that which Giaever used, but argued that the probability of two electrons tunnelling simultaneously would be very small, so that any effects might be unobservable. This plausible argument is now known not to be valid. However, in view of it I turned my attention to a different possiblity, that the *normal* currents through the barrier might be modified by the phase difference. An argument in favour of the existence of such an effect was the fact that matrix elements for processes in a superconductor are modified from those for the corresponding processes in a normal metal by the so-called coherence factors, (3) which are in turn dependent on $\Delta\Phi$ (through the u_k's and v_k's of the BCS theory). At this time there was no theory available to calculate the tunnelling current, apart from the heuristic formula of Giaever, (7) which was in agreement with experiment but could not be derived from basic theory. I was able, however, to make a qualitative prediction concerning the time dependence of the current. Gor'kov (4) had noted that the F function in his theory should be time-dependent, being proportional to $e^{-2i\mu t/\hbar}$, where μ is the chemical potential as before. (10) The phase Φ should thus obey the relation

$$\partial\Phi/\partial t = -2\mu/\hbar, \tag{3}$$

while in a two-superconductor system the phase difference obeys the relation

$$\frac{\partial}{\partial t}(\Delta\Phi) = 2eV/\hbar, \tag{4}$$

where V is the potential difference between the two superconducting regions, so that

$$\Delta\Phi = 2eVt/\hbar + \text{const.} \tag{5}$$

Since nothing changes physically if $\Delta\Phi$ is changed by a multiple of 2π, I was led to expect a periodically varying current at a frequency $2eV/h$.

The problem of how to calculate the barrier current was resolved when one day Anderson showed me a preprint he had just received from Chicago, (11) in which Cohen, Falicov and Phillips calculated the current flowing in a superconductor-barrier-normal metal system, confirming Giaever's formula. They introduced a new and very simple way to calculate the barrier current—they simply used conservation of charge to equate it to the time derivative of the amount of charge on one side of the barrier. They evaluated this time derivative by perturbation theory, treating the tunnelling of electrons through the barrier as a perturbation on a system consisting of two isolated subsystems between which tunnelling does not take place.

I immediately set to work to extend the calculation to a situation in which both sides of the barrier were superconducting. The expression obtained was of the form

$$I = I_0(V) + I_1'(V)\cos(\Delta\Phi) + I_1(V)\sin(\Delta\Phi). \quad (6)$$

At finite voltages the linear increase with time of $\Delta\Phi$ implies that the only contribution to the dc current comes from the first term, which is the same as Giaever's prediction, thus extending the results of Cohen et al. to the two-superconductor case. The second term had a form consistent with my expectations of a $\Delta\Phi$ dependence of the current due to tunnelling of quasi-particles. The third term, however, was completely unexpected, as the coefficient $I_1(V)$, unlike $I_0(V)$, was an *even* function of V and would not be expected to vanish when V was put equal to zero. The $\Delta\Phi$ dependent current at zero voltage had the obvious interpretation of a supercurrent, but in view of the qualitative argument mentioned earlier I had not expected a contribution to appear to the same order of magnitude as the quasiparticle current, and it was some days before I was able to convince myself that I had not made an error in the calculation.

Since sin $(\Delta\Phi)$ can take any value from -1 to $+1$, the theory predicted a value of the critical supercurrent of $I_1(0)$. At a finite voltage V an 'ac supercurrent' of amplitude

$$\sqrt{\{[I_1(V)]^2 + [I_1'(V)]^2\}}$$

and frequency $2eV/h$ was expected. As mentioned earlier, the only contribution to the dc current in this situation $(V \neq 0)$ comes from the $I_0(V)$ term, so that a two-section current-voltage relation of the form indicated in Fig. 2 is expected.

I next considered the effect of superimposing an oscillatory voltage at frequency ν on to a steady voltage V. By assuming the effect of the oscillatory voltage to be to modulate the frequency of the ac supercurrent I concluded that constant-voltage steps would appear at voltages V for which the frequency of the unmodulated ac supercurrent was an integral multiple of ν, i.e. when $V = nh\nu/2e$ for some integer n.

The embarrassing feature of the theory at this point was that the effects predicted were too large! The magnitude of the predicted supercurrent was proportional to the normal state conductivity of the barrier, and of the same order of magnitude as the jump in current occurring as the voltage passes through that at which production of pairs of quasi-particles becomes possible. Examination of the literature showed that possibly dc supercurrents of this magnitude had been observed, for example in the first published observation of tunnelling between two evaporated-film superconductors by Nicol, Shapiro and Smith (12) (fig. 3). Giaever (13) had made a similar observation, but ascribed the supercurrents seen to conduction through metallic shorts through the barrier layer. As supercurrents were not always seen, it seemed that the explanation in terms of shorts might be the correct one, and the whole theory might have been of mathematical interest only (as was indeed suggested in the literature soon after).

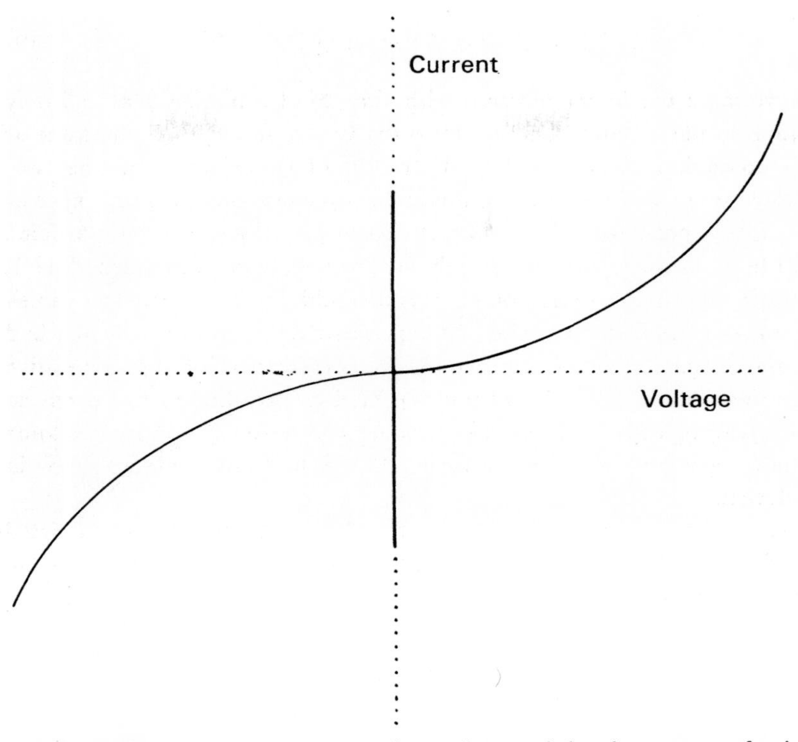

Fig. 2. Predicted two-part current-voltage characteristic of a superconducting tunnel junction.

Then a few days later Phil Anderson walked in with an explanation for the missing supercurrents, which was sufficiently convincing for me to decide to go ahead and publish my calculation, (14) although it turned out later not to have been the correct explanation. He pointed out that my relation between the critical supercurrent and the normal state resistivity depended on the assumption of time-reversal symmetry, which would be violated if a magnetic field were present. I was able to calculate the magnitude of the effect by using the Ginzburg-Landau theory to find the effect of the field on the phase of the *F* functions, and concluded that the Earth's field could have a drastic effect on the supercurrents.

Brian Pippard then suggested that I should try to observe tunneling supercurrents myself, by measuring the characteristics of a junction in a compensated field. The result was negative—a current less than a thousandth of the predicted critical current was sufficient to produce a detectable voltage across the junction. This experiment was at one time to be written up in a chapter of my thesis entitled 'Two Unsuccessful Experiments in Electron Tunnelling between Superconductors'.

Eventually Anderson realized that the reason for the non-observation of dc supercurrents in some specimens was that electrical noise transmitted down the measuring leads to the specimen could be sufficient in high-resistance

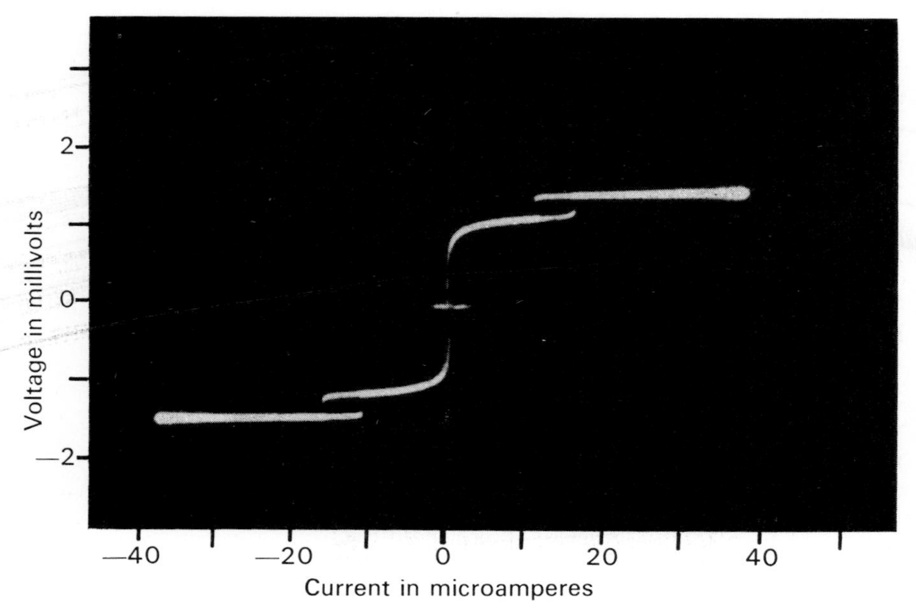

Fig. 3. The first published observation of tunnelling between two evaporated-film superconductors (Nicol, Shapiro and Smith, reference 6). A zero-voltage supercurrent is clearly visible. It was not until the experiments of Anderson and Rowell (reference 15) that such supercurrents could be definitely ascribed to the tunnelling process.

specimens to produce a current exceeding the critical current. Together with John Rowell he made some low resistance specimens and soon obtained convincing evidence (15) for the existence of tunnelling supercurrents, shown particularly by the sensitivity to magnetic fields, which would not be present in the case of conduction through a metallic short. In one specimen they found a critical current of 0.30 mA in the Earth's magnetic field. When the field was compensated, the critical current increased by more than a factor of two, to 0.65 mA, while a field of 2mT was sufficient to destroy the zero-voltage supercurrents completely. Later Rowell (16) investigated the field dependence of the critical current in detail, and obtained results related to the diffraction pattern of a single slit, a connection first suggested by J. C. Phillips (unpublished). This work was advanced by Jaklevic, Lambe, Silver and Mercereau, (17) who connected two junctions in parallel and were able to observe the analogue of the Young's slit interference experiment. The sensitivity of the critical current to applied magnetic field can be increased by increasing the area enclosed between the two branches of the circuit, and Zimmerman and Silver (18) were able to achieve a sensitivity of 10^{-13} T.

Indirect evidence for the ac supercurrents come soon after. Shapiro (19) shone microwaves on to a junction and observed the predicted appearance of steps in the current-voltage characteristics. The voltages at which the steps occurred changed as the frequency of the microwaves was changed, in the manner expected. In 1966, Langenberg, Parker and Taylor (20) measured the ratio of voltage to frequency to 60 parts per million and found agreement with the value of $h/2e$ then accepted. Later they increased their accuracy sufficiently to be able to discover errors in the previously accepted values of the fundamental constants and derive more accurate estimates, (21, 22), thus carrying out to fruition an early suggestion of Pippard (unpublished). The ac supercurrent is now used to compare voltages in different standards laboratories without the necessity for the interchange of banks of standard cells. If two laboratories irradiate specimens with radiation of the same frequency, constant-voltage steps appear at identical voltages. The intercomparison of frequencies can be carried out in a straightforward manner by transmission of radio signals.

At the end of 1963, the evidence for the existence of the ac supercurrent was only indirect. John Adkins and I tried to observe the effect by coupling together two junctions by a short ($\sim$ 0.2 mm.) thin-film transmission line. The idea was that radiation emitted by one junction would modify the characteristics of the other. The experiment, planned to form the second part of the thesis chapter referred to above, was unsuccessful, for reasons which are still unclear. Later, Giaever (23) was able to observe the ac supercurrent by a similar method to the one we had considered, and then Yanson, Svistunov and Dmitrenko (24) succeeded in observing radiation emitted by the ac supercurrent with a conventional detector.

Finally, I should like to describe the SLUG, (25) developed in the Royal Society Mond Laboratory by John Clarke while he was a research student. John was attempting to make a high-sensitivity galvanometer using the previously described magnetic interferometers with two junctions connected in parallel. One day Paul Wraight, who shared a room with John, observed that the fact that one cannot solder niobium using ordinary solder must mean that if one allows a molten blob of solder to solidify in contact with niobium there must be an intermediate layer of oxide, which might have a suitable thickness to act as a tunnelling barrier. This proved to be the case. However, in John's specimens, in which a niobium wire was completely surrounded by a blob of solder, the critical current through the barrier proved to be completely insensitive to externally applied magnetic fields. It was, however, found to be sensitive to the magnetic field produced by passing a current through the central wire. This fact led to the development of a galvanometer with sensitivity of 10^{-14} volts at a time constant of 1 s.

There have been many other developments which I have not had time to describe here. I should like to conclude by saying how fascinating it has been for me to watch over the years the many developments in laboratories over the world, which followed from asking one simple question, namely what is the physical significance of broken symmetry in superconductors?

REFERENCES AND FOOTNOTES

1. Anderson, P. W., *Concepts in Solids,* W. A., Benjamin, Inc., New York, 1963.
2. Anderson, P. W. Phys. Rev. 112, 1900 (1958).
3. Cooper, L. N., *Les Prix Nobel en 1972,* Nobel Foundation 1973, p. 64.
4. Gor'kov, L. P., J. Exptl. Theoret. Phys. (USSR) 36, 1918 (1959); translation: Soviet Phys. JETP 9, 1364 (1959).
5. Meissner, H. Phys, Rev. 117, 672 (1960).
6. Smith, P. H., Shapiro, S., Miles, J. L. and Nicol, J. Phys. Rev. Letters 6, 686 (1961).
7. Giaever, I., Phys. Rev. Letters 5, 464 (1960).
8. Giaever, I., *Les Prix Nobel en 1973,* Nobel Foundation 1974, p.
9. Pippard, A. B., *Proceedings of the VIIth International Conference on Low Temperature Physics,* ed. Graham, G. M. and Hollis Hallett, A. C., North-Holland, Amsterdam 1961, p. 320.
10. Gor'kov's result may be extended to finite temperatures by the following argument. The density operator of a system in equilibrium has the form $Z^{-1} \exp\{-\beta(H-\mu N)\}$ where must contain a small symmetrybreaking term in order to set up an ensemble in which Φ has a definite value. Since this operator commutes with H-μN, all quantities are time-independent if the Hamiltonian of the system is taken to be H-μH. Transition to a situation where the time dependence is given by the true Hamiltonian H can be accomplished by means of a gauge transformation, and consideration of the effect of this transformation on the electron operators gives immediately Gor'kovs result $F \propto \exp(-2\, i\, \mu\, t/\hbar)$.
11. Cohen, M. H., Falicov, L. M. and Phillips, J. C., Phys. Rev. Letters 8, 316 (1962).
12. Nicol, J., Shapiro, S. and Smith, P. H., Phys. Rev. Letters 5, 461 (1960).
13. Giaever, I., Phys. Rev. Letters 5, 464 (1960).
14. Josephson, B. D., Phys. Letters 1, 251 (1962).
15. Anderson, P. W. and Rowell, J. M., Phys. Rev. Letters 10, 230 (1963).
16. Rowell, J. M., Phys. Rev. Letters 11, 200 (1963).
17. Jaklevic, R. C., Lambe, J., Silver, A. H. and Mercereau, J. E., Phys. Rev. Letters 12, 159 (1964).
18. Zimmerman, J. E., and Silver, A. H., Phys. Rev. 141, 367 (1966).
19. Shapiro, S. Phys. Rev. Letters 11, 80 (1963).
20. Langenberg, D. N., Parker, W. H. and Taylor, B. N., Phys. Rev. 150, 186 (1966).
21. Parker, W. H., Taylor, B. N. and Langenberg, D. N., Phys. Rev. Letters 18, 287 (1967).
22. Taylor, B. N., Parker, W. H. and Langenberg, D. N., *The Fundamental Constants and Quantum Electrodynamics,* Academic Press, New York and London, 1969.
23. Giaever, I., Phys. Rev. Letters 14, 904 (1965).
24. Yanson, I. K., Svistunov, V. M. and Dmitrenko, I. M., J. Exptl. Theoret. Phys. (USSR) 21, 650 (1965); translation: Soviet Phys. JETP 48, 976 (1965).
25. Clarke, J., Phil. Mag. 13, 115 (1966).

Physics 1974

MARTIN RYLE and ANTONY HEWISH

for their pioneering research in radio astrophysics: Ryle for his observations and inventions, in particular of the aperture synthesis technique, and Hewish for his decisive role in the discovery of pulsars

THE NOBEL PRIZE FOR PHYSICS

Speech by professor HANS WILHELMSSON of the Royal Academy of Sciences
Translation from the Swedish text

Your Majesty, Your Royal Highnesses, Ladies and Gentlemen,
The subject of the Nobel Prize in Physics this year is the science of Astrophysics, the Physics of the stars and galactic systems.

Problems concerning our Universe on a large scale, its constitution and evolution, play an essential rôle in present day scientific discussions.

We are curious about the behaviour of our Universe. In order to draw reliable conclusions regarding cosmological models it is necessary to gather detailed information about conditions in the remote parts of the Cosmos.

Radio-astronomy offers unique possibilities for studying what is taking place, or in reality what occurred very long ago, at enormous distances from Earth, as far out as thousands of millions of lightyears from us. The radio waves now reaching us have been travelling for thousands of millions of years at the speed of light to reach our Earth from those very remote sources.

It is indeed a thrilling fact that the radio signals we record today here on Earth left their cosmic sources at a time when hardly any flowers or living creatures, and certainly no physicists, existed on Earth.

New and epoch-making discoveries have been made in the field of Radio-astrophysics during the last decade, discoveries that are also exceedingly important contributions to modern Physics, for example in establishing through radio-astronomical observations the presence of matter in a superdense state.

One single cubic centimeter of this superdense matter has a weight of thousands of millions of tons. It consists of tightly-packed neutrons. A neutron star appears as a consequence of a star explosion, a so-called supernova event. Neutron stars, with a diameter of about 10 kilometers, are from a cosmic point of view extremely small objects. They represent the final state in the evolution of certain stars.

This year's Nobel Prize winners in Physics, Martin Ryle and Antony Hewish, developed new radio-astronomical techniques. Their observations of cosmic radio sources represent extremely noteworthy research results.

In order to collect radio waves from cosmic radio sources one utilizes radio-telescopes. It is important that a radio-telescope should have a large area, both for highest possible sensitivity and for the high angular resolution that is needed to discriminate among the various cosmic sources of radio radiation.

For observation of exceedingly small sources it is, however, no longer possible to build a single radio-telescope of sufficient size. Ryle and his collaborators therefore developed the method of aperture synthesis. Instead of making one huge aerial, a number of small aerials are used in this method, and the signals received by them are combined in such a way as to provide the necessary extreme accuracy.

Instead of many small aerials, Ryle in fact made use of a few aerials that could be moved successively to different positions on the ground. Ryle also invented the extremely elegant and powerful technique utilizing the rotation of the Earth to move his radio-telescopes. With this technique he obtained a resolution in his observations that corresponded to an aerial of enormous size.

Ryle's measurements enable us to conclude that a steady-state model of the Universe can not be accepted. The Cosmos on a large scale has to be described by dynamic, evolutionary models.

In his latest construction in Cambridge, Ryle obtained an angular resolution permitting the mapping of cosmic radio sources with an error less than one second of arc!

The radio-astronomical instruments invented and developed by Martin Ryle, and utilized so successfully by him and his collaborators in their observations, have been one of the most important elements of the latest discoveries in Astrophysics.

Antony Hewish and his collaborators in Cambridge, in the Autumn of 1967, made a unique and unexpected discovery that has revolutionized Astrophysics. They had constructed new aerials and instruments to study the influence of the outer corona of the Sun on the radiation detected from remote point sources. A special receiver capable of extremely rapid response had been built.

The fast receiver provided a result quite different from its intended purpose. By chance the receiver detected short pulses of radio signals that were repeated periodically about every second, and with exceedingly high precision in the pulse repetition rates.

It was concluded that the radiation originated from cosmic sources of previously unknown type. These sources were subsequently named pulsars.

One has come to the conclusion that the central part of a pulsar consists of a neutron star. The pulsars are also accompanied by magnetic fields many millions of times stronger than those found in laboratories on Earth. The neutron star is surrounded by an electrically-conducting gas or plasma. Each pulsar rotates and emits beams of radiation in the Universe, resembling those from a light-house. The beams strike the Earth periodically with high precision.

These pulsars are indeed the world clocks which our Nobel Prize winner Harry Martinson mentions in his poetry.

Allow me to quote this poet of space:

"World clocks tick and space gleams
everything changes place and order".

Early in the history of pulsar research it was suspected that neutron star matter existed in the centres of supernovas. Radio-telescopes were aimed towards the centre of the Crab nebula, a magnificent glaring gaseous remnant of a supernova event that is known, from Chinese annals, to have occurred in 1054 A.D., and indeed, they detected a pulsar! This pulsar emits not only radio pulses, as expected from a pulsar, but pulses of light and x-rays as well. It is comparatively young, rotates rapidly and is in fact exceptional among pulsars.

Antony Hewish played a decisive rôle in the discovery of pulsars. This discovery, which is of extraordinary scientific interest, opens the way to new methods for studying matter under extreme physical conditions.

The contributions of Ryle and Hewish represent an important step forward in our knowledge of the Universe. Thanks to their work new fields of research have become part of Astrophysics. The gigantic laboratory of the Universe offers rich possibilities for future research.

Sir Martin,

Some of the most fundamental questions in Physics have been elucidated as a result of your brilliant research. Your inventions and observations have brought new foundations for our conception of the Universe.

Professor Antony Hewish,

The discovery of pulsars, for which you played a decisive rôle, is a most outstanding example of how in recent years our knowledge of the Universe has been dramatically extended. Your research has contributed greatly to Astrophysics and to Physics in general.

On behalf of the Royal Academy of Sciences I wish to express our admiration and to convey to you our warmest congratulations.

The Royal Academy of Sciences regrets that Sir Martin Ryle is not here today.

May I now ask you, Professor Hewish, to receive your prize and also the prize awarded to Sir Martin Ryle from the hands of His Majesty the King.

Antony Hewish

ANTHONY HEWISH

I was born in Fowey, Cornwall, on 11 May 1924, the youngest of three sons and my father was a banker. I grew up in Newquay, on the Atlantic coast and there developed a love of the sea and boats. I was educated at King's College, Taunton and went to the University of Cambridge in 1942. From 1943–46 I was engaged in war service at the Royal Aircraft Establishment, Farnborough and also at the Telecommunications Research Establishment, Malvern. I was involved with airborne radar-counter-measure devices and during this period I also worked with Martin Ryle.

Returning to Cambridge in 1946 I graduated in 1948 and immediately joined Ryle's research team at the Cavendish Laboratory. I obtained my Ph.D. in 1952, became a Research Fellow at Gonville and Caius College where I had been an undergraduate, and in 1961 transferred to Churchill College as Director of Studies in Physics. I was University Lecturer during 1961–69, Reader during 1969–71 and Professor of Radio Astronomy from 1971 until my retirement in 1989. Following Ryle's illness in 1977 I assumed leadership of the Cambridge radio astronomy group and was head of the Mullard Radio Astronomy Observatory from 1982–88.

My decision to begin research in radio astronomy was influenced both by my wartime experience with electronics and antennas and by one of my teachers, Jack Ratcliffe, who had given an excellent course on electromagnetic theory during my final undergraduate year and whom I had also encountered at Malvern. He was head of radiophysics at the Cavendish Laboratory at that time.

My first research was concerned with propagation of radiation through inhomogeneous transparent media and this has remained a lifelong interest. The first two radio "stars" had just been discovered and I realised that their scintillation, or "twinkling", could be used to probe conditions in the ionosphere. I developed the theory of diffraction by phase-modulating screens and set up radio interferometers to exploit my ideas. Thus I was able to make pioneering measurements of the height and physical scale of plasma clouds in the ionosphere and also to estimate wind speeds in this region. Following our Cambridge discovery of interplanetary scintillation in 1964 I developed similar methods to make the first ground-based measurements of the solar wind and these were later adopted in the USA, Japan and India for long term observations. I also showed how interplanetary scintillation could be used to obtain very high angular resolution in radio astronomy, equivalent to an interferometer with a baseline of 1000 km — something which had not then been achieved in this field. It was to exploit this technique on a large sample of radio galaxies that I conceived the idea of a giant phased-array antenna for a major sky survey. This required instrumental capabilities quite different from those of any existing radio telescope, namely very high sensitivity at long wavelengths, and a multi-beam capability for repeated whole-sky surveys on a day to day basis.

I obtained funds to construct the antenna in 1965 and it was completed in 1967. The sky survey to detect all scintillating sources down to the sensitivity threshold began in July. By a stroke of good fortune the observational requirements were precisely those needed to detect pulsars. Jocelyn Bell joined the project as a graduate student in 1965, helping as a member of the construction team and then analysing the paper charts of the sky survey. She was quick to spot the week to week variability of one scintillating source which I thought might be a radio flare star, but our more detailed observations subsequently revealed the pulsed nature of the signal.

Surprisingly, the phased array is still a useful research instrument. It has been doubled in area and considerably improved over the years and one of my present interests is the way our daily observations of scintillation over the whole sky can be used to map large-scale disturbances in the solar wind. At present this is the only means of seeing the shape of interplanetary weather patterns so our observations make an useful addition to *in-situ* measurements from spacecraft such as Ulysses, now (1992) on its way to Jupiter.

Looking back over my forty years in radio astronomy I feel extremely privileged to have been in at the beginning as a member of Martin Ryle's group at the Cavendish. We were a closely-knit team and besides my own research programmes I was also involved in the design and construction of Ryle's first antennas employing the novel principle of aperture synthesis.

Teaching physics at the University, and more general lecturing to wider audiences has been a major concern. I developed an association with the Royal Institution in London when it was directed by Sir Lawrence Bragg, giving one of the well known Christmas Lectures and subsequently several Friday Evening Discourses. I believe scientists have a duty to share the excitement and pleasure of their work with the general public, and I enjoy the challenge of presenting difficult ideas in an understandable way.

I have been happily married since 1950. My son is a physicist and obtained his Ph.D. for neutron scattering in liquids, while my daughter is a language teacher.

Honours and Awards

Hamilton Prize, Cambridge (1952); Eddington Medal, Royal Astronomical Society (1969); Charles Vernon Boys Prize, Institute of Physics (1970); Dellinger Medal, International Union of Radio Science (1972); Michelson Medal, Franklin Institute (1973); Hopkins Prize, Cambridge Philosophical Society (1973); Holwech Medal and Prize, Societé Françaisе de Physique (1974); Nobel Prize in Physics (1974); Hughes Medal, Royal Society (1976).

Honorary ScD.s from the Universities of Leicester (1976), Exeter (1977), Manchester (1989) and Santa Maria, Brazil (1989).

Fellow of the Royal Society (1968), Foreign Honorary Member, American Academy of Arts and Sciences (1977), Foreign Fellow, Indian National Science Academy (1982), Honorary Fellow, Indian Institution of Electronics and Telecommunication Engineers (1985), Associate Member, Belgian Royal Academy (1989).

PULSARS AND HIGH DENSITY PHYSICS

Nobel Lecture, December 12, 1974

by

ANTONY HEWISH
University of Cambridge, Cavendish Laboratory, Cambridge, England

DISCOVERY OF PULSARS

The trail which ultimately led to the first pulsar began in 1948 when I joined Ryle's small research team and became interested in the general problem of the propagation of radiation through irregular transparent media. We are all familiar with the twinkling of visible stars and my task was to understand why radio stars also twinkled. I was fortunate to have been taught by Ratcliffe, who first showed me the power of Fourier techniques in dealing with such diffraction phenomena. By a modest extension of existing theory I was able to show that our radio stars twinkled because of plasma clouds in the ionosphere at heights around 300 km, and I was also able to measure the speed of ionospheric winds in this region (1).

My fascination in using extra-terrestrial radio sources for studying the intervening plasma next brought me to the solar corona. From observations of the angular scattering of radiation passing through the corona, using simple radio interferometers, I was eventually able to trace the solar atmosphere out to one half the radius of the Earth's orbit (2).

In my notebook for 1954 there is a comment that, if radio sources were of small enough angular size, they would illuminate the solar atmosphere with sufficient coherence to produce interference patterns at the Earth which would be detectable as a very rapid fluctuation of intensity. Unfortunately the information then available showed that the few sources known were more than one hundred times too large to produce this effect, and I did not pursue the idea. This was sad because the phenomenon was discovered by chance, about eight years later, by Margaret Clarke long after I had forgotten all about my comment. She was involved with a survey of radio sources at Cambridge and noticed that three particular sources showed variations of intensity. She pointed out that two of the sources were known to have angular sizes of less than 2″ and estimated that a scintillation mechanism required plasma irregularities at distances of thousands of km but she concluded that the fluctuations were an unsolved mystery (3). During a group discussion I suddenly remembered my earlier conclusion and realised that, if the radio sources subtended an angle of less than 1″, they might show the predicted intensity scintillation caused by plasma clouds in the interplanetary medium. With the assistance of Scott and Collins special observations of 3C 48 and other quasi-stellar radio sources were made and the scintillation phenomenon was immediately confirmed (4).

Since interplanetary scintillation, as we called this new effect, could be detected in any direction in space. I used it to study the solar wind, which had by then been discovered by space probes launched into orbits far beyond the magnetosphere. It was interesting to track the interplanetary diffraction patterns as they raced across England at speeds in excess of 300 km s^{-1}, and to sample the behaviour of the solar wind far outside the plane of the ecliptic where spacecraft have yet to venture (5).

The scintillation technique also provided an extremely simple and useful means of showing which radio sources had angular sizes in the range 0″.1—1″.0. The first really unusual source to be uncovered by this method turned up in 1965 when, with my student Okoye, I was studying radio emission from the Crab Nebula. We found a prominent scintillating component within the nebula which was far too small to be explained by conventional synchotron radiation and we suggested that this might be the remains of the original star which had exploded and which still showed activity in the form of flare-type radio emission (6). This source later turned out to be none other than the famous Crab Nebula Pulsar.

In 1965 I drew up plans for a radio telescope with which I intended to carry out a large-scale survey of more than 1000 radio galaxies using interplanetary scintillation to provide high angular resolution. To achieve the required sensitivity it was necessary to cover an area of 18,000 m^2 and, because scintillation due to plasmas is most pronounced at long wavelengths, I used a wavelength of 3.7 m. The final design was an array containing 2048 dipole antennas. Lather that year I was joined by a new graduate student, Jocelyn Bell, and she become responsible for the network of cables connecting the dipoles. The entire system was built with local effort and we relied heavily upon the willing assistance of many members of the Cambridge team.

The radio telescope was complete, and tested, by July 1967 and we immediately commenced a survey of the sky. Our method of utilising scintillation for the quantitative measurement of angular sizes demanded repeated observations so that every source could be studied at many different solar elongations. In fact we surveyed the entire range of accessible sky at intervals of one week. To maintain a continuous assessment of the survey we arranged to plot the positions of scintillating radio sources on a sky-chart, as each record was analysed, and to add points as the observations were repeated at weekly intervals. In this way genuine sources could be distinguished from electrical interference since the latter would be unlikely to recur with the same celestial coordinates. It is greatly to Jocelyn Bell's credit that she was able to keep up with the flow of paper from the four recorders.

One day around the middle of August 1967 Jocelyn showed me a record indicating fluctuating signals that could have been a faint source undergoing scintillation when observed in the antisolar direction. This was unusual since strong scintillation rarely occurs in this direction and we first thought that the signals might be electrical interference. So we continued the routine survey. By the end of September the source had been detected on several occasions, although it was not always present, and I suspected that we had located a flare star, perhaps similar to the M-type dwarfs under investigation by Lovell.

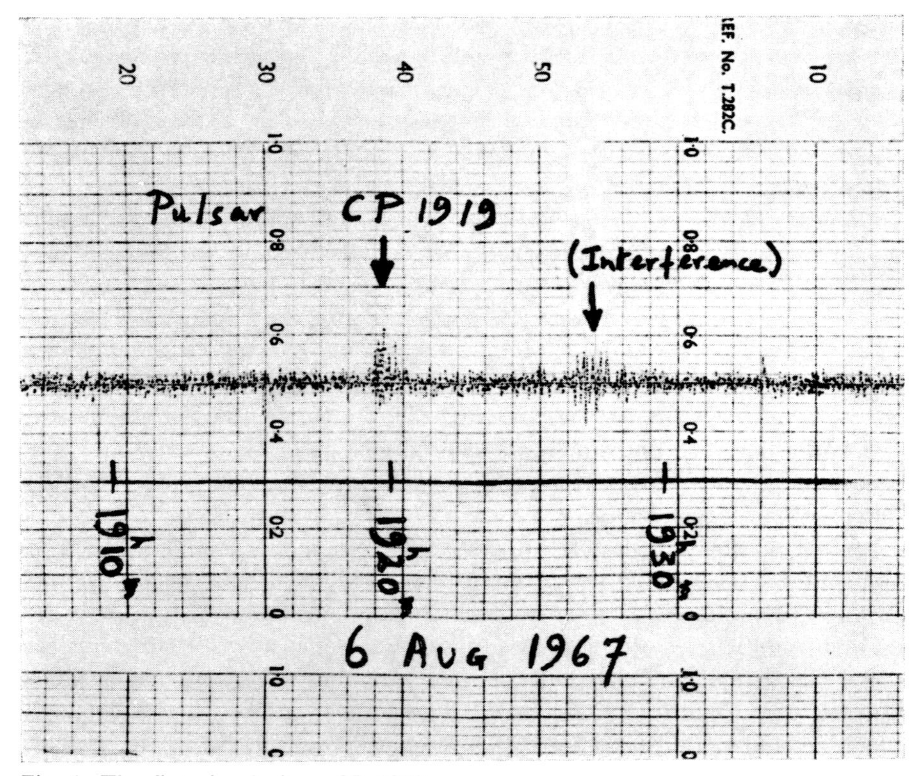

Fig. 1. The first signals from CP 1919.

But the source also exhibited apparent shifts of right ascension of up to 90 seconds which was evidence against a celestial origin. We installed a high-speed recorder to study the nature of the fluctuating signals but met with no success as the source intensity faded below our detection limit. During October

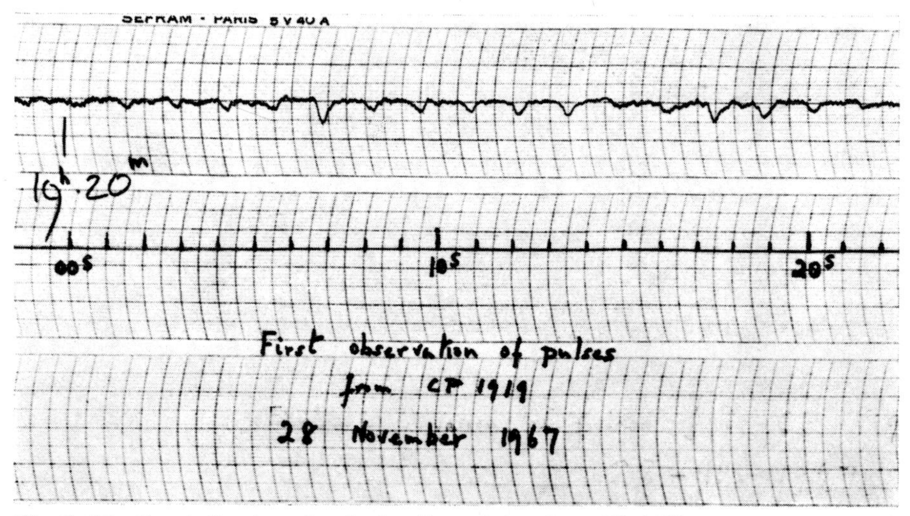

Fig. 2. The first indication of pulsed radio emission from CP 1919.

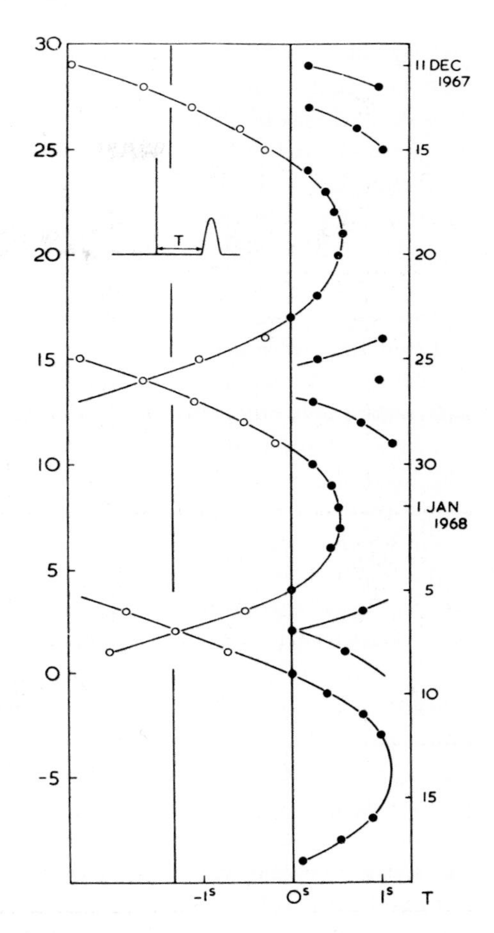

Fig. 3. Timing measurements showing Doppler shift due to the orbital motion of the Earth.

this recorder was required for pre-arranged observations of another source, 3C 273, to check certain aspects of scintillation theory, and it was not until November 28th that we obtained the first evidence that our mysterious source was emitting regular pulses of radiation at intervals of just greater than one second. I could not believe that any natural source would radiate in this fashion and I immediately consulted astronomical colleagues at other observatories to enquire whether they had any equipment in operation which might possibly generate electrical interference at a sidereal time near $19^h\ 19^m$.

In early December the source increased in intensity and the pulses were clearly visible above the noise. Knowing that the signals were pulsed enabled me to ascertain their electrical phase and I reanalysed the routine survey records. This showed that the right ascension was constant. The apparent variations had been caused by the changing intensity of the source. Still sceptical, I arranged a device to display accurate time marks at one second intervals broadcast from the MSF Rugby time service and on December 11th began daily timing measurements. To my astonishment the readings fell in a regular pattern, to within the observational uncertainty of 0.1s, showing that the pulsed source kept time to better than 1 part in 10^6. Meanwhile my colleagues Pilkington, and Scott and Collins, found by quite independent methods that the signal exhibited a rapidly sweeping frequency of about -5 MHz s^{-1}. This showed that the duration of each pulse, at one particular radio frequency, was approximately 16 ms.

Having found no satisfactory terrestrial explanation for the pulses we now began to believe that they could only be generated by some source far beyond the solar system, and the short duration of each pulse suggested that the radiator could not be larger than a small planet. We had to face the possibility that the signals were, indeed, generated on a planet circling some distant star, and that they were artificial. I knew that timing measurements, if continued for a few weeks, would reveal any orbital motion of the source as a Doppler shift, and I felt compelled to maintain a curtain of silence until this result was known with some certainty. Without doubt, those weeks in December 1967 were the most exciting in my life.

It turned out that the Doppler shift was precisely that due to the motion of the Earth alone, and we began to seek explanations involving dwarf stars, or the hypothetical neutron stars. My friends in the library at the optical observatory were surprised to see a radio astronomer taking so keen an interest in books on stellar evolution. I finally decided that the gravitational oscillation of an entire star provided a possible mechanism for explaining the periodic emission of radio pulses, and that the fundamental frequency obtainable from white dwarf stars was too low. I suggested that a higher order mode was needed in the case of a white dwarf, or that a neutron star of the lowest allowed density, vibrating in the fundamental mode, might give the required periodicity. We also estimated the distance of the source on the assumption that the frequency sweep was caused by pulse dispersion in the interstellar plasma, and obtained a value of 65 parsec, a typical stellar distance.

While I was preparing a coherent account of this rather hectic research, in January 1968, Jocelyn Bell was scrutinising all our sky-survey recordings with her typical persistence and diligence and she produced a list of possible additional pulsar positions. These were observed again for evidence of pulsed radiation and before submitting our paper for publication, on February 8th, we were confident that three additional pulsars existed although their parameters were then only crudely known. I well remember the morning when Jocelyn came into my room with a recording of a possible pulsar that she had made during the previous night at a right ascension 09^{h} 50^{m}. When we spread the chart over the floor and placed a metre rule against it a periodicity of 0.25s was just discernible. This was confirmed later when the receiver was adjusted to a narrower bandwidth, and the rapidity of this pulsar made explanations involving white dwarf stars increasingly difficult.

The months that followed the announcement (7) of our discovery were busy ones for observers and theoreticians alike, as radio telescopes all over the world turned towards the first pulsars and information flooded in at a phenomenal rate. It was Gold (8) who first suggested that the rotation of neutron stars provided the simplest and most flexible mechanism to explain the pulsar clock, and his prediction that the pulse period should increase with

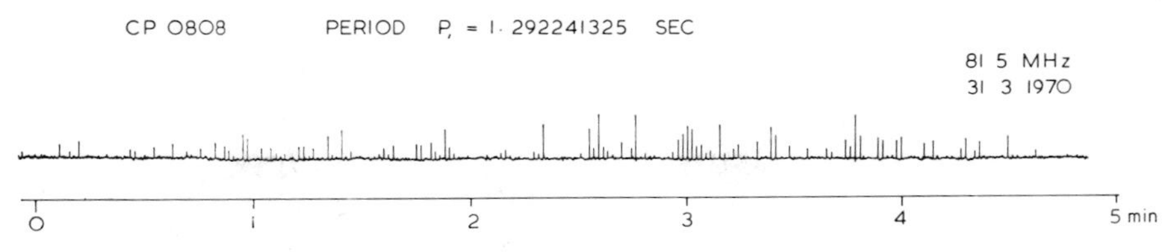

Fig. 4. Radiation from a typical pulsar.

time soon received dramatic confirmation with the discovery of the pulsar in the Crab Nebula (9, 10). Further impressive support for the neutron star hypothesis was the detection of pulsed light from the star which had previously been identified as the remnant of the original explosion. This, according to theories of stellar evolution, is precisely where a young neutron star should be created. Gold also showed that the loss of rotational energy, calculated from the increase of period for a neutron star model, was exactly that needed to power the observed synchrotron light from the nebula.

Now, in 1974, with more than 130 pulsars charted in the heavens, there is overwhelming evidence that the neutron star "lighthouse" model is correct. No other star could spin fast enough, without fragmenting, to account for the most rapid pulsars yet periods ranging from 33 ms to 3.5 s are readily accommodated by the rotation theory. At the same time there is unfortunately no satisfactory theory to account for the radio emission generated by these tiny stars which have radii of only 10 km.

High Density Physics inside Neutron Stars

The prediction that matter at the almost unimaginable density of 10^{18} kg m^{-3} might be formed under gravitational compression inside stars was first made by Baade and Zwicky (11) in 1934, soon after Chadwick's discovery of the neutron. At this density only a small fraction of the original protons and electrons could exist and matter would consist predominantly of neutrons. It is the denegeracy pressure arising from the neutrons, which obey Fermi statistics, that balances further gravitational compression, alhough finally the Fermi energy becomes relativistic and further gravitational collapse ensues. Since complex nuclei are generated by nuclear fusion inside hot stars, where there is a large thermal pressure, the degenerate neutron state can only be found when fusion ceases and we deal with the cooling "ashes" of stellar evolution. The stars that give rise to neutron stars are more massive than the Sun, and it is believed that the formation of neutron stars is associated with supernova explosions.

Since the discovery of pulsars there has been great activity amongst solid-state physicists around the world because neutron matter, at any temperature less than about 10^9 K, behaves rather like ordinary matter close to the absolute zero of temperature. The generally agreed model of a neutron star consists of concentric shells with very different physical properties as reviewed by Ruderman (12).

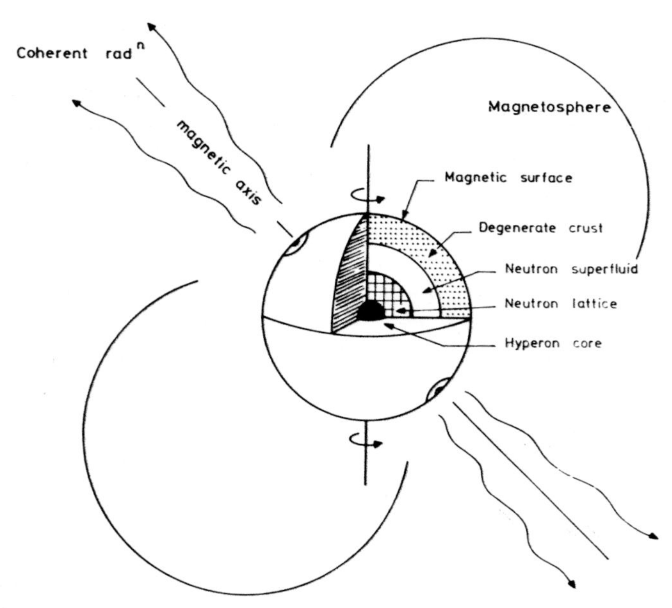

Fig. 5. Model of a neutron star.

At the surface of the star it is likely that there exists a shell of iron since ^{56}Fe is the most stable nucleus. The atoms would be normal if no magnetic field were present. In astrophysics it is unwise to ignore magnetic phenomena and gravitational collapse following a supernova explosion probably compresses the original stellar magnetic flux to produce surface field strengths of 10^8 T or more. In fields of this magnitude the radius of gyration of electrons in atomic energy levels becomes smaller than the Bohr radius and the electronic wave functions adopt a cylindrical shape. It is far harder to ionize distorted atoms of this kind and this is of importance when considering the generation of a magnetosphere surrounding the neutron star.

Beneath the iron skin the increasing compression forces electrons into higher energy states until they are entirely freed from the positive nuclei. The unscreened nuclei then settle into a rigid lattice having a melting temperature of about 10^9 K. At greater depths the electron energies become relativistic and they begin to combine with protons in the nuclei, thus adding to the neutron population. This is the process of inverse β decay. At a sufficient depth nearly all the electrons and protons have disappeared and the nuclei have been converted to a sea of neutrons.

The energy gap for neutron pairing is of the order of several MeV, corresponding to a superfluid transition temperature of 10^9—10^{10} K, and since young neutron stars cool rapidly to temperatures below 10^9 K, the neutron sea is expected to behave like a quantum superfluid. The few remaining protons will similarly pair and enter a superconducting state, while the residual electrons will behave normally. The bulk motion of the neutron superfluid must be irrotational, but an effective solid body rotation can be simulated with a distribution of quantised vortex lines containing a small fraction of normal fluid neutrons.

At yet deeper levels the neutron-neutron interaction may result in the creation of a solid neutron lattice, although this possibility is under debate, and finally there is the question of a material composed of stable hyperons.

Evidence that neutron stars do indeed have a structure similar to the predicted models has been obtained from extended timing observations of pulsars (13). These show that the systematic increase of period, corresponding to a steady loss of rotational energy from the spinning star, is sometimes interrupted by discontinuous changes. Most pulsars are observed to be slowing down on a typical timescale of 10^6—10^7 years, althouh the most rapid pulsars, in the Crab and Vela supernovae, have timescales of only 10^3 and 10^4 years respectively. The discontinuities often show an abrupt decrease of period, followed by a recovery to a slightly reduced value with a characteristic relaxation time.

For the Crab pulsar this effect can be explained by a rigid crust-liquid core model. Young neutron stars are likely to be spinning rapidly at birth, with angular velocities up to 10^4 radian s^{-1}, and they will therefore have a spheroidal shape. As a star slows down it will tend to become less spheroidal and the rigid crust will fracture at irregular intervals as the increasing strain overcomes rigidity. When this occurs the crust will momentarily spin more rapidly, but later the increased angular momentum will be coupled into the fluid interior, where the bulk of the mass resides. The observed time constant for coupling is in good agreement with the superfluid model, and would be far smaller in the case of a normal fluid interior. It is remarkable that a crust shrinkage of only 10 μm is sufficient to explain the period anomalies for the Crab pulsar. When similar reasoning is applied to the Vela pulsar, for which the anomalies are larger, it is found necessary to invoke a solid neutron lattice core in which strains imposed when the star was young are intermittently relaxed.

Plasma Physics outside neutron stars

It is strange that there appears to be more understanding of the interior of neutron stars, than of their atmospheres wherein is generated the radiation which makes them detectable. Ginzburg and Zheleznyakov (14) have summarised the electrodynamic problems in detail. The model upon which theorists are concentrating most attention is that of an oblique magnetic rotator, in which the pulsar may be regarded as a dynamo, powered by the initial store of rotational kinetic energy, and converting this into radiation together with a flux of relativistic particles by means of the large magnetic field. The oblique rotator model was first considered by Pacini (15) before pulsars had been found, and it was Gold (8) who suggested that an extended corotating magnetosphere played a vital role.

Goldreich and Julian (16) showed that electrical forces arising from unipolar induction would be sufficient to drag charges from the stellar surface

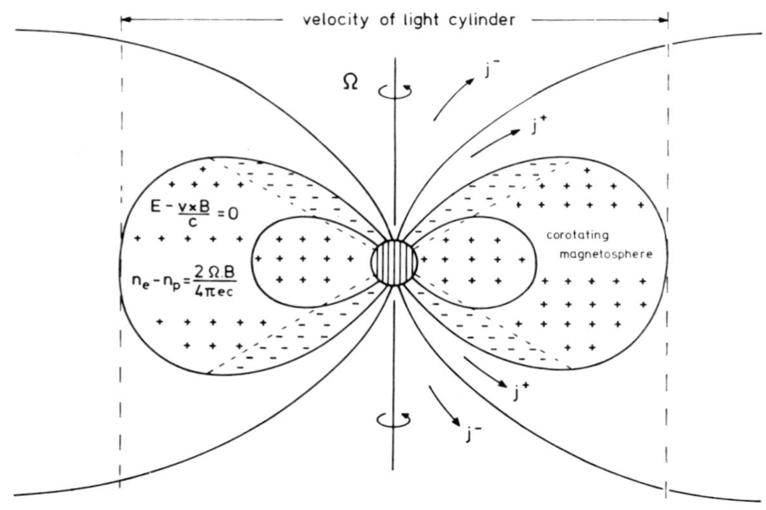

Fig. 6. The neutron star magnetosphere for an aligned magnetic rotator.

and then distribute them in a corotating magnetosphere. It is not yet known whether such a distribution is stable, and the plasma differs from laboratory plasmas in that almost complete charge separation occurs. Inertial forces must dominate when the corotation velocity approaches c, and beyond the velocity of light cylinder the plasma breaks away to create a stellar wind. In such models the polar regions are believed to play a crucial role since particles can escape along 'open' field lines.

Within such an overall framework exists the ordered motion of the charges which generate the beamed radio waves that we observe, and also those regions which emit light and X-rays for the youngest pulsar in the Crab. The fascinating richness of the phenomena involving polarisation, pulse shapes, radio spectra, intensity variations, and complex secondary periodicities, must eventually provide vital evidence to resolve our present uncertainties. There is good reason to believe that the general outline is correct. Simple dynamics shows that the surface magnetic field strength B_0 is proportional to $\left(P\frac{dP}{dt}\right)^{1/2}$ were P is the pulsar period, and observations of many pulsars give $B_0 \sim 10^8$ T when conventional neutron star models are assumed. Further evidence comes from pulsar ages which are aproximately $P\left(\frac{dP}{dt}\right)^{-1}$.

Typical ages are 10^6—10^7 years although 10^3 years is obtained for the Crab pulsar, in good agreement with the known age of the supernova.

Conclusion

In outlining the physics of neutron stars, and my good fortune in stumbling upon them, I hope I have given some idea of the interest and rewards of extending physics beyond the confines of laboratories. These are good times in which to be an astrophysicist. I am also deeply aware of my debt to all my colleagues in the Cavendish Laboratory. Firstly to Sir Martin Ryle for his unique flair in creating so congenial and stimulating a team in which to work. Secondly to Jocelyn Bell for the care, diligence and persistence that led to our discovery so early in the scintillation programme, and finally to my friends who contributed so generously in many aspects of the work.

References

1. Hewish, A., Proc. Roy. Soc. (London), *214,* 494, (1952).
2. Hewish, A. and Wyndham, J. D., Mon. Not. R. astr. Soc., *126,* 469, (1963).
3. Clarke, M. E., Ph.D. Thesis. Cambridge (1964).
4. Hewish, A., Scott, P. F. and Wills, D., Nature, *203,* 1214, (1964).
5. Dennison, P. A. and Hewish, A., Nature, *213,* 343, (1967).
6. Hewish, A. and Okoye, S. E., Nature, *207,* 59, (1965).
7. Hewish, A., Bell, S. J., Pilkington, J. D. H., Scott, P. F. and Collins, R. A., Nature *217,* 709, (1968).
8. Gold, T., Nature, *218,* 731, (1968).
9. Staelin, D. H. and Reifenstein, E. C., Science, *162,* 1481, (1968).
10. Comella, J. M., Craft, H. D., Lovelace, R. V. E., Sutton, J. M. and Tyler, G. L., Nature, *221,* 453, (1969).
11. Baade, W., and Zwicky, F., Proc. Nat. Acad. Sci. *20,* 255, (1934).
12. Ruderman, M., Ann. Rev. Astron. Astrophys., *10,* 427, (1972).
13. Pines, D., Shaham, J., and Ruderman, M., I.Au.c Symposium No 53, (1973).
14. Ginzburg, V. L. and Zheleznyakov, V. V., Ann. Rev. Astron. Astrophys., *13.* (1975) in press.
15. Pacini, F., Nature, *219,* 145, (1968).
16. Goldreich, P. and Julian, W. H., Astrophys. J., *157,* 869, (1969).

Martin Ryle

MARTIN RYLE

I was born on September 27, 1918, the second of five children. My father John A. Ryle was a doctor who, after the war, was appointed to the first Chair of Social Medicine at Oxford University.

I was educated at Bradfield College and Oxford, where I graduated in 1939. During the war years I worked on the development of radar and other radio systems for the R.A.F. and, though gaining much in engineering experience and in understanding people, rapidly forgot most of the physics I had learned.

In 1945 J. A. Ratcliffe, who had been leading the ionospheric work in the Cavendish Laboratory, Cambridge before the war, suggested that I apply for a fellowship to join his group to start an investigation of the radio emission from the Sun, which had recently been discovered accidentally with radar equipment.

During these early months, and for many years afterwards both Ratcliffe and Sir Lawrence Bragg, then Cavendish Professor, gave enormous support and encouragement to me. Bragg's own work on X-ray crystallography involved techniques very similar to those we were developing for "aperture synthesis", and he always showed a delighted interest in the way our work progressed.

In 1948 I was appointed to a Lectureship in Physics and in 1949 elected to a Fellowship at Trinity College. At this time Tony Hewish joined me, and in fact four other members of our present team started their research during the period 1948—52.

In 1959 the University recognized our work by appointing me to a new Chair of Radio Astronomy.

During 1964—7 I was president of Commission 40 of the International Astronomical Union, and in 1972 was appointed Astronomer Royal.

In 1947 I married Rowena Palmer, and we have two daughters, Alison and Claire, and a son, John. We enjoy sailing small boats, two of which I have designed and built myself.

Awards

1952 Fellow of Royal Society of London.
1954 Hughes Medal, Royal Society of London.
1955 Halley Lecturer, University of Oxford.
1958 Bakerian Lecturer, Royal Society of London.

1963 Van der Pol Medal, U.R.S.I.

1964 Gold Medal, Royal Astronomical Society, London.

1965 Henry Draper Medal, U.S. National Academy of Sciences; Holweck Prize, Société Française de Physique.

1968 Elected Foreign Member of the Royal Danish Academy of Sciences and Letters.

1970 Elected Foreign Honorary Member of American Academy of Arts and Sciences.

1971 Elected Foreign Member of U.S.S.R. Academy of Sciences.

Morris N. Liebmann Award; Institution of Electrical & Electronic Engineers.

Faraday Medal, Institution of Electrical Engineers.

Popov Medal, USSR Academy of Sciences.

Michelson Medal, Franklin Institute, U.S.A.

1973 Royal Medal, Royal Society of London.

1974 Bruce Medal, Astronomical Society of the Pacific.

Foreign Member, Deutsche Akademie der Naturforscher, Leopoldina.

Honorary D.Sc. of the Universities of Strathclyde (1968), Oxford (1969) and Nicholas Copernicus University of Torun (1973).

Sir Martin Ryle died in 1984.

RADIO TELESCOPES OF LARGE RESOLVING POWER

Nobel Lecture, December 12, 1974

by Martin Ryle
University of Cambridge, Cavendish Laboratory, Cambridge, England

I think that the event which, more than anything else, led me to the search for ways of making more powerful radio telescopes, was the recognition, in 1952, that the intense source in the constellation of Cygnus was a distant galaxy—1000 million light years away. This discovery showed that some galaxies were capable of producing radio emission about a million times more intense than that from our own Galaxy or the Andromeda nebula, and the mechanisms responsible were quite unknown. It seemed quite likely that some of the weaker sources already detected with the small radio telescopes then available might be similar in character; if so they would be at distances comparable with the limits of observation of the largest optical telescopes. It was therefore possible that more powerful radio telescopes might eventually provide the best way of distinguishing between different cosmological models. It was not until 1958 (1) that it could be shown with some certainty that most of the sources were indeed powerful extragalactic objects, but the possibilities were so exciting even in 1952 that my colleagues and I set about the task of designing instruments capable of extending the observations to weaker and weaker sources, and of exploring their internal structure.

The early observations were severely limited both by the poor angular resolution and by the limited sensitivity. It was usually impossible to obtain any information about the structure of a source, and adjacent sources could often not be properly separated, whilst attempts to relate the radio sources to optically visible objects were often prevented by the poor positional accuracy. The use of interferometers allowed better positions to be obtained, and sometimes made it possible to derive simple models for the source structure. Few of the sources were found to have an angular size greater than 2—3 minutes of arc.

The problem of making detailed maps of such sources arises simply from the fact that the wavelengths used are some million times greater than optical wavelengths—so that even to obtain a radio picture with the same resolution as that of the unaided human eye ($\sim 1'$ arc) we would need a telescope having a diameter of about 1 km operating at a wavelength of 50 cm. At the same time the instrument will be effective only if the surface accuracy is good enough to make a proper image, corresponding to errors of $\leqslant \lambda/20$ or a few cm; the engineering problems of building such an instrument are clearly enormous.

With the development, around 1960, of masers and parametric amplifiers capable of providing receiving systems of good sensitivity at wavelengths of a few cm, it became possible to build telescopes of diameter 10—100 m with

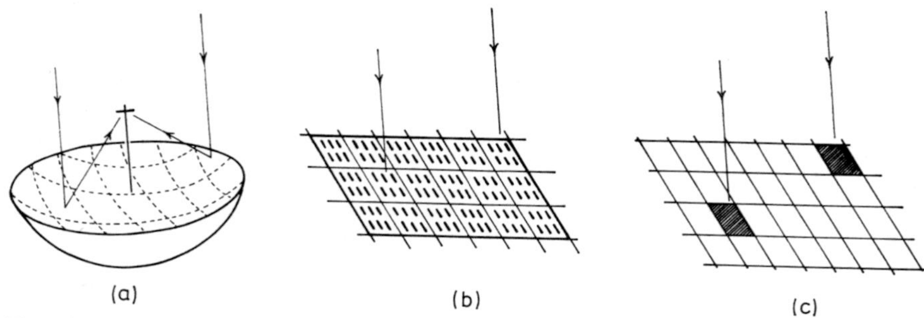

Fig. 1. The use of (a) a paraboloid, (b) an array of dipoles or (c) the sequential sampling of the wavefront by small aerial elements to achieve a high resolving power by combining the signals from a large part of the incident wavefront.

sufficient sensitivity and with angular resolutions of $\sim 1'$ arc; even with such instruments the engineering problems of constructing a rigid enough surface are considerable, and it is likely to be difficult to build a conventional paraboloid capable of angular resolutions much better than $1'$ arc.

I would like now to describe an entirely different approach to the problem in which small aerial elements are moved to occupy successively the whole of a much larger aperture plane. The development and use of "aperture synthesis" systems has occupied much of our team in Cambridge over the past 20 years.

The principle of the method is extremely simple. In all methods used to obtain a large resolving power, that is to distinguish the wavefront from a particular direction and ignore those from adjacent directions, we arrange to combine the field measured over as large an area as possible of the wavefront. In a paraboloid we do this by providing a suitably shaped reflecting surface, so that the fields incident on different parts of the sampled wavefront are combined at the focus (Fig. 1(a)); the voltage produced in the receiving dipole represents the sum of these fields. We can achieve the same result if we use an array of dipoles connected together through equal lengths of cable (Fig. 1(b)).

Suppose now that only a small part of the wavefront is sampled, but that different parts are sampled in turn (Fig. 1(c)); could we combine these signals to produce the same effect? Since in general, we do not know the phase of the incident field at different times this would not normally be possible but if we continue to measure *one* of the samples while we measure the others we can use the signal from this one as a phase reference to correct the values measured in other parts of the wavefront. In this way, by using two small aerial elements, we can again add the fields over the wavefront—the area of which is now determined by the range of relative positions taken by the two aerial elements.

It might be thought that this method would be extremely slow, for if we are to sample an area of side D using elements of side d, it is necessary to observe with $\frac{2D^2}{d^2}$ different relative positions of the two aerial elements. In practice, however, the method is not significantly slower than the use of the large equivalent instrument for although a large number of observations must be made, the results may be combined in a computer using additional phase differences, which correspond to many different wave directions (as in a phased array or dipoles), so that with the one set of observations an *area* of sky may be mapped which is limited only by the diffraction pattern of the small elements themselves; there are in fact some $\frac{D^2}{d^2}$ different directions which can be scanned in this way, and which would have had to be explored sequentially by a conventional instrument, so that the total observing time of the two methods is nearly the same.

It can also be seen that the sensitivity of the system is much better than would be associated with the small elements, for the signal from a particular point in the sky is contributing to that point on the map for the *whole* observing period; the resulting signal-to-noise is in fact equivalent to the use of an instrument having a collecting area $2d^2 \sqrt{\frac{2D^2}{d^2}}$ $\sim 3Dd$, a figure which may be much greater than that of the elements themselves, and although it is not as great as if the full instrument of area D^2 had been built, it may exceed that of any instrument which *can* be built.

Unlike a paraboloid or array, in which both the sensitivity and resolving power are fixed as soon as the wavelength is decided, the value of d may be chosen so that the sensitivity, for any particular wavelength and type of observation, is matched to the resolution.

The method of aperture synthesis avoids the severe structural problems of building very large and accurate paraboloids or arrays, and allows both high resolving power and large effective collecting area to be obtained with a minimum of engineering structure and therefore cost. Provision must be made for the relative movement of the small elements, and their relative positions and electrical connecting paths must be known with an accuracy equal to the surface accuracy of the equivalent instrument ($\leqslant \lambda/20$). Automatic computing is needed to carry out the Fourier inversion involved in combining the observations to provide a map of the sky.

Historically, the forerunners fo this type of instrument were realized in the early days when observations in both Australia and England with aerial elements having a range of separations were used to determine the distribution of radio brightness across the solar disc. In the earliest observations the Sun was assumed to show spherical symmetry, and no measurements of phase were necessary so that a precise knowledge of the relative positions of the elements, and of the electrical path lengths to the receiver were unnecessary. A similar technique was used to establish the profile of radio brightness across the plane of the Galaxy (2).

The first synthesis instrument capable of mapping an abritrary distribution of sources was built at Cambridge in 1954 by John Blythe (3); it consisted of a long thin element covering, in effect, a whole row of Fig. 1(c) used in con-

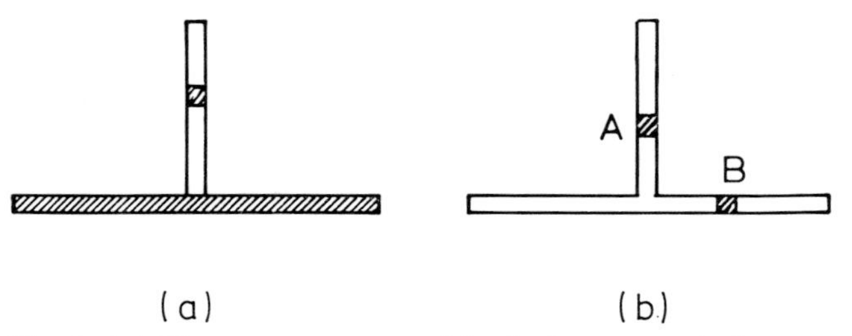

Fig. 2. (a) The arrangement used in the instrument built in 1954 by J. H. Bythe. (b) The equivalent instrument using two small elements.

junction with a smaller element moved to 38 different positions along a perpendicular line (Fig. 2(a)) to synthesise a square instrument giving a resolution of 2°.2. This instrument provided the first detailed maps of the galactic emission at a long radio wavelength (7.9 m).

Larger instruments using this same configuration were built at Cambridge during the succeeding years, including an instrument of high sensitivity and 45′ arc resolution also at $\lambda = 7.9$ m (4) and a second operating at $\lambda = 1.7$ m with 25′ arc resolution which was used by Paul Scott and others to locate nearly 5000 sources in the northern sky (5, 6).

These instruments used a very cheap form of construction; for $\lambda > 1$ m an efficient reflecting surface may be provided by thin (~ 1 mm diameter) wires 5—10 cm apart. In the case of the $\lambda = 1.7$ m instrument, wires stretched across simple parabolic frames of welded steel tube provided a cylindrical paraboloid 450 m long and 20 m wide (Fig. 3) at a cost of about £2 per m^2.

Fig. 3. Photograph of the east—west arm of the $\lambda = 1.7$ m instrument built in 1957 with which nearly 5000 sources were located.

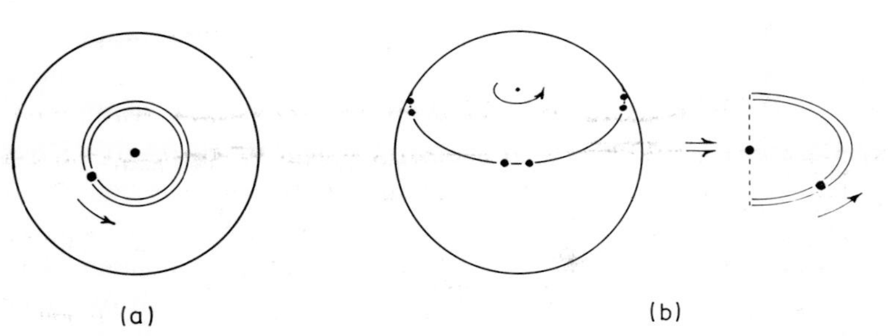

Fig. 4. (a) Two aerial elements mounted near the North Pole observing throughout the day are equivalent to one ring of a much larger instrument.
(b) The elements may be used at other latitudes if arranged on an east—west line and used to track the chosen point for 12^h.

With the need for still greater resolving power, we realized that physically larger systems operating at metre wavelengths would no longer prove successful, because of the limitation imposed by irregularities of electron density in the ionosphere. But at shorter wavelengths where these are unimportant it becomes difficult to make efficient reflectors by using stretched wires, both because of their deflection by the wind, and because with the closer spacing needed there is difficulty with them twisting together. For operating wavelengths of < 50 cm a much more rigid supporting structure must be used, and the engineering costs of building a long element become very great.

The obvious solution is to use the system illustrated in Fig. 1(c), in which the engineering structure is confined to two small elements—where much higher costs per m^2 are acceptable. The method for altering the relative positions of the two elements presents some practical problems; suppose that the elements are mounted on two railway tracks at right angles (Fig. 2(b)), so that for each position of A on the N—S track B is moved to every position along the E—W track. For values of $\frac{D}{d} \sim 50$, there are then 5000 different arrangements, and if B is moved each day, the observations will take 5000 days and although a map will then be available for the whole strip of sky, the period is too long for a graduate student's thesis!

Alternatively B could be moved rapidly—so that several positions could be fitted into the time during which the area of sky remains in the beam of the small elements. This will reduce the total time of the observations, at the expense of observing only parts of the strip of sky. We can clearly extend this period, and so allow more relative positions of A and B each day, if we arrange for the elements to track the chosen point in the sky for an extended period.

As soon as we do this, we realize that the rotation of the earth is itself providing us with a relative motion of A and B as seen from the source, without our having to move them on the surface of the earth at all. Suppose, for example, we have our two elements mounted near the North Pole and we use them to observe an area of sky centred on the Celestial Pole; in this case we do not even have to arrange for them to track. Over a 24^h period, one will have traced out a circular path about the other (Fig. 4(a)), and the signals recorded during this time can be combined to provide the same response as that of the equivalent ring aerial; by simply altering the separation along a line on say 50 successive days a complete aperture can then be synthesized. Miss Ann Neville and I set up an experimental system in 1960—61 to test the method and develop the computing; we used it to map a region 8° in diameter round the North Celestial Pole at a wavelength of 1.7 m (7). We connected up different 14 m sections of the long cylindrical paraboloid (Fig. 3) with some other small aerials to simulate the use of two 14 m diameter elements at different spacings. The effective diameter of the synthesized instrument was 1 km and it provided an angular resolution of 4'.5 arc.

As well as showing that the method really worked, it provided some interesting astronomical results—in particular by allowing the detection of sources some 8 times weaker than had been observed before; even though the area of sky covered was only some 50 square degrees the results were useful in our cosmological investigations.

In practice only 12^h observations are needed because of the symmetry of the system and observations need not be made *from* the North Pole or limited *to* the Celestial Pole, provided that the elements are situated on an East—West axis, and each is able to track the required region of sky for 12^h (Fig. 4(b)). At low declinations the synthesized instrument becomes elliptical with the north-south aperture reduced by sin δ. The engineering simplicity of moving the elements along a line, and the consequent great saving in the area of land needed are, however, such great advantages that we eventually built three large instruments in Cambridge with equivalent instrumental diameters of 0.8, 1.6 and nearly 5 km.

These instruments are known as the Half-Mile, the One-Mile and (because its construction coincided with the early negotiations for the entry of Britain into the European Community), the 5 km Telescopes! The One-Mile telescope was the first to be built, and this started observations in 1964.

It is interesting that as early as 1954 we had discussed the possibility of building a high resolution instrument on exactly these principles, and I have recently found two entries in an old note-book:—

"*8.6.1954* *Possible research student and other projects.*
... 3(f). North Polar Survey on 81.5 Mc/s. Effective gain area $\sim 25 \times 1500 = 37{,}500$ sq. ft. Effective resolving power area $\simeq 10^6$ sq. ft."
(The entry included a diagram of the proposed aerial element)

"*29.6.1954*
Do 3(f) in all directions where 180° rotation available? above about 20° might be possible by directing aerials in successively different directions—i.e. observation not on meridian."

A third entry on 22.7.1954 discusses the east—west rail track to be used for the latter programme with two 30 ft aerials mounted on it, the arrangement of cabling needed to compensate for the different path lengths to the two aerials when observing off the meridian, and the selection of directions of observation "to give uniform cover of Fourier terms".

Why then, with its obvious simplicity and economy, did we not build this instrument in 1954? The answer is that at this time there were no computers with sufficient speed and storage capacity to do the Fourier inversion of the data. EDSAC I, which was the first stored-programme computer, was built by Dr. M. V. Wilkes of the Cambridge University Mathematical Laboratory, and came into operation in 1949. It was used for reducing John Blythe's observations and took some 15^h of computing to do the 38-point transform for every 4^m of the 24^h scan of the sky. It would not have been practicable to use it for the 2-dimensional inversion needed for the earth-rotation synthesis. By 1958 the completion of the much faster EDSAC II, and the development by Dr. David Wheeler of the Mathematical Laboratory of the fast fourier transform (incidentally some six years before these methods came into general use) made possible the efficient reduction of the 7.9 m and 1.7 m surveys, and also enabled the trials of the 1.7 m earth-rotation synthesis to be made in 1961; even with EDSAC II, however, the reduction for the small area of sky covered in the latter survey took the whole night.

During the early stages in the design of the One-Mile telescope in 1961, I discussed with Maurice Wilkes the considerably greater problems of reducing the data from this instrument, but by then the replacement of EDSAC II was planned and the new TITAN computer, which came into operation in 1963, was easily capable of dealing with the output of the One-Mile telescope. The development of aperture synthesis has therefore been very closely linked to the development of more and more powerful computers, and it is interesting to speculate how our work in Cambridge would have proceeded if, for example, computer development had been five years behind its actual course.

The two programmes in my 1954 note-book subsequently formed the basis of two Ph.D. theses in 1964 and 1965.

Now I return to the design of the large instruments whose layout is shown in Fig. 5. The One-Mile telescope consists of three 18 m dishes, two fixed at 0.8 km spacing, the third mounted on a 0.8 km rail-track (Fig. 6); this ar-

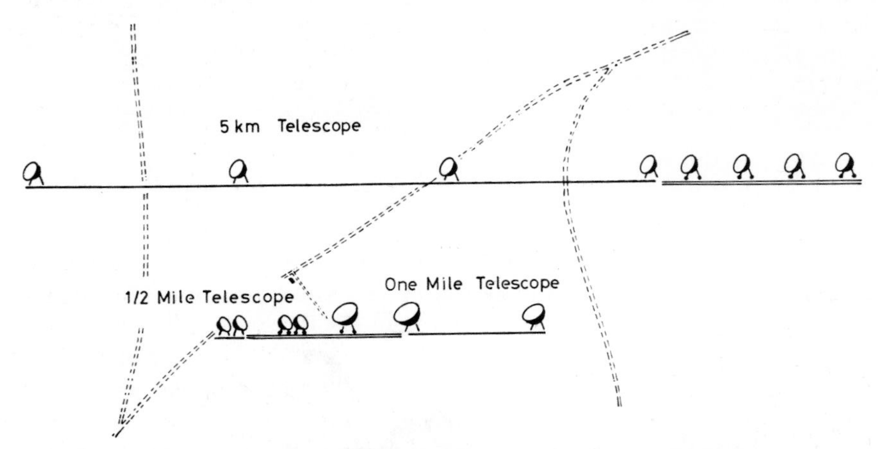

Fig. 5. Sketch map showing the arrangement of One-Mile, Half-Mile and 5 km telescopes.

Fig. 6. The One-Mile telescope, showing the west, railmounted, dish in the foreground, with the two fixed dishes behind.

rangement was cheaper than building the longer rail track and it also provided two spacings at a time. It was designed for two main programmes: (a) The detection of much fainter and therefore more distant sources (see Fig. 7) in order to explore the early history of the Universe, and so try and distinguish between different cosmological models, and (b) To make radio maps of individual sources in an attempt to understand the physical mechanisms within them; most of the sources studied have been powerful extragalactic objects, but the remnants of supernova explosions are perhaps physically as important.

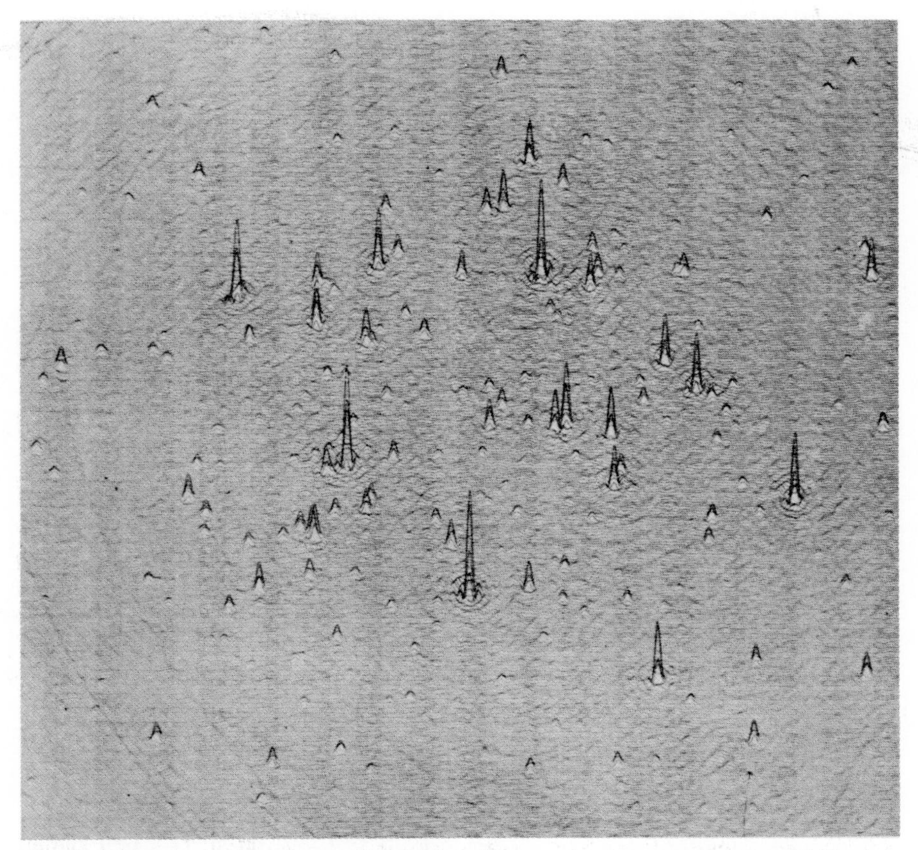

Fig. 7. Map obtained with the One-Mile telescope showing sources about 100 times fainter than had been observed before.

The problem of the physics of radio galaxies and quasars and the cosmological problem are strangely linked; we appear to be living in an evolving Universe, so that very distant sources which, due to the signal travel time, we observe as they were when the Universe was younger, may be systematically different from a sample of nearby sources. But the intrinsically most powerful sources are so rare that there *are* no nearby ones, whilst the weak sources cannot be detected at great distances. If we are to understand *how* the Universe is evolving, we may first have to solve the physical problem of the individual source—so that we can infer the differences in its evolution at earlier cosmological epochs.

The Half-Mile telescope was built later by John Shakeshaft and John Baldwin. It was actually built very cheaply because as can be seen from Fig. 5, it made use of the same rail track, and we were able to get the four 9 m dishes at scrap-metal prices from a discontinued radio link service, and only the mounts had to be built. It has been used mainly with a radio spectrometer covering the 21 cm wavelength band of neutral hydrogen to map the distribution of density and velocity of the hydrogen in a number of nearby galaxies,

Fig. 8. The 5 km telescope, with the movable dishes in the foreground.

and forms part of a programme concerning the formation and evolution of galaxies.

The 5 km telescope was completed in 1971, and because it represents a rather more advanced design I will describe it in more detail. It was designed solely for the purpose of mapping individual sources, and besides its larger overall size, the individual dishes are more accurate to allow operation at wavelengths as short as 2 cm. As a result the angular resolution is $\sim 1''$ arc, a figure comparable with the resolution attained by large optical telescopes on good mountain sites. It is at present being used on a wavelength of 6 cm, where the resolution is $2''$ arc.

In order to improve the speed of observation, four fixed and four movable elements mounted on a rail-track are used, as shown in Fig. 5; this arrangement provides 16 spacings simultaneously, and a single 12^h observation produces a $2''$ arc main response with circular grating responses separated by $42''$ arc. Sources of smaller extent than $42''$ arc can therefore be mapped with a single 12^h observation; more extensive fields of view require further observations with intermediate positions of the movable elements to suppress the grating responses.

For operation at these short wavelengths the positioning of the elements, and the electrical cable connections, must be stable and measured with an accuracy better than 1 mm. Conventional surveying methods allowed each element to be located to $\pm$ 10 mm, and the final alignment had to be based entirely on radio observations; the distance between the two outer fixed elements (on which the scale of declination is based) was found in this way to be 3430828.7 ± 0.25 mm, and no changes outside this error have been found over a 2-year period. The combination of azimuth and longitude, on which the measurement of right-ascension depends, was established by observing the bright fundamental star Algol, which is a weak and variable radio source.

The telescope is controlled by an on-line computer which continually updates the position of the selected map-centre for precession, aberration etc., and uses this to compute the path differences (corrected for atmospheric refraction) to each pair of elements; these values are then used to control electrical delays in the signals from each element before they are combined in the receivers. The outputs of the receivers are sampled by the computer and stored on a magnetic disc, so that at the completion of the observation they may be combined to form a map of the area observed. The map is then drawn on a curve-plotter controlled by the computer.

This instrument has been used in a wide range of astronomical programmes from the study of ionized hydrogen clouds in our Galaxy to distant quasars. Following the accurate calibration survey it became evident that as an astrometric instrument—to establish a coordinate system across the sky—its measuring accuracy was comparable with the best optical methods, whilst overcoming some of the difficulties in optical work such as the measurement of large angles. Bruce Elsmore is involved in a collaborative programme with optical observers to relate the positions of quasars—(some of which are compact sources at both optical and radio wavelengths)—as measured by radio means, to those derived from the fundamental stars, in order to determine any large-scale non-uniformities which may exist in the present astrometric systems. He also showed how this type of instrument may be used for the direct measurement of astronomical time—without the need for collaborative observations at different longitudes to correct for polar motion—again with an accuracy comparable with optical methods (~ 5 mS in a 12^h observation).

Another programme is concerned with a study of the birth of stars; when a cloud of gas condenses to form a star, the dust which it contains provides such an effective screen that newly-formed stars, with their surrounding regions of ionized hydrogen, can never be seen optically; only after this dust cloud has dispersed does the star appear. The dust introduces no appreciable absorption at radio wavelengths, so that radio observations allow these regions to be studied at the earliest stages.

NGC 7538 is an example of such a region, and the upper part of Fig. 9 shows the radio emission as mapped with the One-Mile telescope. The large diffuse component corresponds almost exactly with the optical nebulosity, and represents the cloud of gas ionized by one or more O-stars formed about a million years ago, with the dust sufficiently dispersed to allow the light to be seen. The compact lower component corresponds to gas ionized by much younger stars, which are still embedded in dust too dense for any optical emission to escape, and it is invisible on the photograph. When this southern component was mapped with the higher resolution of the 5 km telescope, the

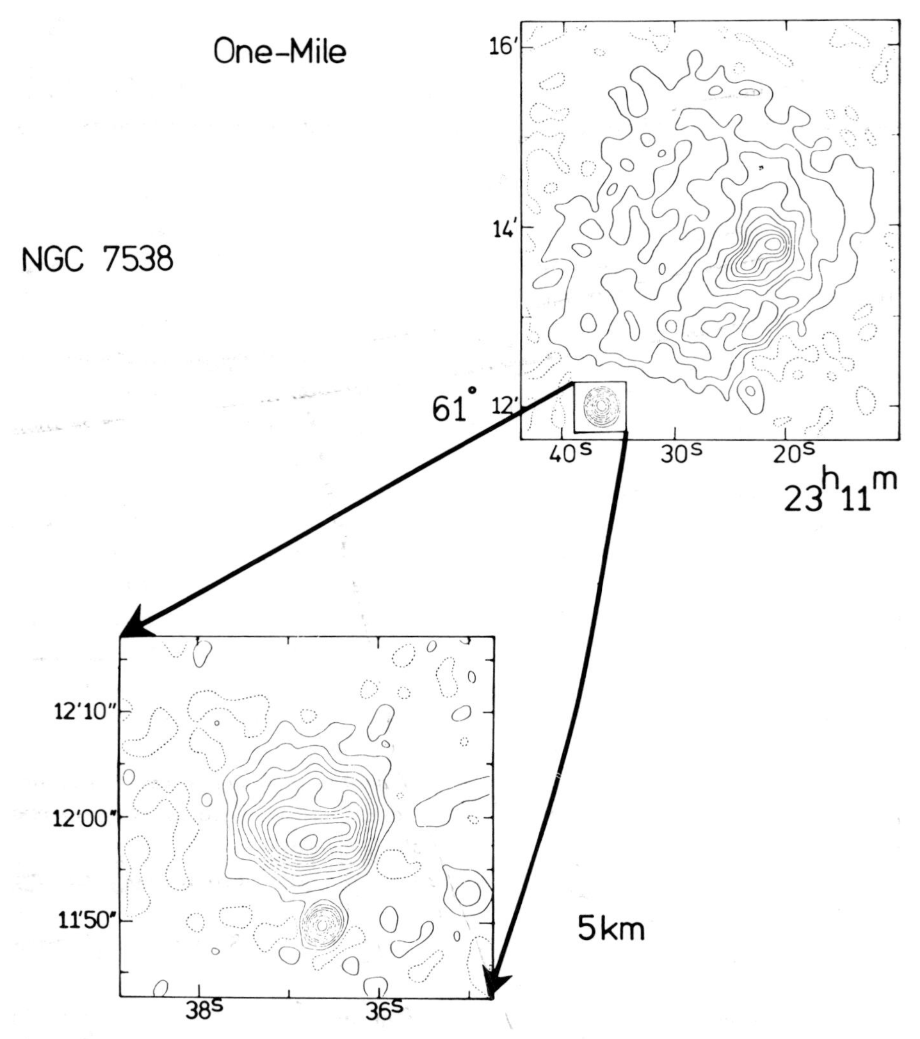

Fig. 9. The ionized hydrogen cloud NGC 7538. The upper radio map shows the large cloud associated with the optical emission, and another, compact, component to the south. This compact component is shown with greater resolution below.

lower map was obtained, showing that there is an ionized cloud some 10″ arc in diameter, probably produced by the radiation from a star of spectral type O8, and an even more compact cloud to the south of this, produced by a still younger star, only a few thousand years old. The dust surrounding these two compact regions is heated by the young stars they contain, and both have been detected by their infra-red emission (8).

But the most extensive programme has been the mapping of extragalactic sources—the radio galaxies and quasars; galaxies which, during a brief fraction of their lives, produce some 10^{60} ergs of energy, equivalent to the total annihilation of the matter in about a million suns, by a mechanism which is not understood.

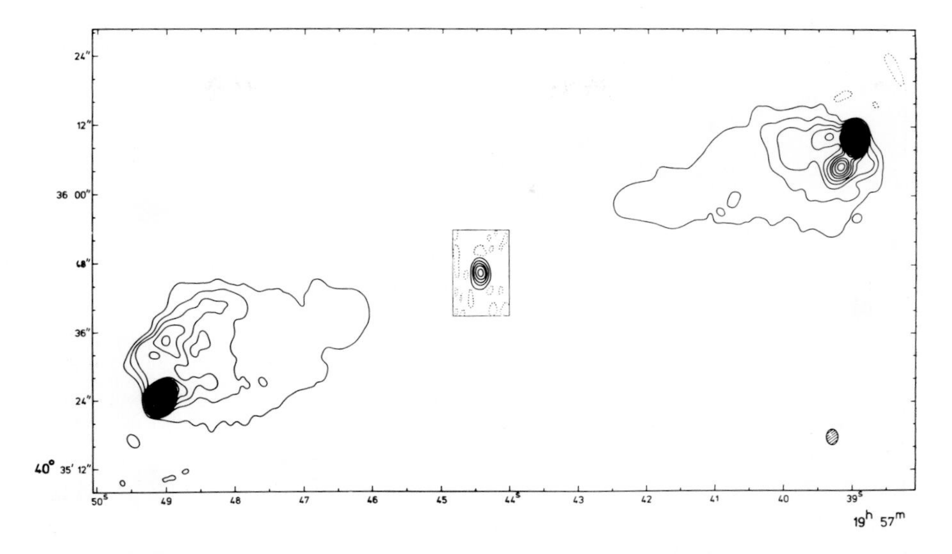

Fig. 10. The powerful radio galaxy in the constellation of Cygnus mapped with the 5 km telescope. The compact outer components are exceedingly bright—(31 and 41 contours). The central component—which corresponds to the nucleus of the optical galaxy is very weak and is drawn with contours spaced at 1/5 the interval.

Fig. 10 shows the new radio map of the source in the constellation of Cygnus—the first powerful radio galaxy to be recognized. The distribution of polarized emission from the north component is shown in Fig. 11, giving in-

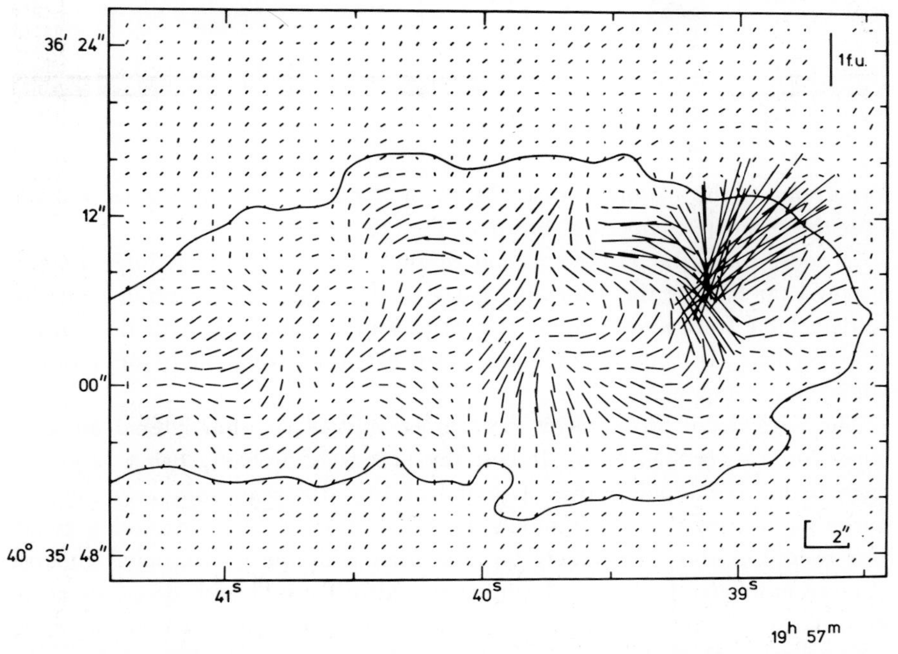

Fig. 11. The polarization of the emission from the north component of the Cygnus source which shows the magnetic field to be turbulent on a scale $\sim 10^4$ light-years.

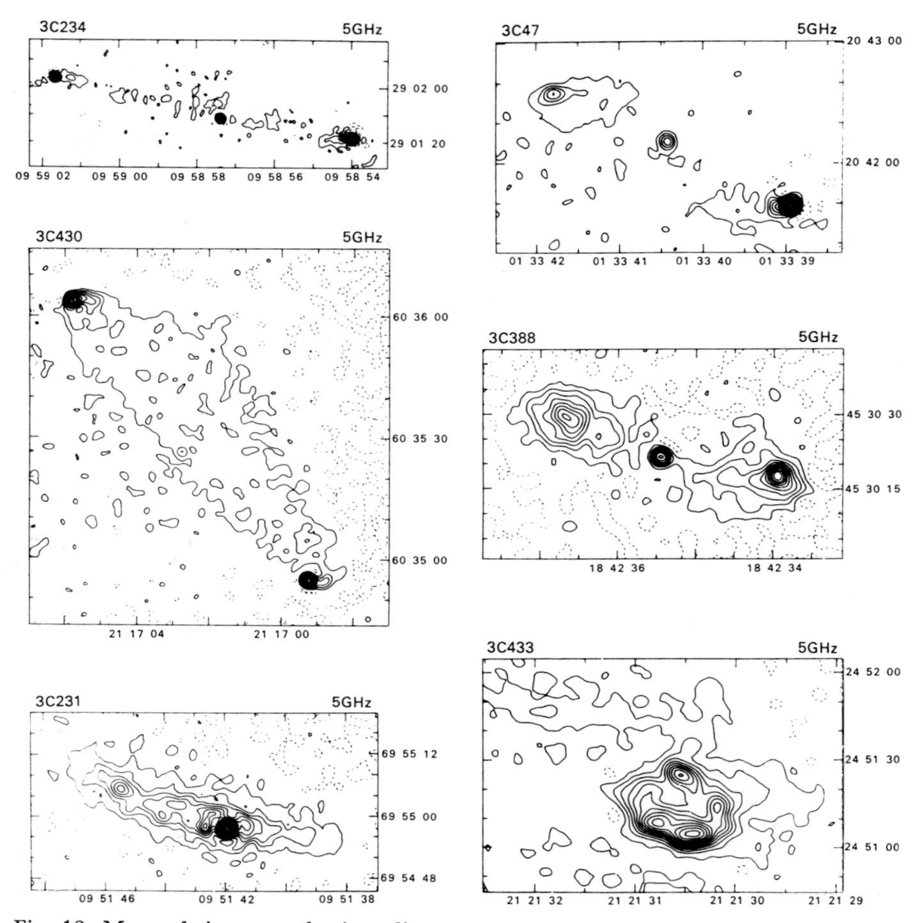

Fig. 12. Maps of six extragalactic radio sources.

formation on the magnetic field. Maps of a number of other sources made with the 5 km telescope are shown in Fig. 12.

In most cases the radio emission originates mainly in two huge regions disposed far outside the associated galaxy—although weak emission may also be detectable from a very compact central source coincident with the nucleus of the galaxy. In some cases much more extensive components or a bridge linking the components occur.

The finer detail provided by the 5 km telescope has already enabled some important conclusions to be drawn; the energy is probably being produced more or less continuously over a period of 10^7—10^8 years in a very compact nucleus and not, as was originally thought, in some single explosive event. The source of this energy may be associated with the gravitational collapse of large numbers of stars, in the manner which Tony Hewish describes in his lecture, or by material falling into a much more massive collapsed object at the nucleus of the galaxy. The mechanism for transmitting this energy to the compact heads of the main components (e.g. Fig. 10) is not understood, but may involve a narrow beam of low frequency electromagnetic waves or relativistic particles (9, 10). The interaction of this beam with the surrounding intergalactic medium might then accelerate the electrons responsible for the radio emission from the compact heads, and their subsequent diffusion into the region behind the heads can probably explain the general shape of the extensive components.

While much remains unanswered, the present conclusions were only reached when detailed maps became available; the physical processes relating the nucleus, the compact heads, and the extensive tails or bridges can clearly only be investigated when the relationship between these structural components is known.

What can we expect in the future? In 1954, the first aperture synthesis telescope provided maps with a resolution of $2^\circ.2$; today we have maps with a resolution of $2''$ arc. Can we foresee a continuing development with radio pictures having much *better* resolution than the optical ones? The technical problems of increasing the aperture or decreasing the operating wavelength are severe, but they do not appear to be as serious as the limitations imposed by the earth's atmosphere; in optical observations atmospheric turbulence on a scale of ~ 10 cm in the lower atmosphere introduces irregularities in the incident wavefront which normally limits the resolution to $\sim 1''$ arc. At radio wavelengths the contribution of these small-scale irregularities is not important, but there are also irregularities of refractive index on a much larger scale in the troposphere. Two distinct types have been found in a series of observations with the One-Mile and 5 km telescopes; neither can be attributed to variations of air density, and both are probably due to non-uniformity in the partial pressure of water-vapour, which makes an important contribution to the refractive index at radio wavelengths. One class has a typical scale size of ~ 0.7 km and is attributed to turbulence in the troposphere due to solar heating of the ground in the same way that fair-weather cumulus clouds develop. These irregularities, however, are often detected in clear air conditions without the formation of cumulus clouds; they only occur during day-time and are more severe during summer months. The second class,—which shows only slight diurnal or annual variation, has a much larger scale size, typically 10—20 km, and there may be still larger scales which have not yet been recognized. The origin of these disturbances is not known, and it is therefore not possible to predict how they might depend on geographical position.

Under very good conditions—representing about 1 % of the total time, the atmospheric irregularities are extremely small and correspond to a distortion of the incident wavefront by < 0.2 mm over 5 km; under these conditions, operation at a wavelength of 4 mm or less would be possible and should provide maps with a resolution better than $0''.2$ arc. These excellent observing conditions have only been encountered during periods of widespread winter fog when the atmosphere is extremely stable, a result which illustrates the differing requirements in seeking good sites for optical and radio observatories!

For most of the time the atmospheric irregularities are considerably worse, and although there is insufficient information on scale sizes > 20 km, the use of instruments much larger than this will introduce difficulties associated with the curvature of the atmosphere. One might guess that it should be possible to build instruments which would give a resolution better than $0''.5$ arc for perhaps 50 % of the winter months.

To reach a greater resolution new techniques capable of correcting for the atmospheric effects will be necessary. One simple, though expensive, solution would be to build a second dish alongside each element, so that observations of a reference point source close to the area to be mapped, could be made simultaneously at every spacing; the observed phase errors for this reference source could then be used to provide a continuous correction for the signals from the area being mapped.

Such techniques can clearly be extended to the interferometers having baselines of many thousands of km (VLBI) which have been made possible by the development of atomic frequency standards. These instruments have shown the existence of very small components, $\sim 0''.001$ arc in some sources. The use of a comparison source for eliminating both atmospheric and instrumental phase was first used at Jodrell Bank in the special case of sources of the OH maser line at $\lambda = 18$ cm, where different components within the primary beam can be distinguished by their frequency; if one is used as a phase reference the relative positions of the others can be found (11).

For continuum sources a reference outside the primary beam of the instrument must, in general, be used and two elements at each location are needed. This technique has been used in the U.S.A. to reduce both instrumental and atmospheric phase variations in measurements of the gravitational deflection of radio waves by the sun (12); one pair of elements was used to observe a source close to the Sun, while the other pair observed a reference source about 10° away.

The accuracy of the correction, and hence the shortest wavelength at which mapping could be achieved, would depend on the angular separation between the area to be mapped and a reference source sufficiently intense and of sufficiently small angular size. But even if adequate phase stability can be attained in this way, there is a serious practical difficulty in making maps with resolution $\sim 0''.001$ arc, due to the inevitable poor sampling of the aperture plane. Even with 5 or 6 stations distributed across one hemisphere of the world, and using every possible combination of the signals from them, with observing periods lasting several hours, the fraction of the aperture plane which can be filled is still very small, so that the field of view which can be mapped without ambiguity from secondary responses is unlikely to exceed $\sim 0''.02$ arc. Whilst there seems little hope of deriving complete maps of most sources with this resolution, there are certainly some central components where such a map could provide very important information.

But I think it may also be important for our understanding of the mechanisms operating in the main components of radio sources, to obtain complete maps with intermediate resolution; for this work extensions of the present synthesis techniques, while retaining good filling of the aperture plane, are needed.

The last 25 years have seen a remarkable improvement in the performance of radio telescopes, which has in turn led to a much greater understanding of the strange sources of "high-energy astrophysics" and of the nature of the Universe as a whole.

I feel very fortunate to have started my research at a time which allowed me and my colleagues to play a part in these exciting developments.

References

1. Ryle, M. (1958) Proc. Roy. Soc. A., *248,* 289.
2. Scheuer, P. A. G. & Ryle, M. (1953) Mon. Not. R. astr. Soc., *113,* 3.
3. Blythe, J. H. (1957) Mon. Not. R. astr. Soc., *117,* 644.
4. Costain, C. H. & Smith, F. G. (1960) Mon. Not. R. astr. Soc., *121,* 405.
5. Pilkington, J. D. H. & Scott, P. F. (1965) Mem. R. astr. Soc., *69,* 183.
6. Gower, J. F. R., Scott, P. F. & Wills, D. (1967) Mem. R. astr. Soc., *71,* 49
7. Ryle, M. & Neville, A. C. (1962) Mon. Not. R. astr. Soc., *125,* 39.
8. Wynn-Williams, C. G., Becklin, E. E. & Neugebauer, G. (1974) Ap. J., *187,* 473.
9. Rees, M. (1971) Nature, *229,* 312.
10. Scheuer, P. A. G. (1974) Mon. Not. R. astr. Soc., *166,* 513.
11. Cooper, A. J., Davies, R. D. & Booth, R. S. (1971) Mon. Not. R. astr. Soc., *152,* 383
12. Counselmann, C. C., Kent, S. M., Knight, C. A., Shapiro, I. I., Clarke, T. A., Hinteregger, H. F., Rogers, A. E. E. & Whitney, A. R. (1974) Phys. Rev. Lett.. *33,* 1621.

Physics 1975

AAGE BOHR, BEN R MOTTELSON and JAMES RAINWATER

for the discovery of the connection between collective motion and particle motion in atomic nuclei and the development of the theory of the structure of the atomic nucelus based on this connection

THE NOBEL PRIZE FOR PHYSICS

Speech by professor SVEN JOHANSSON of the Royal Academy of Sciences
Translation from the Swedish text

Your Majesties, Your Royal Highnesses, Ladies and Gentlemen,
At the end of the 1940's, nuclear physics had advanced to a stage where a more detailed picture of the structure of the atomic nucleus was beginning to emerge and it was becoming possible to calculate its properties in a quantitative way. One knew that the nucleus consists of protons and neutrons, the so-called nucleons. They are kept together by nuclear forces, which give rise to a potential well, in which the nucleons move. The details of the nuclear structure were, however, unknown and one had to a great extent to rely upon models. These models were rather incomplete and partly contradictory. The oldest is the drop model in which the nucleus is regarded as a liquid drop, the nucleons corresponding to the molecules of the liquid. This model could be used with a certain success for a description of the mechanism of nuclear reactions, in particular for fission. On the other hand, one could not find any excited states of the nucleus corresponding to rotations or vibrations of the drop. Neither could certain other properties of the nucleus, particularly those associated with the "magic nubers", be explained by means of the drop model. These show that individual nucleons in a decisive way affect the behaviour of the nucleus. This discovery, which is systematized in the shell model, was awarded the 1963 Nobel Prize for Physics.

It was soon found that the nucleus has properties, which cannot be explained by these models. perhaps the most striking one was the very marked deviation of the charge distribution from spherical symmetry, which was observed in several cases. It was also pointed out that this might indicate that certain nuclei are not spherical but are deformed as an elipsoid, but no one could give a reasonable explanation of this phenomenon.

The solution of the problem was first presented by James Rainwater of Columbia University, New York, in a short paper submitted for publication in April 1950. In this, he considers the interaction between the main part of the nucleons, which form an inner core, and the outer, the valence nucleons. He points out that the valence nucleons can influence the shape of the core. Since the valence nucleons move in a field which is determined by the distribution of the inner nucleons, this influence is mutual. If several valence nucleons move in similar orbits, this polarizing effect on the core can be so great that the nucleus as a whole becomes permanently deformed. Expressed very simply, it can be said that as a result of their motion, certain nucleons expose the "walls" of the nucleus to such high centrifugal pressure that it becomes deformed. Rainwater also attempted to calculate this effect and got results that agreed with experimental data on the charge distributions.

Aage Bohr, working in Copenhagen, but at this time on a visit to Columbia University, had, independently of Rainwater, been thinking along the same lines. In a paper, submitted for publication about a month after Rainwater's, he formulates the problem of the interaction of a valence nucleon with the core in a general way.

These relatively vague ideas were further developed by Bohr in a famous work from 1951, in which he gives a comprehensive study of the coupling of oscillations of the nuclear surface to the motion of the individual nucleons. By analysing the theoretical formula for the kinetic energy of the nucleus, he could predict the different types of collective excitations: vibration, consisting of a periodic change of the shape of the nucleus around a certain mean value, and rotation of the whole nucleus around an axis perpendicular to the symmetry axis. In the latter case, the nucleus does not rotate as a rigid body, but the motion consists of a surface wave propagating around the nucleus.

Up to this point, the progress made had been purely theoretical and the new ideas to a great extent lacked experimental support. The very important comparison with experimental data was done in three papers, written jointly by Aage Bohr and Ben Mottelson and published in the years 1952—53. The most spectacular finding was the discovery that the position of energy levels in certain nuclei could be explained by the assumption that they form a rotational spectrum. The agreement between theory and experiment was so complete that there could be no doubt of the correctness of the theory. This gave stimulus to new theoretical studies, but, above all, to many experiments to verify the theoretical predictions.

This dynamic progress very soon led to a deepened understanding of the structure of the atomic nucleus. Even this further development towards a more refined theory was inspired and influenced in a decisive way by Bohr and Mottelson. For example, they showed together with Pines that the nucleons have a tendency to form pairs. A consequence of this is that nuclear matter has properties reminiscent of superconductors.

Drs Bohr, Mottelson and Rainwater,

In your pioneering works you have laid the foundation of a theory of the collective properties of atomic nuclei. This has been an inspiration to an intensive research activity in nuclear structure physics. The further development in this field has in a striking way confirmed the validity and great importance of your fundamental investigations.

On behalf of the Royal Academy of Sciences I wish to convey to you our warmest congratulations and I now ask you to receive your prize from the hands of His Majesty the King.

AAGE BOHR

I was born in Copenhagen on June 19, 1922, as the fourth son of Niels Bohr and Margrethe Bohr (née Nørlund). During my early childhood, my parents lived at the Institute for Theoretical Physics (now the Niels Bohr Institute), and the remarkable generation of scientists who came to join my father in his work became for us children Uncle Kramers, Uncle Klein, Uncle Nishina, Uncle Heisenberg, Uncle Pauli, etc. When I was about ten years old, my parents moved to the mansion at Carlsberg, where they were hosts for widening circles of scholars, artists, and persons in public life.

I went to school for twelve years at Sortedam Gymnasium (H. Adler's fællesskole) and am indebted to many of my teachers, both in the humanities and in the sciences, for inspiration and encouragement.

I began studying physics at the University of Copenhagen in 1940 (a few months after the German occupation of Denmark). By that time, I had already begun to assist my father with correspondence, with his writing of articles of a general epistemological character, and gradually also in connection with his work in physics. In those years, he was concerned partly with problems of nuclear physics and partly with problems relating to the penetration of atomic particles through matter.

In October 1943, my father had to flee Denmark to avoid arrest by the Nazis, and the whole family managed to escape to Sweden, where we were warmly received. Shortly afterwards, my father proceeded to England, and I followed after him. He became associated with the atomic energy project and, during the two years until we returned to Denmark, in August 1945, we travelled together spending extensive periods in London, Washington, and Los Alamos. I was acting as his assistant and secretary and had the opportunity daily to share in his work and thoughts. We were members of the British team, and my official position was that of a junior scientific officer employed by the Department of Scientific and Industrial Research in London. In another context, I have attempted to describe some of the events of those years and my father's efforts relating to the prospects raised by the atomic weapons[1].

On my return to Denmark, I resumed my studies at the University and obtained a master's degree in 1946. My thesis was concerned with some aspects of atomic stopping problems.

For the spring term of 1948, I was a member of the Institute for Advanced Study in Princeton. On a visit during that period to Columbia University

[1] Niels Bohr. His life and work as seen by his friends and colleagues, p. 191. Ed. by S. Rozental, North-Holland Publishing Company, Amsterdam 1967.

and through discussions with professor I. I. Rabi, I became interested in a newly discovered effect in the hyperfine structure in deuterium. This led on to my association with Columbia University from January 1949 to August 1950. As described in my lecture, this was for me a very fruitful association.

Soon after my return to Copenhagen, I began the close cooperation with Ben Mottelson which has continued ever since. The main direction of our work is described in the lectures included in the present volume. During the last fifteen years, a major part of our efforts has been connected with the attempt to present the status of our understanding of nuclear structure in a monograph, of which Volume I (Single-Particle Motion) appeared in 1969, and Volume II (Nuclear Deformations) in 1975. We feel that in our cooperation, we have been able to exploit possibilities that lie in a dialogue between kindred spirits that have been attuned through a long period of common experience and jointly developed understanding. It has been our good fortune to work closely together with colleagues at the Niels Bohr Institute and Nordita, including the many outstanding scientists who have come from all parts of the world and have so greatly enriched the scientific atmosphere and personal contacts.

I have been connected with the Niels Bohr Institute since the completion of my university studies, first as a research fellow and from 1956 as a professor of physics at the University of Copenhagen. After the death of my father in 1962, I followed him as director of the Institute until 1970.

For our whole circle, it has been a challenge to exploit the opportunities provided by the traditions of the Institute, of which I would like especially to mention two aspects. One concerns the fruitful interplay between experimental and theoretical investigations. The other concerns the promotion of international cooperation as a vital factor in the development of science itself and also as a means to strengthen the mutual knowledge and understanding between nations.

In 1957, Nordita (Nordisk Institut for Teoretisk Atomfysik) was founded on the premises of the Niels Bohr Institute, and the two institutes operate in close association. I have been a member of the Board of Nordita from 1957 until 1975, and since then director of this institute.

In March 1950, in New York City, I was married to Marietta Soffer. We have three children, Vilhelm, Tomas, and Margrethe. Both for my wife and myself, the personal friendships that have grown out of scientific contacts with colleagues from many different countries have been an important part of our lives, and the travels we have made together in connection with the world-wide scientific co-operation have given us rich treasures of experiences.

ROTATIONAL MOTION IN NUCLEI

Nobel Lecture, December 11, 1975

by

AAGE BOHR
The Niels Bohr Institute and Nordita
Copenhagen, Denmark

The exploration of nuclear structure over the last quarter century has been a rich experience for those who have had the privilege to participate. As the nucleus has been subjected to more and more penetrating probes, it has continued to reveal unexpected facets and to open new perspectives. The preparation of our talks today has been an occasion for Ben Mottelson and myself to relive the excitement of this period and to recall the interplay of so many ideas and discoveries coming from the worldwide community of nuclear physicists, as well as the warmth of the personal relations that have been involved.

In this development, the study of rotational motion has had a special role. Because of the simplicity of this mode of excitation and the many quantitative relations it implies, it has been an important testing ground for many of the general ideas on nuclear dynamics. Indeed, the response to rotational motion has played a prominent role in the development of dynamical concepts ranging from celestial mechanics to the spectra of elementary particles.

EARLY IDEAS ON NUCLEAR ROTATION

The question of whether nuclei can rotate became an issue already in the very early days of nuclear spectroscopy (1, 2). Quantized rotational motion had been encountered in molecular spectra (3), but atoms provide examples of quantal systems that do not rotate collectively. The available data on nuclear excitation spectra, as obtained for example from the fine structure of *a* decay, appeared to provide evidence against the occurrence of low-lying rotational excitations, but the discussion was hampered by the expectation that rotational motion would either be a property of all nuclei or be generally excluded, as in atoms, and by the assumption that the moment of inertia would have the rigid-body value, as in molecular rotations. The issue, however, took a totally new form with the establishment of the nuclear shell model (4).

Just at that time, in early 1949, I came to Columbia University as a research fellow and had the good fortune of working in the stimulating atmosphere of the Pupin Laboratory where so many great discoveries were being made under the inspiring leadership of I.I. Rabi. One of the areas of great activity was the study of nuclear moments, which was playing such a crucial role in the development of the new ideas on nuclear structure.

To-day, it is difficult to fully imagine the great impact of the evidence for nuclear shell structure on the physicists brought up with the concepts of the liquid-drop and compound-nucleus models, which had provided the basis for interpreting nuclear phenomena over the previous decade (5)[1]. I would like also to recall my father's reaction to the new evidence, which presented the sort of dilemma that he would respond to as a welcome opportunity for deeper understanding. In the summer of 1949, he was in contact with John Wheeler on the continuation of their work on the fission process, and in this connection, in order to "clear his thoughts", he wrote some tentative comments on the incorporation of the contrasting evidence into a more general picture of nuclear constitution and the implications for nuclear reactions (7). These comments helped to stimulate my own thinking on the subject, which was primarily concerned with the interpretation of nuclear moments[2].

The evidence on magnetic moments, which at the time constituted one of the most extensive quantitative bodies of data on nuclear properties, presented a special challenge. The moments showed a striking correlation with the predictions of the one-particle model (9, 4), but at the same time exhibited major deviations indicative of an important missing element. The incomparable precision that had been achieved in the determination of the magnetic moments, as well as in the measurement of the hyperfine structure following the pioneering work of Rabi, Bloch, and Purcell, was even able to provide information on the distribution of magnetism inside the nucleus (10, 11).

A clue for understanding the deviations in the nuclear coupling scheme from that of the single-particle model was provided by the fact that many nuclei have quadrupole moments that are more than an order of magnitude larger than could be attributed to a single particle[3]. This finding directly implied a sharing of angular momentum with many particles, and might seem to imply a break-down of the one-particle model. However, essential features of the single-particle model could be retained by assuming that the average nuclear field in which a nucleon moves deviates from spherical symmetry (15). This picture leads to a nuclear model resembling that of a molecule, in which the nuclear core possesses vibrational and rotational degrees

[1] The struggle involved in facing up to the new evidence is vividly described by Jensen (6). Our discussions with Hans Jensen over the years concerning many of the crucial issues in the development provided for us a special challenge and inspiration.

[2] The interplay between individual-particle and collective motion was also at that time taken up by John Wheeler. Together with David Hill, he later published the extensive article on "Nuclear Constitution and the Interpretation of Fission Phenomena" (8), which has continued over the years to provide inspiration for the understanding of new features of nuclear phenomena.

[3] The first evidence for a non-spherical nuclear shape came from the observation of a quadrupole component in the hyperfine structure of optical spectra (12). The analysis showed that the electric quadrupole moments of the nuclei concerned were more than an order of magnitude greater than the maximum value that could be attributed to a single proton and suggested a deformation of the nucleus as a whole (13). The problem of the large quadrupole moments came into focus with the rapid accumulation of evidence on nuclear quadrupole moments in the years after the war and the analysis of these moments on the basis of the shell model (14).

of freedom. For the rotational motion there seemed no reason to expect the classical rigid-body value; however, the large number of nucleons participating in the deformation suggested that the rotational frequency would be small compared with those associated with the motion of the individual particles. In such a situation, one obtains definite limiting coupling schemes (see Fig. 1) which could be compared with the empirical magnetic moments and the evidence on the distribution of nuclear magnetism, with encouraging results (15, 17)[4].

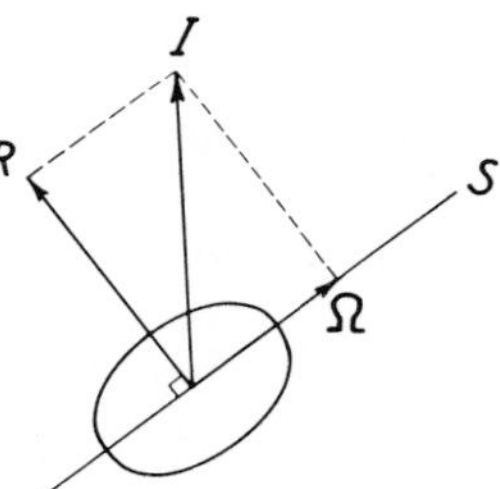

Fig. 1. Coupling scheme for particle in slowly rotating spheroidal nucleus. The intrinsic quantum number Ω represents the projection of the particle angular momentum along the nuclear symmetry axis S, while R is the collective angular momentum of the nuclear core and is directed perpendicular to the symmetry axis, since the component along S which is a constant of the motion, vanishes in the nuclear ground state. The total angular momentum is denoted by I. The figure is from (16).

In the meantime and, in fact, at nearly the same point in space, James Rainwater had been thinking about the origin of the large nuclear quadrupole moments and conceived an idea that was to play a crucial role in the following development. He realized that a non-spherical equilibrium shape would arise as a direct consequence of single-particle motion in anisotropic orbits, when one takes into account the deformability of the nucleus as a whole, as in the liquid-drop model (19).

On my return to Copenhagen in the autumn of 1950, I took up the problem of incorporating the coupling suggested by Rainwater into a consistent dynamical system describing the motion of a particle in a deformable core. For this coupled system, the rotational motion emerges as a low-frequency component of the vibrational degrees of freedom, for sufficiently strong coupling. The rotational motion resembles a wave travelling across the nuclear surface and the moment of inertia is much smaller than for rigid rotation (see Fig. 2).

Soon, I was joined by Ben Mottelson in pursuing the consequences of the interplay of individual-particle and collective motion for the great variety of nuclear phenomena that was then coming within the range of experimental

[4] The effect on the magnetic moments of a sharing of angular momentum between the single particle and oscillations of the nuclear surface was considered at the same time by Foldy and Milford (18).

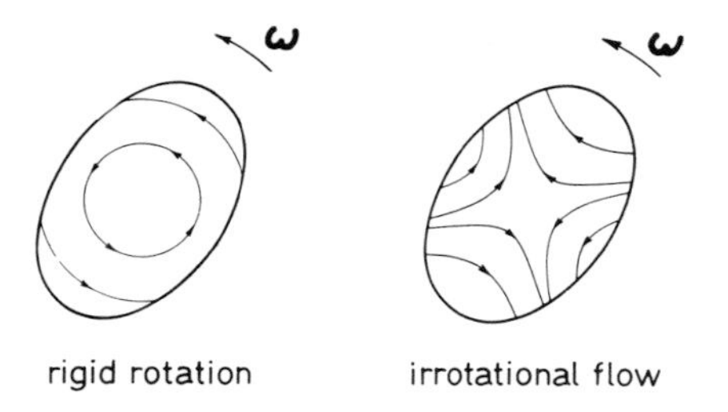

Fig. 2. Velocity fields for rotational motion. For the rotation generated by irrotational flow, the velocity is proportional to the nuclear deformation (amplitude of the travelling wave). Thus, for a spheroidal shape, the moment of inertia is $\mathcal{J} = \mathcal{J}_{rig}\ (\Delta R/R)^2$, where $\mathcal{J}_{rig}$ is the moment for rigid rotation, while R is the mean radius and ΔR (assumed small compared with R) is the difference between major and minor semi-axes. The figure is from (16).

studies (20). In addition to the nuclear moments, important new evidence had come from the classification of the nuclear isomers (21) and beta decay (22) as well as from the discovery of single-particle motion in nuclear reactions (23, 24). It appeared that one had a framework for bringing together most of the available evidence, but in the quantitative confrontation with experiment, one faced the uncertainty in the parameters describing the collective properties of the nucleus. It was already clear that the liquid-drop description was inadequate, and one lacked a basis for evaluating the effect of the shell structure on the collective parameters.

THE DISCOVERY OF ROTATIONAL SPECTRA

At this point, one obtained a foothold through the discovery that the coupling scheme characteristic of strongly deformed nuclei with the striking rotational band structure was in fact realized for an extensive class of nuclei. The first indication had come from the realization by Goldhaber and Sunyar that the electric quadrupole transition rates for the decay of low-lying excited states in even-even nuclei were, in some cases, much greater than could be accounted for by a single-particle transition and thus suggested a collective mode of excitation (21). A rotational interpretation (25) yielded values for the nuclear eccentricity in promising agreement with those deduced from the spectroscopic quadrupole moments.

Soon after, the evidence began to accumulate that these excitations were part of a level sequence with angular momenta $I = 0, 2, 4 \ldots$ and energies proportional to $I\ (I+1)$ (26, 27); examples of the first such spectra are shown in Fig. 3. For ourselves, it was a thrilling experience to receive a prepublication copy of the 1953 compilation by Hollander, Perlman, and Seaborg (29) with its wealth of information on radioactive transitions, which made it possible to identify so many rotational sequences.

The exciting spring of 1953 culminated with the discovery of the Coulomb excitation process (30, 31), which opened the possibility for a systematic study of rotational excitations (30, 32). Already the very first experiments by

Huus and Zupančič (see Fig. 4) provided a decisive quantitative test of the rotational coupling scheme in an odd nucleus, involving the strong coupling between intrinsic and rotational angular momenta[5].

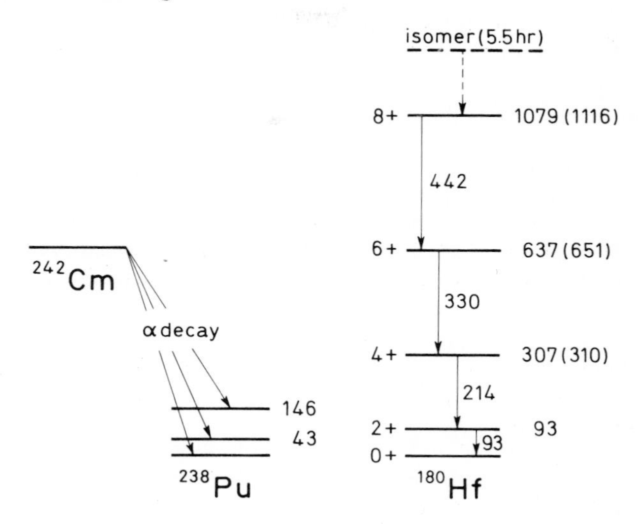

Fig. 3. Rotational spectra for ^{238}Pu and ^{180}Hf. The spectrum of ^{180}Hf (from (26)) was deduced from the observed γ lines associated with the decay of the isomeric state (28). The energies are in keV, and the numbers in parenthesis are calculated from the energy of the first excited state, assuming the energies to be proportional to $I(I+1)$.

The spectrum of ^{238}Pu was established by Asaro and Perlman (27) from measurements of the fine structure in the α decay of ^{242}Cm. Subsequent evidence showed the spin-parity sequence to be 0+, 2+, 4+, and the energies are seen to be closely proportional to $I(I+1)$.

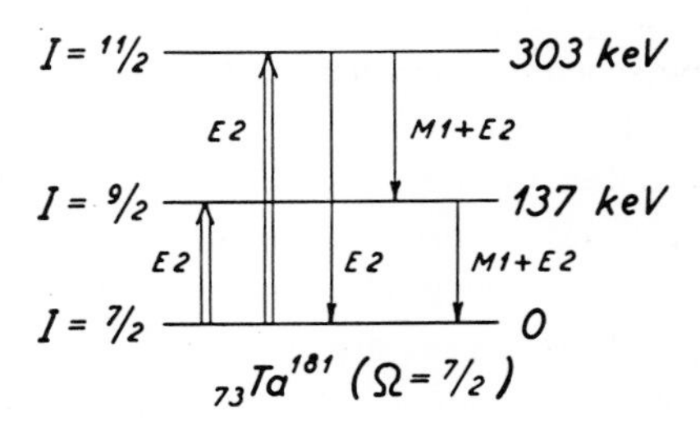

Fig. 4. Rotational excitations in ^{181}Ta observed by Coulomb excitation. In an odd-A nucleus with intrinsic angular momentum Ω (see Fig. 1), the rotational excitations involve the sequence $I = \Omega,\ \Omega+1,\ \Omega+2, \ldots$, all with the same parity. In the Coulomb excitation process, the action of the electric field of the projectile on the nuclear quadrupole moment induces E2 (electric quadrupole) transitions and can thus populate the first two rotational excitations. The observed energies (30) are seen to be approximately proportional to $I(I+1)$.

The excited states decay by E2 and M1 (magnetic dipole) transitions, and the rotational interpretation implies simple intensity relations. For example, the reduced E2 matrix elements within the band are proportional to the Clebsch-Gordan coefficient $< I_i \Omega\ 20 | I_f \Omega >$, where I_i and I_f are the angular momenta of initial and final states. The figure is from (16).

[5] The quantitative interpretation of the cross sections could be based on the semi-classical theory of Coulomb excitation developed by Ter-Martirosyan (33) and Alder and Winther (34).

This was a period of almost explosive development in the power and versatility of nuclear spectroscopy, which rapidly led to a very extensive body of data on nuclear rotational spectra. The development went hand in hand with a clarification and expansion of the theoretical basis.

Fig. 5 shows the region of nuclei in which rotational band structure has so far been identified. The vertical and horizontal lines indicate neutron and proton numbers that form closed shells, and the strongly deformed nuclei are seen to occur in regions where there are many particles in unfilled shells that can contribute to the deformation.

The rotational coupling scheme could be tested not only by the sequence of spin values and regularities in the energy separations, but also by the intensity relations that govern transitions leading to different members of a rotational band (37, 38, 39). The leading order intensity rules are of a purely geometrical character depending only on the rotational quantum numbers and the multipolarity of the transitions (see the examples in Fig. 4 and Fig. 10).

The basis for the rotational coupling scheme and its predictive power were greatly strengthened by the recognition that the low-lying bands in odd-A nuclei could be associated with one-particle orbits in the deformed potential (40, 41, 42). The example in Fig. 6 shows the spectrum of ^{235}U with its high level density and apparently great complexity. However, as indicated, the states can be grouped into rotational bands that correspond uniquely to those expected from the Nilsson diagram shown in Fig. 7.

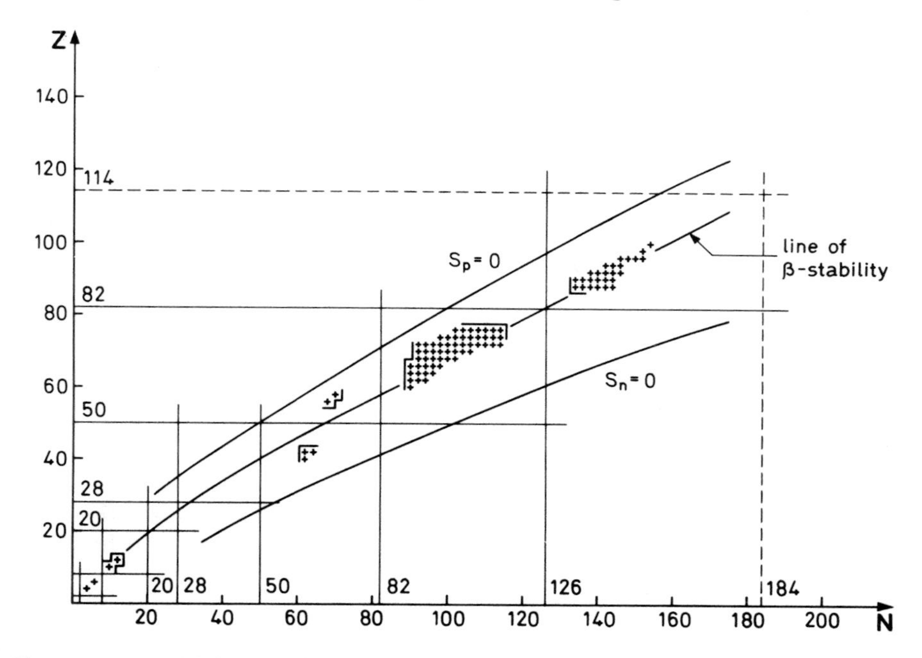

Fig. 5. Regions of deformed nuclei. The crosses represent even-even nuclei, whose excitation spectra exhibit an approximate $I(I+1)$ dependence, indicating rotational band structure. The figure is from (35) and is based on the data in (36). The curves labelled $S_n = 0$ and $S_p = 0$ are the estimated borders of instability with respect to neutron and proton emission.

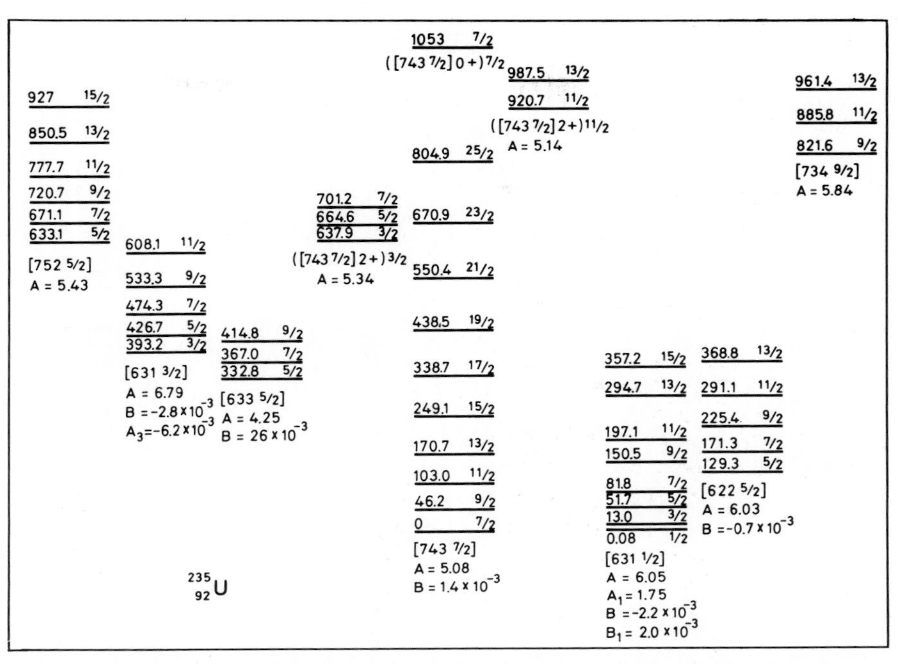

Fig. 6. Spectrum of ^{235}U. The figure is from (35) and is based on the experimental data from Coulomb excitation (43), ^{239}Pu α decay (43a), one-particle transfer (44), and the ^{234}U (n γ) reaction (45). All energies are in keV. The levels are grouped into rotational bands characterized by the spin sequence, energy dependence, and intensity rules. The energies within a band can be represented by a power series expansion of the form $E(I) = AI(I+1) + BI^2(I+1)^2 + \ldots \; (-1)^{I+\Omega} \; (I+\Omega)! \; ((I-\Omega)!)^{-1} \; (A_{2\Omega}+B_{2\Omega} \; I(I+1)+\ldots)$, with the parameters given in the figure. The low-lying bands are labelled by the quantum numbers of the available single-particle orbits (see Fig. 7), with particle-like states drawn to the right of the ground-state band and hole-like states to the left. The bands beginning at 638, 921, and 1053 keV represent quadrupole vibrational excitations of the ground-state configuration.

The regions of deformation in Fig. 5 refer to the nuclear ground-state configurations; another dimension is associated with the possibility of excited states with equilibrium shapes quite different from those of the ground state. For example, some of the closed-shell nuclei are found to have strongly deformed excited configurations[6]. Another example of sharpe isomerism with associated rotational band structure is encountered in the metastable, very strongly deformed states that occur in heavy nuclei along the path to fission (50, 51).

[6] The fact that the first excited states in ^{16}O and ^{40}Ca have positive parity, while the low-lying single-particle excitations are restricted to negative parity, implies that these states involve the excitation of a larger number of particles. It was suggested (47) that the excited positive parity states might be associated with collective quadrupole deformations. The existence of a rotational band structure in ^{16}O was convincingly established as a result of the ^{12}C ($\alpha\alpha$) studies (48) and the observation of strongly enhanced E2-transition matrix elements (49).

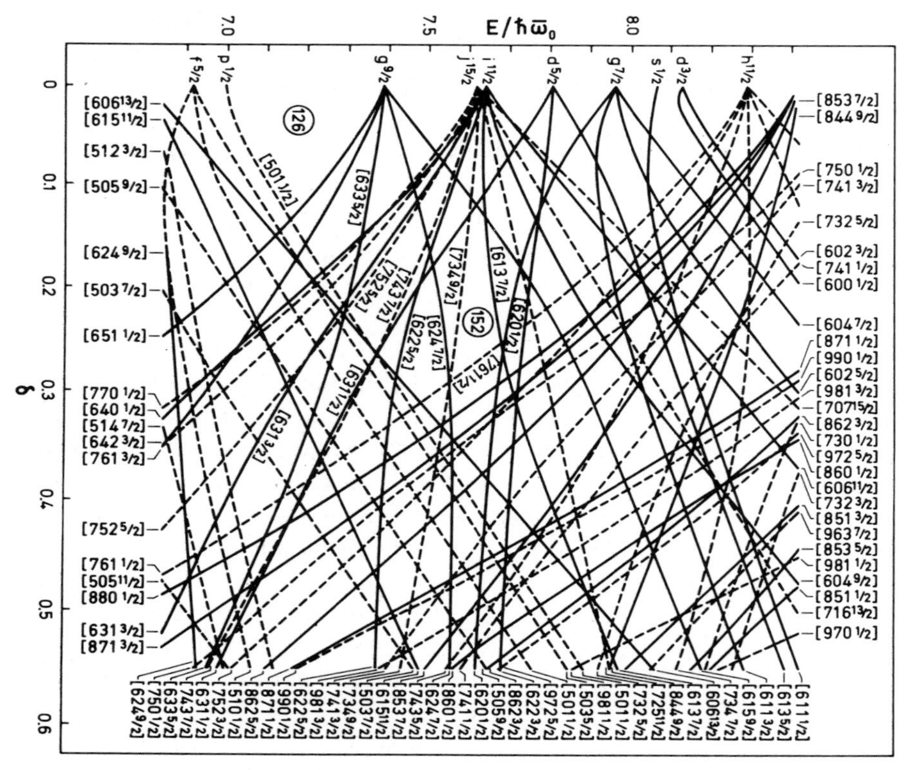

Fig. 7. Neutron orbits in prolate potential. The figure (from (35)) shows the energies of single-particle orbits calculated in an appropriate nuclear potential by Gustafson, Lamm, Nilsson, and Nilsson (46). The single-particle energies are given in units of $\hbar\bar{\omega}$, which represents the separation between major shells and, for ^{235}U, has the approximate value 6.6 MeV. The deformation parameter δ is a measure of the nuclear eccentricity; the value determined for ^{235}U, from the observed E2 transition moments, is $\delta \approx 0.25$. The single-particle states are labelled by the "asymptotic" quantum numbers $[\mathcal{N} n_3\ \Lambda\ \Omega]$. The last quantum number Ω, which represents the component j_3 of the total angular momentum along the symmetry axis, is a constant of the motion for all values of δ. The additional quantum numbers refer to the structure of the orbits in the limit of large deformations, where they represent the total number of nodal surfaces ($\mathcal{N}$), the number of nodal surfaces perpendicular to the symmetry axis (n_3), and the component of orbital angular momentum along the symmetry axis (Λ). Each orbit is doubly degenerate ($j_3 = \pm\Omega$), and a pairwise filling of orbits contributes no net angular momentum along the symmetry axis. For ^{235}U, with neutron number 143, it is seen that the lowest two configurations are expected to involve an odd neutron occupying the orbits [743 7/2] or [631 1/2], in agreement with the observed spectrum (see Fig. 6). It is also seen that the other observed low-lying bands in ^{235}U correspond to neighbouring orbits in the present figure.

New possibilities for studying nuclear rotational motion were opened by the discovery of marked anisotropies in the angular distribution of fission fragments (52), which could be interpreted in terms of the rotational quantum numbers labelling the individual channels through which the fissioning nucleus passes the saddle-point shape (53). Present developments in the experimental tools hold promise of providing detailed information about band structure in the fission channels and thereby on rotational motion under circumstances radically different from those studied previously.

CONNECTION BETWEEN ROTATIONAL AND SINGLE-PARTICLE MOTION

The detailed testing of the rotational coupling scheme and the successful classification of intrinsic spectra provided a firm starting point for the next step in the development, which concerned the dynamics underlying the rotational motion.

The basis for this development was the bold idea of Inglis (54) to derive the moment of inertia by simply summing the inertial effect of each particle as it is dragged around by a uniformly rotating potential (see Fig. 8). In this approach, the potential appears to be externally "cranked", and the problems concerning the self-consistent origin for the rotating potential and the limitations of such a semi-classical description have continued over the years to be hotly debated issues. The discussion has clarified many points concerning the connection between collective and single-particle motion, but the basic idea of the cranking model has stood its tests to a remarkable extent (55, 35).

The evaluation of the moments of intertia on the basis of the cranking model gave the unexpected result that, for independent-particle motion, the moment would have a value approximately corresponding to rigid rotation (56). The fact that the observed moments were appreciably smaller than the rigid-body values could be qualitatively understood from the effect of the residual interactions that tend to bind the particles into pairs with angular momentum zero. A few years later, a basis for a systematic treatment of the moment of inertia with the inclusion of the many-body correlations associated with the pairing effect was given by Migdal (57) and Belyaev (58),

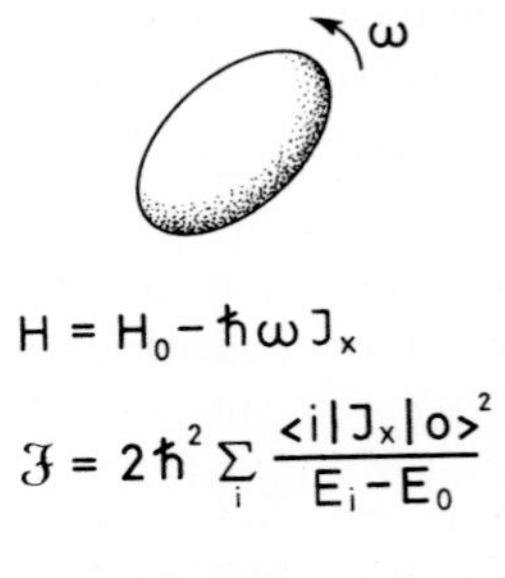

$$H = H_0 - \hbar\omega J_x$$

$$\mathcal{J} = 2\hbar^2 \sum_i \frac{\langle i|J_x|0\rangle^2}{E_i - E_0}$$

cranking model

Fig. 8. Nuclear moment of inertia from cranking model. The Hamiltonian H describing particle motion in a potential rotating with frequency ω about the x axis is obtained from the Hamiltonian H_0 for motion in a fixed potential by the addition of the term proportional to the component $\mathcal{J}_x$ of the total angular momentum, which represents the Coriolis and centrifugal forces acting in the rotating co-ordinate frame. The moment of inertia is obtained from a second-order perturbation treatment of this term and involves a sum over the excited states i. For independent-particle motion, the moment of inertia can be expressed as a sum of the contributions from the individual particles.

exploiting the new concepts that had in the meantime been developed for the treatment of electronic correlations in a superconductor (59); see also the following talk (60).

The nuclear moment of inertia is thus intermediate between the limiting values corresponding to rigid rotation and to the hydrodynamical picture of irrotational flow that was assumed in the early models of nuclear rotation. Indeed, the classical pictures involving a local flow provide too limited a framework for the description of nuclear rotation, since, in nuclear matter, the size of the pairs (the coherence length) is greater than the diameter of the largest existing nuclei. Macrosopic superflow of nuclear matter and quantized vortex lines may occur, however, in the interior of rotating neutron stars (61).

While these developments illuminated the many-body aspects of nuclear rotation, appropriate to systems with a very large number of nucleons, a parallel development took its starting point from the opposite side. Shell-model calculations exploiting the power of group-theoretical classification schemes and high-speed electronic computers could be extended to configurations with several particles outside of closed shells. It was quite a dramatic moment when it was realized that some of the spectra in the light nuclei that had been successfully analyzed by the shell-model approach could be given a very simple interpretation in terms of the rotational coupling scheme[7].

The recognition that rotational features can manifest themselves already in configurations with very few particles provided the background for Elliott's discovery that the rotational coupling scheme can be given a precise significance in terms of the SU_3 unitary symmetry classification, for particles moving in a harmonic oscillator potential (65). This elegant model had a great impact at the time and has continued to provide an invaluable testing ground for many ideas concerning nuclear rotation. Indeed, it has been a major inspiration to be able, even in this limiting case, to see through the entire correlation structure in the many-body wave function associated with the collective motion. Thus, for example, the model explicitly exhibits the separation between intrinsic and collective motion and implies an intrinsic excitation spectrum that differs from that of independent-particle motion in a deformed field by the removal of the "spurious" degrees of freedom that have gone into the collective spectrum.

This development also brought into focus the limitation to the concept of rotation arising from the finite number of particles in the nucleus. The rotational spectrum in the SU_3 model is of finite dimension (compact symmetry group) corresponding to the existence of a maximum angular momentum that can be obtained from a specified shell-model configuration. For low-lying bands, this maximum angular momentum is of the order of magnitude

[7] In this connection, a special role was played by the spectrum of ^{19}F. The shell-model analysis of this three-particle configuration had been given by Elliott and Flowers (62) and the rotational interpretation was recognized by Paul (63); the approximate identity of the wave functions derived by the two approaches was established by Redlich (64).

of the number of nucleons A and, in some of the light nuclei, one has, in fact, obtained evidence for such a limitation in the ground-state rotational bands[8]. However, the proper place of this effect in nuclear rotations is still an open issue due to the major deviations from the schematized SU_3 picture.

GENERAL THEORY OF ROTATION

The increasing precision and richness of the spectroscopic data kept posing problems that called for a framework, in which one could clearly distinguish between the general relations characteristic of the rotational coupling scheme and the features that depend more specifically on the internal structure and the dynamics of the rotational motion[9]. For ourselves, an added incentive was provided by the challenge of presenting the theory of rotation as part of a broad view of nuclear structure. The view-points that I shall try to summarize gradually emerged in this prolonged labour (70, 71, 35).

In a general theory of rotation, symmetry plays a central role. Indeed, the very occurrence of collective rotational degrees of freedom may be said to originate in a breaking of rotational invariance, which introduces a "deformation" that makes it possible to specify an orientation of the system. Rotation represents the collective mode associated with such a spontaneous symmetry breaking (Goldstone boson).

The full degrees of freedom associated with rotations in three-dimensional space come into play if the deformation completely breaks the rotational symmetry, thus permitting a unique specification of the orientation. If the deformation is invariant with respect to a subgroup of rotations, the corresponding elements are part of the intrinsic degrees of freedom, and the collective rotational modes of excitation are correspondingly reduced, disappearing entirely in the limit of spherical symmetry.

The symmetry of the deformation is thus reflected in the multitude of states that belong together in rotational families and the sequence of rotational quantum numbers labelling these states, in a similar manner as in the symmetry classification of molecular rotational spectra. The nuclear rotational spectra shown in Figs. 3, 4, and 6 imply a deformation with axial symmetry and invariance with respect to a rotation of 180° about an axis perpendicular to the symmetry axis (D_∞ symmetry group). It can also be inferred from the observed spectra that the deformation is invariant with respect to space and time reflection.

[8] The evidence (66, 67) concerns the behaviour of the quadrupole transition rates, which are expected to vanish with the approach to the band termination (65). This behaviour reflects the gradual alignment of the angular momenta of the particles and the associated changes in the nuclear shape that lead eventually to a state with axial symmetry with respect to the angular momentum and hence no collective radiation (68), (35).

[9] In this development, a significant role was played by the high-resolution spectroscopic studies (69) which led to the establishment of a generalized intensity relation in the E2 decay of the γ-vibrational band in ^{156}Gd.

The recognition of the deformation and its degree of symmetry breaking as the central element in defining rotational degress of freedom opens new perspectives for generalized rotational spectra associated with deformations in many different dimensions including spin, isospin, and gauge spaces, in addition to the geometrical space of our classical world. The resulting rotational band structure may involve comprehensive families of states labelled by the different quantum numbers of the internally broken symmetries. Relations between quantum numbers belonging to different spaces may arise from invariance of the deformation with respect to a combination of operations in the different spaces[10].

The Regge trajectories that have played a prominent role in the study of hadronic properties have features reminiscent of rotational spectra, but the symmetry and nature of possible internal deformations of hadrons remain to be established. Such deformations might be associated with boundaries for the regions of quark confinement.

The condensates in superfluid systems involve a deformation of the field that creates the condensed bosons or fermion pairs. Thus, the process of addition or removal of a correlated pair of electrons from a superconductor (as in a Josephson junction) or of a nucleon pair from a superfluid nucleus constitutes a rotational mode in the gauge space in which particle number plays the role of angular momentum (73). Such pair rotational spectra, involving families of states in different nuclei, appear as a prominent feature in the study of two-particle transfer processes (74). The gauge space is often felt as a rather abstract construction but, in the particle-transfer processes, it is experienced in a very real manner.

The relationship between the members of a rotational band manifests itself in the simple dependence of matrix elements on the rotational quantum numbers, as first encountered in the $I(I+1)$ dependence of the energy spectra and in the leading-order intensity rules that govern transitions leading to different members of a band. The underlying deformation is expressed by the occurrence of collective transitions within the band.

For sufficiently small values of the rotational quantum numbers, the analysis of matrix elements can be based on an expansion in powers of the angular momentum. The general structure of such an expansion depends on the symmetry of the deformation and takes an especially simple form for axially symmetric systems. As an example, Fig. 9 shows the two lowest bands observed in ^{166}Er. The energies within each band have been measured with enormous precision and can be expressed as a power series that converges rather rapidly for the range of angular momentum values included in the figure. Similar expansions can be given for matrix elements of tensor operators representing

[10] A well-known example is provided by the strong-coupling fixed-source model of the pion-nucleon system, in which the intrinsic deformation is invariant with respect to simultaneous rotations in geometrical and isospin spaces resulting in a band structure with $I = T$ (72, 35).

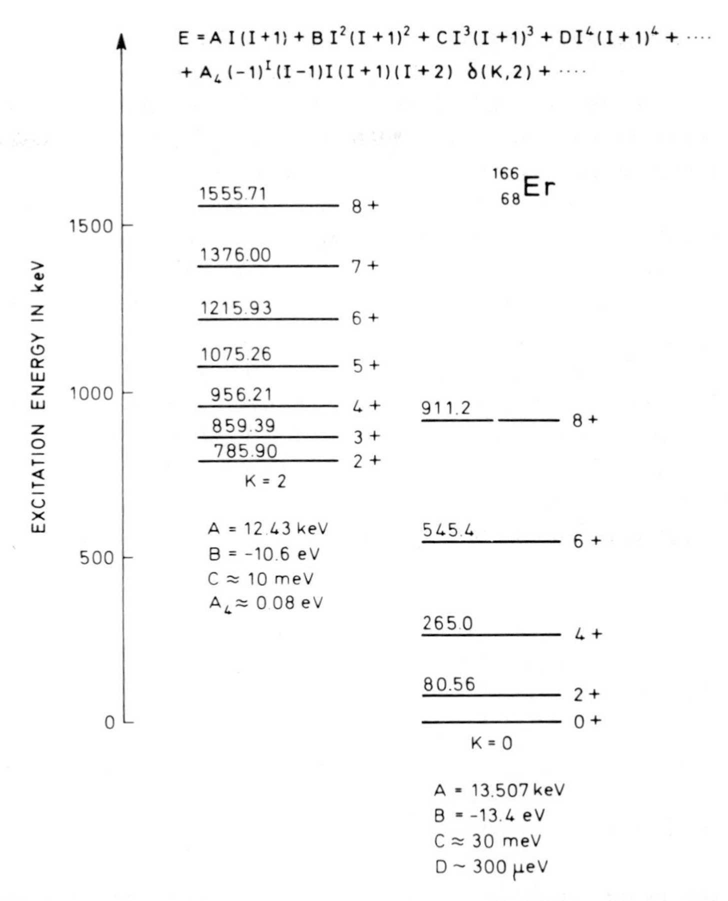

Fig. 9. Rotational bands in ^{166}Er. The figure is from (35) and is based on the experimental data by Reich and Cline (75). The bands are labelled by the component K of the total angular momentum with respect to the symmetry axis. The $K = 2$ band appears to represent the excitation of a mode of quadrupole vibrations involving deviations from axial symmetry in the nuclear shape.

electromagnetic transitions, β decay, particle transfer, etc. Thus, extensive measurements have been made of the E2 transitions between the two bands in ^{166}Er, and Fig. 10 shows the analysis of the empirical transition matrix elements in terms of the expansion in the angular momentum quantum numbers of initial and final states.

Such an analysis of the experimental data provides a phenomenological description of the rotational spectra in terms of a set of physically significant parameters. These parameters characterize the internal structure of the system with inclusion of the renormalization effects arising from the coupling to the rotational motion.

A systematic analysis of these parameters may be based on the ideas of the cranking model, and this approach has yielded important qualitative insight into the variety of effects associated with the rotational motion. However, in

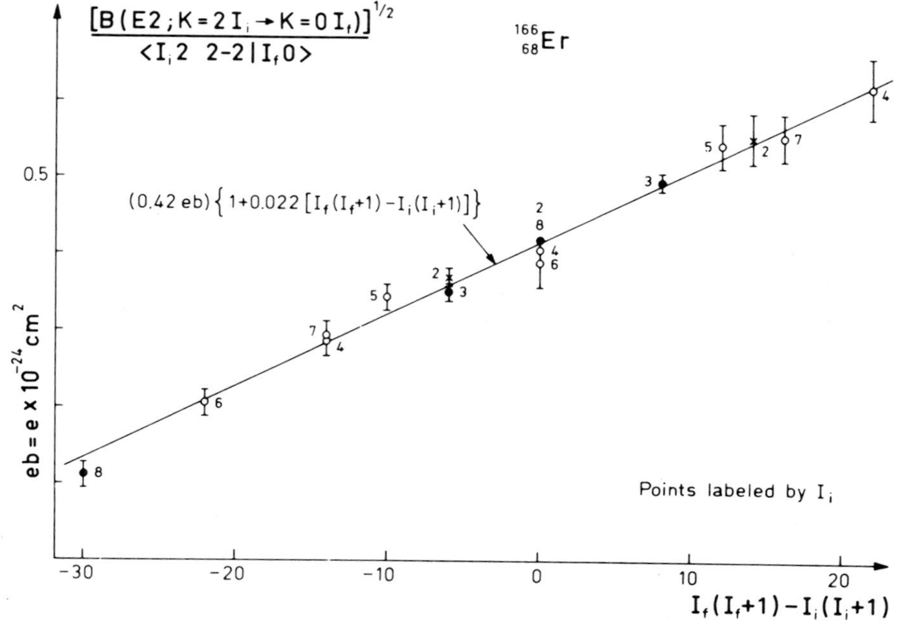

Fig. 10. Intensity relation for E2 transitions between rotational bands. The figure, which is from (35) and is based upon experimental data in (76), shows the measured reduced electric quadrupole transition probabilities B(E2) for transitions between members of the $K = 2$ and $K = 0$ bands in ^{166}Er (see Fig. 9). An expansion similar to that of the energies in Fig. 9, but taking into account the tensor properties of the E2 operator, leads to an expression for $(B(E2)^{1/2}$ which involves a Clebsch-Gordan coefficient $< I_i\, K_i\, 2-2|I_f\, K_f >$ (geometrical factor) multiplied by a power series in the angular momenta of I_i and I_f of the initial and final states. The leading term in this expansion is a constant, and the next term is linear in $I_f(I_f+1)-I_i(I_i+1)$; the experimental data are seen to be rather well represented by these two terms.

this program, one faces significant unsolved problems. The basic coupling involved in the cranking model can be studied directly in the Coriolis coupling between rotational bands in odd-A nuclei associated with different orbits of the unpaired particle (77). The experiments have revealed, somewhat shockingly that, in many cases, this coupling is considerably smaller than the one directly experienced by the particles as a result of the nuclear rotation with respect to the distant galaxies (78). It is possible that this result may reflect an effect of the rotation on the nuclear potential itself (57, 79, 80, 35), but the problem stands as an open issue.

CURRENT PERSPECTIVES

In the years ahead, the study of nuclear rotation holds promising new perspectives. Not only are we faced with the probiem already mentioned of a more deep-going probing of the rotational motion, which has become possible with the powerful modern tools of nuclear spectroscopy, but new frontiers are opening up through the possibility of studying nuclear states with very large values of the angular momentum. In reactions induced by heavy ions, it is in fact now possible to produce nuclei with as much as a hundred units of angular momentum. We thus encounter nuclear matter under quite novel conditions, where centrifugal stresses may profoundly affect the structure of the nucleus. The challenge of this new frontier has strongly excited the imagination of the nuclear physics community.

A schematic phase diagram showing energy versus angular momentum for a nucleus with mass number $A \approx 160$ is shown in Fig. 11. The lower curve representing the smallest energy, for given angular momentum, is referred to as the yrast line. The upper curve gives the fission barrier, as a function of angular momentum, estimated on the basis of the liquid-drop model (81). For $I \approx 100$, the nucleus is expected to become unstable with respect to fission, and the available data on cross sections for compound-nucleus formation in heavy ion collisions seem to confirm the approximate validity of this estimate of the limiting angular momentum (82).

Present information on nuclear spectra is confined almost exclusively to a small region in the left-hand corner of the phase diagram, and a vast extension of the field is therefore coming within range of exploration. Special interest attaches to the region just above the yrast line, where the nucleus, though highly excited, remains cold, since almost the entire excitation energy is concentrated in a single degree of freedom. One thus expects an excitation spectrum with a level density and a degree of order similar to that near the ground state. The extension of nuclear spectroscopy into this region may therefore offer the opportunity for a penetrating exploration of how the nuclear structure responds to the increasing angular momentum.

In recent years, it has been possible to identify quantal states in the yrast region up to $I \approx 20$ to 25, and striking new phenomena have been observed.

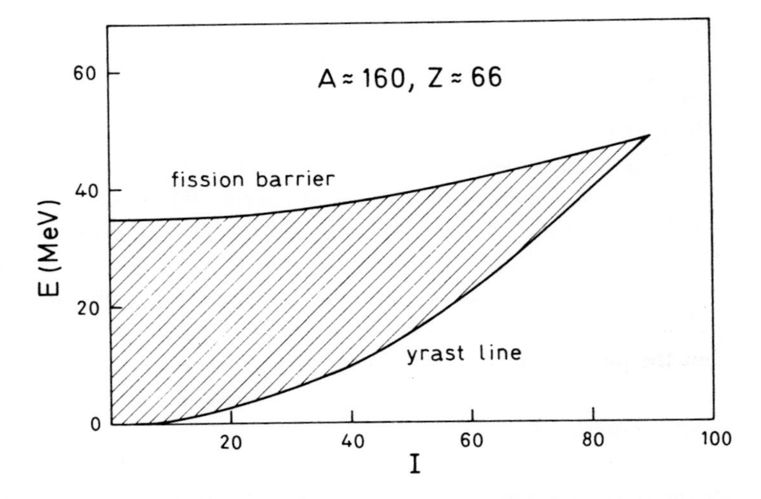

Fig. 11. Nuclear phase diagram for excitation energy versus angular momentum. The yrast line and the fission barrier represent estimates, due to Cohen, Plasil, and Swiatecki (81), based on the liquid-drop model, with the assumption of the rigid-body value for the moment of inertia.

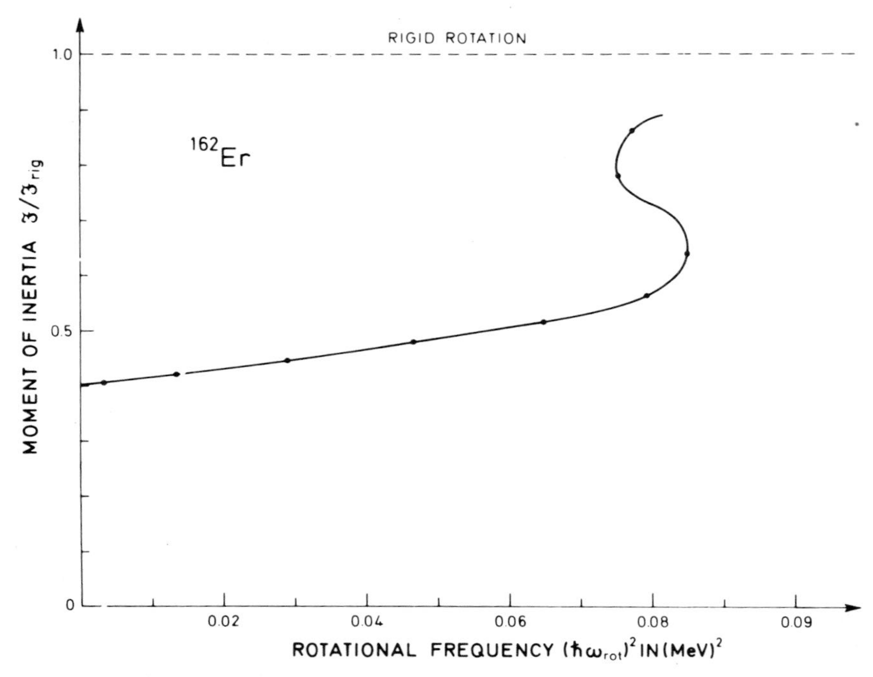

Fig. 12. Moment of inertia as function of rotational frequency. The figure is from (83) and is based on the experimental data of Johnson, Ryde, and Hjorth (84). The rotational frequency is defined as the derivative of the rotational energy with respect to the angular momentum and is obtained by a linear interpolation in the variable $I(I+1)$ between the quantal states. The moment of inertia is defined in the usual manner as the ratio between the angular momentum and the rotational frequency.

An example is shown in Fig. 12, in which the moment of inertia is plotted against the rotational frequency. This "back-bending"effect was discovered here in Stockholm at the Research Institute for Atomic Physics, and has been found to be a rather general phenomenon.

In the region of angular momenta concerned, one is approaching the phase transition from superfluid to normal nuclear matter, which is expected to occur when the increase in rotational energy implied by the smaller moment of inertia of the superfluid phase upsets the gain in correlation energy (85). The transition is quite analogous to the destruction of superconductivity by a magnetic field and is expected to be associated with an approach of the moment of inertia to the rigid-body value characteristic of the normal phase.

The back-bending effect appears to be a manifestation of a band crossing, by which a new band with a larger moment of inertia and correspondingly smaller rotational frequency for given angular momentum, moves onto the yrast line. Such a band crossing may arise in connection with the phase transition, since the excitation energy for a quasiparticle in the rotating potential may vanish, even though the order parameter (the binding energy of the correlated pairs) remains finite, in rather close analogy to the situation in gapless superconductors (86). In fact, in the rotating potential, the angular

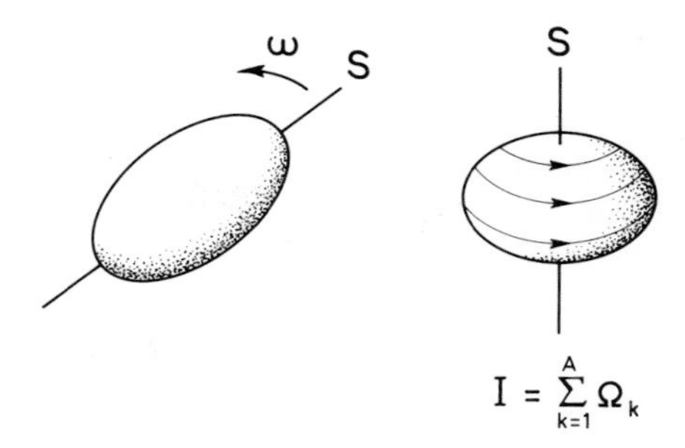

Fig. 13. Collective rotation contrasted with alignment of particle angular momenta along a symmetry axis.

momentum carried by the quasiparticle tends to become aligned in the direction of the axis of rotation. The excitation of the quasiparticle is thus associated with a reduction in the angular momentum and, hence, of the energy that is carried by the collective rotation (87).

It must be emphasized that, as yet, there is no quantitative interpretation of the striking new phenomena, as exemplified by Fig. 12. One is facing the challenge of analyzing a phase transition in terms of the individual quantal states.

For still larger values of the angular momentum, the centrifugal stresses are expected to produce major changes in the nuclear shape, until finally the system becomes unstable with respect to fission. The path that a given nucleus follows in deformation space will depend on the interplay of quantal effects associated with the shell structure and classical centrifugal effects similar to those in a rotating liquid drop. A richness of phenomena can be envisaged, but I shall mention only one of the intriguing possibilities.

The classical centrifugal effects tend to drive the rotating system into a shape that is oblate with respect to the axis of rotation, as is the case for the rotating earth. An oblate nucleus, with its angular momentum along the symmetry axis, will represent a form for rotation that is entirely different from that encountered in the low-energy spectrum, where the axis of rotation is perpendicular to the symmetry axis (see Fig. 13). For a nucleus spinning about its symmetry axis, the average density and potential are static, and the total angular momentum is the sum of the quantized contributions from the individual particles. In this special situation, we are therefore no longer dealing with a collective rotational motion characterized by enhanced radiative transitions, and the possibility arises of yrast states with relatively long lifetimes (88). If such high-spin metastable states (super-dizzy nuclei) do in fact occur, the study of their decay will provide quite new opportunities for exploring rotational motion in the nucleus at very high angular momenta.

Thus, the study of nuclear rotation has continued over the years to be alive and to reveal new, challenging dimensions. Yet, this is only a very special aspect of the broader field of nuclear dynamics that will be the subject of the following talk.

REFERENCES

1. Thibaud, J., Comptes rendus *191,* 656 (1930)
2. Teller, E., and Wheeler, J. A., Phys. Rev. *53,* 778 (1938)
3. Bjerrum, N. in *Nernst Festschrift,* p. 90, Knapp, Halle 1912
4. Mayer, M. G., Phys. Rev. *75,* 209 (1949); Haxel, O., Jensen, J. H. D., and Suess, H. E., Phys. Rev. *75,* 1766 (1949)
5. Bohr, N., Nature *137,* 344 (1936); Bohr, N., and Kalckar, F., Mat. Fys. Medd. Dan. Vid. Selsk. *14,* no, 10 (1937); Weisskopf, V. F., Phys. Rev. *52,* 295 (1937); Meitner, L., and Frich, O. R., Nature *143,* 239 (1939); Bohr, N., and Wheeler, J. A., Phys. Rev. *56,* 426 (1939); Frenkel, J., J. Phys. (USSR) *1,* 125 (1939); see also Phys. Rev. *55,* 987 (1939)
6. Jensen, J. H. D., Les Prix Nobel en 1963, p. 153 Imprimerie Royale P. A. Norstedt & Söner, Stockholm 1964
7. Bohr, N., *Tentative Comments on Atomic and Nuclear Constitution.* Manuscript dated August 1949. Niels Bohr Archives, The Niels Bohr Institute, Copenhagen
8. Hill, D. L., and Wheeler, J. A., Phys. Rev. *89,* 1102 (1953)
9. Schmidt, T., Z Physik *106,* 358 (1937)
10. Bitter, F., Phys. Rev. *75,* 1326 (1949); *76,* 150, (1949)
11. Bohr, A., and Weisskopf, V. F., Phys. Rev. *77,* 94 (1950)
12. Schüler, H., and Schmidt, Th., Z. Physik *94,* 457 (1935)
13. Casimir, H. B. G., *On the Interaction Between Atomic Nuclei and Electrons,* Prize Essay, Taylor's Tweede Genootschap, Haarlem (1936); see also Kopfermann, H., Kernmomente, Akademische Verlagsgesellschaft, Leipzig 1940
14. Townes, C. H., Foley, H. M., and Low, W., Phys. Rev. *76,* 1415 (1949)
15. Bohr, A., Phys. Rev. *81,* 134 (1951)
16. Bohr, A., *Rotational States of Atomic Nuclei,* Munksgård, Copenhagen 1954
17. Bohr, A., Phys. Rev. *81,* 331 (1951)
18. Foldy, L. L., and Milford, F. J., Phys. Rev. *80,* 751 (1950)
19. Rainwater, J., Phys. Rev. *79,* 432 (1950)
20. Bohr, A., and Mottelson, B. R., Mat. Fys. Medd. Dan. Vid. Sselsk. *27,* no. 16 (1953)
21. Goldhaber, M., and Sunyar. A. W., Phys. Rev. *83,* 906 (1951)
22. Mayer, M. G., Moszkowski, S. A., and Nordheim, L. W., Rev. Mod. Phys. *23,* 315 (1951) Nordheim, L. W., Rev. Mod. Phys. *23,* 322 (1951)
23. Barschall, H. H., Phys. Rev. *86,* 431 (1952)
24. Weisskopf, V. F., Physica *18,* 1083 (1952)
25. Bohr, A., and Mottelson, B. R., Phys. Rev. *89,* 316 (1953)
26. Bohr, A., and Mottelson, B. R., Phys. Rev. *90,* 717 (1953)
27. Asaro, F., and Perlman, I., Phys. Rev. *91,* 763 (1953)
28. Burson, S. B., Blair, K. W., Keller, H. B., and Wexler, S., Phys. Rev. *83,* 62 (1951)
29. Hollander, J. M., Perlman, I., and Seaborg, G. T., Rev. Mod. Phys. *25,* 469 (1953)
30. Huus, T. and Zupančič, Č., Mat. Fys. Medd. Dan. Vid. Selsk. *28,* no. 1 (1953)
31. McClelland, C. L., and Goodman, C., Phys. Rev. *91,* 760 (1953)
32. Heydenburg, N. P., and Temmer, G. M., Phys. Rev. *94,* 1399 (1954) and *100,* 150 (1955)
33. Ter-Martirosyan, K. A., Zh. Eksper. Teor. Fiz. *22,* 284 (1952)
34. Alder, K., and Winther, A. Phys. Rev. *91,* 1578 (1953)
35. Bohr, A., and Mottelson, B. R., *Nuclear Structure,* Vol. II. Benjamin, W. A. Inc., Reading, Mass. 1975
36. Sakai, M., Nuclear Data Tables *A8,* 323 (1970) and *A10,* 511 (1972)
37. Alaga, G., Alder, K., Bohr, A., and Mottelson, B. R., Mat. Fys. Medd. Dan. Vid. Selsk. *29,* no. 9 (1955)
38. Bohr, A., Fröman, P. O., and Mottelson, B. R., Mat. Fys. Medd. Dan. Vid. Selsk. *29,* no. 10 (1955)

39. Satchler, G. R., Phys. Rev. *97,* 1416 (1955)
40. Nilsson, S. G., Mat. Fys. Medd. Dan. Vid. selsk. *29,* no. 16 (1955)
41. Mottelson, B. R., and Nilsson, S. G., Phys. Rev. *99,* 1615 (1955)
42. Gottfried, K.,.Phys. Rev. *103,* 1017 (1956)
43. Stephens, F. S., Holtz, M. D., Diamond, R. M., and Newton, J. O., Nuclear Phys. *A115,* 129 (1968)
43a. Cline, J. E., Nuclear Phys. *A106,* 481 (1968)
44. Elze, Th. W., and Huizenga, J. R., Nuclear Phys, *A133,* 10 (1969); Braid. T. H., Chasman, R. R., Erskine. J. R., and Friedman, A. M., Phys. Rev. *C1,* 275 (1970)
45. Jurney, E. T., *Neutron Capture Gamma-Ray Spectroscopy,* Proceedings of the International Symposium held in Studsvik, p. 431, International Atomic Energy Agency, Vienna 1969
46. Gustafson, C., Lamm. I. L., Nilsson, B., and Nilsson, S. G., Arkiv Fysik *36,* 613 (1967)
47. Morinaga, H., Phys. Rev. *101,* 254 (1956)
48. Carter, E. B., Mitchell, G. E., and Davies, R. H., Phys. Rev. *133,* B1421 (1964)
49. Gorodetzky, S., Mennrath, P., Benenson, W., Chevallier, P., and Scheibling, F., J. phys. radium *24,* 887 (1963)
50. Polikanov, S. M., Druin, V. A., Karnaukhov, V. A., Mikheev, V. L., Pleve, A. A., Skobelev, N. K., Subbotin, V. G., Ter-Akop'yan, G. M., and Fomichev, V. A., J. Exptl. Theoret. Phys. (USSR) *42,* 1464 (1962); transl. Soviet Physics JETP *15,* 1016
51. Specht, H. J., Weber, J., Konecny, E., and Heunemann, D., Phys. Letters *41B,* 43 (1972)
52. Winhold, E. J., Demos, P. T., and Halpern, I., Phys. Rev. *87,* 1139 (1952)
53. Bohr, A. in *Proc. Intern. Conf. on the Peaceful Uses of Atomic Energy,* vol. 2, p. 151, United Nations, New York, 1956
54. Inglis, D. R., Phys. Rev. *96,* 1059 (1954)
55. Thouless, D. J., and Valatin, J. G., Nuclear Phys. *31,* 211 (1962)
56. Bohr, A., and Mottelson, B. R., Mat. Fys. Medd. Dan. Vid. Selsk. *30,* no. 1 (1955)
57. Migdal, A. B., Nuclear Phys. *13,* 655 (1959)
58. Belyaev, S. T., Mat. Fys. Medd. Dan. Selsk. *31,* no. 11 (1959)
59. Bardeen, J., Cooper, L. N., and Schrieffer, J. R., Phys. Rev. *106,* 162 and *108,* 1175 (1957)
60. Mottelson, B. R., this volume, p. 82.
61. Ruderman, M., Ann. Rev. Astron. and Astrophys. *10,* 427 (1972)
62. Elliott, J. P., and Flowers, B. H., Proc. Roy. Soc. (London) *A229,* 536, (1955)
63. Paul, E. B., Phil. Mag. *2,* 311 (1957)
64. Redlich, M. G., Phys. Rev. *110,* 468 (1958)
65. Elliott, J. P., Proc. Roy. Soc. (London) *A245,* 128 and 562 (1958)
66. Jackson, K. P., Ram, K. B., Lawson, P. G., Chapman N. G., and Allen, K. W., Phys. Letters *30B,* 162 (1969)
67. Alexander, T. K., Häusser, O., McDonald, A. B., Ferguson, A. J., Diamond, W. T., and Litherland, A. E., Nuclear Phys. *A179,* 477 (1972)
68. Bohr, A. in *Intern. Nuclear Phys. Conf.,* p. 489, Ed.-in-Chief R. L. Becker, Academic Press, New York 1967
69. Hansen, P. G., Nielsen, O. B., and Sheline, R. K., Nuclear Phys. *12,* 389 (1959)
70. Bohr, A., and Mottelson, B. R., Atomnaya Energiya *14,* 41 (1963)
71. Bohr, A. in *Symmetry Properties of Nuclei,* Proc. of 15th Solvay Conf. on Phys. 1970, p. 187. Gordon and Breach Science Publ., London 1974
72. Henley, E. M., and Thirring, W., *Elementary Quantum Field Theory,* McGraw Hill, New York 1962
73. Anderson, P. W., Rev. Mod. Phys. *38,* 298 (1966)
74. Middleton, R., and Pullen, D. J., Nuclear Phys. *51,* 77 (1964); see also Broglia, R.

A., Hansen, O., and Riedel, C. *Advances in Nuclear Physics 6,* 287, Plenum Press, New York 1973
75. Reich, C. W., and Cline, J. E., Nuclear Phys. *A159,* 181 (1970)
76. Gallagher, C. J., Jr., Nielsen, O. B., and Sunyar, A. W., Phys. Letters *16,* 298 (1965); Günther, C., and Parsignault, D. R., Phys. Rev. *153,* 1297 (1967); Domingos, J. M., Symons, G. D., and Douglas, A. C., Nlclear Phys. *A180,* 600 (1972)
77. Kerman, A. K., Mat. Fys. Medd. Dan. Vid. Selsk. *30,* no. 15 (1956)
78. Stephens, F. (1960), quoted by Hyde, E., Perlman, I., and Seaborg, G. T. in *The Nuclear Properties of the Heavy Elements,* Vol. II, p. 732, Prentice Hall, Englewood Cliffs, N.J. 1964; Hjorth, S. A., Ryde, H., Hagemann, K. A., Løvhøiden, G., and Waddington, J. C., Nuclear Phys. *A144,* 513 (1970); see also the discussion in (35)
79. Belyaev, S. T., Nuclear Phys. *24,* 322 (1961)
80. Hamamoto, I., Nuclear Phys. *A232,* 445, (1974)
81. Cohen, S., Plasil, F., and Swiatecki, W. J., Ann, Phys. *82,* 557 (1974)
82. Britt, H. C., Erkilla, B. H., Stokes, R. H., Gutbrod, H. H., Plasil, F., Ferguson, R. L., and Blann, M., Phys. Rev. *C* in press; Gauvin, H., Guerrau, D., Le Beyec, Y., Lefort, M., Plasil, F., and Tarrago, X., Phys. Letters, *58B,* 163 (1975)
83. Bohr, A., and, Mottelson, B. R., *The Many Facets of Nuclear Structure.* Ann. Rev. Nuclear Science *23,* 363 (1973)
84. Johnson, A., Ryde, H., and Hjorth, S. A., Nuclear Phys. *A179,* 753 (1972)
85. Mottelson, B. R., and Valatin, J. G., Phys. Rev. Letters *5,* 511 (1960)
86. Goswami, A., Lin, L., and Struble, G. L., Phys. Letters *25B,* 451 (1967)
87. Sephens, F. S., and Simon, R. S., Nuclear Phys. *A183,* 257 (1972)
88. Bohr, A., and Mottelson, B. R., Physica Scripta *10A,* 13 (1974)

Ben R. Mottelson

BEN R. MOTTELSON

I was born in Chicago, Illinois, on July 9, 1926, the second of three children of Goodman Mottelson and Georgia Mottelson (née Blum). My father held a university degree in engineering. My childhood home was a place where scientific, political and moral issues were freely and vigorously discussed. I attended primary school and high school in the village of La Grange, Illinois.

Graduating from high school during the second world war, I was sent by the U.S. Navy to Purdue University for officers training (V12 program) and remained there to receive a Bachelor of Science degree in 1947. My graduate studies were at Harvard University and my PhD work on a problem in nuclar physics was directed by Professor Julian Schwinger and completed in 1950.

Receiving a Sheldon Traveling Fellowship from Harvard University I chose to spend the year (1950—51) at the Institute for Theoretical Physics in Copenhagen (later the Niels Bohr Institute) where so much of modern physics had been created and where there were such special traditions for international cooperation. A fellowship from the U.S. Atomic Energy Commission permitted me to continue my work in Copenhagen for two more years after which I held a research position in the CERN (European Organization for Nuclear Research) theoretical study group that was formed in Copenhagen. With the founding of the Nordic Institute for Theoretical Atomic Physics in Copenhagen (1957) I received a position as professor which I have held since. The spring term of 1959 was spent as visiting professor in the University of California at Berkeley.

The close scientific collaboration with Aage Bohr was begun in 1951 and has continued ever since. We feel that in this cooperation we have been able to exploit possibilities that lie in a dialogue between kindred spirits that have been attuned through a long period of common experience and jointly developed understanding. The lectures that are published in this volume attempt a discussion of the main influences that we have built on and the viewpoints that have been developed in this collaboration. It has been our good fortune to work closely together with colleagues at the Niels Bohr Institute and Nordita, including the many outstanding scientists who have come from all parts of the world and have so enriched the scientific atmosphere and personal contacts.

In 1948 I married Nancy Jane Reno and we had three children. The family became Danish citizens in 1971.

ELEMENTARY MODES OF EXCITATION IN THE NUCLEUS

Nobel Lecture, December 11, 1975

by
BEN R. MOTTELSON
NORDITA, Copenhagen, Denmark

In the field of nuclear dynamics a central theme has been the struggle to find the proper place for the complementary concepts referring to the independent motion of the individual nucleons and the collective behaviour of the nucleus as a whole. This development has been a continuing process involving the interplay of ideas and discoveries relating to all different aspects of nuclear phenomena. The multi-dimensionality of this development makes it tempting to go directly to a description of our present understanding and to the problems and perspectives as they appear today. However, an attempt to follow the evolution of some of the principal ideas may be instructive in illustrating the struggle for understanding of many-body systems, which have continued to inspire the development of fundamental new concepts, even in cases where the basic equations of motion are well established. Concepts appropriate for desribing the wealth of nuclear phenomena have been derived from a combination of many different approaches, including the exploration of general relations following from considerations of symmetry, the study of model systems, sometimes of a grossly oversimplified nature, and, of course, the clues provided by the experimental discoveries which have again and again given the development entirely new directions.[1]

The situation in 1950, when I first came to Copenhagen, was characterized by the inescapable fact that the nucleus sometimes exhibited phenomena characteristic of independent-particle motion, while other phenomena, such as the fission process and the large quadrupole moments, clearly involved a collective behaviour of the whole nucleus.

It was also clear from the work of Rainwater that there was an important coupling between the motion of the individual particles and the collective deformation, and one was thus faced with the problem of exploring the properties of a dynamical system involving such coupled degrees of freedom (1, 2, 3, 4).

$$\begin{aligned} H &= H_{\text{vib}} + H_{\text{part}} + H_{\text{coupl}} \\ H_{\text{vib}} &= \tfrac{1}{2} C_\lambda \sum_\mu |\alpha_{\lambda\mu}|^2 + \tfrac{1}{2} D_\lambda \sum_\mu |\dot{\alpha}_{\lambda\mu}|^2 \\ H_{\text{coupl}} &= \sum_p k(r_p) \sum_\mu \alpha_{\lambda\mu} Y_{\lambda\mu}(\varphi_p \theta_p) \end{aligned} \tag{1}$$

[1] We would like to take this opportunity to pay tribute to the ingenuity and resourcefulness of the generation of experimentalists whose untiring efforts have created the basis for the development sketched in our reports today.

where $\alpha_{\lambda\mu}$ are the amplitudes of the nuclear deformation expanded in spherical harmonics and (r_p,θ_p,φ_p) are the coordinates of the particles considered. The coupling term represents the effect of the deformation on the one-particle potential.

I remember vividly the many lively discussions in these years reflecting the feeling of unease, not to say total disbelief, of many of our colleagues concerning the simultaneous use of both collective and single-particle coordinates to describe a system that we all agreed was ultimately built out of the neutrons and protons themselves. Niels Bohr participated very actively in these discussions. Something of the flavour of this contribution can perhaps be gathered from the exhange recorded in the Proceedings of the CERN International Physics Conference in Copenhagen from June 1952; I had given a report on our work, and in the discussion Rosenfeld "asked how far this model is based on first principles". N. Bohr "answered that it appeared difficult to define what one should understand by first principles in a field of knowledge where our starting point is empirical evidence of different kinds, which is not directly combinable".

I would like to take this opportunity to acknowledge the tremendous inspiration it has been for me to have had the privilege to work for the entire period covered by this report within the unique scientific environment created by Niels Bohr.

INTERPRETATION OF LOW-ENERGY NUCLEAR EXCITATION SPECTRA

In the beginning of the 1950'ies, there existed very little evidence on nuclear spectra, which could be used to test these idea. In the following years, however, a dramatic development of nuclear spectroscopy took place. The new data made possible the identification of the characteristic patterns of rotational spectra (5) and shortly afterwards the recognition by Scharff-Goldhaber and Weneser (6) that a significant class of spectra exhibit patterns corresponding to quadrupole vibrations about a spherical equilibrium[2]. The existence of the static deformations in certain classes of nuclei received further decisive confirmation in the successful classification of the intrinsic states of these spectra in terms of one-particle motion in an appropriately deformed potential (5).

A striking feature in the developing picture of nuclear excitation spectra was the distinction between a class of nuclei with spherical shape and others with large deformations. The clue to the origin of this distinction came, rather unexpectedly, from the analysis of the moments of intertia of the rotational spectra. The cranking model of Inglis (14) had provided a starting point for a microscopic interpretation of the rotational motion, and the analysis

[2] This step followed the recognition of striking regularities in the low-energy spectra of even-even nuclei, including the spins and parities (7, 8), energy systematics (8, 9, 10, 11), and selection rules (12).

showed that significant deviations from independent-particle motion were required to account for the observed magnitude of the moments of inertia. These correlations could be attributed to the residual interactions that tend to bind the nucleons into pairs with angular momentum zero. Such a pair is spherically symmetric, and this nucleonic correlation could therefore, at the same time, be seen to provide an effect tending to stabilize the spherical shape (15).

Thus, quite suddenly the way was opened to a qualitative understanding of the whole pattern of the low-energy excitation spectra in terms of a competition between the pairing effect and the tendency toward deformations implied by the anisotropy of the single-particle orbits. The outcome of this competition depends on the number of particles in unfilled shells; for few particles, the deformation in the absence of interactions is relatively small and can easily be dominated by the tendency to form spherical pairs; but with increasing number of particles, the spherical equilibrium shape becomes less stable, and eventually a transition takes place to a deformed equilibrium shape. These considerations are illustrated by the potential energy surfaces shown in Fig. 1.

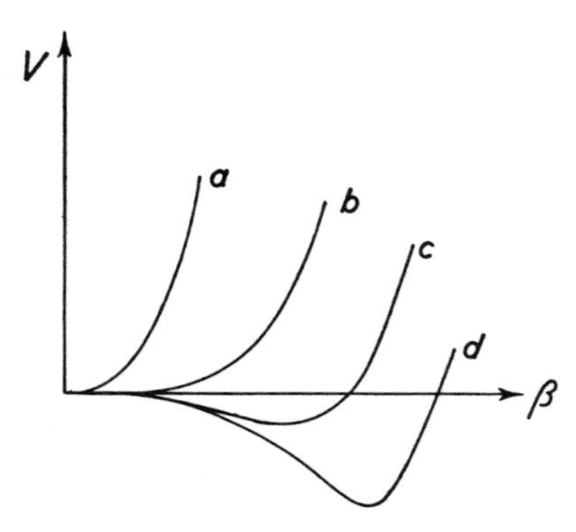

Fig. 1. Nuclear potential energy function. The figure, taken from (13), gives a schematic representation of the nuclear energy as function of the deformation β. The curve *a* represents a configuration with only relatively few particles outside of closed shells. As particles are added, the restoring force for the spherical shape ($\beta = 0$) decreases (curve *b*). Still further from the closed shells, the spherical shape may become unstable (curves *c* and *d*) and the nucleus acquires a non-spherical equilibrium shape.

MICROSCOPIC DESCRIPTION OF COLLECTIVE MOTION

This qualitative interpretation of the nuclear coupling schemes could soon be given a firmer basis in terms of many-body wave functions that describe the correlation effects governing the low-energy nuclear spectra.

A step towards a microscopic understanding of the deformation effect resulted from the discovery of rotational spectra in light nuclei[3]. For these nuclei, even a few particles represent a significant fraction of the total and can

[3] Rotational band structure and the classification of the intrinsic states for *(sd)*-shell nuclei was first establisched in 1955 following the extensive series of experiments at Chalk River (see the survey by A. Litherland et al. (16)). For the classification of the *p*-shell nuclei in terms of the rotational coupling scheme, see Kurath and Pičman (17). For our own understanding of the special flavour of these very light nuclei, the discussions over the years with Tom Lauritsen were a continuing challenge and source of inspiration.

give rise to deformations that are among the largest observed. The spectra of some of these nuclei had previously been successfully analyzed in terms of shell-model configurations (5). Thus, for the first time one had a many-body wave function with rotational relationships and one could see explicitly that the main effect of the rather complicated finite range interactions employed in the shell-model calculations had been to generate a deformed average potential.

The essence of this development was brought into focus by Elliott's discovery that the SU_3 classification scheme for particles in a harmonic oscillator potential leads to multiplets with rotational relations (18). The effective two-body interaction that is invariant under SU_3 symmetry (when acting within the configurations of a major shell) and thus leads to the rotational coupling scheme, is given by the scalar product of the quadrupole moments of each pair of particles.

$$V_{\mathrm{eff}} = \tfrac{1}{2} H \sum_{ij} (q(i)q(j))_0 \qquad (2)$$
$$q_\mu(i) = r_i^2\, Y_{2\mu}(\varphi_i \theta_i)$$

Such a two-body force is equivalent to the interaction of each particle with the total quadrupole moment of the system and thus to the effect of an ellipsoidal deformation in the average potential.

In retrospect, the important lesson of this development was the recognition that the aligned wave function

$$\Psi = A\,(\psi_{k1}(x_1)\,\psi_{k2}(x_2) \ldots . \psi_{kA}(x_A)) \qquad (3)$$

obtained as a simple product of single-particle states in a self-consistent deformed potential provides a starting point for the full many-body wave function[4]. This view-point had indeed been implied by the establishment of the classification based on the Nilsson scheme, but the revelation of the exact SU_3 solution, even in such an oversimplified model, contributed greatly to the confidence in this approach.

The second major development involved the many-particle interpretation of the nuclear pairing effect. As we have seen, this problem had become a crucial one for the quantitative analysis of collective motion in the nucleus, but the story of the pairing effect goes back much further, to the very earliest days of nuclear physics (21). The discovery of the neutron made it possible to

[4] The wave function given by Eq. (3) represents the intrinsic state in the absence of rotation, and can be directly employed in obtaining the leading-order intensity relations. The I-dependent terms, such as the rotational energy, are obtained by including the rotational perturbations in the intrinsic motion, as in the cranking model. The SU_3 coupling scheme represents a special case in which the total function, with the inclusion of rotational effects can be expressed as a projection of the intrinsic wave function onto a state of specified angular momentum (18). (Such projected wave functions had been employed earlier (19); see also the discussion in (20).)

interpret the accumulated systematics concerning the differences in stability of odd and even nuclei in terms of an additional binding associated with even numbers of protons or neutrons (22). This effect later provided the basis for understanding the striking difference in the fission of the odd and even isotopes of uranium (23). The pairing effect also played an important role in the development of the shell model since it provided the basis for the interpretation of many of the properties of odd-A nuclei in terms of the binding states of the last odd particle (24, 25, 26).

The key to understanding the correlation effect underlying the odd-even differences came from the discovery by Bardeen, Cooper, and Schrieffer of the profound new concepts for treating the electronic correlations in superconductors (27)[5]. It was a marvellous thing that the correlations, which might appear to be associated with such complexity, could be simply expressed in terms of a generalized one-body problem in which the particles move in a potential which creates and annihilates pairs of particles giving rise to the quasiparticles that are superpositions of particles and holes (30, 31). It could also be seen that the many-body wave function represented a generalization of Racah's seniority coupling scheme (32) which had been exploited in the interpretation of the one-particle model in nuclei.

One thus had available the basic tools for a microscopic analysis of the coupling schemes encountered in the low-energy nuclear spectra. These tools were rapidly exploited to treat the moments of inertia of rotating nuclei (33, 34, 35, 36), the potential energy surfaces and inertial parameters for the vibrations of spherical nuclei (33, 37), as well as the effects of pair correlations on a variety of nuclear processes (38, 39, 40, 41).

This was indeed a period of heady development in the understanding of many-body problems with a fruitful interplay of experience gained from the study of so many different systems that nature had provided, including the "elementary particles" that had stimulated the development of the powerful tools of relativistic field theory. An important clarification in the description of collective motion was the new way of viewing the normal modes of vibration as built out of correlated two-quasiparticle (or particle-hole) excitations. The significant part of the interactions creates and annihilates two such basic excitations, and the vibrations can thus be obtained from the solution of a generalized two-body problem (42). This approach not only comple-

[5] It was a fortunate circumstance for us that David Pines spent a period of several months in Copenhagen in the summer of 1957, during which he introduced us to the exciting new developments in the theory of superconductivity. Through the discussions with him, the relevance of these concepts to the problem of pair correlations in nuclei became apparent (28). An important component in these discussions was the fact that the experimental evidence had been accumulating for the existence of an energy gap in the excitation spectra of nuclei reminiscent of that observed in superconductors (15, 28). (For the recognition of the odd-even difference in nuclear excitation spectra, striking evidence had come from the high-resolution spectroscopic studies of ^{182}W and ^{183}W made possible by the bent crystal spectrometer (29).)

mented the previously applied adiabatic treatment of nuclear collective motion, bus also gave a broader scope to the concept of vibration that was to be important for the subsequent development.

The whole picture of nuclear physics at this stage in the development is beautifully expressed by Weisskopf in his summary talk at the Kingston Conference in 1960, where the recurring theme is his comment again and again: "It works surprisingly well."[6]

THE GREAT VARIETY OF COLLECTIVE MODES

While the low-frequency spectra are dominated by transitions of particles within the partly filled shells, new aspects of nuclear dynamics are associated with the excitation of the closed shells. The classic example of a collective excitation of this type is the "giant dipole resonance" which was discovered in the study of the photo-processes soon after the war (43), and which could be given an interpretation in terms of collective motion of the neutron and proton fluids with respect to each other (44, 45).

After the development of the shell model, attempts were made to describe the photo-absorption in terms of single-particle excitations (46), but one encountered the problem that the one-particle excitations that should carry the main part of the dipole strength appeared in a part of the spectrum quite distinct from that in which the strong dipole absorption was observed (see Fig. 2). This led to a period of lively discussions, and for a time it was felt that the single-particle and collective descriptions represented opposite and mutually exclusive interpretations (47).

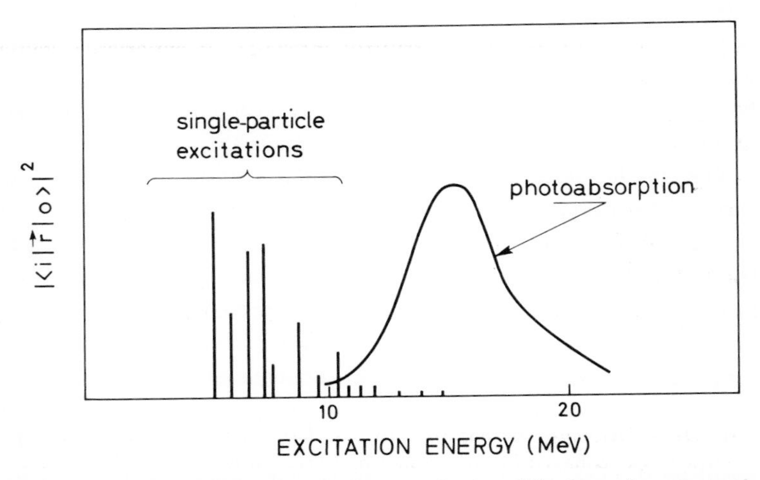

Fig. 2. Frequency distribution of nuclear electric dipole excitations. The figure is a schematic representation (for $A \approx 100$) of the dipole strength for single-particle excitations as compared with the observed frequency distribution of the photo-absorption cross section.

[6] We would like to acknowledge the deep importance for us of the close personal contact with Viki Weisskopf who has provided inspiration for a whole generation of nuclear physicists.

A step in the resolution of the problem resulted from a study of the interaction effects in the single-particle excitations of the closed-shell configuration of ^{16}O, which revealed a strong tendency towards the formation of linear combinations of different particle-hole configurations collecting the major part of the dipole strength and shifting it to higher energy (48). A highly simplified model based on degenerate single-particle excitations, as in the harmonic oscillator potential, again provided valuable insight by exhibiting exact solutions, in which the total dipole strength was collected into a single high-frequency excitation (49, 50). These schematic models could soon be seen in the more general framework of the normal modes treatment referred to above.

In carrying through this program, one faced the uncertainty in the effective forces to be employed, but it was found possible to represent the interactions by an oscillating average potential acting with opposite sign on neutrons and protons, the strength of which could be related to the isovector component in the static central potential that is present in nuclei with a neutron excess (51, 20).[7] Indeed, it appears that all the collective nuclear modes that have been identified can be traced back to average fields of specific symmetry generated by the effective interaction.

The new insight into the manner in which the vibrations are generated by the interactions in the various channels of particle excitations opened a whole new perspective, since one became liberated from the classical picture of vibrations and could begin to imagine the enormously greater variety of vibrational phenomena that are characteristic of quantal systems. This perspective became apparent 10 to 15 years ago, but there was at that time very little experimental evidence on which to build. The understanding of some of the features in this rich fabric of possibilities has been the result of a gradual process (which added a decade to the gestation of Vol. II of our work on Nuclear Structure) and which is still continuing. A few examples may give an impression of the scope of the new phenomena.

The dipole mode is of isovector character and each quantum of excitation carries unit isospin. It is thus a component of a triplet, which also includes excitations that turn neutrons into protons and *vice versa.* In a nucleus with equal numbers of neutrons and protons, and total isospin $T_0 = 0$ in the ground state, the triplet of excitations represents an isobaric multiplet and the different states are therefore directly related in terms of rotations in isospace. However, in a nucleus with neutron excess and total isospin $T_0 \neq 0$ *in the ground* state, the dipole excitations with charge exchange may be very different from those with zero component of the isopin (see Fig. 3). The resulting dipole excitation spectrum is schematically illustrated in Fig. 4 and

[7] The close similarity of the results of the hydrodynamic and the microscopic treatments is a special feature of the dipole mode (20), associated with the fact that the single-particle response function for this channel is concentrated in a single frequency region (see Fig. 2).

presents an example of symmetry breaking resulting from the lack of isobaric isotropy of the "vacuum" (the nuclear ground state). Some of the features in the pattern indicated in Fig. 4 have been experimentally confirmed, but the major part of this rich structure remains to be explored[8].

Another dimension to the vibrational concept is associated with the possibility of collective fields that create or annihilate pairs of particles, in

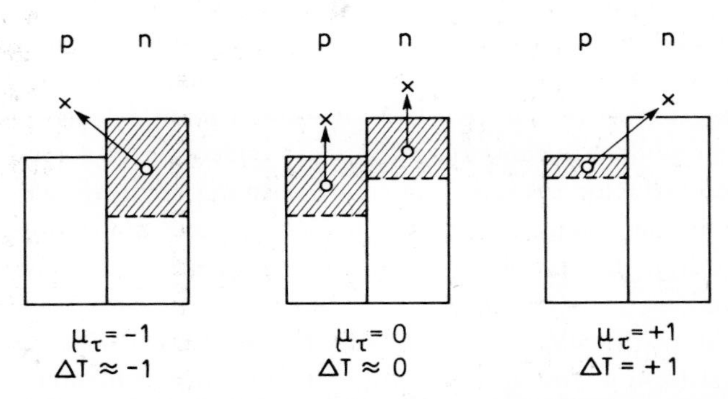

Fig. 3. Single-particle dipole excitations in a nucleus with neutron excess. The boxes represent the occupied proton (p) and neutron (n) orbits and the hatched domains correspond to the particle orbits that can be excited by the isovector dipole field with different components, μ_τ. For large values of the neutron excess, the excitations lead to a change ΔT in the total isospin quantum number equal to μ_τ. The figure is from (20).

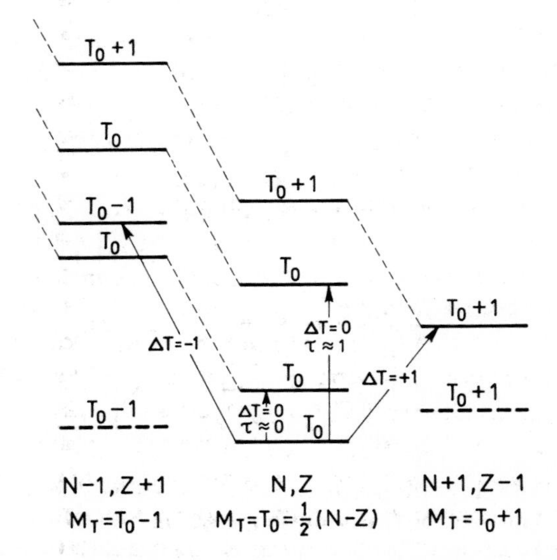

Fig. 4. Isospin of vibrational excitations in nucleus with neutron excess. The ground state of such a nucleus has a total isospin component $M_T = \frac{1}{2}(N-Z)$ and total isospin $T_0 = M_T$. The figure gives a schematic illustration of the pattern of states formed by adding vibrational quanta with isospin $\tau = 0$ and $\tau = 1$. Isobaric analogue states are connected by thin broken lines. The ground states of the isobaric nuclei with $M_T = T_0 \pm 1$ are indicated by dashed lines. The figure is from (20).

[8] For a summary of this development, see Fallieros (52) and reference (20).

contrast to the field associated with the dipole mode that creates particle-hole pairs and therefore conserves particle number. The new fields are connected with the pairing component in the nuclear interactions which tend to bind pairs into a highly correlated state of angular momentum zero. The addition of such a pair to a closed shell constitutes an excitation that can be repeated and which can thus be viewed as a quantum of a vibrational mode. Fig. 5 shows the pair-vibrational spectrum with the two modes associated with addition and removal of the neutrons from the closed-shell configuration of ^{208}Pb. One thus encounters a vibrational band in which the members belong to different nuclei. In systems with many particles outside closed shells the ground state can be viewed as a condensate of correlated pairs as in the superconductor (54). Such a condensate can be expressed as a static *deformation* in the magnitude of the pair field, and the addition and removal of pairs from the condensate constitute the associated rotational mode of excitation.

The clarification of the dynamical role of pair fields in the nucleus has resulted from a close interplay of experimental and theoretical work[9]. From the

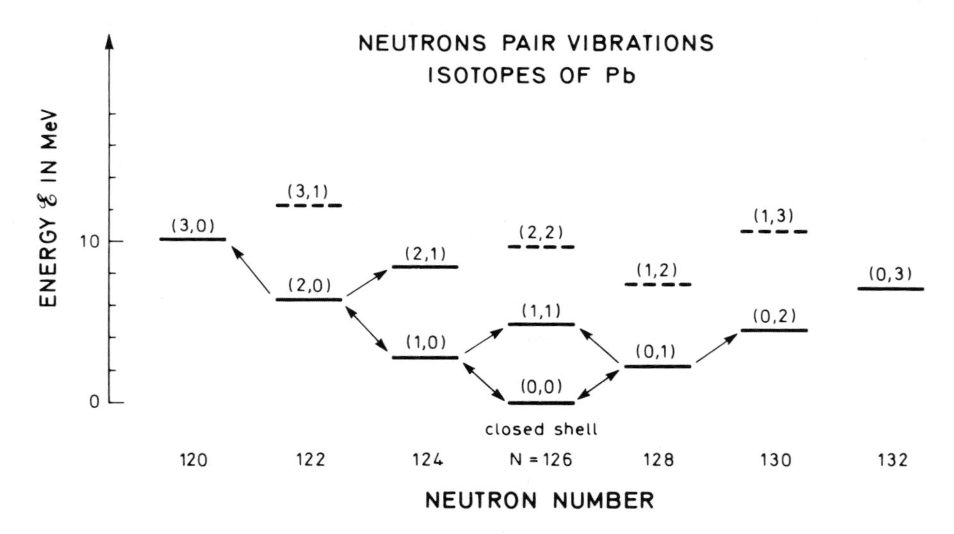

Fig. 5. Neutron monopole pair vibrations based on ^{208}Pb. The levels in the pair-vibrational spectrum are labelled by the quantum numbers (n_-, n_+) where $n_\pm$ corresponds to the number of correlated $\mathcal{J} = 0$ pairs that have been added to or removed from the closed-shell configuration of ^{208}Pb. Thus, the levels $(n_-, 0)$ and $(0, n_+)$ correspond to the ground states of the even Pb isotopes. The observed levels are indicated by solid lines, while the dashed lines indicate the predicted positions of additional levels. The strong two-neutron transfer processes ((pt) and (tp)) that have so far been observed are indicated by arrows. The figure is from (53).

[9] The concept of pair vibrations in nuclei evolved through the discussions of Högaasen-Feldman (55), early versions of ref. (20) (see, for example, (56)), and Bès and Broglia (57). Excited states of pair vibrational type were identified in the region of ^{208}Pb by Bjerregaard, Hansen, Nathan, and Hinds (58).

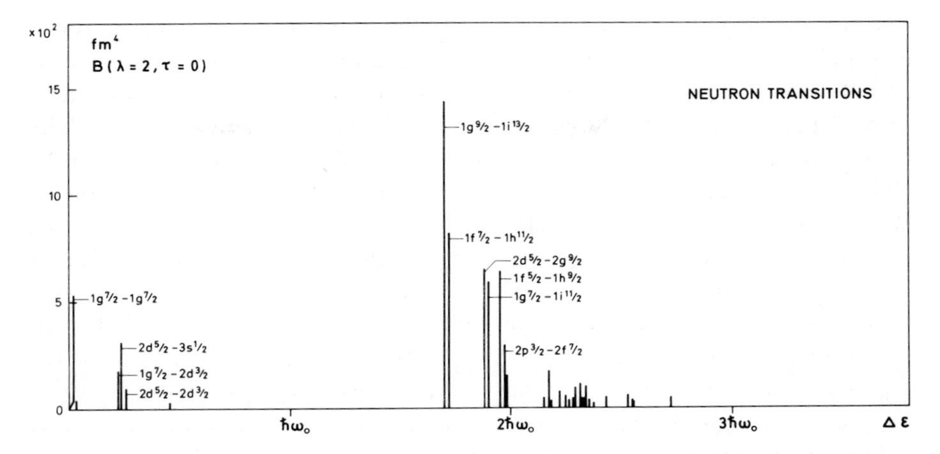

Fig. 6. Single-particle response function for quadrupole excitations. The figure gives the strength of the transitions produced by the quadrupole operator $r^2 Y_{2\mu}$ acting on a nucleus with neutron number $N = 60$. The single-particle spectrum has been obtained from a potential represented by a harmonic oscillator with the addition of spin-orbit coupling and anharmonic terms reflecting the flatter bottom and steeper sides of the nuclear potential. The excitation energies are plotted in terms of the oscillator frequency ω_0, and for the nucleus considered $h\omega \approx 8.7$ MeV. The figure is from (20).

experimental side, the decisive contribution came from the study of reactions in which a correlated pair of nucleons is added or removed from the nucleus as in the (tp) or (pt) reactions (59).

The new views of vibrations also lead to important insight concerning shape oscillations. While the early considerations were guided by the classical picture provided by the liquid-drop model (60, 4), the lesson of the microscopic theory has been that one must begin the analysis of the collective modes by studying the single-particle excitations produced by fields of the appropriate symmetry.

For quadrupole excitations, an example of such a single-particle response function is shown in Fig. 6 and reveals that the quadrupole excitations involve two very different frequency regions. The first is associated with transitions within the partially filled shells and gives rise to the low-frequency quadrupole mode discussed above. The second frequency region in the quadrupole response function is associated with transitions between orbits separated by two major shells and contains most of the oscillator strength. This group of transitions generates a high-frequency collective mode which has been eagerly expected for many years (61); a few years ago, the study of inelastic electron scattering led to the identification of this mode (62) (see Fig. 7), which has since been found as a systematic feature in a wide variety of inelastic scattering experiments (63). This discovery opens the possibility for a deeper probing of one of the fundamental degress of freedom in the nucleus.

Returning of the quadrupole response function, the low-frequency excitations reflect a degeneracy in the single-particle spectrum, which is responsi-

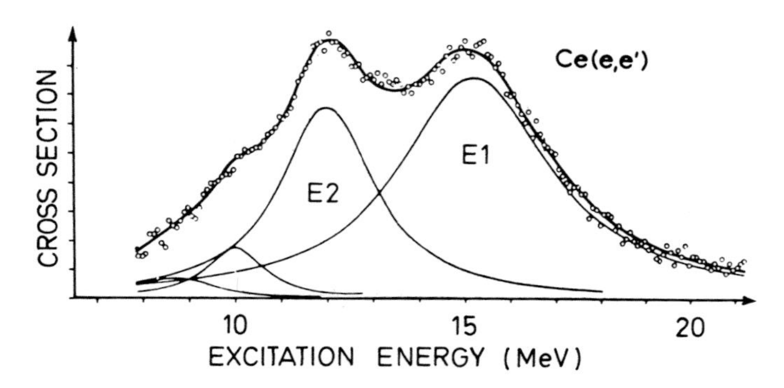

Fig. 7. Inelastic electron scattering on Ce. The highest energy resonance line corresponds with the well-known isovector dipole resonance observed in photo-absorption, while the resonance at an excitation energy of about 12 MeV is identified with the isoscalar quadrupole mode. The figure is from (62).

ble for the tendency to break away from spherical symmetry and to form a spheroidal equilibrium shape (Jahn-Teller effect). One may ask: What underlies this degeneracy in the single-particle spectrum? Apparently this question was never seriously asked until the discovery of the fission isomers (64) revealed the occurrence of important shell-structure effects in potentials that deviate in a major way from spherical symmetry (65) (saddle-point shape). These developments posed in an acute way the question of the general conditions for the occurrence of significant degeneracies in the eigenvalue spectrum for the wave equation. It has been possible to relate this question to the occurrence of degenerate families of periodic orbits in the corresponding classical problem (66, 67, 20) and the instabilities that arise for partially filled shells directly reflect the geometry of these classical orbits. Thus, the observed quadrupole deformations in nuclei can be associated with the elliptical orbits for particle motion in an harmonic oscillator potential (see Fig. 8). The nuclear potential in heavy nuclei also supports orbits of triangular symmetry, and indeed there is evidence for an incipient octupole instability in heavy nuclei.

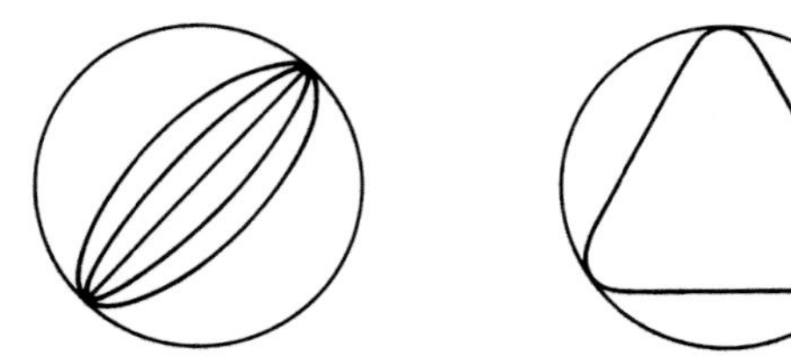

Fig. 8. Periodic orbits in nuclear potential. For small values of angular momentum the motion resembles the elliptical orbits in the oscillator potential. For larger values of angular momentum the effects of the rather sharp nuclear surface can give rise to approximately triangular orbits.

MODERN VIEW OF PARTICLE-VIBRATION COUPLING

The picture of nuclear dynamics that has emerged from these developments thus involves a great variety of different collective excitations that are as elementary as the single-particle excitations themselves, in the sense that they remain as approximately independent entities in the construction of the nuclear excitation spectrum. Examples of the superposition of elementary modes of excitation are given in Fig. 9 (see also Fig. 5).

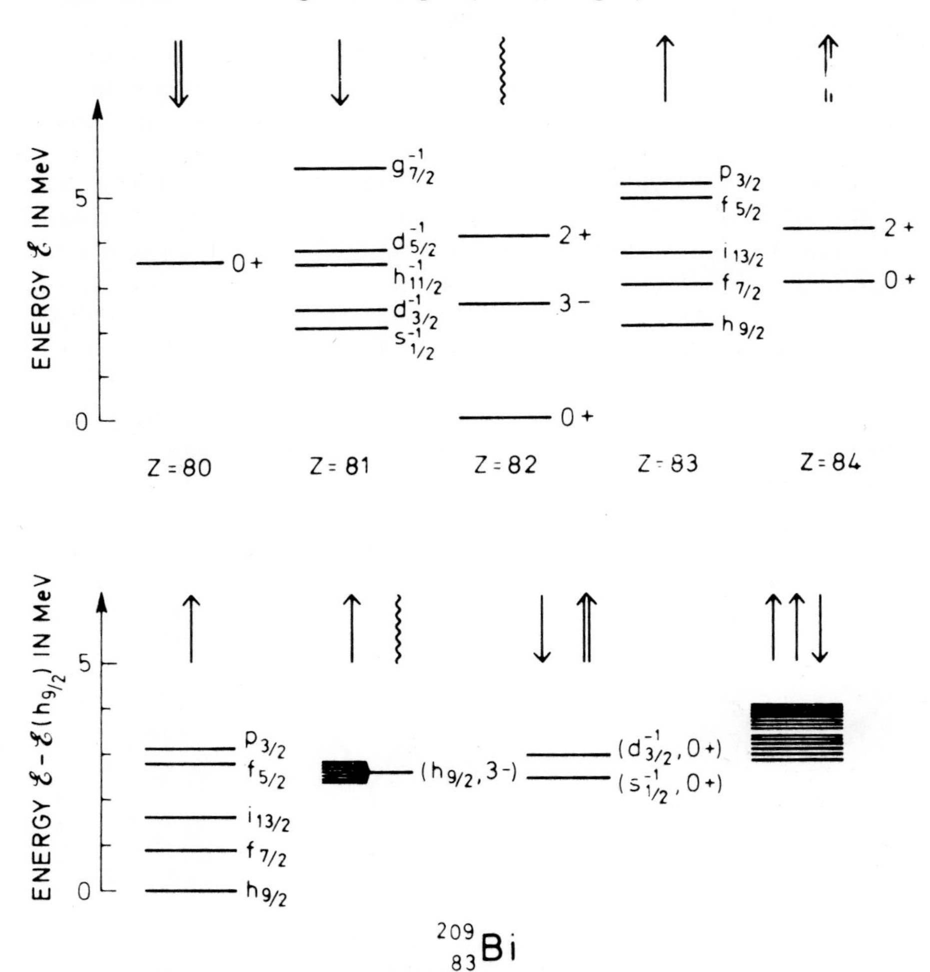

Fig. 9. Elementary excitations based on the closed shell of ^{208}Pb. The upper part of the figure shows fermion excitations involving the addition or removal of a single proton ($\Delta Z = +1$ or $\Delta Z = -1$), and boson excitations involving correlated pairs of protons ($\Delta Z = \pm 2$) as well as collective shape oscillations (particle-hole excitations) in ^{208}Pb itself.

The lower part of the figure gives the observed spectrum of $^{209}_{83}$Bi, which comprises partly the single-proton states, and partly states involving the combinations of a single particle or a single hole with a collective boson. The configuration ($h_{9/2}$, 3−) gives rise to a septuplet of states with $I = 3/2, 5/2 \ldots 15/2$ which have all been identified within an energy region of a few hundred keV (see figure 12). At an excitation energy of about 3 MeV, a rather dense spectrum of two-particle one-hole states sets in, as indicated to the right in the figure. The figure is from (53).

In the analysis of the elementary modes and their interactions, a central element is the particle-vibration coupling which expresses the variations in the average potential associated with the collective vibrational amplitude. This coupling is the organizing element that generates the self-consistent collective modes out of the particle excitations. At the same time it gives rise to interactions that provide the natural limitation to the analysis in terms of elementary modes.

Information about the particle-vibration coupling comes from a variety of sources. For some modes, such as the shape oscillations, the coupling can be related to observed static potentials. More generally, the couplings directly manifest themselves in inelastic scattering processes and indirectly in the properties of the modes and their interactions.

The average one-particle potentials appearing in the particle-vibration coupling are of course ultimately related to the underlying nucleonic interactions. Indeed, many of our colleagues would stress the incompleteness in a description that is not explicitly based on these interactions. However, we would emphasize that the potentials are physically significant quantities in terms of which one can establish relationships between a great variety of nuclear phenomena.[10]

It is of course a great challenge to exploit the extensive and precise information available on the two-body forces and the structure of hadrons in order to shed light on the average nuclear potentials. The problem is a classical one in nuclear physics and has continued to reveal new facets, not only because of the complexity of the nuclear forces, but also due to the many subtle correlations that may contribute to the effective interactions in the nuclear medium.

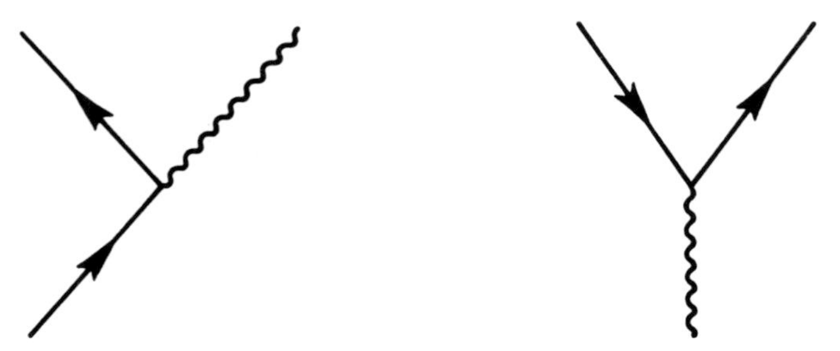

Fig. 10. Basic diagrams for particle-vibration coupling. The solid lines represent particles and the wavy line a phonon of a collective excitation. The particle-vibration coupling creates or annihilates vibrational quanta and at the same time either scatters a particle (or hole) or creates a particle-hole pair.

[10] This issue appears to be endemic in all strongly interacting many-body systems ranging from condensed matter to elementary particles. The approach described here is closely related to that of the Fermi-liquid theory os developed by Landau (68). This formulation operates with a phenomenological effective interaction between the quasiparticles from which the coupling between the particles and the collective modes can be derived. The description of nuclear dynamics in terms of the concepts employed in the theory of Fermi liquids has been developed by Migdal (69).

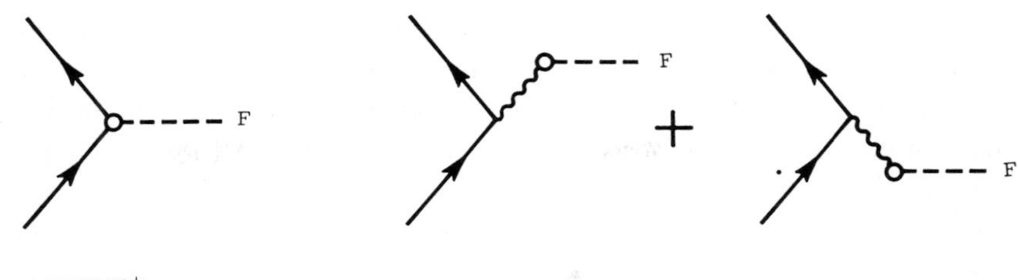

Fig. 11. Renormalization of the matrix elements of a single-particle moment resulting from particle-vibration coupling. The moment F may be any operator that acts on the degrees of freedom of a single particle, such as an electric or magnetic moment, β-decay transition moment, etc.

The basic matrix elements of the particle-vibration coupling can be represented by the diagrams in Fig. 10, which form the basis for a nuclear field theory based on the elementary modes of fermion and boson type. In lowest order, the coupling gives rise to a renormalization of the effective moments of a particle illustrated by the diagrams in Fig. 11. This renormalization is a major effect in the transitions between low-lying single-particle states and provides the answer to the old dilemma concerning the distribution of the strength between the particle excitations and the collective modes. Thus, for example, for the dipole mode, the one-particle excitations carry a very small admixture of the collective mode, which is sufficient to almost cancel the dipole moment of the bare particle[11].

Acting in higher order, the particle-vibration coupling gives rise to a wealth of different effects, including interactions between the different elementary modes, anharmonicities in the vibrational motion, self-energy effects, etc. An example is provided by the interaction between a single particle and a phonon in ^{209}Bi (see Fig. 12)[12]. The lowest single-proton state h 9/2 can be superposed on the octupole excitation observed in ^{208}Pb and gives rise to a septuplet with I = 3/2, . . . 15/2. The splitting of the septuplet receives

[11] While the renormalization of the electric quadrupole operator followed directly from the coupling to the deformation of the nuclear surface (13, 20), the occurrence of large deviations in the magnetic moments for configurations with a single particle outside of closed shells was felt as an especially severe challenge to the description in terms of particles coupled to surface oscillations (see, for example, (4)). The clue to the understanding of this effect came from the recognition that special kinds of configuration mixings could give rise to large first-order effects in the magnetic moments (70, 71). Later, it was recognized that this was a manifestation of the particle-vibration coupling involving collective modes of spin-flip type ($\lambda\pi = 1+$) (61, 20). Experimental evidence for the occurrence of such modes in heavy nuclei came only at a much later time (72). The interpretation of the strong M1 transitions in light nuclei was discussed by Kurath (73) in terms of an intermediate situation between *(LS)* and *(jj)* coupling.

[12] The discovery of the weak-coupling multiplet in ^{209}Bi (75) was a major incentive to the exploration of the scope of the particle-vibration coupling (76, 20).

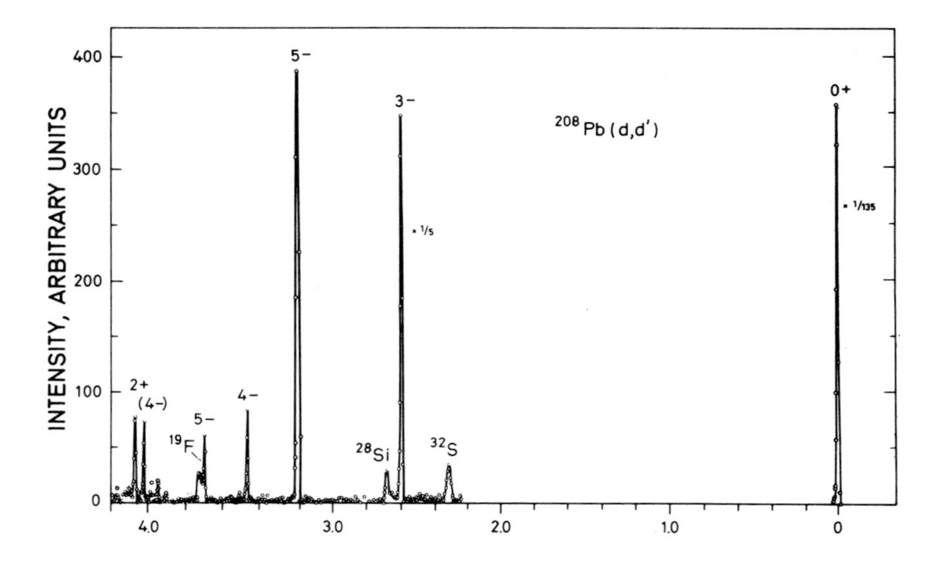

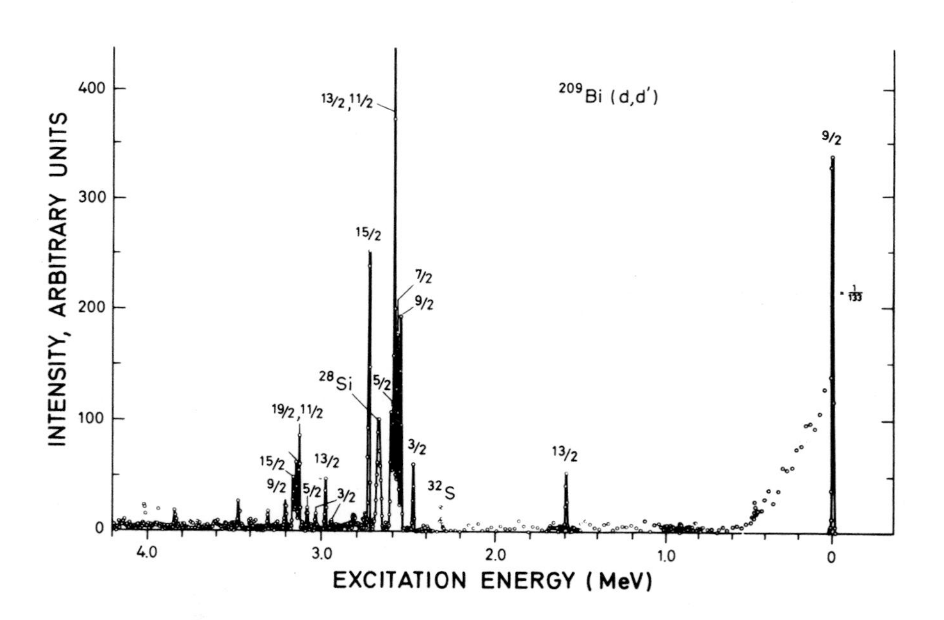

Fig. 12. Energy spectra of deuterons scattered from ^{208}Pb and ^{209}Bi. The prominent inelastic group in ^{208}Pb corresponds to the excitation of an octupole vibrational phonon ($I\pi = 3-$; $h\omega_3 = 2.6$ MeV). In ^{209}Bi the ground state has $I = 9/2$, corresponding to a single $h_{9/2}$ proton outside the closed-shell configurations. The excitation of the octupole quantum in ^{209}Bi leads to a septuplet of states in the neighbourhood of 2.6 MeV with $I = 3/2, 5/2 \ldots 15/2$. The figure is from (20) and is based on data from (74).

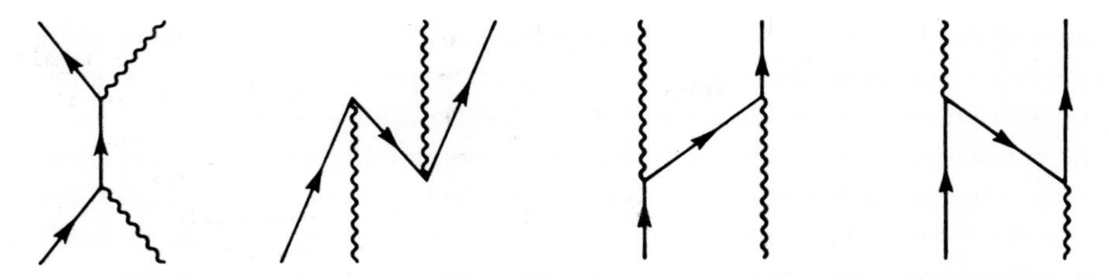

Fig. 13. Second-order diagrams contributing to the energy of a particle-phonon multiplet.

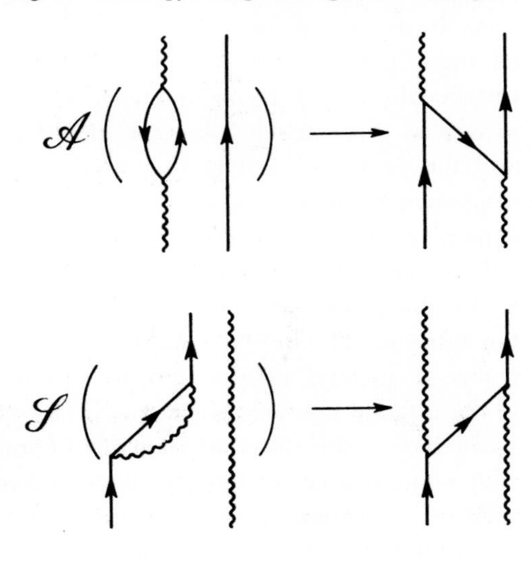

Fig. 14. Linked diagrams associated with symmetrization of particle plus phonon states.

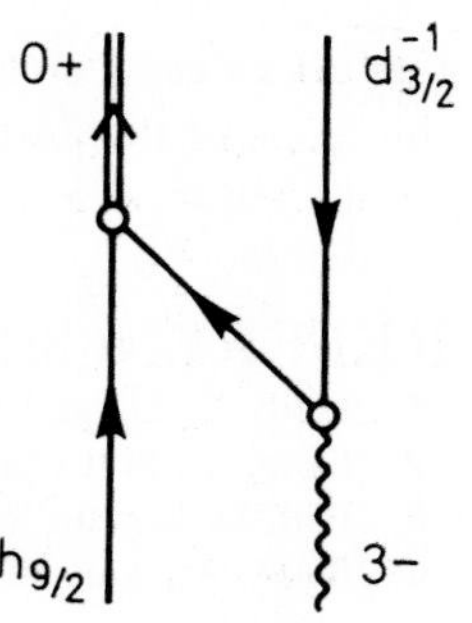

Fig. 15. Coupling between the configurations ($h_{9/2}$ 3−)3/2+ and ($d^{-1}_{3/2}$ 0+)3/2+, based on particle-vibration vertices.

contributions from the octupole coupling, which can be estimated from the second-order diagrams shown in Fig. 13 (and which are seen to correspond to those of the Compton effect in electrodynamics).

It is an important feature of this calculation that the interactions contain the effect of the antisymmetry between the single particle considered and the particles out of which the vibration is built. This effect is contained in the last diagram in Fig. 13, as schematically indicated in Fig. 14. In a similar manner, the third diagram in Fig. 13 contains the effect of the Bose symmetry of the two identical octupole quanta.

The particle-vibration coupling also leads to the interaction between "crossed" channels, such as illustrated in Fig. 15, which exhibits the coupling of the $I = 3/2$ member of the septuplet in ^{209}Bi to the state obtained by superposing a quantum consisting of a pair of protons coupled to angular momentum zero (as in the ground state of ^{210}Po) and a single-proton hole in the configuration $d^{-1}_{3/2}$ (as observed in the spectrum of ^{207}Tl). The treatment of this diagram takes proper account of the fact that the two configurations considered are not mutually orthogonal, as must be expected quite generally in a description that exploits simultaneously the quanta of particle-hole type as well as those involving two particles or holes.

As illustrated by these examples, it appears that the nuclear field theory based upon the particle-vibration coupling provides a systematic method for treating the old problems of the overcompleteness of the degrees of freedom, as well as those arising from the identity of the particles appearing explicitly and the particles participating in the collective motion (20, 77). This development is one of the active frontiers in the current exploration of nuclear dynamics.

Looking back over this whole development one cannot help but be impressed by the enormous richness and variety of correlation effects exhibited by the nucleus. This lesson coincides with that learned in so many other domains of quantal physics and reflects the almost inexhaustible possibilities in the quantal many-body systems. The connections between the problems encountered in the different domains of quantal physics dealing with systems with many degrees of freedom have become increasingly apparent, and have been of inspiration, not least to the nuclear physicists who find themselves at an intermediate position on the quantum ladder. Looking forward, we feel that the efforts to view the various branches of quantal physics as a whole may to an even greater extent become a stimulus to a deeper understanding of the scope of this broad development.

REFERENCES:

1. Foldy, L. L., and Milford, F. J., Phys Rev. *80,* 751 (1950)
2. Bohr, A., Mat Fys. Medd. Dan. Vid, Selsk. *26* no. 14 (1952)
3. Hill, D. L., and Wheeler, J. A., Phys. Rev. *89,* 1102 (1953)
4. Bohr, A., and Mottelson, B. R., Mat. Fysk. Medd. Dan. Vid. Selsk. *27,* no. 16 (1953)
5. see A. Bohr, preceding lecture
6. Scharff-Goldhaber, G., and Weneser, J., Phys. Rev. *98,* 212 (1955)
7. Goldhaber, M., and Sunyar, A. W., Phys. Rev. *83,* 906 (1951)
8. Scharff-Goldhaber, G., Physica *18,* 1105 (1952)
9. Stähelin, P., andr Preiswerk, P., Helv. Phys. Acta *24,* 623 (1951)
10. Rosenblum, S., and Valadares, M., Compt, rend. *235,* 711 (1952)
11. Asaro, F., and Perlman, I., Phys. Rev. *87,* 393 (1953)
12. Kraushaar, J. J., and Goldhaber, M., Phys. Rev. *89,* 1081 (1953)
13. Alder, K., Bohr, A., Huus, T., Mottelson, B., and Winther, A., Rev. Mod. Phys. *28,* 432 (1956)
14. Inglis, D. R., Phys. Rev. *96,* 1059, (1954)
15. Bohr, A., and Mottelson, B. R., Mat. Fys. Medd. Dan. Vid. Selsk. *30,* no. 1 (1955)
16. Litherland, A. E., McManus, H., Paul, E. B., Bromley, D. A., and Gove, H. E., Can. J. Phys. *36,* 378 (1958)

17. Kurath D., and Pičman, L., Nuclear Phys. *10,* 313 (1959)
18. Elliott, J. P., Proc. Roy. Soc. (London) *A245,* 128 and 562 (1958)
19. Hill, D. L., and Wheeler, J. A., Phys. Rev. *89,* 1102 (1953); Peierls, R. E., and Yoccoz, J., Proc. Phys. Soc. (Lodon) *A70,* 381 (1957); Yoccoz, J. Proc. Phys. Soc. (London) *A70,* 388 (1957);Villars, F., Ann. Rev. Nuclear Sci. *7,* 185 (1957)
20. Bohr, A., and Mottelson, B. R., Nuclear Structure, Vol. II, Benjamin, W. A. Inc., Reading, Mass. 1975
21. Rutherford, E., Chadwick, J., and Ellis, C. D., *Radiations from Radioactive Substances,* Cambridge 1930
22. Heisenberg, W., Z. Physik *78,* 156 (1932)
23. Bohr N. and Wheeler, J. A. Phys. Rev. *56,* 426 (1939)
24. Mayer, M. G. Phys. Rev. *78,* 22 (1950)
25. Racah, G. in *Farkas Memorial Volume,* p. 294 eds. A. Farkas and E. P. Wigner, Research Council of Israel, Jerusalem 1952
26. Racah, G. and Talmi, I. Phys. Rev. *89,* 913 (1953)
27. Bardeen, J., Cooper, L. N., Schrieffer, J. R., Phys. Rev. *106,* 162 and *108,* 1175 (1957)
28. Bohr, A., Mottelson, B. R., and Pines, D., Phys. Rev. *110,* 936 (1958)
29. Murray, J. J., Boehm, F., Marmier, P., and Du Mond, J. W., Phys. Rev. *97,* 1007 (1955)
30. Bogoliubov, N. N., J. Exptl, Theoret. Phys. (USSR) *34,* 58 (1958)
31. Valatin, J. G., Nuovo cimento *7,* 843 (1958)
32. Racah, G., Phys. Rev. *63,* 367 (1943)
33. Belyaev, S. T., Mat. Fys. Medd. Dan. Vid. Selsk. *31,* no. 11 (1959)
34. Migdal, A. B., Nuclear Phys. *13,* 644 (1959)
35. Griffin, J. J., and Rich, M., Phys. Rev. *118,* 850 (1960)
36. Nilsson, S. G., and Prior, O., Mat. Fys. Medd. Dan. Vid. Selsk. *32,* no. 16 (1961)
37. Kisslinger, L. S., and Sorensen, R. A., Mat. Fys. Medd. Vid. Selsk. *32,* no. 9 (1960)
38. Soloviev, V G., Mat. Fys. Skr. Dan. Vid. Selsk. *1,* no. 11 (1961) and Phys. Letters *1,* 202 (1962)
39. Yoshida, S., Phys. Rev. *123,* 2122 (1961) and Nuclear Phys. *33,* 685 (1962)
40. Mang, H. J., and Rasmussen, J. O., Mat. Fys. Skr. Dan. Vid. Selsk. 2, no. 3 (1962)
41. Cohen, B. L., and Price, R. E., Phys. Rev. *121,* 1441 (1961)
42. Glassgold, A. E., Heckrotte, W., and Watson, K. M., Ann. Phys. 6, 1 (1959); Ferrell, R. A., and Fallieros, S., Phys. Rev. *116,* 660 (1959); Goldstone, J., and Gottfried, K., Nuovo cimento *13,* 849 (1959); Takagi, S., Progr. Theoret. Phys. (Kyoto) *21,* 174 (1959); Ikeda, K., Kobayasi, M., Marumori, T., Shiozaki, T., and Takagi, S., Progr. Theoret. Phys. (Kyoto *22,* 663 (1959); Arvieu, R., and Vénéroni, M., Compt, rend. *250,* 992 and 2155 (1960) and *252,* 670 (1961); Baranger, M., Phys. Rev. *120,* 957 (1960); Kobayasi, M., and Marumori, T., Progr. Theoret. Phys. (Kyoto) *23,* 387 (1960); Marumori, T., Progr. Theoret. Phys. (Kyoto) *24,* 331 (1960); Thouless, D. J., Nuclear Phys. *22,* 78 (1961)
43. Baldwin, G. C., and Klaiber, G. S., Phys. Rev. *71,* 3 (1947) and *73,* 1156 (1948)
44. Goldhaber, M., and Teller, E., Phys. Rev. *74,* 1046 (1948)
45. Steinwedel, H., and Jensen, J. H. D., Z. Naturforsch. *5a,* 413 (1950)
46. see especially Wilkinson, D. H., Physica *22,* 1039 (1956)
47. see, for example, the discussion in *Proc. Glasgow Conf. on Nuclear and Meson Physics,* eds. Bellamy, E. H., andr Moorhouse, R. G., Pergamon Press, London 1955
48. Elliott, J. P., and Flowers, B. H., Proc. Roy. Soc. (London) *A242,* 57 (1957)
49. Brink, D. M., Nuclear Phys, *4,* 215 (1957)
50. Brown, G. E., and Bolsterli, M., Phys. Rev. Letters *3,* 472 (1959)
51. Bohr, A., and Mottelson, B. R. in *Neutron Capture Gamma-Ray Spectroscopy,* p. 3, International Atomic Energy Agency, Vienna 1969

52. Fallieros, S. in *Intern. Conf. on Photonuclear Reactions and Applications*, Vol. I, p. 401, ed. Berman, B. L., U.S. Atomic Energy Commission, Oak Ridge, Tenn. 1973
53. Bohr, A., and Mottelson, B., Ann. Rev. Nuclear Sci. *23,* 363 (1973)
54. Anderson, P. W. in *Lectures on the Many-Body Problem 2,* 113, ed. Caianiello, E. R., Academic Press, New York 1964
55. Högaasen-Feldman, J., Nuclear Phys, *28,* 258 (1961)
56. Bohr A. in *Compt. rend. du Congrès Intern. de Physique Nucléaire,* Vol. 1, p. 487, ed. Gugenberger, P., C.N.R.S. Paris 1964 and *Nuclear Structure,* Dubna Symposium, p. 179, International Atomic Energy Agency, Vienne 1968
57. Bès, D. R., and Broglia, R. A., Nuclear Phys. *80,* 289 (1966)
58. Bjerregaard, J. H., Hansen, O., Nathan, O., and Hinds, S., Nuclear Phys. *89,* 337 (1966)
59. see the review by Broglia, R. A. Hansen, O. and Riedel, C. *Advances in Nuclear Phys. 6,* 287, Plenum Press, New York 1973
60. Bohr, N., and Kalckar, F., Mat. Fys. Medd. Dan Vid. Selsk. *14,* no. 10 (1937)
61. Mottelson, B. in *Proc. International Conf. on Nuclear Structure,* p. 525, eds. Bromley, D. A., and Vogt, E. W., Univ. of Toronto Press, Toronto 1960
62. Pitthan, R., and Walcher, Th., Phys. Letters *36B,* 563 (1971) and Z. Naturforsch. *27a,* 1683 (1972)
63. see, for example, the review by Satchler, G. R., Phys. Reports *14,* 97 (1974)
64. Polikanov, S. M., Druin, V. A., Karnaukhov, V. A., Mikheev, V. L. Pleve, A. A., Skobelev, N. K., Subbotin, V. G., Ter-Akop'yan, G. M., and Fomichev, V. A., J. Exptl. Theoret. Phys. (USSR) *42,* 1464 (1962)
65. Strutinsky, V. M., Nuclear Phys. *A95,* 420 (1967)
66. Balian, R., and Bloch, C. Ann. Phys. *69,* 76 (1971)
67. Wheeler, J. A. in *Atti del Convegno Mendeleeviano,* p. 189, ed. M. Verde, Accad. delle Scienze di Torino. Torino 1971; Swiatecki, W. J., private communication (1971)
68. Landau, L. D., J. Exptl. Theoret. Phys. (USSR) *30,* 1058 (1956) and *32,* 59 (1957)
69. Migdal, A. B., *Theory of Finite Fermi Systems and Applications to Atomic Nuclei,* Wiley (Interscience) New York 1967
70. Arima, A., and Horie, H., Progr. Theoret. Phys. (Kyoto) *12,* 623 (1954)
71. Blin-Styole, R. J., and Perks, M. A., Proc. Phys. Soc. (London) *67A,* 885 (1954)
72. Berman, B. L., Kelly, M. A., Bramblett, R. L., Caldwell. J. T. Davis, H. S., and Fultz, S. C., Phys. Rev. *185,* 1576 (1969); Pitthan, R., and Walcher, Th. Z. Naturforsch. *27A,* 1683 (1972)
73. Kurath, D., Phys. Rev. *130,* 1525 (1963)
74. Ungrin, J. Diamond, R. M., Tjøm, P. O., and Elbek, B., Mat. Fys. Medd. Dan. Vid Selsk. *38,* no. 8 (1971)
75. Alster, J., Phys. Rev *141,* 1138 (1966); Hafele, J. C., and Woods, R., Phys. Letters *23B* 579 (1966)
76. Mottelson, B. R. in *Proc. Intern. Conf on Nuclear Structure* p. 87, ed. J. Sanada, Phys. Soc. of Japan (1968); Hamamoto, I., Nuclear Phys. *A126,* 545 (1969) *A141* 1, and *A155,* 362 (1970); Bès, D. R., and Broglia, R. A. Phys. Rev. *3C,* 2349 and 2389 (1971); Broglia, R. A., Paar, V., and Bès, D. R., Phys. Letters *37B,* 159 (1971)
77. Bès, D. R., Dussel, G. G., Broglia, R. A., Liotta, R., and Mottelson, B. R., Phys. Letters *52B,* 253 (1974)

James Rainwater

JAMES RAINWATER

I was born December 9, 1917 in a small town in Idaho (Council) where my parents had moved to from California to operate a general store. My father, who had previously been a civil engineer, died in the great influenza epidemic of 1918. My mother then moved with me and her mother to Hanford, Calif. in the San Joaquin Valley of California, where she was re-married to George Fowler a few years later. In my schooling through high school, I excelled mainly in chemistry, physics and mathematics. Due mainly to my record on an open chemistry competition given by Cal Tech, I was admitted, graduating in 1939 as a physics major. Carl David Anderson was my physics group recitation instructor when he received his Nobel Prize and Milliken was the President of the Institute. I had a short biology course taught by Thomas Hunt Morgan. In 1939 I began graduate study in physics as a teaching assistant at Columbia University where I have remained. During the first two years, I had courses under I. I. Rabi, Enrico Fermi, Edward Teller and J. R. Dunning. Fermi was working on neutron moderator assemblies which led to the first working nuclear "pile" after his group was moved to Chicago. Dunning, Booth, Slack, and Von Grosse held the basic patent on the gaseous diffusion process for ^{235}U enrichment and were working on its development. This evolved into the Oak Ridge enrichment plants and the present U.S. technology for ^{235}U enrichment.

In March 1942, I married Emma Louise Smith. We have three sons, James, Robert and William who are all now adults. We also had a daughter, Elizabeth Ann, who died while young.

During W. W. II, I worked with W. W. Havens, Jr. and C. S. Wu under Dr. Dunning (Manhattan Project) mainly doing pulsed neutron spectroscopy using the small Columbia cyclotron. I received my Ph. D after my thesis was de-classified in 1946. I continued at Columbia, first as an instructor, reaching the rank of full professor in 1952. About 1946 funding was obtained from the Office of Naval Research to build a synchrocyclotron which became operational in early 1950. I was involved with the facility development from the beginning and my research has used that facility ever since. The research included neutron resonance spectroscopy, the angular distribution of pion elastic and inelastic scattering on nuclei with optical model fitting. Best known are the muonic-atom-x-ray studies starting with the pioneering 1953 paper with Val Fitch which first established the smaller proton charge radii of nuclei.

Starting in 1948, I taught an advanced nuclear physics graduate course. The Maria Mayer shell model suggestion in 1949 was a great triumph and fitted my belief that a nuclear shell model should represent a proper approach to understanding nuclear structure. Combined with developments of Weizsaker's semi-empirical explanation of nuclear binding, and the Bohr-Wheeler 1939 paper on nuclear fission, emphasizing distorted nuclear shapes, I was prepared to see an explanation of large nuclear quadrupole moments. The full concept came to me in late 1949 when attending a colloquium by Prof. C. H. Townes who described the experimental situation for nuclear quadrupole moments. It was a fortuitous situation made even more so by the fact that I was sharing an office with Aage Bohr that year. We had many discussions of the implications, subsequently very successfully exploited by Bohr, Mottelson, and others of the Copenhagen Institute.

Since I joined the Columbia Physics Dept., in 1939, it has been my privilege to have as teachers and/or colleagues many previous Nobel Laureates in Physics: E. Fermi, I. I. Rabi, H. Bethe (Visiting Prof.), P. Kusch, W. Lamb, C. H. Townes, T. D. Lee and L. Cooper in addition to R. A. Milliken, C. D. Anderson, and T. H. Morgan (Biology) while I was an undergraduate at Cal Tech.

Organization Membership, etc.

Fellow: American Physical Society, Institute of Electrical and Electronic Engineers, New York Academy of Sciences, American Association for the Advancement of Sciences.

Member: National Academy of Sciences, Optical Society of America, American Association of Physics Teachers

Recipient: Ernest Orlando Lawrence Award for Physics, 1963.

Dr. Rainwater died in 1986.

BACKGROUND FOR THE SPHEROIDAL NUCLEAR MODEL PROPOSAL

Nobel Lecture, December 11, 1975
by JAMES RAINWATER
Columbia University, New York, N.Y., USA

The conceptual developments on which my award is based occurred to me about twenty-six years ago in late 1949. I shall attempt, as accurately as I can remember, to reconstruct how I viewed the situation of the nuclear shell model and non-spherical nuclear shape at that time.

In a sense the subject began in 1910 when Ernest Rutherford's α particle scattering experiments (1) showed that the nuclear size is $\leqslant 10^{-12}$ cm radius, although the atom size is $\sim 10^{-8}$ cm. This led to Niels Bohr's (2) 1913 theory of the hydrogen atom in terms of quantized electron orbits about the nucleus. This was extended by many workers, especially via the Wilson-Sommerfeld quantization rule that $\int p_i dq_i = n_i h$ for each degree of freedom, where q_i and p_i are the generalized coordinates and momenta of an electron in its orbit about the nucleus. The proposal in 1925 by Goudsmit and Uhlenbeck (3) of the concept of spin 1/2 for the electron and the statement by Pauli (4) of the exclusion principle for electrons, later generalized to all spin 1/2 particles, led to an understanding of the Periodic Table of the Elements, using the old quantum theory, in terms of filling electron shells.

The development of quantum mechanics in 1926 placed the subject on a proper foundation and led to an explosion of the development of atomic physics as is evident from a perusal of the 1935 treatise by E. U. Condon and G. H. Shortley, *The Theory of Atomic Spectra*, Cambridge University Press (1935 and 1951). In the case of the electron orbits or shells about the nucleus, the potential is dominated by the central coulomb attraction of the nucleus, thus permitting treatment of angular momentum as a good quantum number to a good approximation. The coulomb force law was completely known. For the nucleus, early attempts to treat it as composed of protons and electrons were unsatisfactory. When the neutron was discovered by Chadwick in 1932, the picture shifted to a nucleus composed of neutrons and protons bound by strong short range forces. Measurements of nuclear spins soon established that the neutron and proton should probably be taken to have spin 1/2 and to obey Dirac Theory and the Pauli exclusion principle, thus providing a basis for a nuclear shell model. My own detailed introduction to the subject was mainly provided by Bethe's massive review of Nuclear Physics (5) in the 1936, 1937 issues of Reviews of Modern Physics.

The subject of attempts at a nuclear shell model are reviewed by Bethe and Bacher. (5) I was particularly familiar with the 1937 article by Feenberg and Phillips, (6) "On the Structure of Light Nuclei", where the Hartree method was used with a simplified assumed potential to investigate possible spin orbit Russell Saunders coupling states in filling the first $l = 1$ shell between

^4He and ^{16}O, to explain the behavior of ground and excited nuclear states, etc. A model of particles in a spherical box has the first 1 *s* ($l = 0$) state filled by 2 neutrons (N) and 2 protons (Z) at ^4He. This nucleus is certainly exceptionally stable, having a binding energy of over 20 MeV for the last nucleon. The first *p* shell ($l = 1$) then begins, which is closed at ^{16}O. It is interesting that the mass $A = 5$ system is unable to bind the last nucleon and appears as a resonance for neutron or proton scattering on helium. The third shell holds the second *s* and the first *d* ($l = 2$) shell and is filled at ^{40}Ca ($Z = N = 20$) which is also unusually bound. It is the heaviest stable nucleus having $N = Z$. Beyond this the predicted shell closings disagreed with experiment. The basic force law between nucleons was poorly known.

Before 1940 it was known that the nuclear volume and total nuclear binding both increased roughly linearly with A, the number of nucleons. The range of the nuclear force between nucleons was known to be $\approx 2 \times 10^{-13}$ cm and to be deep enough to give the single bound *s* ground state for the deuteron when n and p spins were parallel, but not when they are antiparallel. A major question involved the reason for "saturation of nuclear forces: i.e., why binding did not increase as $A(A-1)$, the number of possible pairings with a "collapsed" nucleus having radius $\approx 10^{-13}$ cm. This was "answered" by Heisenberg, Wigner, Majorana and others in an ad-hoc fashion by assuming "exchange forces" which were attractive or repulsive depending on the wave function exchange properties. Only after 1950 did Jastrow introduce the concept of a short range repulsion which is now accepted as the reason.

In 1935, Weizsacker introduced his semi-emperical binding energy formula (7) including volume, surface, isotope, coulomb, and "odd-even" or pairing terms to explain the general trend of nuclear binding. The surface term noted that surface nucleons were less bound, giving a decrease in binding proportional to $A^{2/3}$ for the radius proportional to $A^{1/3}$. This gives less binding for light nuclei and partially explains why maximum stability occurs near ^{56}Fe. The isotope term is easily understood on a shell model basis or using a Fermi-Thomas statistical model. The number of filled space states increases as $(Z/2)$ or $(N/2)$ for protons and neutrons. For a given A, minimum kinetic energy occurs for $N = Z$. For $N > Z$, one must change $(N-Z)/2$ protons to neutrons of higher kinetic energy, with the average kinetic eneugy change per transferred nucleon proportional to $(N-Z)$ for a total kinetic energy increase proportional to $(N-Z)^2$. This favors $N = Z$ for stability. This is balanced by the coulomb repulsion energy of the protons which is proportional to $Z(Z-1)/R$. This favors having only neutrons. The stability balance for stable nuclei has an increasingly large fraction of the nucleons as neutrons as A becomes large. This term also gives reduced binding per nucleon beyond ^{56}Fe and leads to instability against α decay beyond $A \approx 208$ with not too long lifetimes for the ^4He fragment to penetrate the coulomb barrier. It was observed that even N, even Z (e,e) nuclei were unusually stable relative to odd, odd (0,0) nuclei, such that after ^{14}N the stable nuclei for even A all were (e,e), often having two stable even Z values for each even $A \geqslant 36$. For odd A, there is almost always only one naturally occurring stable Z value, with (e,o) and (o,e) equally favored. This extra binding, $+\delta$ for ee, zero for A odd, and $-\delta$ for (0,0) has $\delta \sim 1$ to 3 MeV, decreasing as A increases approximately as $12A^{-1/2}$ MeV. (See Bethe and Bacher (5), p. 104.) It is also observed that the ground states of even A nuclei have net spin zero, indicating a space pairing (potential energy) for strongest interaction to cancel the angular momentum contributions. Figure 1, from the Bohr—Mottelson text (8), plots the observed binding per nucleon for beta stable nuclei, vs A, with a best fit semi-emperical curve for comparison. The deviations of the experimental bindings from the smooth curve give hints of shell structure effects.

In the early 1930's, the energy dependence of the interaction cross section for reactions involving neutrons or protons incident on nuclei was treated by what is now referred to as an optical model approach. The incident nucleon-nucleus interaction was treated using a smoothed interaction potential for the

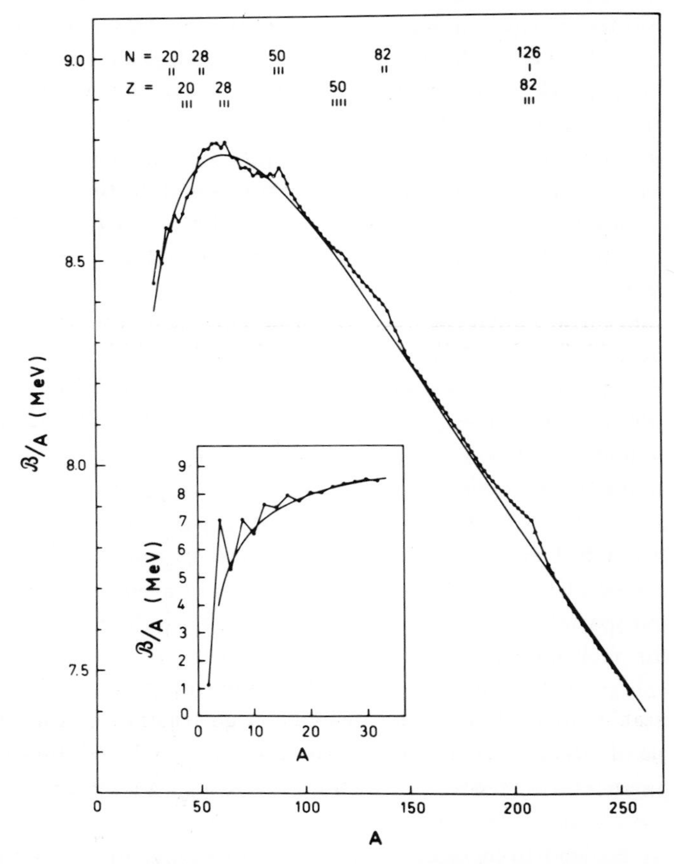

Fig. 1. The average binding energy per nucleon is plotted for nuclei stable against β decay. It is compared with the semi-emperical formula $B/A = [15.56-17.23\ A^{-1/3}-23.28(\mathcal{N}-\mathcal{Z})^2/A^2]\text{MeV}-3\mathcal{Z}^2e^2/5R_cA$, with $R_c = 1.24A^{1/3}$ fm. This figure is from Ref. 8, courtesy of W. A. Benjamin, Inc.

nucleon inside the nucleus. This model predicted "shape" resonances with huge resonance widths and spacings. Early experiments (5) using slow neutrons revealed cross section (compound nucleus) resonances for medium-heavy nuclei $\sim$ 10 to 100 eV apart, with $<$ 1 eV resonance widths. This led N. Bohr to suggest a liquid drop model (9) of the nucleus where the incoming nucleon, as for a molecule hitting a liquid drop, is absorbed near the surface and loses its identity. This is not necessarily incompatible with a shell model, since the shell model refers mainly to the lowest states of a set of fermions in the nuclear "container". However, when combined with the discouragingly poor fits with experiment of detailed shell model predictions (6), the situation $\sim$ 1948 was one of great discouragement concerning a shell model approach.

In the first part of 1949, three groups presented different "explanations" of nuclear shell structure (10) in the same issue of Physical Review. Of these, that of Maria Mayer became the now accepted model. A similar proposal by J. H. D. Jensen and colleagues at the same time led to the Nobel Award in Physics to Mayer and Jensen in 1963. From 1948 to $\sim$ 1962, I taught a course in "Advanced Nuclear Physics" for graduate students at Columbia. I was also, as an experimental physicist, working on the completion of the Columbia University Nevis Synchrocyclotron which first became operational in March, 1950. During the 1949—50 academic year, I shared an office, Room 910 Pupin, with Aage Bohr who was visiting Columbia that year. I was particularly excited about the Mayer shell model which suddenly made understandable a vast amount of experimental data on spins, magnetic moments, isomeric states, β decay systematics, and the "magic numbers" at Z, $N = 2,8$, 20 (28), 50, 82, 126. I reviewed this material at a seminar at Columbia that year.

For over a year previously, I had felt that shell model aspects should have a large degree of validity for nuclei for the following reason. When one considers forming the nuclear wave functions, in $3A$ dimensional coordinate space, for A nucleons in a spherical box the size of the nucleus, the shell model states result in lowest kinetic energy. The effective potential energy and the shell model kinetic energy (for $r < R$) are both quite large compared with the net binding energy ($\sim$ 8 MeV) for the least bound nucleons. This is illustrated in Fig. 2 (from Ref. 8). The single particle state energies vs A have as the "valence" nucleon that with E_n about —8 MeV. If one attempts to use ψ functions wherein the spatial behavior for each nucleon is very different from that predicted by the shell model, the effect is equivalent to mixing in large amounts of higher energy states having compatible symmetry properties. This mixture of high curvature ψ states would greatly increase the $< T >$ for the least bound nucleons. I pictured the net ψ fundtion *not* as a pure Hartree product of single particle ψ functions, but as being nearly so for the long wave length Fourier aspects of the functions. The short range nucleon-nucleon attractive force would lead to local distortions and clusterings in 3 A dimensional space such as of deuterons and of α particle structures, etc., but low energy studies would emphasize the long wavelength Fourier aspects which are suggested by the shell model. I was thus delighted by the success of the Mayer model. (I was

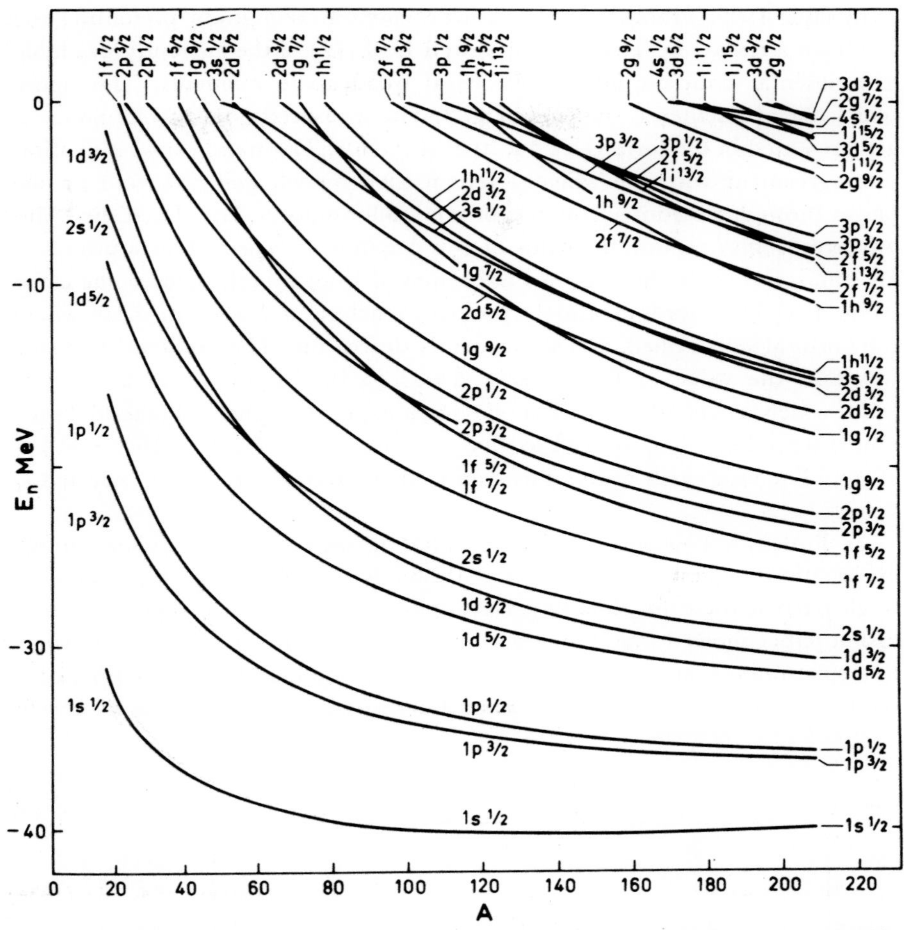

Fig. 2. Energies of neutron orbits using a model of C. J. Veje (from Ref. 8, Vol. I, p. 239). Courtesy of W. A. Benjamin, Inc. The least bound nucleons have energy ~ (−8 MeV) which is small compared with their potential or kinetic energies inside the nucleus.

not then aware of Jensen's work.) The N. Bohr liquid drop model for nuclear reactions and fission did not seem to me to contradict the shell model since the concept of scattering is meaningless for a many fermion ground state, but not for an incident continuum state particle which is not inhibited by the Pauli principle from knocking bound nucleons to excited (unoccupied) states. The compound nucleus states emphasized by Bohr involved an eventual sharing of the excitation by many nucleons so ~ 10 eV level spacing for medium A nuclei plus slow $l = 0$ neutrons could result. Since ~ 1941, I had been using the small Columbia cyclotron to carry out slow neutron time of flight spectroscopy studies in collaboration with W. W. Havens, Jr., and C. S. Wu, under Professor J. R. Dunning. We were quite aware of the famous 1939 paper of N. Bohr and J. A. Wheeler on the theory of nuclear fission (11) which emphasized that excited nuclei need not be spherical.

In later 1949, Professor C. H. Townes gave a colloquium presenting the results of a review by Townes, Foley, and Low (12) of the currently available experimental data on nuclear electrical quadrupole moments. The figure which they presented is shown in Fig. 3. The measured quadrupole moments are presented in the form $Q/(1.5\times10^{-13}\ A^{1/3}\ \text{cm})^2$. The trend shows a qualitative agreement with the Mayer—Jensen shell model, going to zero as one passes through closed neutron and proton shell numbers. For closed shell plus one extra high l proton, the value of Q is negative as expected for a proton in an equatorial orbit. As nucleons are removed from a high l closed shell, the value of Q becomes increasingly positive, reaching a maximum near where the l orbital is half filled, and subsequently decreasing. The problem expressed was that the value of Q/R^2, using $R = 1.5\times10^{-13}A^{1/3}$ cm, reaches 10 for ^{176}Lu which is over 30 times what one might expect for spherical potential shell model wave functions coupled to give a 7^- state ($Z = 71$, $N = 105$, $\tau = 4\times10^{10}$y). The rare earth nuclei particularly show much larger than expected Q values.

As Professor Townes was talking, what seemed like the obvious simple explanation suggested itself to me. Although the Mayer shell model used single particle wave functions based on a spherical potential, the Bohr—Wheeler fission paper showed that, if energetically favorable, the nucleus would distort to a spheroidal shape. For small values of the fractional difference β between the major and minor axes, for constant nuclear volume, the surface area term

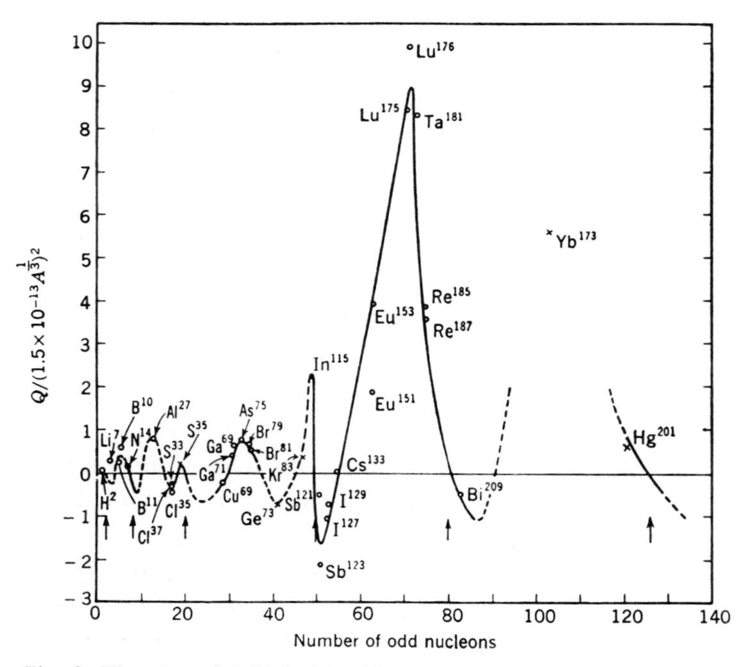

Fig. 3. The plot of $Q/(1.5\ A^{1/3}\times10^{-13}\ \text{cm})^2$ for known nuclear quadrupole moments as presented by Townes, Foley, and Low (1949, Ref. 12).

increases as β^2, with the decrease in coulomb energy compensating in part (for high Z). My picture assumed constant well depth, but with a distortion where R increased to $(1+2\beta/3)R_0$ in the z direction and decreased to $(1-\beta/3)R_0$ in the x and y directions (or to $R_0 e^{2\beta/3}$ and $R_0 e^{-\beta/3}$). If one uses trial ψ functions which are identically distorted, the potential energy $< V >$ is the same, but the kinetic energy $< T_z > = (1+2\beta/3)^{-2} < T_z >_0$ and $< T_x >$ and $< T_y >$ become $(1-\beta/3)^{-2}$ as large as before. For high $|m|$ states, the orbits are nearly equatorial and $< T >$ is nearly proportional to R_x^{-2} or R_y^{-2}, with $< T_x >_0 \approx < T_y >_0 \gg < T_z >_0$. This clearly favors β *negative*, or a bulge at the equator to disk (oblate) shape. Each 1 % increase in equator radius (R_x and R_y) gives about 2 % decrease in $< T >$, or $\delta T/T \approx +2\beta/3$. For a closed shell, $< T_x >_0 = < T_y >_0 = < T_z >_0$ averaged over all l_z $(= m)$ for high l, so there is zero net linear term in the change in total kinetic energy with the distortion parameter β. For a high l closed shell *minus* equatorial (high $|m|$) orbitals, the net nuclear angular momentum is the negative of the contribution of the missing nucleons (holes) and the contribution to the kinetic energy term linear in β is equal and opposite to that of the missing equatorial orbit nucleons. The important point is that this yields a term linear in β favoring $|\beta| \neq 0$, while the restoring terms are quadratic in β. The expected equlibrium β is thus $\neq 0$ and is proportional to the coefficient of the linear term for not too large deviations of β from unity. This gives a prolate (cigar) shaped distortion.

The next step was to attempt a more quantitative evaluation of the β^2 restoring term. For this, I found the 1939 paper (13, 7) by E. Feenberg useful. He noted that the surface energy increased as $E_s = E_s^0 [1+ \frac{8}{45} \beta^2 \ldots]$ and the coulomb energy decreased as $E_c = E_c^0 [1- \frac{4}{45} \beta^2 \ldots]$ which requires $F = 2E_s^0/E_c^0$ $(\approx 42.6\ A/Z^2) > 1$ for a net positive restoring β^2 term. This predicted zero net β^2 restoring term for $Z \sim 125$ for beta stable nuclei (no resistance to fission). The net term was β^2 $[2.74\ A^{2/3} - 0.054\ Z^2 A^{-1/3}]$ MeV. Using this value gave [14] $Q/R^2 = -11$ for a single high l nucleon above closed shell for a ficticious case of $A \sim 176$. The picture, if anything, seemed capable of giving even larger Q/R^2 values than were observed experimentally.

For a prolate spheroidal potential, with the distortion axis in the z direction, the φ dependence of the single particle ψ for $l_z = m$ is still $e^{im\varphi}$. However l_x and l_y and $\vec{l}^2$ cannot be good quantum numbers. The core must somehow share the net angular momentum. This consideration helps when one considers the deviations of the observed magnetic moments from the Schmidt limits predicted by the simple shell model.

Aage Bohr pointed out to me at the time (14) that if the nucleus is a spheroid with an "intrinsic" quadrupole moment Q_0 relative to its distortion axis, and total angular momentum is I, the maximum "observed" Q is reduced by a factor $I(2I-1)/(I+1)(2I+3) = 1/10,\ 2/7,\ 5/12$, and $28/55$ for $I = 1, 2, 3, 4$. This emphasizes that $Q = 0$ for $I = 0$ or $1/2$, but Q_0 may not be zero. Bohr,

Mottelson and colleagues (15) subsequently treated the situation for coulomb excitation cross sections for low lying rotational states. The excitation cross sections uniquely establish the intrinsic quadrupole moment Q_0 for the ground states of distorted even-even nuclei as well as for odd A nuclei. Figure 4 was prepared by Professor Townes ~ 1957 for a review article on measured quadru-

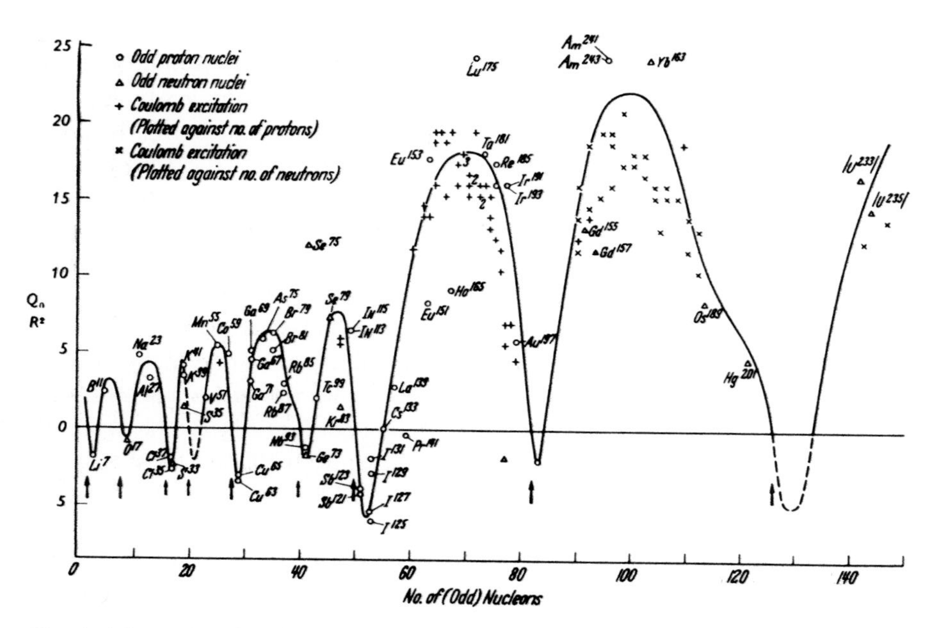

Fig. 4. A later plot of the intrinsic quadrupole moments, Q_0/R^2, prepared by C. H. Townes (Ref. 16), using $R = 1.2\ A^{1/3} \times 10^{-13}$ cm. This figure supercedes Fig. 3. It emphasizes the large size of the quadrupole moments relative to values $|Q_0/R^2| < 1$ expected for a spherical nucleus shell model.

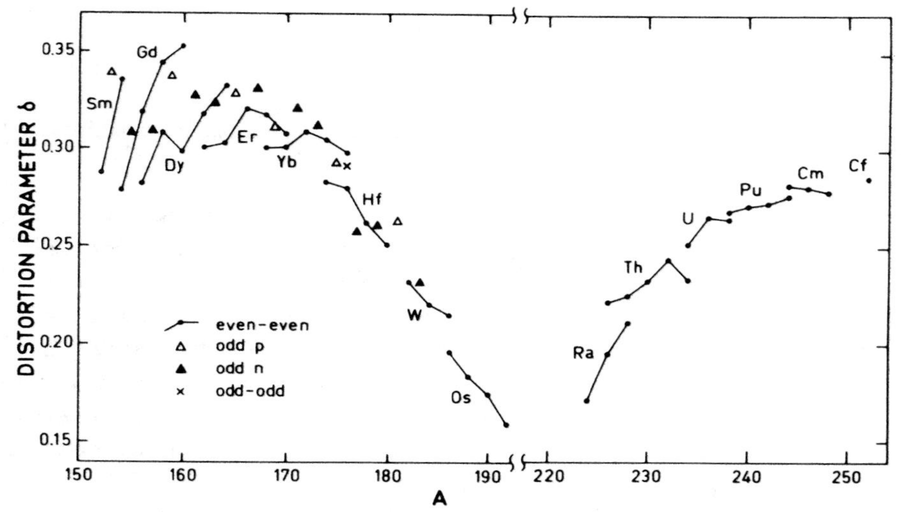

Fig. 5. A plot of the experimental distortion parameter δ ($\approx \beta$ of this paper) in the rare earth region and beyond mass ~ 220. From Ref. 8, Vol. II.

pole moments (16). The largest intrinsic quadrupole moments occur for the rare earth region before the double closed shell $Z = 82$, $N = 126$, and beyond $A \sim 230$ where even higher j single particle states are involved. Figure 5 shows a recent plot from the just released Vol. II of Bohr—Mottelson, *Nuclear Structure* (8). The distortion parameter δ is nearly the same as the parameter β discussed above. It is seen, as was evident from Professor Townes' 1949 colloquium (12), that many nuclei deviate quite strongly from spherical shape so it does not make sense to use a spherical nuclear model in these regions of atomic size.

After Professor Townes' colloquium, Dr. Bohr and I had many discussions of my concept. He was particularly interested in the dynamical aspects. The distortion bulge could in principle vibrate or move around to give the effect of rotational levels. The first result was his January 1951 paper (17), "On the Quantization of Angular Momenta in Heavy Nuclei". The subsequent exploitation of the subject by Bohr, Mottelson and their colleagues is now history and the main reason for our presence here at this time.

I should mention that the program of evaluating the energies of single particle states in distorted nuclei was subsequently carried out in proper form by Mottelson and Nilsson and by Nilsson alone in the form of "Nilsson diagrams" such as in Fig. 6, which is for proton single particle states beyond

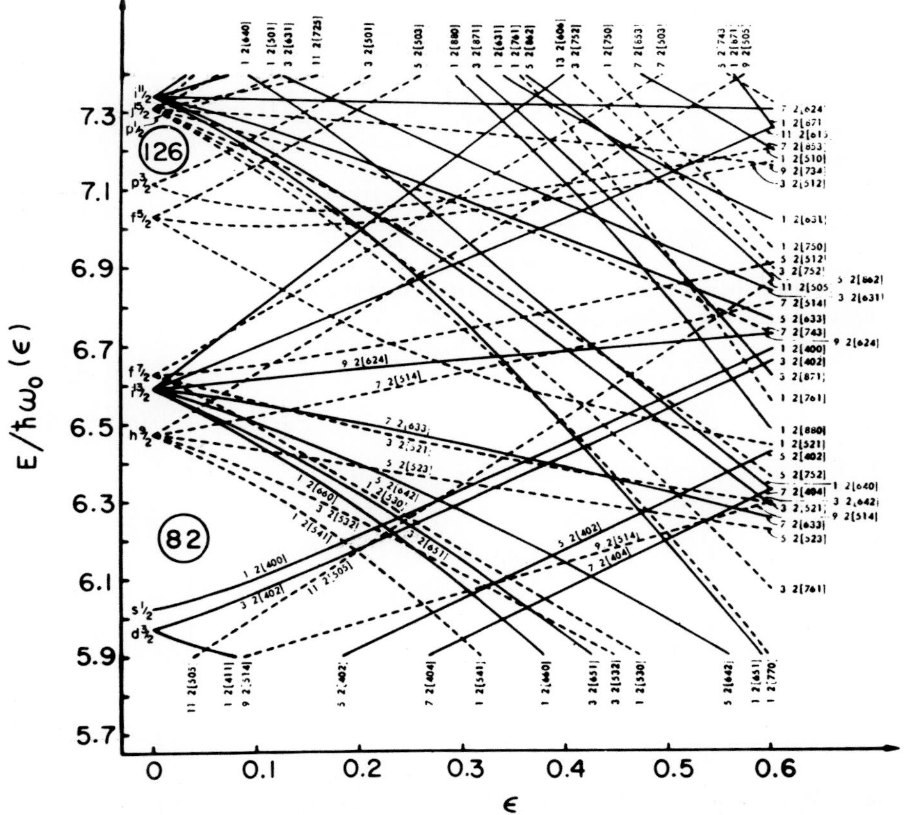

Fig. 6. Nilsson Diagram of single particle shell model proton states vs distortion for $Z > 82$.

$Z = 82$ vs the distortion. They have also made detailed comparisons, with experimental values of the predicted distortions, etc. with generally excellent results (8). It has also been established that some nuclei have appreciable octopole electric moments and distortions, a generalization of the concept.

One interesting feature of the distorted nucleus shell model is that as the distortion increases, the net energy may go through a minimum and then increase until the energy of an initially higher energy orbital, which decreases faster with deformation, crosses below the previous last filled orbital and subsequently becomes the defining least bound filled state. The net energy may then decrease and show a second minima, etc. vs distortion. This is shown in Fig. 7 which is Fig. 25 of Dr. S. A. Moszkowski's review article (18). This effect seems to be present in sub-threshold nuclear fission where the barrier shape has two minima as shown in Fig. 8 (from Ref. 8, Vol. II, p. 633). This was suggested by Dr. V. M. Strutinski (19) in 1967.

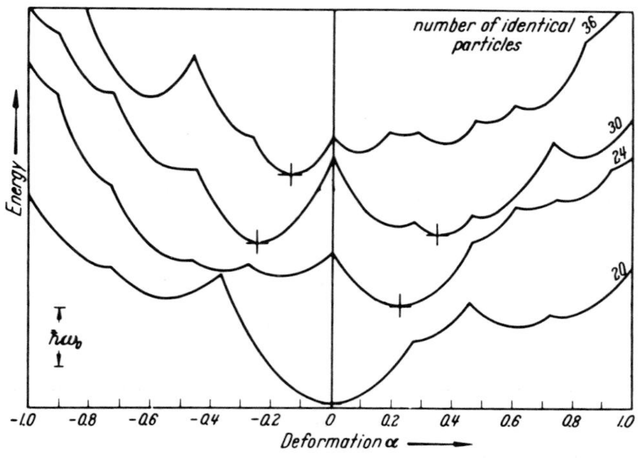

Fig. 7. Deformation potentials for various stages of shell filling-spheroidal harmonic oscillator binding potential. S. A. Moszkowski, Ref. 18, Fig. 25.

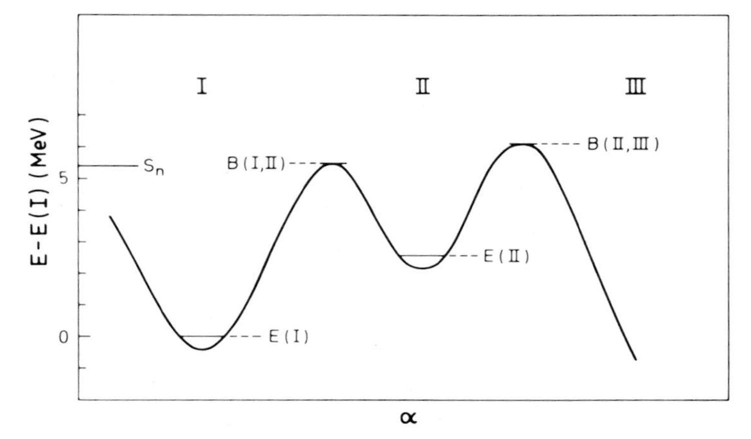

Fig. 8. Double hump energy vs distortion proposed by Strutinski to explain the observed features in sub-threshold nuclear fission. From Ref. 8, Vol. II.

There is one additional effect which I have not yet mentioned which favors spherical shape. If reference is made to the 1937 paper by Feenberg and Phillips (6) on the relative binding of different configurations having two or more $l = 1$ nucleons beyond the ^{4}He core which are combined to form various total L and S ($L-S$ coupling) states for a short range attractive only force, it is seen that the overlap is sensitive to how this is done. As an example, for $A = 6$, the two p nucleon wave functions take on the form $(x+iy)f(r)/\sqrt{2}$, $(x-iy)f(r)/\sqrt{2}$, and $zf(r)$. The combination $(x_1x_2+y_1y_2+z_1z_2)/\sqrt{3}$ for $L = 0$ is more strongly bound than such choices as z_1z_2, or $(x_1x_2+y_1y_2)/\sqrt{2}$ which are favored by a spheroidal potential but do not correspond to an eigenstate of $\vec{L}^2$. Such an effect may inhibit the distortion for small distortions until the gain from the distortion is more overwhelming relative to such symmetry effects on the interaction potential energy.

Since 1950, I have been mainly concerned with experimental physics research using the Nevis Synchrocyclotron. I have been an admiring spectator of the developments of the theory by the Copenhagen group. My main other (experimental) contribution was in the muonic atom x-ray studies started with Val Fitch (20) in 1953 where we first established the smaller charge radii for nuclei. When I made my proposal for use of a spheroidal nuclear model (14), it seemed to be an obvious answer which would immediately be simultaneously suggested by all theorists in the field. I do not understand why it was not. I was also surprised and dismayed to hear one or more respected theorists announce in every Nuclear Physics Conference which I attended through $\sim$ 1955 some such comment as, "Although the Nuclear Shell Model seems emperically to work very well, there is at present no theoretical justification as to why it should apply." Fortunately, such opinions are no longer expressed.

Although my consideration of the "forcing term" for spheroidal nuclear distortion considered the dependence of the single particle kinetic energy on the distortion, I have never seen a description of my work elsewhere in those terms. A common equivalent phrasing is the "centrifugal force exerted on the barrier" by the orbit. Another method is to compute the increase in the potential energy interaction on distortion. This is equivalent, since for a single particle eigenstate, there is zero rate of change of energy with distortions of ψ. Thus $< V >$ and $< T >$ must give equal but opposite contributions to the term linear in β.

I wish to thank the Physical Review, W. A. Benjamin, Inc., and Springer-Verlag for permission to use the various figures.

REFERENCES

1. Rutherford, E., Cambridge Phil. Soc. Proc. *15*, 465 (1910).
2. Bohr, N., Phil. Mag. *26*, 1 (1913).
3. Uhlenbeck and Goudsmit, Naturweis *13*, 593 (1925), Nature *117*, 264 (1926).
4. Pauli, W., Zeits für Phys. *31*, 765 (1925).
5. Bethe, H. A. and Bacher, R. F., Rev. Mod. Phys. *8*, 82–229 (1936). Bethe, H. A., Rev. Mod. Phys. *9*, 69–224 (1937); Livingston, M. S. and Bethe, H. A., Rev. Mod. Phys. *9*, 245–390 (1937).
6. Feenberg, E. and Phillips, M. L., Phys. Rev. *51*, 597 (1937).
7. Feenberg, c.f. E., "Semi-Emperical Theory of the Nuclear Energy Surface", Rev. Mod. Phys. *19*, 239 (1947).
8. Bohr, A. and Mottelson, B., *Nuclear Structure*, Vol. I (1969) and II (1975). Benjamin, W. A., Inc.
9. Bohr, N., Nature *137*, 344 (1936). Bohr and Kalckar, Mat. Fys. Medd. Dan. Vid. Selsk. *14*, #10 (1937).
10. Maria Mayer, Phys. Rev. *75*, 1969 (1949). Feenberg, E. and Hammack, K. C., Phys. Rev. *75*, 1877 (1949). Nordheim, L., Phys. Rev. *75*, 1894 (1949). Also, see Mayer, M. G., and Jensen, J. H. D., *Elementary Theory of Nuclear Shell Structure* (1955), Wiley, New York, N.Y.
11. Bohr, N. and Wheeler, J. A., Phys. Rev. *56*, 426 (1939).
12. Townes, C. H., Foley, H. and Low, W., Phys. Rev. *76*, 1415 (1949).
13. Feenberg, E., Phys. Rev. *55*, 504 (1939).
14. Rainwater, J., "Nuclear Energy Level Argument for a Spheroidal Nuclear Model", Phys. Rev. *79*, 432 (1950).
15. Alder, K., Bohr, A., Huus, T., Mottelson, B., Winther, A., Rev. Mod. Phys. *28*, 432 (1956).
16. Townes, C. H., "Determination of Nuclear Quadrupole Moments", Handbuch der Physick Vol. *38*, (1959), edited by Flügge, S., Springer-Verlag, Berlin.
17. Bohr, A., Phys. Rev. *81*, 134 (1951).
18. Moszkowski, S. A., "Models of Nuclear Structure", Handbuch der Physick Vol. *39*, (1957), edited by Flügge, S., Springer-Verlag, Berlin.
19. Strutinski, V. M., Nucl. Phys. A *95*, 420 (1967).
20. Fitch, V. and Rainwater, J., Phys. Rev. *92*, 789 (1953).

Physics 1976

BURTON RICHTER and SAMUEL C C TING

for their pioneering work in the discovery of a heavy elementary particle of a new kind

THE NOBEL PRIZE FOR PHYSICS

Speech by professor GÖSTA EKSPONG of the Royal Academy of Sciences
Translation from the Swedish text

Your Majesties, Your Royal Highnesses, Ladies and Gentlemen,
By decision of the Royal Swedish Academy of Sciences, this year's Nobel Prize for physics has been awarded to Professor Burton Richter and to Professor Samuel Ting for their pioneering work in the discovery of a heavy elementary particle of a new kind.

This discovery has opened new vistas and given rise to great activity in all laboratories around the world where resources are available. It brings with it the promise of a deeper understanding of all matter and of several of its fundamental forces.

Elementary particles are very small compared to our human dimensions. They are smaller than viruses and molecules and atoms, even smaller than the tiny nucleus of most atoms. They are of great importance when it comes to understanding the basic structure and the basic forces of the material world. In some cases they can even be of importance to society. A basic philosophy is that the material units on any level of subdivision derive their properties from the levels below.

Seventy years ago the first elementary particle was involved in a Nobel Prize. This was at a time when no valid picture of atoms had been formulated. In his Nobel lecture in 1906, J. J. Thompson spoke about his discovery of the electron as one of the bricks of which atoms are built up. Today we know that the electron plays a decisive role in many sciences and technologies and through them in many walks of life—it binds together the molecules of our own bodies, it carries the electricity which makes our lamps shine and it literally draws up the pictures on the TV-screens.

Forty years ago Carl David Andersson was awarded a Nobel Prize for the discovery of the positron—which is the antiparticle to the electron. In the presentation of the award in 1936, it was mentioned that twins of one electron and one positron could be born out of the energy coming from radiation. The reverse can also happen. If the two opposite types of particle meet they can disappear and the energy, which can never be destroyed, shows up as radiation. Only in recent years has this description been enriched through experiments at higher energies, where, among many researchers, both Richter and Ting have contributed.

It is with these two particles that the Nobel laureates Ting and Richter have again experimented in most successful ways. Ting discovered the new particle when he investigated how twins of one electron and one positron are born at very high energies. Richter arranged for electrons and positrons to meet in head-on collisions and the new particle appeared when conditions were exactly right. Both have carried out their researches at laboratories with large particle accelerators and other heavy equipment, which take the place of microscopes when it comes to investigating the smallest structures of matter. Ting and his team of researchers from Massachusetts Institute of Technology set up their cleverly designed apparatus at the Brookhaven National Laboratory on Long Island. Richter and his teams from Stanford and Berkeley built their sophisticated instrumentation complex at the Stanford Linear Accelerator Center in California. In the two different laboratories and with very different methods both found almost simultaneously a clear signal that a new, heavy particle was involved—born in violent collisions and dying shortly afterwards. The letter J was chosen as name at Brookhaven, the greek letter ψ (psi) at Stanford.

The multitude of elementary particles can be beautifully grouped together in families with well-defined boundaries. Missing members have been found in many cases, in some cases they still remain to be found. All seem to derive their properties from a deeper level of subdivision where only a few building bricks, called quarks, are required.

The unique thing about the J-ψ particle is that it does not belong to any of the families as they were known before 1974. Further particles have been discovered resembling the J-ψ one. The reappraisal of particle family structures now required has already begun in terms of a new dimension, corresponding to the new fourth quark already suggested in other contexts.

Most of the recently found particles of normal type can be described as hills of varying height and width in the energy landscape of the physicists, not too unlike pictures of the mounds, barrows and pyramids which the archeologists take an interest in. In the landscape of particles the new J-ψ surprised physicists by being more than twice as heavy as any comparable particle and yet a thousand times more narrow. One can perhaps better imagine the surprise of an explorer in the jungle if he suddenly were to discover a new pyramid, twice as heavy as the largest one in Tikal and yet a thousand times narrower and thus higher. After checking and rechecking that he is not the victim of an optical illusion he would certainly claim that such a remarkable mausoleum must entail the existence of a hidden culture.

Professor Richter, Professor Ting,

I have compared you to explorers of almost unknown territory in which you have discovered new startling structures. Like many great explorers you have had with you teams of skilful people. I would like you to convey to them our congratulations upon these admirable achievements. Your own unrelenting efforts in the field of electron-positron research over a large number of years and your visions have been of outstanding importance and have now culminated in the dramatic discovery of the J-ψ particle. You have greatly influenced and enriched your research field: the physics of elementary particles after November 1974 is recognized to be different from what it was before.

I have the pleasure and the honour on behalf of the Academy to extend to you our warmest congratulations and I now invite you to receive your prizes from the hands of His Majesty the King.

Burton Richter

BURTON RICHTER

I was born on 22 March 1931 in New York, the elder child of Abraham and Fanny Richter. In 1948 I entered the Massachusetts Institute of Technology, undecided between studies of chemistry and physics, but my first year convinced me that physics was more interesting to me. The most influential teacher in my undergraduate years was Professor Francis Friedman, who opened my eyes to the beauty of physics.

In the summer following my junior year, I began work with Professor Francis Bitter in MIT's magnet laboratory. During that summer I had my introduction to the electron-positron system, working part-time with Professor Martin Deutsch, who was conducting his classical positronium experiments using a large magnet in Bitter's laboratory. Under Bitter's direction, I completed my senior thesis on the quadratic Zeeman effect in hydrogen.

I entered graduate school at MIT in 1952, continuing to work with Bitter and his group. During my first year as a graduate student, we worked on a measurement of the isotope shift and hyperfine structure of mercury isotopes. My job was to make the relatively short-lived mercury-197 isotope by using the MIT cyclotron to bombard gold with a deuteron beam. By the end of the year I found myself more interested in the nuclear- and particle-physics problems to which I had been exposed and in the accelerator I had used, than in the main theme of the experiment. I arranged to spend six months at the Brookhaven National Laboratory's 3-GeV proton accelerator to see if particle physics was really what I wanted to do. It was, and I returned to the MIT synchrotron laboratory. This small machine was a magnificent training ground for students, for not only did we have to design and build the apparatus required for our experiments, but we also had to help maintain and operate the accelerator. My Ph.D. thesis was completed on the photoproduction of pi-mesons from hydrogen, under the direction of Dr. L. S. Osborne, in 1956.

During my years at the synchrotron laboratory, I had become interested in the theory of quantum electrodynamics and had decided that what I would most like to do after completing my dissertation work was to probe the short-distance behavior of the electromagnetic interaction. So I sought a job at Stanford's High-Energy Physics Laboratory where there was a 700 MeV electron linear accelerator. My first experiment there, the study of electron-positron pairs by gamma-rays, established that quantum electrodynamics was correct to distances as small as about 10^{-13} cm.

In 1960, I married Laurose Becker. We have two children, Elizabeth, born in 1961, and Matthew, born in 1963.

In 1957, G. K. O'Neill of Princeton had proposed building a colliding beam machine that would use the HEPL linac as an injector, and allow electron-electron scattering to be studied at a center-of-mass energy ten times larger than my pair experiment. I joined O'Neill and with W. C. Barber and B. Gittelman we began to build the first colliding beam device. It took us about six years to make the beams behave properly. This device was the ancestor of all of the colliding beam storage rings to follow. The technique has been so productive that all high-energy physics accelerators now being developed are colliding beam devices.

In 1965, after we had finally made a very complicated accelerator work and had built the needed experimental apparatus, the experiment was carried out, with the result that the validity of quantum electrodynamics was extended down to less than 10^{-14} cm.

Even before the ring at HEPL was operating, I had begun to think about a high-energy electron-positron colliding-beam machine and what one could do with it. In particular, I wanted to study the structure of the strongly interacting particles. I had moved to SLAC in 1963, and with the encouragement of W. K. H. Panofsky, the SLAC Director, I set up a group to make a final design of a high-energy electron-positron machine. We completed a preliminary design in 1964 and in 1965 submitted a request for funds to the Atomic Energy Commission. That was the beginning of a long struggle to obtain funding for the device, during which I made some excursions into other experiments. My group designed and built part of the large magnetic spectrometer complex at SLAC and used it to do a series of pi- and K-meson photoproduction experiments. Throughout this time, however, I kept pushing for the storage ring and kept the design group alive. Finally, in 1970, we received funds to begin building the storage ring (now called SPEAR) as well as a large magnetic detector that we had designed for the first set of experiments. In 1973 the experiments finally began, and the results were all that I had hoped for. The discovery for which I have been honored with the Nobel Prize and the experiments that elucidated exactly what that discovery implied are described in the accompanying lecture. Much more has been done with the SPEAR storage ring, but that is another story.

I spent the academic year, 1975–76, on sabbatical leave at CERN, Geneva. During that year I began an experiment on the ISR, the CERN 30 by 30 GeV proton storage rings, and worked out the general energy scaling laws for high-energy electron-positron colliding-beam storage rings. My motive for this last work was two-fold — to solve the general problems and to look specifically at the parameters of a collider in the 100–200 GeV c.m. energy range that would, I thought, be required to better understand the weak interaction and its relation to the electromagnetic interaction. That study turned into the first-order design of the 27 km circumference LEP project at CERN that was so brilliantly brought into being by the CERN staff in the 1980's.

An interesting sidelight to the LEP story is the attempt by Professor Guy von Dardel of Lund and Chairman of the European Committee for Future Accelerators and I to turn LEP into an inter-regional project. We failed because we couldn't interest either the American or European high-energy physics communities in a collaboration even on as large a scale as LEP. The time was not right, but it surely must be sooner or later.

The general scaling laws for storage rings showed that the size and cost of such machines increased as the square of the energy. LEP, though very large, was financially feasible, but a machine of ten times the energy of LEP would not be. I began to think about alternative approaches with more favorable scaling laws and soon focused on the idea of the linear collider where electron and positron beams from separate linear accelerators were fired at each other to produce the high-energy interactions. The key to achieving sufficient reaction rate to allow interesting physics studies at high energies was to make the beam extremely small at the interaction point, many orders of magnitude less in area than the colliding beams in the storage rings.

In 1978 I met A. N. Skrinsky of Novosibirsk and Maury Tigner of Cornell at a workshop we were attending on future possibilities for high energy machines. We discovered that we had all been thinking along the same general lines and at that workshop we derived, with the help of others present, the critical equations for the design of linear colliders. On returning from the workshop I got a group of people together at the Stanford Linear Accelerator Center and we began to investigate the possibility of turning the two-mile-long SLAC linac into a linear collider. It would be a hybrid kind of machine, with both electrons and positrons accelerated in the same linear accelerator, and with an array of magnets at the end to separate the two beams and then bring them back into head-on collisions. The beams had to have a radius of approximately two microns at the collision point to get enough events to be interesting as a physics research tool, roughly a factor of 1000 less in area than the colliding beams in a storage ring. Construction of SLAC Linear Collider began in 1983, and was finished in late 1987. The first physics experiments began in 1990. Probably the most lasting contribution that this facility makes to particle physics will be the work on accelerator physics and beam dynamics that has been done with the machine and which forms the basis of very active R&D programs aimed at TeV-scale linear colliders for the future. The R&D program is being pursued in the U.S., Europe, the Soviet Union and Japan. Perhaps this will be the inter-regional machine that von Dardel and I tried to make of LEP in the later 1970's.

Along the way I succumbed to temptation and became a scientific administrator first as Technical Director of the Stanford Linear Accelerator Center from 1982 to 1984, and then Director from 1984 to the present. The job of a laboratory director is much different from the job of a physicist, particularly in a time of tight budgets. It is much easier to do physics when someone else gets the funds than it is to get the funds for others to do the research.

Writing this brief biography had made me realize what a long love affiar I have had with the electron. Like most love affairs, it has had its ups and downs, but for me the joys have far outweighed the frustrations.

FROM THE PSI TO CHARM—THE EXPERIMENTS OF 1975 AND 1976

Nobel Lecture, December 11, 1976

by

BURTON RICHTER
Stanford University, Stanford, California, USA

1. INTRODUCTION

Exactly 25 months ago the announcement of the ψ/J particle by Professor Ting's and my groups [1, 2] burst on the community of particle physicists. Nothing so strange and completely unexpected had happened in particle physics for many years. Ten days later my group found the second of the ψ's, [3] and the sense of excitment in the community intensified. The long awaited discovery of *anything* which would give a clue to the proper direction in which to move in understanding the elementary particles loosed a flood of theoretical papers that washed over the journals in the next year.

The experiments that I and my colleagues carried through in the two years after the discovery of the ψ have, I believe, selected from all the competing explanations the one that is probably correct. It is these experiments that I wish to describe. The rapid progress is a consequence of the power of the electron-positron colliding-beam technique, and so I also want to describe this technique and tell something of my involvement in it.

2. COLLIDING BEAMS

I completed my graduate studies at M.I.T. in 1956, and in the Fall of that year I took a position at the High-Energy Physics Laboratory (HEPL) at Stanford University. My main research interest at that time was in exploring the high momentum-transfer or short-distance behavior of quantum electrodynamics (QED). My original plan for a QED experiment had been to use the 700-MeV electron linac at HEPL in a study of electron-electron scattering. Within a short time, however, I came to realize that a different experiment would be both technically simpler to carry out and would also probe QED more deeply (though somewhat differently). During my first year at HEPL I did this latter experiment, which involved the photoproduction of electron-positron pairs in which one of the members of the pair emerged at a large angle. This experiment succeeded in establishing the validity of QED down to distances of about 10^{-13} cm.

2.1 *The Stanford-Princeton Electron-Electron Storage Rings*

In 1957 the idea of an electron-electron scattering experiment came alive again, although in a much different form. This happened when G. K. O'Neill of Princeton University informally proposed the construction at HEPL of a figure 8-shaped set of rings capable of storing counter-rotating beams of electrons at energies up to 500 meV for each beam. In this plan the HEPL linac was to act as the injector for the rings, and the circulating electron beams would collide in the common straight section between the two rings. O'Neill's aim was not only to demonstrate the feasibility of colliding electron beams, but also to carry out electron-electron scattering at an energy that could significantly extend the range of validity of QED.

The potential of such an e^-e^-, colliding-beam experiment, with its total center-of-mass energy of 1000 MeV, was much greater than the ~ 50 MeV that would have been available to test QED in my original e^-e^- scattering idea. Thus when O'Neill asked me to join in this work, I accepted enthusiastically and became an accelerator builder as well as an experimenter. With two other collaborators, W. C. Barber and B. Gittelman, we set out in 1958 to build the first large storage ring, and we hoped to have our first experimental results in perhaps three years. These results were not in fact forthcoming until seven years later, for there was much to learn about the behavior of beams in storage rings; but what we learned during that long and often frustrating time opened up a new field of particle physics research. [4]

2.2 *A Moment of Realization*

Let me digress here for a moment to recount a formative experience. In 1959, as the work on the HEPL rings progressed, I was also trying to learn something about how to calculate cross sections in QED under the tutelage of Stanford theorist J. D. Bjorken. One of the problems Bjorken gave me was to calculate the cross section for the production of a pair of point-like particles having zero spin (bosons) in electron-positron annihilation. I carried out this calculation, but I was troubled by the fact that no point-like bosons were known to exist. The only spin-zero bosons I knew about were pions, and the strong interactions to which these particles were subject gave them a finite size. I realized that the structure function of the particle would have to enter into the cross section to account for this finite size. The structure function for the pion could be measured in an experiment in which e^+e^- annihilation resulted in the production of pion pairs. Further, the structures of any of the family of strongly interacting particles (hadrons) could be determined by measuring their production cross sections in e^+e^- annihilation. It's certain that many people had realized all this before, but it came as a revelation to me at that time, and it headed me firmly on the course that eventually led to this platform.

2.3 *The Electron-Positron Annihilation Process*

This connection between e^+e^- annihilation and hadrons is worth a brief elaboration here, since it is central to the experimental results I shall describe later. The method by which new particles are created in electron-positron collisions is a particularly simple one that I have always naively pictured in the following way. The unique annihilation process can occur only in the collision between a particle and its antiparticle. The process proceeds in two steps:

1. The particle and antiparticle coalesce, and all the attributes that give them their identities cancel. For a brief instant there is created a tiny

electromagnetic fireball of enormous energy density and precisely defined quantum numbers: $J^{PC} = 1^{--}$; all others cancel out to zero.

2. The energy within the fireball then rematerializes into *any* combination of newly created particles that satisfies two criteria: (a) the total mass of the created particles is less than or equal to the total energy of the fireball; (b) the overall quantum numbers of the created particles are the same as those of the fireball. There is no restriction on the individual particles that comprise the final state, only on their sum.

The formation of the fireball or virtual-photon intermediate state in e^+e^- annihilation is described in QED, a theory whose predictions have so far been confirmed by every experimental test. Since we therefore understand Step 1, the creation of the fireball, we are in a sense using the known e^+e^- annihilation process to probe the unknown hadrons that are produced in Step 2 of the process. Our ignorance is thus limited to the structure of the final-state hadrons and to the final-state interactions that occur when particles are created close together. And while that is a great deal of ignorance, it is much less than that of any other particle-production process. In addition, the quantum numbers of the final state in e^+e^- annihilation are simple enough so that we can hope to calculate them from our theoretical models. This is in sharp contrast, for example, to high-energy hadron-hadron collisions, in which very many different angular-momentum states may be involved and thus must be calculated.

2.4 *The SPEAR Electron-Positron Storage Ring*

In 1961, while work on the e^-e^- rings at HEPL continued, I began with D. Ritson of Stanford some preliminary design on a larger e^+e^- storage ring. In 1963 I moved from HEPL to the Stanford Linear Accelerator Center (SLAC), and set up a small group to carry out the final design of the e^+e^- ring. The design energy chosen was 3 GeV (each beam). A preliminary proposal for this colliding-beam machine was completed in 1964, and in 1965 a full, formal proposal was submitted to the U.S. Atomic Energy Commission (now ERDA).

There followed a period of about five years before any funding for this proposed project could be obtained. During this time, other groups became convinced of the research potential of the e^+e^- colliding-beam technique, and several other projects began construction. We watched this other activity enviously, worked at refining our own design, and tried to appropriate any good ideas the others had come up with. Finally, in 1970, funds were made available for a reduced version of our project, now called "SPEAR", and we all fell to and managed to get it built in record time—some 21 months from the start of construction to the first beam collisions [5].

The SPEAR storage ring is located in a part of the large experimental area at the end of the 3-kilometer-long SLAC linac. The facility is shown schematically in Fig. 1. Short pulses of positrons, then electrons, are injected from the SLAC accelerator through alternate legs of the Y-shaped magnetic injection channel into the SPEAR ring. The stored beams actually consist of only a single short bunch of each kind of particle, and the bunches colllide

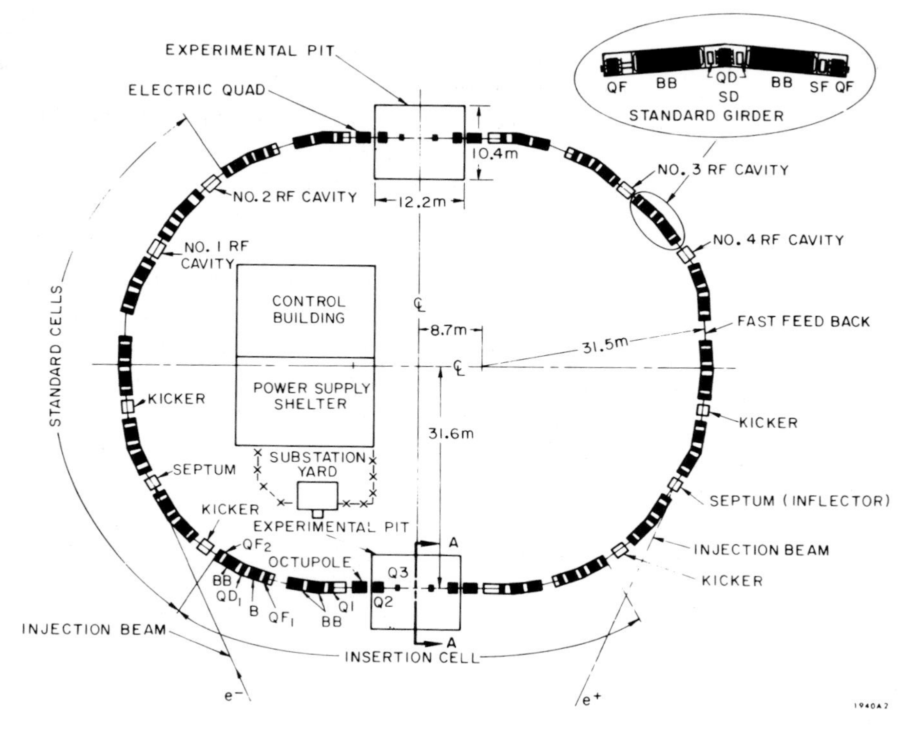

1. Schematic of the SPEAR storage ring.

only at the mid-points of the two straight interaction areas on opposite sides of the machine. Special focusing magnets are used to give the beams a small cross-sectional area at these two interaction points. The time required to fill the ring with electrons and positrons is typically 15—30 minutes, while the data-taking periods between successive fillings are about 2 hours. To achieve this long lifetime, it is necessary to hold a pressure of about 5×10^{-9} torr in the vacuum chamber. The center-of-mass (c.m.) energy of the colliding e^+e^- system can be varied from 2.6 to 8 GeV. The radiofrequency power required to compensate for synchrotron radiation losses rises to 300 kilowatts at the maximum operating energy. The volume within which the e^+e^- collisions occur is small and well-defined ($\sigma_x \times \sigma_y \times \sigma_z = 0.1 \times 0.01 \times 5$ cm^3), which is a great convenience for detection.

2.5 *The Mark I Magnetic Detector*

While SPEAR was being designed, we were also thinking about the kind of experimental apparatus that would be needed to carry out the physics. In the 1965 SPEAR proposal, we had described two different kinds of detectors: the first, a non-magnetic detector that would have looked only at particle multiplicities and angular distributions, with some rather crude particle-identification capability; the second, a magnetic detector that could add accurate momentum measurement to these other capabilities. When the early results in 1969, from the ADONE storage ring at Frascati, Italy, indicated that hadrons were being produced more copiously than expected, I decided that it would be very important to learn more about the final states than could be done with the non-magnetic detector.

Confronted thus with the enlarged task of building not only the SPEAR facility itself but also a large and complex magnetic detector, I began to face up to the fact that my group at SLAC had bitten off more than it could reasonably chew, and began to search out possible collaborators. We were soon joined by the groups of M. Perl, of SLAC; and W. Chinowsky, G. Goldhaber and G. Trilling of the University of California's Lawrence Radiation Laboratory (LBL). This added manpower included physicists, graduate students, engineers, programmers and technicians. My group was responsible for the construction of SPEAR and for the inner core of the magnetic detector, while our collaborators built much of the particle-identification apparatus and also did most of the programming work that was necessary to find tracks and reconstruct events.

This collaborative effort results in the Mark I magnetic detector, shown schematically in Fig. 2. The Mark I magnet produces a solenoidal field, coaxial with the beams, of about 4 kilogauss throughout a field volume of about 20 cubic meters. Particles moving radially outward from the beam-interaction point pass successively through the following elements: the beam vacuum pipe; a trigger counter; 16 concentric cylinders of magnetostrictive wire spark chambers that provide tracking information for momentum measurements; a cylindrical array of 48 scintillators that act as both trigger and time-of-flight counters; the one-radiation-length thick aluminium magnet coil; a cylindrical array of 24 lead-scintillator shower counters that provide

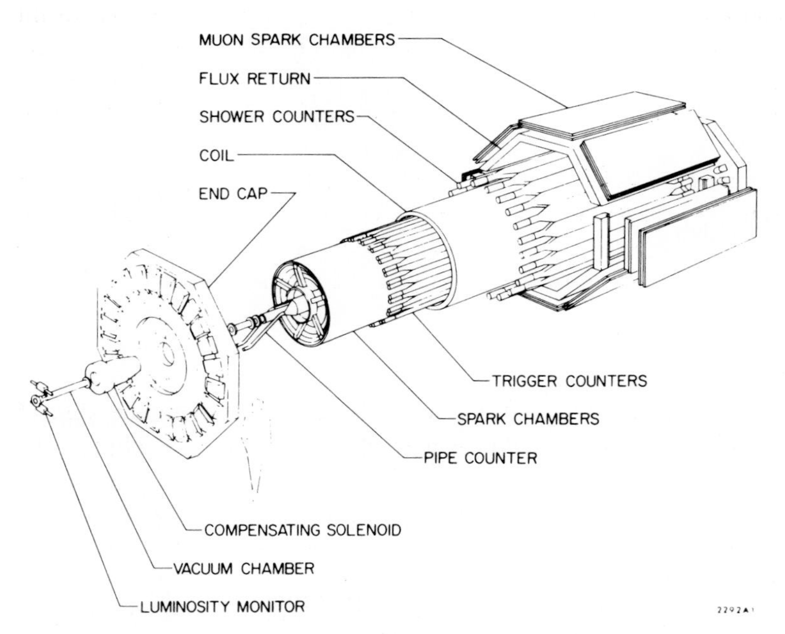

2. An exploded view of the SLAG-LBL magnetic detector.

electron identification; the 20-cm-thick iron flux-return plates of the magnet; and finally an additional array of plane spark chambers used to separate muons from hadrons.

The Mark I magnetic detector was ready to begin taking data in February 1973. During the fall of 1977 it will be replaced at SPEAR by a generally similar device, the Mark II, that will incorporate a number of important improvements. During its career, however, the Mark I has produced a remarkable amount of spectacular physics [6].

3. EARLY EXPERIMENTAL RESULTS

I would like to set the stage for the description of the journey from the ψ's to charm by briefly reviewing here the situation that existed just before the discovery of the new particles. The main international conference in high-energy physics during 1974 was held in July in London. I presented a talk at the London Conference [7] in which I tried to summarize what had been learned up until that time about the production of hadrons in e^+e^- annihilation. This information, shown in Fig. 3, will require a little bit of explanation.

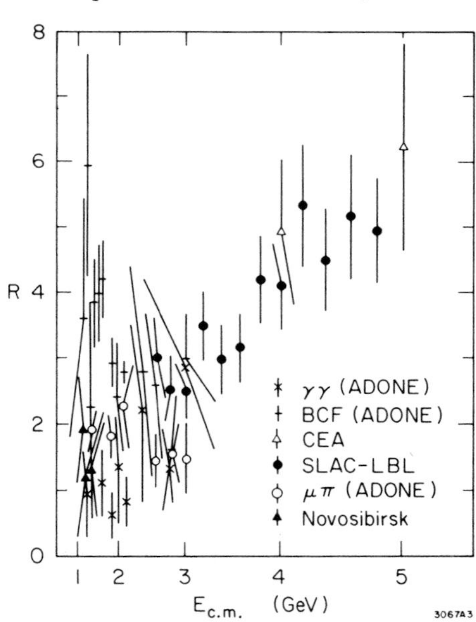

3. The ratio R as of July 1974.

3.1. *The Hadron/Muon-Pair Ratio*

Measurements of the process $e^+e^- \rightarrow$hadrons can be presented straightforwardly in a graph which plots the hadron-production cross section against the c.m. energy of the colliding e^+e^- system. For reasons that I shall explain later, it has become common practice to replace the hadron-production cross section in such graphs by the following ratio:

$$R = \frac{\text{cross section for } e^+e^- \rightarrow \text{hadrons}}{\text{cross section for } e^+e^- \rightarrow \mu^+\mu^-}. \tag{1}$$

It is that ratio R that is plotted vs. c.m. energy in Fig. 3. Historically, the earliest measurements of R were made at the ADONE ring at Frascati; these occupy the lower-energy region of the graph, and they indicate values of R ranging from less than 1 to about 6. These were followed by two important measurements of R made at the storage ring that had been created by rebuilding the Cambridge Electron Accelerator (CEA) at Harvard; the CEA measurements gave an R value of about 5 at $E_{c.m.}$ of 4 GeV, and $R \simeq 6$ at 5 GeV. The early experimental results from the SLAC-LBL experiment at SPEAR filled in some of the gap between the ADONE and CEA results, and between the two CEA points, in a consistent manner; that is, the SPEAR data appear to join smoothly onto both the lower and higher energy data from ADONE and from CEA. With the exception of the experimental points at the very lowest energies, the general picture conveyed by Fig. 3 is that the value of R seems to rise smoothly from perhaps 2 to 6 as $E_{c.m.}$ increases from about 2 to 5 GeV.

3.2. *The Theoretical Predictions*

During the same London Conference in 1974, J. Ellis of CERN [8] undertook the complementary task of summarizing the process $e^+e^- \rightarrow$hadrons from a theoretical point of view. Once again, the predictions of many different theories could most conveniently be expressed in terms of the hadron/muon-pair ratio R rather than directly as hadron-production cross sections. The most widely accepted theory of the hadrons at that time gave the prediction that $R = 2$; but there were many theories. Let me illustrate this by reproducing here, as Table I, the compilation of R predictions that Ellis included in his London Talk. As this table shows, these predictions of the hadron/muon-pair ratio ranged upward from 0.36 to ∞, with many a stop along the way.

I included this table to emphasize the situation that prevailed in the Summer of 1974—vast confusion. The cause of the confusion lay in the paucity of e^+e^- data and the lack of experimental clues to the proper direction from elsewhere in particle physics. The clue lay just around the next corner, but that corner itself appeared as a totally unexpected turn in the road.

Table I. Table of Values of R from the Talk by J. Ellis at the 1974 London Conference [8] (references in table from Ellis's talk)

Value	*Model*	
0.36	Bethe-Salpeter bound quarks	Bohm *et al.*, Ref .42
2/3	Gell-Mann-Zweig quarks	
0.69	Generalized vector meson dominance	Renard, Ref. 49
~ 1	Composite quarks	Raitio, Ref. 43
10/9	Gell-Mann-Zweig with charm	Glashow *et al.*, Ref. 31
2	Colored quarks	
2.5 to 3	Generalized vector meson dominance	Greco, Ref. 30
2 to 5	Generalized vector meson dominance	Sakurai, Gounaris, Ref. 47

3—1/3	Colored charmed quarks	Glashow *et al.*, Ref. 31
4	Han-Nambu quarks	Han and Nambu, Ref. 32
5.7 ± 0.9	Trace anomaly and ρ dominance	Terazawa, Ref. 27
$5.8 {+3.2 \atop -3.5}$	Trace anomaly and ε dominance	Orito *et al.*, Ref. 25
6	Han-Nambu with charm	Han and Nambu, Ref. 32
6.69 to 7.77	Broken scale invariance	Choudhury, Ref. 18
8	Tati quarks	Han and Nambu, Ref. 32
8 ± 2	Trace anomaly, and ε dominance	Eliezer, Ref. 26
9	Gravitational cut-off, Universality	Parisi, Ref. 40
9	Broken scale invariance	Nachtmann, Ref. 39
16	$Su_{12} \times SU_2$) gauge models	Fritzsch & Minkowski, Ref. 34
35—1/3	$SU_{16} \times SU_{16}$) gauge models	Fritzsch & Minkowski, Ref. 34
~5000	High Z quarks	Yock, Ref. 73
70,383	Schwinger's quarks	Yock, Ref. 73
∞	∞ of partons	Cabibbo and Karl, Ref. 9 Matveev and Tolkachev, Ref. 35 Rozenblit, Ref. 36

4. THE PSI PARTICLES

4.1. *Widths of the Psi Resonances*

Figure 4 shows the cross section for hadron production at SPEAR on a scale where all of the data can be plotted on a single graph. This figure is clearly dominated by the giant resonance peaks of the ψ and the ψ'. The extreme

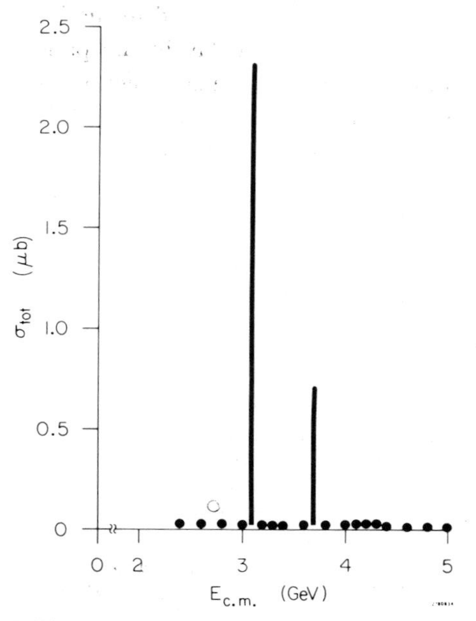

4. The total cross section of hadron production *vs.* center-of-mass energy.

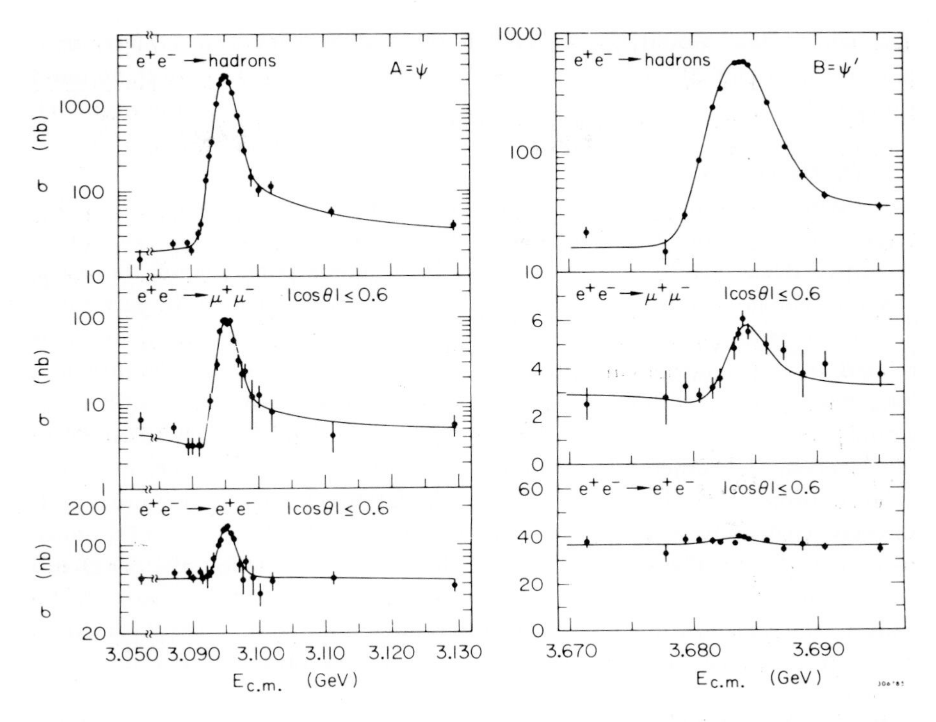

5. Hadron, $\mu^+\mu^-$ and e^+e^- pair production cross section in the regions of the ψ and ψ'. The curves are fits to the data using the energy spread in the colliding beams as the determinant of the widths.

narrowness of the peaks implies that these two states are very long-lived, which is the principal reason why they could not be accounted for by the previously successful model of hadronic structure. In Fig. 5 we show the ψ and ψ' peaks on a greatly expanded energy scale, and also as they are measured for three different decay modes: ψ, $\psi'\rightarrow$hadrons; ψ, $\psi'\rightarrow\mu^+\mu^-$; and ψ, $\psi'\rightarrow e^+e^-$. In this figure the ψ and ψ' peaks can be seen to have experimental widths ot about 2 MeV and 3 MeV, respectively. These observed widths are just about what would be expected from the intrinsic spread in energies that exists within the positron and electron beams *alone*, which means that the true widths of the two states must be very much narrower. The true widths can be determined accurately from the areas that are included under the peaks in Fig. 5 and are given by the following expression:

$$\int \sigma_i \, dE = \frac{6\pi^2}{M^2} B_e B_i \; \Gamma \tag{2}$$

where σ_i is the cross section to produce final state i, B_i is the branching fraction to that state, B_e is the branching fraction to e^+e^-, M is the mass of the state, and Γ is its total width. The analysis is somewhat complicated by radiative corrections but can be done, with the result that [9]

$$\begin{aligned} \Gamma\psi &= 69\pm13 \text{ keV} \\ \Gamma\psi' &= 225\pm56 \text{ keV} \end{aligned} \tag{3}$$

The widths that would be expected if the psi particles were conventional hadrons are about 20% of their masses. Thus the new states are several thousand times narrower than those expected on the basis of the conventional model.

4.2. *Psi Quantum Numbers*

The quantum numbers of the new psi states were expected to be $J^{PC} = 1^{--}$ because of their direct production in e^+e^- annihilation and also because of the equal decay rates to e^+e^- and $\mu^+\mu^-$. In so new a phenomenon, however, anything can go, and so that assumption needed to be confirmed. In particular, one of the tentative explanations of the psi particles was that they might be related to the hypothetical intermediate vector boson, a particle that had long been posited as the carrier of the weak force. Such an identification would permit the psi's to be a mixture of $J^{PC} = 1^{--}$ and 1^{+-}. These quantum numbers can be studied by looking for an interference effect between on- and off-peak production of muon pairs, since the latter is known to be pure 1^{--}. If the new particles were also 1^{--}, then an interference should occur and produce two recognizable effects: a small dip in the cross section below the peak, and an apparent shift in the position of the peak relative to that observed in the hadron channels. In addition, any admixture of 1^{+-} could be expected to show up as a forward/backward asymmetry in the observed angular distribution.

This analysis was carried out as soon as there were sufficient data available for the purpose. The postulated interference effect was in fact observed, as shown in Fig. 6, while no angular asymmetry was seen [8, 9]. Thus both of the psi states were firmly established as $J^{PC} = 1^{--}$.

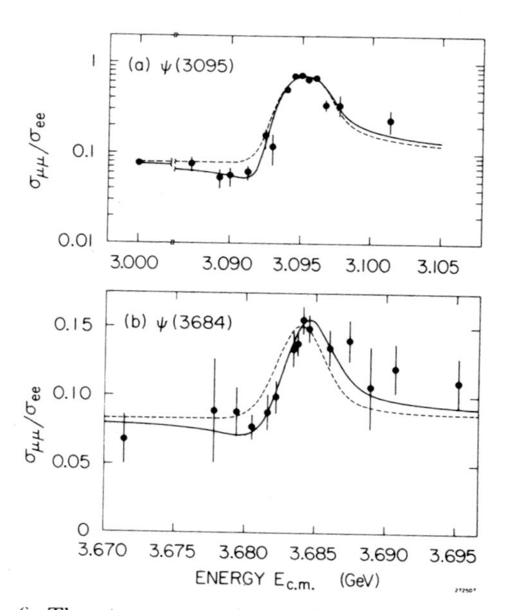

6. The $\mu^+\mu^-$ cross sections at the ψ and the ψ'. The solid curves show the results expected if both states are $J^{PC} = 1^{--}$ and hence interfere with the non-resonant $\mu^+\mu^-$ production. The dashed curves assume no interference.

4.3. *Psi Decay Modes*

We also studied the many decay modes of the ψ and ψ'. In these studies it was important to distinguish between direct and "second-order" decay processes, a point that is illustrated in Fig. 7. This figure shows the following processes:

$$\begin{array}{lll} \text{(a)}\ e^+e^- \to \gamma \to \psi \to \text{hadrons} & \text{(direct decay)} & \\ \text{(b)}\ e^+e^- \to \gamma \to \psi \to \gamma \to \text{hadrons} & \left.\begin{array}{l}\text{(second-order electro-}\\ \text{magnetic decay)}\end{array}\right\} & (4) \\ \text{(c)}\ e^+e^- \to \gamma \to \psi \to \gamma \to \mu^+\mu^- & & \end{array}$$

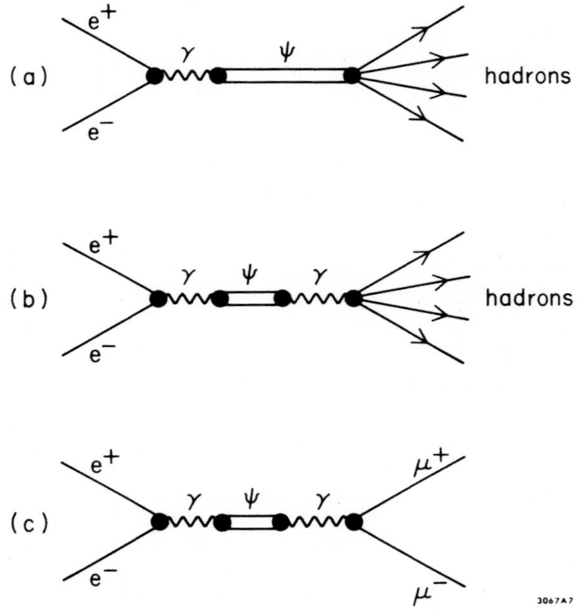

7. Feynman diagrams for ψ production and (a) direct decay to hadrons, (b) second-order electromagnetic decay to hadrons, and (c) second-order electromagnetic decay to $\mu^+\mu^-$.

In processes (b) and (c), hadrons and muon-pairs are produced by virtual photons in exactly the same way that they are produced at off-resonance energies. If the observed hadrons were produced *only* through second-order electromagnetic decay, then the hadron/muon-pair production ratio, R, would be the same on-resonance as off. This is decidedly not the case. Since R is much larger on-resonance than off, both ψ and ψ' do have direct hadronic decays.

More branching fractions for specific hadronic channels have been measured for the ψ and ψ' than for any other particles. Most of these are of interest only to the specialist, but a few have told us a good deal about the psi particles. Since the second-order electromagnetic decays also complicate these analyses, we must again make on- and off-resonance comparisons between muon-pair production and the production of specific hadronic final states. In Fig. 8 we show such a comparison plotted against the number of pions observed in the final state [10]. Even numbers of pions observed are consistent with what is

8. The ratio of the ratios of hadron to $\mu^+\mu^-$ production on and off the ψ resonance *vs.* the number of π mesons in the final state.

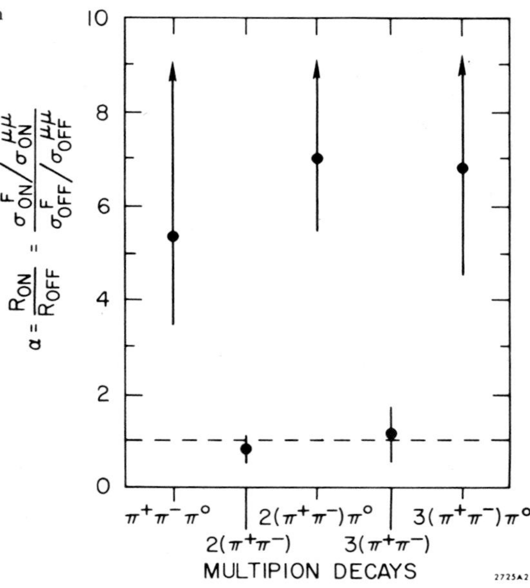

expected from second-order electromagnetic decays, while the observed odd-pion decays are much enhanced. The ψ decays appear, from these data, to be governed by a certain selection rule (G-parity conservation) that is known to govern only the behavior of hadrons, thus indicating that the ψ itself is a hadron.

There are certain specific decay modes that, if observed, provide definite evidence on the isospin of the psi particles. Such modes are

$$\psi \text{ or } \psi' \rightarrow \pi^+\pi^-\pi^\circ\ \Lambda\bar{\Lambda},\ p\bar{p}. \tag{5}$$

Each of these decay modes has in fact been seen, thus establishing $I^G J^{PC} = 0^- 1^{--}$ for both particles.

4.4. *Search for Other Narrow Resonances*

By operating the SPEAR storage ring in a "scanning" mode, we have been able to carry out a systematic search for any other very narrow, psi-like resonances that may exist. In this scanning mode, the ring is filled and set to the initial energy for the scan; data are taken for a minute or two; the ring energy is increased by about an MeV; data are taken again; and so forth. Figure 9 shows these scan data from c.m. energies of about 3.2 to 8 GeV [11, 12]. No statistically significant peaks (other than the ψ' that was found in our first scan) were observed in this search, but this needs two qualifications. The first is that the sensitivity of the search extends down to a limit on possible resonances that have a cross section $\times$ width of about 5% to 10% of that of the ψ. The second qualification is that the particular method of search is sensitive only to extremely narrow resonances like the ψ and ψ'; other, much broader resonances have been found at SPEAR, and we shall soon see how these apparently much different states fit into the picture.

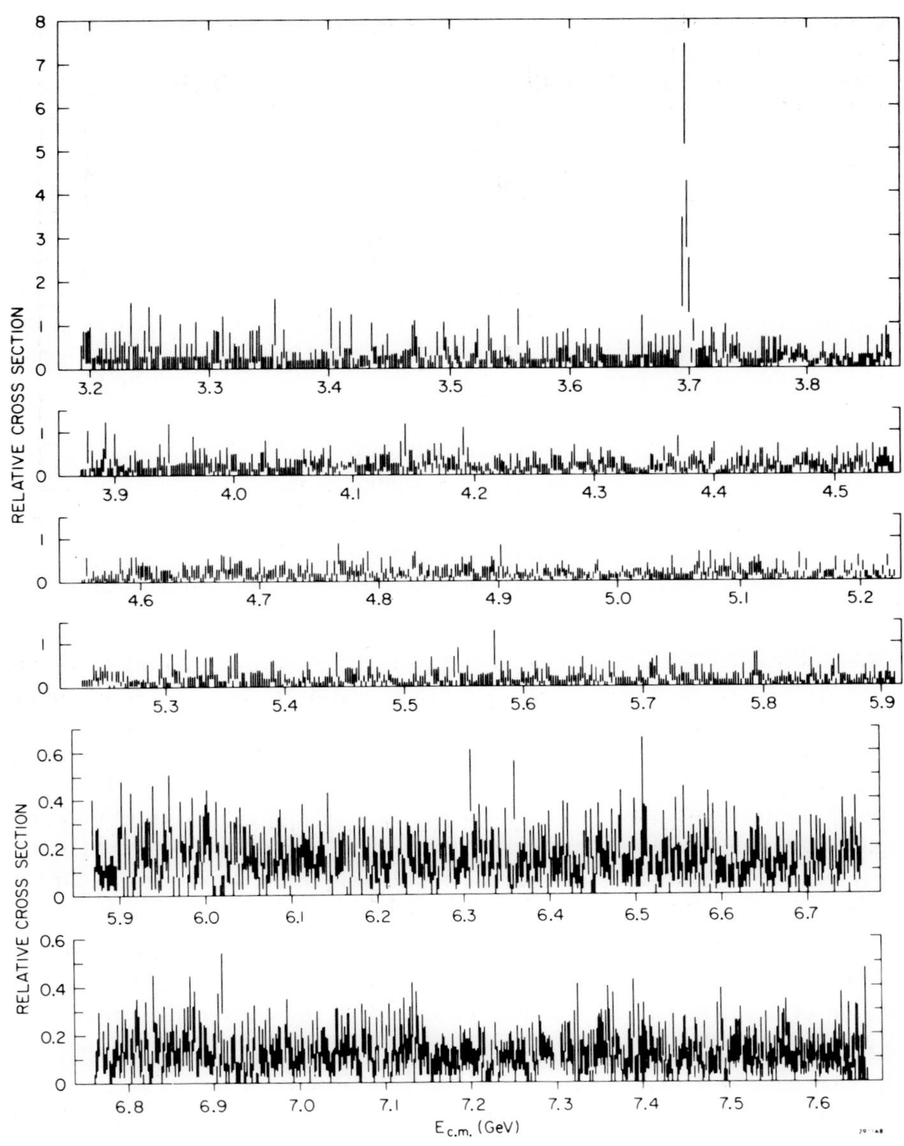

9. The fine-scan data from our search for other narrow ψ-like states. The signal near 3.7 GeV is the ψ'.

5. THE INTERMEDIATE STATES

5.1. *Radiative Transitions*

There are other new states, related to the ψ and ψ' but not directly produced in e^+e^- annihilation, which are observed among the decay products of the two psi particles. More specifically, these new states are produced when either ψ or ψ' decays through the emission of a gamma-ray:

$$\psi \text{ or } \psi' \to \gamma + \text{intermediate state} \tag{6}$$

At least four (perhaps five) distinct intermediate states produced in this way have been observed experimentally.

The first such observation was made by an international collaboration working at the DORIS e^+e^- storage ring at the DESY laboratory in Hamburg [13]. This state was named P_c, and its mass was found to be about 3500 MeV. This same group [14] in collaboration with another group working at DESY later found some evidence for another possible state, which they called X, at about 1800 MeV [15]. At SPEAR, the SLAC-LBL group has identified states with masses of about 3415, 3450 and 3550 MeV, and has also confirmed the existence of the DESY 3500-MeV state. We have used the name χ to distinguish the state intermediate in mass between the $\psi(3095)$ and the $\psi'(3684)$. To summarize these new states:

$$\begin{aligned} &\psi'(3684) \rightarrow \gamma + \chi(3550) \\ &\psi'(3684) \rightarrow \gamma + \chi(3500) \text{ or } P_c \\ &\psi'(3684) \rightarrow \gamma + \chi(3555) \\ &\psi'(3684) \rightarrow \gamma + \chi(3415) \\ &\psi\ (3095) \rightarrow \gamma + X(2800) \quad \text{(not yet firmly established)} \end{aligned} \tag{7}$$

5.2. *Three Methods of Search*

The three methods we have used at SPEAR to search for these intermediate states are indicated schematically in Fig. 10. To begin with, the storage ring is operated at the center-of-mass energy of 3684 MeV that is required for resonant production of the ψ'. In the first search method, Fig. 10(a), ψ' decays to the intermediate state then decays to the ψ through γ-ray emission; and finally the ψ decays, for example, into $\mu^+\mu^-$. The muon-pair is detected along with one or both of the γ-ray photons. This was the method used at DESY to find the 3500-MeV state and also by our group at SLAC to confirm this state [16]. In our apparatus at SPEAR, it will occasionally happen that one of the two γ-ray photons converts into an e^+e^- pair before entering the tracking region of the detector. This allows the energy of the converting γ-ray to be

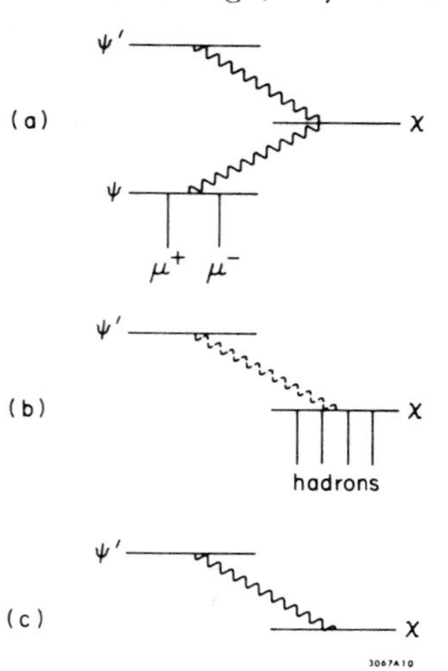

10. Schematics of the three methods of searching for narrow intermediate states.

measured very accurately, and this information can be combined with the measured momenta of the final $\mu^+\mu^-$ pair to make a two-fold ambiguous determination of the mass of the intermediate state. The ambiguity arises from the uncertainty in knowing whether the first or the second gamma-rays in the decay cascade have been detected. It can be resolved by accumulating enough events to determine which assumption results in the narrower mass peak. (The peak associated with the second γ-rays will be Doppler broadened because these photons are emitted from moving sources.) Figure 11 shows the alternate low- and high-mass solutions for a sample of our data [17]. There appears to be clear evidence for states at about 3.45, 3.5 and 3.55 GeV.

The second search method we have used, Fig. 10(b), involves measuring the momenta of the final-state hadrons and reconstructing the mass of the intermediate state [18]. Figure 12 shows two cases in which the effective mass of the final-state hadrons recoils against a missing mass of zero (that is, a γ-ray). In the case where 4 pions are detected, peaks are seen at about 3.4, 3.5 and 3.55 GeV. In contrast, the 2-pion *or* 2-kaon case shows only one clear peak at 3.4 GeV, with perhaps a hint of something at 3.55 GeV. The appearance of the 2-pion or 2-kaon decay modes indicates that the quantum numbers of the states in question must be either 0^{++} or 2^{++}.

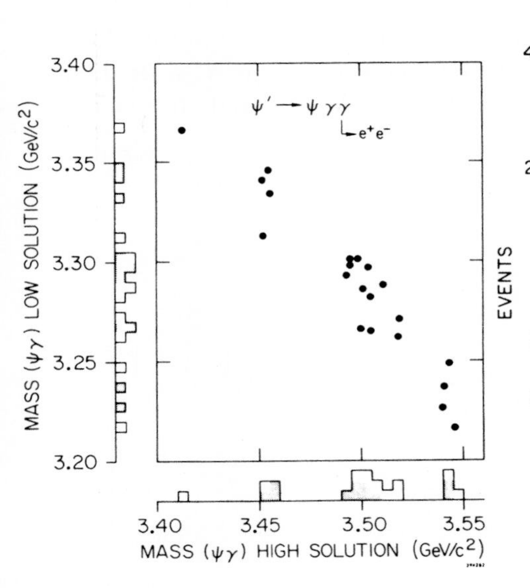

11. The high-resolution ψ—γ mass data. The clustering indicates at least 3 intermediate states.

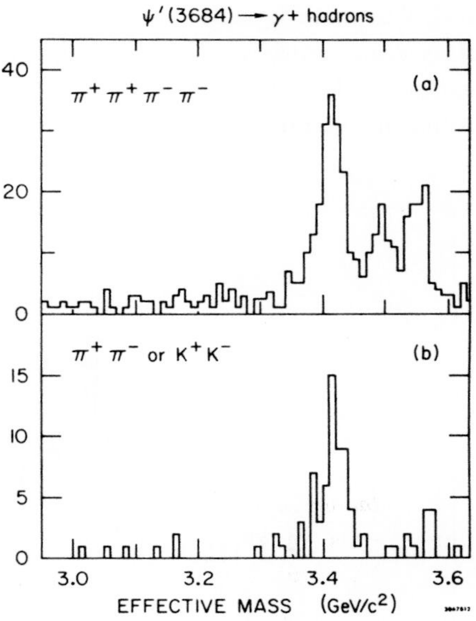

12. The invariant mass of the indicated hadron final states that appear with a γ-ray in ψ' decay. The data show three distinct intermediate states, one of which is not seen in the previous figure.

In the third method of search, Fig. 10(c), only a single γ-ray is detected. The presence of a monoenergetic γ-ray line would signal a radiative transition directly to a specific intermediate state. In our apparatus, this method is difficult to apply because of the severe background problems, but we were able to identify the direct γ-ray transition to the 3.4 GeV state [17]. A different experimental group working at SPEAR (a collaboration among the Universities of Maryland, Princeton, Pavia, Stanford and UC-San Diego) was able to make use of a more refined detection system to observe several of these radiative transitions and to measure the ψ' branching franctions of those states [19].

To summarize, these studies have led to the addition of four (the 2800-MeV state is still marginal) new intermediate state, all with charge-conjugation $C = +1$, to the original ψ and ψ' particles.

6. TOTAL CROSS SECTION AND BROADER STATES

6.1. *Total Cross Section*

So far our discussion of the process $e^+e^- \to$hadrons has been concerned largely with the two psi particles, which are created directly in e^+e^- annihilation, and with the intermediate states, which are not directly created but rather appear only in the decay products of the ψ and ψ'. It is now time to turn our attention to the larger picture of hadron production to see what else can be learned.

Figure 4 presented the total cross section for $e^+e^- \to$hadrons over the full range of c.m. energies accessible to SPEAR. This figure was dominated by the ψ and ψ' resonance peaks, and very little else about the possible structure of the cross section outside of these peaks was observable. We now remedy this situation in Fig. 13, which shows the hadron/muon-pair ratio R, with the dominating ψ and ψ' resonance peaks removed, including their radiative tails.

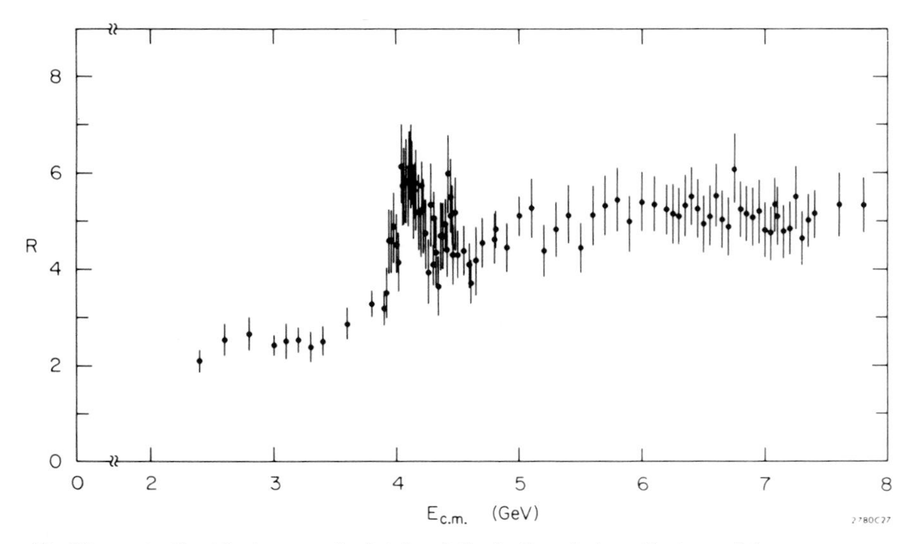

13. The ratio R with the ψ and ψ' deleted (including their radiative tails).

We can characterize the data in the following way. Below about 3.8 GeV, R lies on a roughly constant plateau at a value of $\simeq 2.5$; there is a complex transition region between about 3.8 and perhaps 5 GeV in which there is considerable structure; and above about 5.5 GeV, R once again lies on a roughly constant plateau at a value of $\simeq 5.2$ GeV.

6.2. *Broader (Psi?) States*

The transition region is shown on a much expanded energy scale in Fig. 14. This figure clearly shows that there seem to be several individual resonant states superposed on the rising background curve that connects the lower and upper plateau regions [20]. One state stands out quite clearly at a mass of 3.95 GeV, and another at about 4.4 GeV. The region near 4.1 GeV is remarkably complex and is probably composed of two or more overlapping states; more data will certainly be required to try to sort this out.

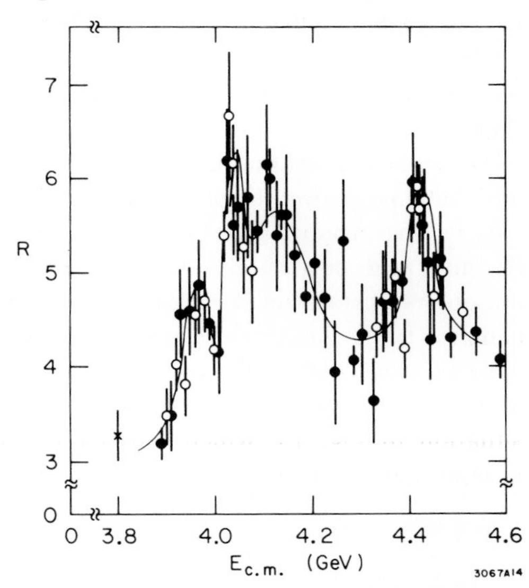

14. An expanded view of R in the transition region around 4 GeV.

The properties of the several states within the transition region are very difficult to determine with any precision. One obvious problem is that these resonances sit on a rapidly rising background whose exact shape is presently neither clear experimentally nor calculable theoretically. Since these new states are, like the ψ's, produced directly in e^+e^- annihilation, they all have $J^{PC} = 1^{--}$ and can therefore interfere with each other, thus distorting the classical resonance shape that would normally be expected from a new particle. Additional shape-distortion might be expected because new particle-production thresholds are almost certainly opening up in the transition region between the lower and upper plateaus. While precise properties can't be given for the new states, we can get some rough numbers from the data. The 3.95-GeV state (ψ'') has a width of about 40—50 MeV. The 4.4-GeV state (ψ'''') seems to be about 30-MeV wide. The 4.1-GeV region (temporarily called ψ''') seems to consist of at least two peaks: one at 4.03 GeV, which is 10—20 MeV wide, and a broad enhancement at 4.1 GeV, about 100-MeV wide.

The widths of all of these states are much greater than the intrinsic energy spread in the e^+e^- beams, and very much greater than the widths of the ψ and ψ'. The suspicion remains, however, that they may still be correctly identified as members of the psi sequence, and that the vast apparent differences between their widths and those of the ψ and ψ' may result simply from the fact that the higher mass states can undergo rapid hadronic decay through new channels that have opened up above the 3684-MeV mass of the ψ'. As with most of the questions in the transition region, this matter will require a good deal more experimental study before it is resolved. In the meantime, however, we shall tentatively add the three or four new psi-like states shown above to the growing list of members of the "psion" family.

7. AN EXCURSION INTO THEORY

Up to this point, we have been cataloguing new particles without much worrying about what it all means. Granting full status to even the several doubtful states, we have a total of 11 new particles. These are grouped together in Fig. 15 in a kind of energy-level diagram, which also includes principal decay modes.

The system shown in Fig. 15, with its radiative transitions, looks remarkably like the energy-level diagram of a simple atom, in fact like the simplest of all "atoms"-positronium, the bound state of an electron and a positron. Although the mass scale for this new positronium is much larger than that of the old, the observed states of the new system can be placed in a one-to-one correspondence with the levels expected for a bound fermion-antifermion system such as e^+e^-. Table II shows these predicted levels together with the most probable assignments of the new particles to the appropriate levels. To gain some insight into the origins of the new positronium system, let's now turn to some specific theoretical models.

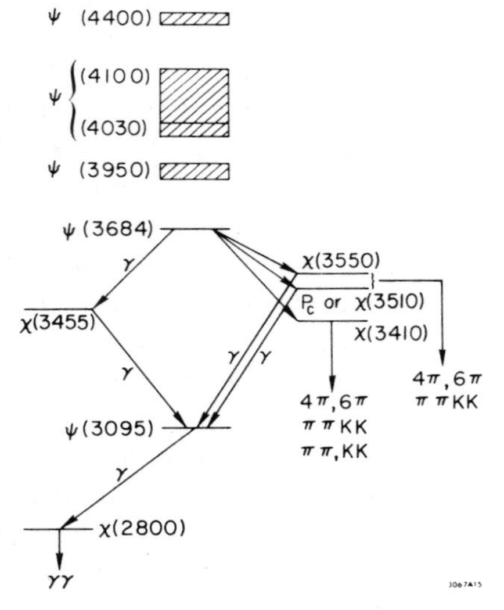

15. An energy-level diagram of the new particles. The many observed decay modes of the psi family have been omitted.

Table II. Some of the low lying bound states of a fermion-antifermion system together with an assignment of the new particle to states with appropriate quantum numbers.

State	L	S	J^{PC}	Particle
1^3S_1	0	1	1^{--}	ψ
2^3S_1	0	1	1^{--}	ψ'
3^3S_1	0	1	1^{--}	ψ'''
1^3D_1	2	1	1^{--}	ψ''
2^3D_1	2	1	1^{--}	ψ''''
1^1S_0	0	0	0^{-+}	X
2^1S_0	0	0	0^{-+}	$\chi(3.45)$
1^3P_0	1	1	0^{++}	$\chi(3.4)$
1^3P_1	1	1	1^{++}	$\chi(3.5)$
1^3P_2	1	1	2^{++}	$\chi(3.55)$

7.1. *The 3-Quark Model*

Some 25 years ago, when only three kinds of hadrons were known (proton, neutron and pi-meson), these particles were universally regarded as simple, indivisible, *elementary* objects. In those days the central task in hadron physics was the effort to understand the strong nuclear force between protons and neutrons in terms of pi-meson exchange. But as the family of hadrons grew steadily larger (they are now numbered in the hundreds), it became increasingly difficult to conceive of them *all* as elementary. In 1963, M. Gell-Mann and G. Zweig independently proposed a solution to this dilema—that *none* of the hadrons was elementary, but rather that all were complex structures in themselves and were built up from different combinations of only three fundamental entities called quarks. These quarks were assumed to carry the familiar 1/2 unit of spin of fermions, but also to have such unfamiliar properties as fractional electric charge and baryon number. A brief listing of the 3 quarks and 3 antiquarks and their properties is given in Table III.

Table III. Properties of the 3 Quarks and 3 Antiquarks

Quarks				Antiquarks			
Symbol	Charge	Baryon Number	Strange-ness	Symbol	Charge	Baryon Number	Strange-ness
u	2/3	1/3	0	$\bar{u}$	−2/3	−1/3	0
d	−1/3	1/3	0	$\bar{d}$	1/3	−1/3	0
s	−1/3	1/3	1	$\bar{s}$	1/3	−1/3	−1

According to this 3-quark model, all mesons were made up of one quark and one antiquark; all baryons, of three quarks; and all antibaryons, of three antiquarks. The quark compositions of some of the better known hadrons are shown here as examples:

$$\pi^+ = u\bar{d},\ K^+ = u\bar{s},\ p = uud,\ \bar{n} = \bar{d}\bar{d}\bar{u}. \tag{8}$$

Prior to 1974, all of the known hadrons could be accommodated within this basic scheme. Three of the possible meson combinations of quark-antiquark ($u\bar{u}$, $d\bar{d}$, $s\bar{s}$) could have the same quantum numbers as the photon, and hence could be produced abundantly in e^+e^- annihilation. These three predicted states had all infact been found; they were the familiar $\varrho(760)$, $\omega(780)$ and $\varphi(1005)$ vector mesons.

7.2. *R in the Quark Model*

The quark model postulated a somewhat different mechanism for the process $e^+e^- \rightarrow$ hadrons than that previously described. For comparison,

Customary View	Quark Model Hypothesis
$e^+e^- \rightarrow \gamma \rightarrow$ hadrons	$e^+e^- \rightarrow \gamma \rightarrow q\bar{q} \rightarrow$ hadrons

where $q\bar{q}$ means any quark-antiquark pair. The quark-model hypothesis is shown schematically in Fig. 16. In this picture the virtual photon intermediate state creates a $q\bar{q}$ pair, which then in turn "clothe" themselves with additional $q\bar{q}$ pairs to form the hadrons that are observed in the final state.

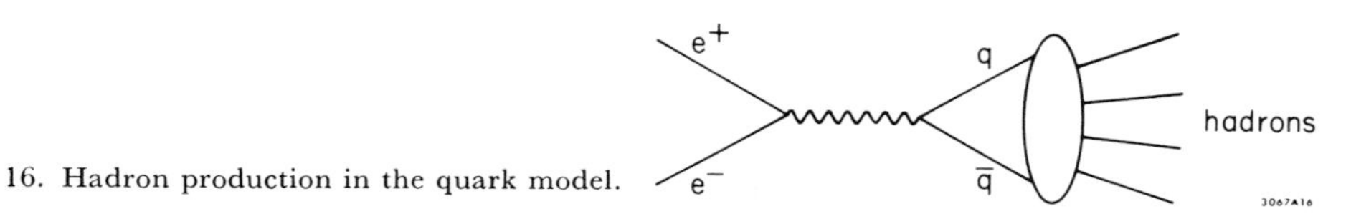

16. Hadron production in the quark model.

Since the quarks are assumed to be elementary, point-like fermions and thus similar to electrons and muons in their electromagnetic properties, it was possible to predict the ratio that should exist between the producton cross sections for quark pairs and muon pairs:

$$\frac{\sigma_{q\bar{q}}}{\sigma_{\mu^+\mu^-}} = q_i^2 \tag{10}$$

where q_i is simply the quark's electric charge. Of course, quarks were supposed to have half-integral spin and fractional charge in the final state, while all hadrons have integral charge and some hadrons have integral spin. In a breathtaking bit of daring it was assumed that the "final-state" interactions between quarks that were necessary to eliminate fractional charge and half-integral spin would have no effect on the basic production cross section. With this assumption the ratio of hadron production to muon-pair production becomes simply

$$R = \sum_{u,d,s} q_i^2. \tag{11}$$

As developed up to 1974, the quark model actually included 3 triplets of quarks, rather than simply 3 quarks, so that with this 3×3 model the hadron/muon-pair ration, R, would be

$$R = (3[2/3)^2 + (-1/3)^2 + (-1/3)^2]) = 2. \tag{12}$$

This beautiful model had great simplicity and explanatory power, but it could not accommodate the ψ and ψ' particles. Nor could it account for the two plateaus that were observed in the measured values of R. The model allowed for excited states of $u\bar{u}$, $d\bar{d}$ and $s\bar{s}$, but the required widths were typically some 20% of the mass of the excited state—more than 1000 times broader than the observed widths of the ψ and ψ'. Before that time there had been a number of suggested modifications or additions to the basic 3-quark scheme. I shall not describe these proposed revisions here except for the one specific model which seems now to best fit the experimental facts.

7.3. *A Fourth Quark*

The first publications of a theory based on 4 rather than 3 basic quarks go all the way back to 1964 [21], only a year or so after the original Gell-Mann/Zweig 3-quark scheme. The motivation at that time was more esthetic than practical, and these models gradually expired for want of an experimental fact that called for more than a 3-quark explanation. In 1970, Glashow, Iliopolous and Maiani [22] breathed life back into the 4-quark model in an elegant paper that dealt with the *weak* rather than the strong interactions. In this work the fourth quark—which had earlier been christened by Glashow the "charmed" quark (c)— was used to explain the non-occurrence of certain weak decays of strange particles in a very simple and straight-forward way. The new c quark was assumed to have a charge of $+2/3$, like the u qark, and also to carry $+1$ unit of a previously unknown quantum number called charm, which was conserved in both the strong and electromagnetic interactions but not in the weak interactions. The c and $\bar{c}$ quarks were also required to have masses somewhat larger than the effective mass of the 3 original quarks, and it was clear that they should be able to combine with the older quarks and antiquarks to form many new kinds of "charmed" hadrons [23].

7.4. "*Charmonium*"

The 4-quark theoretical model became much more compelling with the discovery of the psi particles. This model postulates that the ψ is the lowest mass $c\bar{c}$ system which has the quantum numbers of the photon. The ψ's long life is explained by the fact that the decay of the ψ into ordinary hadrons requires the conversion of *both* c and $\bar{c}$ into other quarks and antiquarks. The positronium-like energy-level states of the psions discussed earlier are also well accounted for by the $c\bar{c}$ system; indeed, 5 specific intermediate states were predicted by Applequist *et al.* [24], and by Eichten *et al.* [25], before they were actually discovered. It was the close analogy with positronium that led Applequist and Politzer to christen the new $c\bar{c}$ system *charmonium*, a name that has caught on.

The 4-quark model also requires two plateaus on R. Above the threshold for charmed-hadron production, the $R = 2$ calculation made above must be modified by the addition of the fourth quark's charge, which results in a prediction of $R = 10/3$ (not enough, but in the right direction). The broad psi-like states at 3.95, 4.1, and 4.4 GeV are accounted for by postulating that the mass of the lightest charmed particle is less than half the mass of the ψ'' (3950) but more than half the mass of the very narrow $\psi'(3684)$, which means that ψ'' can decay strongly to charmed-particle pairs, but ψ' cannot.

To summarize briefly, the 4-quark model of the hadrons seemed to account in at least a qualitative fashion for all of the main experimental information that had been gathered about the psions, and by the early part of 1976 the consensus for charm had become quite strong. The $c\bar{c}$ system of charmonium had provided indirect but persuasive evidence for a fourth, charmed quark, but there remained one very obvious and critically important open question. The particles formed by the $c\bar{c}$ system are not in themselves charmed particles, since charm and anticharm cancel out to zero. But it is necessary to the theory that particles which exhibit charm exist ($c\bar{u}$, $c\bar{d}$, etc.). What was needed, then, was simply the direct experimental observation of charmed particles, and the question was: Where were they [26]?

8. THE DISCOVERY OF CHARM

8.1. *What are We Looking For?*

By early 1976 a great deal had been learned about the properties that the sought-after charmed particles must have. As an example, it was clear that the mass of the lightest of these particles, the charmed D meson, had to fall within the range

$$1843 < m_D < 1900 \text{ MeV}. \tag{13}$$

The lower limit was arrived at by noting once again that the $\psi'(3684)$ was very narrow and therefore could not decay into charmed particles, and also that the upper limit had to be consistent with the begining of the rise of R from its lower to its upper plateau. Since the principal decay product of the c quark was assumed for compelling reasons to be the s quark, then the decay products of charmed particles must preferentially contain strange particles such as the K mesons. The charmed D mesons, for example, could confidently be expected to have the following identifiable decay modes:

$$\begin{aligned} D^\circ &\rightarrow K^-\pi^+ \\ D^\circ &\rightarrow K^-\pi^+\pi^-\pi^+ \\ D^+ &\rightarrow K^-\pi^+\pi^+ \end{aligned} \tag{14}$$

A further point was that, since the charmed quark would decay only through the weak interactions, one might reasonably expect to see evidence of parity violation in the decays of the D mesons.

At SPEAR our collaboration had looked for such charm signatures in the limited data taken before the psi discoveries, but without success. As the post-psi data accumulated throughout 1975, it was evident that we should have another go at it, with particular emphasis on the results obtained at energies close to the expected charm threshold, where the simplest charmed mesons would be produced without serious masking effects from extraneous background. Since I spent the academic year 1975—76 on sabbatical leave at CERN, this chapter of the charmed-particle story belongs to my collaborators.

8.2. *The Charmed Meson*

With the advantages of a much larger data sample and an improvement in the method of distinguishing between pi- and K-mesons in the Mark I detector, a renewed search for charmed particles was begun in 1976. Positive results were not long in coming. The first resonance to turn up in the analysis was one in the mass distribution of the twoparticle system $K^+\pi^+$ in multiparticle events [27]. The evidence for this is shown in Fig. 17. This was the first direct indication of what might be the D meson, for the mass of 1865 MeV was in just the right region. If it was the D°, then presumably the production process was:

$$e^+e^- \rightarrow D^\circ\bar{D}^\circ + X \tag{15}$$

where X represents any other particles. The D° or $\bar{D}^\circ$ would subsequently decay into the observed $K^+\pi^-$ or $K^-\pi^+$ some fraction of the time—the data indicated a branching fraction of about 2% for this charged two-body mode.

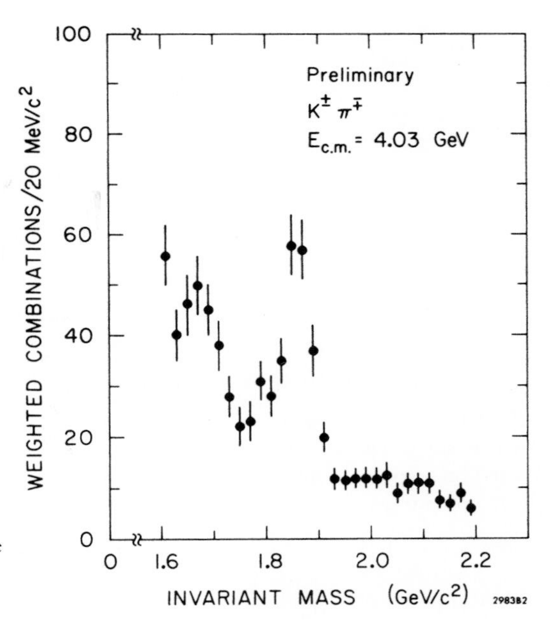

17. The invariant mass distribution of the $K^\pm\pi^\mp$ system in multiparticle final states. The peak at a mass of 1865 MeV is the D° meson.

The branching fraction was a little low compared to the charm-model predictions, but not alarmingly so. The measured width of the resonance was consistent with the resolution of our apparatus, which in this case was determined by the momentum resolution of the detector rather than by the more precise energy resolution of the circulating beams. The measured upper bound on the full width was about 40 MeV; the actual value could well be much smaller, as a weak-interaction decay of the D meson would require.

Continuing analysis of the data yielded two more persuasive findings. The first was a resonance in $K^+\pi^-\pi^+\pi^-$ or $K^-\pi^+\pi^-\pi^+$, which appears to be an alternate decay mode of the D° since the mass is also 1865 MeV. The second was the discovery of the charged companions [28] of the D°, which were observed at the slightly larger mass of 1875 MeV in the following decay channels:

$$\begin{aligned} D^+ &\rightarrow K^-\pi^+\pi^+ \\ D^+ &\rightarrow K^+\pi^-\pi^- \end{aligned} \tag{16}$$

The data for the charged D states are shown in Fig. 18. It is important to note that these states are *not* observed in three-body decay when the pions are oppositely charged:

$$\begin{aligned} D^+ &\rightarrow K^+\pi^-\pi^+ \\ D^- &\rightarrow K^-\pi^+\pi^- \end{aligned} \tag{17}$$

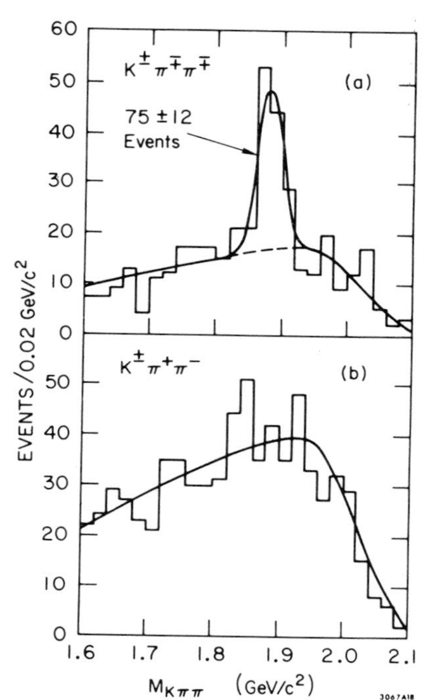

18. The invariant mass distribution of the $K\pi\pi$ system. The $D^{\pm}$ appears in the plot (a) with same-sign pions and not in the plot (b) with opposite-sign pions.

This is precisely what is required by the charmed-quark model. In addition to the clear identification of both neutral and charged D mesons, an excited state [29] of this meson (D*) has also turned up and has been seen to decay to the ground state by both strong and electromagnetic interactions:

$$\begin{aligned} D^* &\rightarrow D+\pi \\ D^* &\rightarrow D+\gamma \end{aligned} \tag{18}$$

Since we have several times mentioned the possibility that the psi-like states having masses above that of the $\psi'(3684)$ may be much broader than ψ and ψ' because they are able to decay strongly into charmed-particle pairs, it is interesting to note that this speculation has now been confirmed in the case of the $\psi'''(4030)$. It now appears, in fact, that the following are the principal decay modes of this particle:

$$\begin{aligned}\psi'''(4030) &\rightarrow D^{\circ}\overline{D}^{*} \\ &\rightarrow D^{*}\overline{D}^{\circ} \\ &\rightarrow D^{*}\overline{D}^{*}\end{aligned} \tag{19}$$

As a final bit of evidence in support of the charmed-meson interpretation of the experimental data, the predicted parity violation in D decay has also been observed. In the decay process $D^{\circ} \rightarrow K^{+}\pi^{-}$, the K and π each have spin-0 and odd intrinsic parity. This means that any spin possessed by the D° must show up as orbital angular momentum in the $K\pi$ system, and thus that the parity of the D° must be given by

$$P = (-1)^{J} \tag{20}$$

where J is the spin of the D°. An analysis of the 3-body decay data, $D^{\pm} \rightarrow$ $\rightarrow K^{-}\pi^{+}\pi^{+}$ or $K^{+}\pi^{-}\pi^{-}$, showed that the parity cannot be the same as that given above, and therefore that parity must be violated in D-meson decay [30].

The experimental data that have been described here are strikingly consistent with the predictions of the 4-quark or charm theory of the hadrons, and there is little doubt that charmed particles have now in fact been found. In addition to these charmed mesons uncovered at SPEAR, there has been recent information from Fermilab that a collaborative group working there under Wonyong Lee has now discovered the first of the charmed baryons [31] actually an antibaryon designed Λ_c to identify it as the charmed counterpart of the Λ.

9. OBSERVATION OF JETS

While this topic is not directly connected with the new particles, it does have a direct bearing on the validity of the quark model. As I noted earlier, the picture of $e^{+}e^{-}$ annihilation that is derived from the quark model indicates that the final-state hadrons do not come directly from the virtual-photon intermediate state, but rather from the quark-antiquark pair that is first created from the electromagnetic fireball and subsequently forms the final hadrons. These hadrons are produced with low transverse momenta with respect to the $q\overline{q}$ direction, and as illustrated in Fig. 19, if the energy is sufficiently high, form two collimated jets of particles whose axes lie along the original $q\overline{q}$ direction.

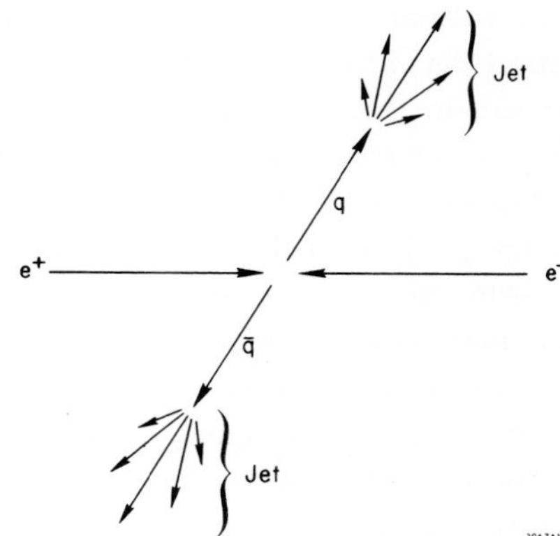

19. Jet production in the quark model.

At SPEAR we have analyzed our highest-energy data [32] by determining for each event those particular axes that minimize the transverse momentum relative to those axes for all of the observed particles. This method of analysis leads to the definition of a quantity we have called "sphericity," which is related to the quadrupole moment of the particle distribution in momentum space. The more jet-like event, the lower the sphericity. Figure 20 shows the data compared to the jet model and to an "isotropic" model with no jet-like characteristics. As the energy increases, the events do become more jet-like as required. The result was excellent agreement, not only in the general sense but also in the finding that the angular distribution of the jet axes was consistent with the $1+\cos^2\vartheta$ distribution that is expected if the jets originate from parent particles of spin-1/2.

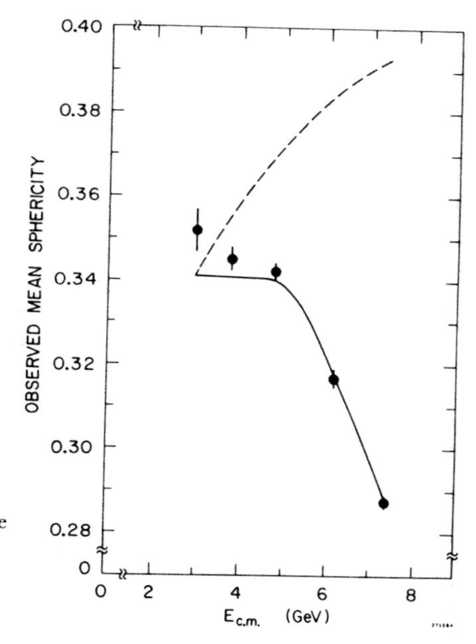

20. The mean sphericity of multihadron events *vs.* center-of-mass energy. The solid curve is that expected of the jet model, while the dashed curve is that expected from an isotropic phase-space model.

In addition, under certain operating conditions the beams in the SPEAR storage ring become polarized, with the electron spin parallel and the positron spin antiparallel to the ring's magnetic bending field. In this polarized condition an azimuthal asymmetry in particle production can appear with respect to the direction of the beams. Jets measured under these conditions also displayed the azimuthal asymmetry that is expected of spin-1/2 particles.

Further, the individual hadrons within the jets also displayed this asymmetry [33]. It will be evident that the greater the momentum of a single hadron, the closer that hadron must lie to the original direction defined by the quark. By looking at pion production in detail, we were able to determine that as the pion momentum approached the maximum value possible for the particular machine energy, so did the azimuthal asymmetry approach the maximum possible asymmetry expected for spin-1/2 particles. This point is illustrated in Fig. 21.

21. The azimuthal asymmetry parameter for pions normalized to the asymmetry in μ-pair production *vs.* the fractional pion momentum.

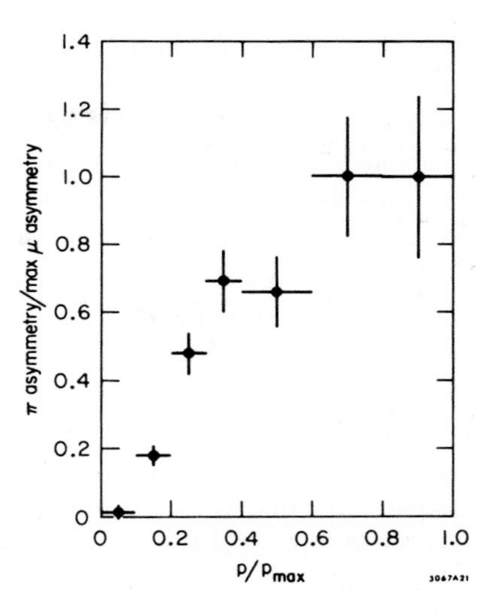

I find it quite remarkable that a collection of hadrons, each of which has integral spin, should display all of the angular-distribution characteristics that are expected for the production of a pair of spin-1/2 particles. Such behavior is possible without assuming the existence of quarks (the final-state helicity must be one along the direction of the particle or jet), but any other explanation seems difficult and cumbersome. In my view the observations of these jet phenomena in e^+e^- annihilation constitute one of the very strongest pieces of evidence for believing that there really is a substructure to the hadrons.

10. CONCLUSIONS AND QUESTIONS

The electron-positron colliding-beam experiments of the past two years have, I believe, settled the question of the significance of the psi particles. The charmonium family, the two plateaus in R, the wide resonances above charm threshold, the charmed particles themselves, the evidence for the weak decays of the charmed particles and the existence of jets—all these support most strongly the ideas of the quark model of hadron substructure and the 4-quark version of that model. To me, one of the most remarkable features of the quark model is that it correctly explains a great deal of data on strongly interacting particles with the most simple-minded of calculations. The charmonium spectrum, for example, is calculated with the nonrelativistic Schrödinger equation using a simple potential. The two plateaus in R and jet structure are explained by assuming that the final-state interactions of strongly interacting particles can be ignored. Why it is all so simple, while at the same time the quarks themselves appear confined to hadrons and are never seen in the free state, is one of the central questions of strong-interaction physics.

We already know, however, that the 4-quark model cannot be the complete story. The colliding-beam experiments are not entirely consistent with this model. The high energy plateau value of R is about 5.1 rather than 3-1/3 as demanded by the charm model. While R = 3-1/3 is only reached in the theory at very high energies, the difference between 3-1/3 and 5.1 are too large to be explained easily. At the same time, there is evidence in our data for a class of events (the μ-e events) which are not easily explained within the framework of 4-quarks and 4 leptons (e^-, v_e, μ^-, v_μ) and which may require an expansion of the lepton family and/or the quark family. These inconsistencies immediately bring up the question of how many quarks and leptons there are.

There are two schools of thought on this question. One school says that the quark system is complete or nearly complete—while there may be a few more quarks to be found, there are a small number of indivisible elements, among which are the present four, and all of the strongly interacting particles are built out of these elementary and indivisible components. The other school says that the quarks themselves are probably built from something still smaller, and that we shall go on forever finding smaller and smaller entities each inside the next larger group.

These and other questions on particle structure may be answered by the next generation of e^+e^- colliding-beam machines now being built at DESY and SLAC which will reach 35 to 40 GeV in the center-of-mass system. Experiments on these machines will begin in 4 to 5 years and should tell us promptly about the existence of new plateaus in R, new "oniums", or new leptons.

An even more fundamental set of questions, which I find more interesting than the number of quarks, will probably not be answered by experiments at any accelerator now in construction. These questions have to do with the possibility of a unified picture of the forces of nature: gravity, the weak interaction, the electromagnetic interaction, and the strong interaction. Weinberg [34] and Salam [35] have made the first models of a unified weak and electromagnetic interaction theory. Attempts have been made at a unified picture of the weak, electromagnetic and strong interactions—more primitive than the Weinberg/Salam model, for the problem is more difficult, but still a beginning. The experimental information required to establish these unified pictures will almost certainly require still higher energies: several hundred GeV in the center-of-mass and again, I believe, in the e^+e^- system. If any of these unified pictures is correct at very high energies, then our only correct field theory, quantum electrondynamics, will necessarily have to break down, and I will have come full circle back to the first experiment I wanted to do as an independent researcher [36].

REFERENCES

1. J. J. Aubert *et al.*, Phys. Rev. Lett. *33*, 1404 (1974).
2. J.-E. Augustin *et al.*, Phys. Rev. Lett. *33*, 1406 (1974).
3. G. S. Abrams *et al.*, Phys. Rev. Lett. *33*, 1453 (1974).
4. The early development of the colliding beams technique was an international effort. The two groups who, in those early days suffered with us through the discovery and conquest of what at times seemed to be an endless series of beam instabilities and technological problems, were those of F. Amman at Frascati and G. I. Budker at Novosibirsk.
5. The success of the SPEAR project is in large measure due to J. Rees who was then my deputy, and to M. Allen, A. M. Boyarski, W. Davies-White, N. Dean, G. E. Fischer, J. Harris, J. Jurow, L. Karvonen, M. J. Lee, R. McConnell, R. Melen, P. Morton, A. Sabersky, M. Sands, R. Scholl and J. Voss.
6. The physicists of the SLAC/LBL group who were responsible for building the detector and for the experiments I will discuss are S. M. Alam, J.-E. Augustin, A. M. Boyarski, M. Breidenbach, F. Bulos, J. M. Dorfan, G. J. Feldman, G. E. Fischer, D. Fryberger, G. Hanson, J. A. Jaros, B. Jean-Marie, R. R. Larsen, D. Lüke, V. Lüth, H. L. Lynch, C. C. Morehouse, J. M. Paterson, M. L. Perl, I. Peruzzi, M. Piccolo, T. P. Pun, P. Rapidis, B. Richter, R. H. Schindler, R. F. Schwitters, J. Siegrist, W. Tanenbaum, and F. Vannucci from SLAC; and G. S. Abrams, D. Briggs, W. C. Carithers, W. Chinowsky, R. G. DeVoe, C. E. Friedberg, G. Goldhaber, R. J. Hollebeek, A. D. Johnson, J. A. Kadyk, A. Litke, B. Lulu, R. J. Madaras, H. K. Nguyen, F. Pierre, B. Sadoulet, G. H. Trilling, J. S. Whitaker, J. Wiss, and J. E. Zipse from LBL.
7. B. Richter, *Proceedings of the XVII International Conference on High Energy Physics, London* (1974).
8. J. Ellis, *Ibid.*
9. A. M. Boyarski *et al.*, Phys. Rev. Lett. *34*, 1357 (1975); V. Lüth *et al.*, Phys. Rev. Lett. *35*, 1124 (1975).
10. B. Jean-Marie *et al.*, Phys. Rev. Lett. *36*, 291 (1976).
11. A. M. Boyarski *et al.*, Phys. Rev. Lett. *34*, 762 (1975).
12. Review paper by R. Schwitters, *Proceedings of the 1975 International Symposium on Lepton and Photon Interactions*, *Stanford University* (1975).
13. W. Braunschweig *et al.*, Phys. Lett. *57B*, 407 (1975).
14. B. H. Wiik, *Proceedings of the 1975 Symposium on Lepton and Photon Interactions*, Stanford University (1975).
15. J. Heintze, *Ibid.*
16. W. Tanenbaum *et al.*, Phys. Rev. Lett. 35, 1323 (1975).
17. S. Whitaker *et al.*, to be published in Phys. Rev. Lett.
18. G. H. Trilling, LBL Report 5535 and *Proceedings of the SLAC Summer Institute on Particle Physics*, Stanford (1976).
19. D. H. Badtke *et al.*, paper submitted to the *XVIII International Conference on High Energy Physics, Tbilisi, USSR* (1976).
20. J. Siegrist *et al.*, Phys. Rev. Lett. *36*, 700 (1976).
21. D. Amati *et al.*, Phys. Lett *11*, 190 (1964);
    J. D. Bjorken and S. L. Glashow, Phys. Lett. *11*, 255 (1964);
    Z. Maki and Y. Chnuki, Prog. Theor. Phys. *32*, 144 (1964);
    Y. Hara, Phys. Rev. *134B*, 701 (1964).
22. S. L. Glashow, J. Iliopolous, and L. Maiani, Phys. Rev. *D2*, 1285 (1970).
23. An excellent review of the status of the charm model at the end of 1974 is that of M. K. Gaillard, B. Lee, and J. L. Rosner, Rev. Mod. Phys. *47*, 277 (1975).

24. T. Applequist *et al.*, Phys. Rev. Lett. 34, 365 (1975).
25. E. Eichten *et al.*, Phys. Rev. Lett. 34, 369 (1975).
26. One possible example of the production of a charmed baryon has been reported by E. G. Cazzoli *et al.*, Phys. Rev. Lett. *34*, 1125 (1975).
27. G. Goldhaber *et al.*, Phys. Rev. Lett. *37*, 255 (1976).
28. I. Peruzzi *et al.* Phys. Rev. Lett. *37*, 569 (1976).
29. Papers on the D* decays and ψ''' decays are in preparation by the SLAC/LBL Group.
30. Strictly speaking, this argument is not airtight. If the D^+ is not in the same isotopic doublet as the D°, the comparison of D^+ and D° decay gives no information on parity violation. The close values of the masses of D^+ and D°, however, make it very probable that they are related.
31. B. Knapp *et al.*, Phys. Rev. Lett. *37*, 822 (1976).
32. G. Hanson *et al.*, Phys. Rev. Lett. *35* 1609 (1975).
33. R. F. Schwitters *et al.*, Phys. Rev. Lett. *35*, 1320 (1975).
34. S. Weinberg, Phys. Rev. Lett. *19*, 1264 (1967).
35. A. Salam, *Proceedings of the 8th Nobel Symposium* (Almquist Wiksells, Stockholm, 1968).
36. I want to acknowledge here those who have been most important in helping me on my circular path in particle physics and whom I have not previously mentioned. They are L. S. Osborne, my thesis adviser; E. Courant and A. Sessler, who helped me understand the mysteries of the behavior of beams in storage rings; S. Drell and J. D. Bjorken, who have been my guides to theoretical physics; M. Sands, who helped me design the storage ring and whose encouragement helped keep me going in the frustrating years of waiting for the funds to build it; W. K. H. Panofsky, who was Director of HEPL and is Director of SLAC, without whose support and desire to see "good physics" done there would be no SPEAR; and Laurose Richter—wife, friend and adviser.

Samuel C. C. Ting

SAMUEL CHAO CHUNG TING

I was born on 27 January 1936 in Ann Arbor, Michigan, the first of three children of Kuan Hai Ting, a professor of engineering, and Tsun-Ying Wang, a professor of psychology. My parents had hoped that I would be born in China, but as I was born prematurely while they were visiting the United States, by accident of birth I became an American citizen. Two months after my birth we returned to China. Owing to wartime conditions I did not have a traditional education until I was twelve. Nevertheless, my parents were always associated with universities, and I thus had the opportunity of meeting the many accomplished scholars who often visited us. Perhaps because of this early influence I have always had the desire to be associated with university life.

Since both my parents were working, I was brought up by my maternal grandmother. My maternal grandfather lost his life during the first Chinese Revolution. After that, at the age of thirty-three, my grandmother decided to go to school, became a teacher, and brought my mother up alone. When I was young I often heard stories from my mother and grandmother recalling the difficult lives they had during that turbulent period and the efforts they made to provide my mother with a good education. Both of them were daring, original, and determined people, and they have left an indelible impression on me.

When I was twenty years old I decided to return to the United States for a better education. My parents' friend, G. G. Brown, Dean of the School of Engineering, University of Michigan, told my parents I would be welcome to stay with him and his family. At that time I knew very little English and had no idea of the cost of living in the United States. In China, I had read that many American students go through college on their own resources. I informed my parents that I would do likewise. I arrived at the Detroit airport on 6 September 1956 with $100, which at the time seemed more than adequate. I was somewhat frightened, did not know anyone, and communication was difficult.

Since I depended on scholarships for my education, I had to work very hard to keep them. Somehow, I managed to obtain degrees in both mathematics and physics from the University of Michigan in three years, and completed my Ph.D. degree in physics under Drs. L. W. Jones and M. L. Perl in 1962.

I went to the European Organization for Nuclear Research (CERN) as a Ford Foundation Fellow. There I had the good fortune to work with Giuseppe Cocconi at the Proton Synchrotron, and I learned a lot of physics from him. He always had a simple way of viewing a complicated problem, did experiments with great care, and impressed me deeply.

In the spring of 1965 I returned to the United States to teach at Columbia University. In those years the Columbia Physics Department was a very stimulating place, and I had the opportunity of watching people such as L. Lederman, T. D. Lee, I. I. Rabi, M. Schwarts, J. Steinberger, C. S. Wu, and others. They all had their own individual style and extremely good taste in physics. I benefitted greatly from my short stay at Columbia.

In my second year at Columbia there was an experiment done at the Cambridge Electron Accelerator on electron-positron pair production by photon collision with a nuclear target. It seemed to show a violation of quantum electrodynamics. I studied this experiment in detail and decided to duplicate it. I contacted G. Weber and W. Jentschke of the Deutsches Elektronen-Synchrotron (DESY) about the possibility of doing a pair production experiment at Hamburg. They were very enthusiastic and encouraged me to begin right away. In March 1966 I took leave from Columbia University to perform this experiment in Hamburg. Since that time I have devoted all my efforts to the physics of electron or muon pairs, investigating quantum electrodynamics, production and decay of photon-like particles, and searching for new particles which decay to electron or muon pairs. These types of experiments are characterized by the need for a high-intensity incident flux, for high rejection against a large number of unwanted background events, and at the same time the need for a detector with good mass resolution.

In order to search for new particles at a higher mass, I brought my group back to the United States in 1971 and started an experiment at Brookhaven National Laboratory. In the fall of 1974 we found evidence of a new, totally unpredicted, heavy particle—the J particle. Since then a whole family of new particles has been found.

In 1969 I joined the Physics Department of the Massachusetts Institute of Technology (MIT). In 1977, I was appointed as the first Thomas Dudley Cabot Institute Professor of Physics at MIT. In recent years it has been my privilege to be associated with M. Deutsch, A. G. Hill, H. Feshbach, W. Jentschke, H. Schopper and G. Weber. All have strongly supported me. In addition, I have enjoyed working with many very outstanding young physicists such as U. Becker, J. Burger, M. Chen, R. Marshall and A. J. S. Smith.

I married Dr. Susan Marks in 1985. We have one son, Christopher, born in 1986 and I have two daughters, Jeanne and Amy, from an earlier marriage.

I have been awarded the Ernest Orlando Lawrence Award from the US government in 1976 and the DeGasperi Award in Science from the Italian government in 1988. I have also received the Eringen Medal awarded by the Society of Engineering Science in 1977, the Golden Leopard Award for Excellence from the town of Taormina, Italy in 1988 and the Gold Medal for Science and Peace from the city of Brescia, Italy in 1988. I am a member of the National Academy of Sciences (US) and the American Physical Society, the Italian Physical Society and the European Physical Society. I have also been elected as a foreign member in Academia Sinica, the Pakistan Academy of Science and the Academy of Science of the USSR (now Russian

Academy of Science). I also hold Doctor Honoris Causa degrees from the University of Michigan, The Chinese University of Hong Kong, Columbia University, the University of Bologna, Moscow State University and the University of Science and Technology in China and am an honorary professor at Jiatong University in Shanghai, China.

THE DISCOVERY OF THE J PARTICLE:

A personal recollection

Nobel Lecture, 11 December, 1976
by
SAMUEL C. C. TING
Massachusetts Institute of Technology, Cambridge, Massachusetts, USA
and
CERN, European Organization for Nuclear Research, Geneva, Switzerland

1. PHOTONS AND HEAVY PHOTONS

The study of the interaction of light with matter is one of the earliest known subjects in physics. An example of this can be found in the *Mo Tsu* [1] (the book of Master Mo, Chou Dynasty, China, 4th century B.C.). In the 20th century, many fundamentally important discoveries in physics were made in connection with the study of light rays. The first Nobel Prize in Physics was awarded to W. C. Röntgen in 1901 for his discovery of X-rays.

In modern times, since the work of Dirac, we realized the possibility of the creation of electron-positron pairs by energetic light quanta. The work of W. E. Lamb and R. C. Retherford provided a critical step in the understanding of interactions between photons and electrons. The elegant formulation of quantum electrodynamics by S. Tomonaga, J. Schwinger and R. Feynman, F. J. Dyson, V. F. Weisskopf and others has led to a procedure for calculating observable effects of the proper electromagnetic field of an electron.

In the last decade, with the construction of giant electron accelerators, with the development of sophisticated detectors for distinguishing electrons from other particles, and finally with the building of electron-positron colliding beam storage rings, much has been learnt about the nature of very high energy light quanta in their interactions with elementary particles. The study of interactions between light and light-like particles (the so-called vector mesons, or heavy photons) eventually led to the discovery of a new family of elementary particles—the first of which is the J particle.

My first knowledge of the concept of light quanta and the role they play in atomic physics came from the classical book "The Atomic Spectra" by Herzberg [2], which I picked up in the summer of 1957 when I was working in New York as a summer student. Just before my graduation from college, I received as a Christmas gift from my father the English translation of the book "Quantum Electrodynamics" by Akhiezer and Berestetskii [2]. During my school years at Michigan I managed to go through this book in some detail and worked out some of the formulas in the book myself. Then, during my years as a junior faculty member at Columbia University, I read with great interest a paper by Drell [2], who pointed out the implications of various tests of quantum electrodynamics at short distances using high-energy electron accelerators. I did a theoretical calculation with Brodsky [3] on how to isolate a certain class of Feynman graphs from the muon production of three muons.

There are basically two ways of testing the theory of interactions between photons, electrons, and muons. The low-energy method, like the Lamb shift or $(g-2)$ experiment, tests the theory to high accuracy at a long distance (or small momentum transfer). For example, the most recent experiment done at CERN by Picasso and collaborators [4] to measure the g-factor anomaly of the muon with a muon storage ring, obtained the result:

$$(g-2)/2 = 0.001165922+0.000000009 \text{ (an accuracy of 10 parts per million)}.$$

This result can be compared with calculations of quantum electrodynamics, including corrections from strong and weak interactions. The theoretical number is

$$(g-2)/2 = 0.001165921 \pm 0.000000010,$$

a most fantastic achievement of both experiment and theory.

The other way of testing quantum electrodynamics involves the study of reactions at large momentum transfers. Using the uncertainty principle $\Delta x \cdot \Delta p \approx h$, this type of experiment, though much less accurate, probes the validity of QED to a large momentum transfer or to a small distance. One such experiment, the process of e^+e^- production by multi-GeV photons in the Coulomb field of the nucleus, has both electromagnetic and strong interaction contributions to the e^+e^- yield. By properly choosing the kinematical conditions we can isolate the contributions from quantum electrodynamics alone and reduce the yield from strong interactions to a few percent level. The momentum transfer to the electron propagator is about 1 GeV; it is related to the effective mass of the e^+e^- pair. The yield of QED pairs is of the order α^3 ($\alpha = 1/137$). Because the yield is third order in α, to obtain a reasonable amount of events the experiment must be able to handle a high intensity of incident flux. A large acceptance detector is necessary not only to collect the events but also to average the steep angular dependence of the yields.

The effective mass of a pair of particles emitted from the same point is obtained by measuring the momentum of each of the particles p_1 and p_2, and the angles θ_1 and θ_2 between their paths and the incident beam direction, and by identifying the two particles simultaneously so that their masses m_1 and m_2 can be determined. The effective mass m of the pair is defined by:

$$m^2 = m_1^2+m_2^2+2[E_1E_2-p_1p_2 \cos(\theta_1+\theta_2)],$$

where E_i = total energy of the particle.

A pair spectrometer has two arms, which measure simultaneously the momenta p_1 and p_2 of the particles and the angles θ_1 and θ_2. Owing to the immense size of the equipment required, the physical position of each arm is often preselected. This restricts θ_1 and θ_2 to a relatively narrow band of possible values. Different effective masses may be explored by varying the accepted momentum of the particles p_1 and p_2.

When the two particles are uncorrelated, the distribution of m is normally a smooth function. A 'narrow' resonance will exhibit a sharp peak above this smooth distribution, while a 'wide' resonance will produce a broader bump.

The identification of particles from the spectrometer is done by

i) measuring the charge and momentum of the particle from its trajectory in a magnetic field;

ii) determining for a given trajectory, or a given momentum, the mass of the particle by measuring its velocity and using the relation $p = m \cdot v$.

The measurement of velocity can be done with Čerenkov counters using the Čerenkov effect. For electrons, their additional property of having only electromagnetic interactions can be used. When an electron enters a dense piece of lead, it loses all its energy by a cascading process which releases photons. The amount of light emitted from a lead-lucite sandwich shower counter (or a lead-glass counter) is thus proportional to the energy of the electron.

In October, 1965, I was invited by W. Jentschke, then Director of the Deutsches-Elektronen Synchrotron (DESY) in Hamburg, Germany, to perform my first experiment on e^+e^- production [5]. The detector we used is shown in Figs. 1a and 1b. It has the following properties that are essential to this type of experiment: i) it can use an incident photon flux of $\sim 10^{11}$/s, with a duty cycle of 2—3 %; ii) the acceptance is very large and is not limited by edges of the magnets or by shielding, being defined by scintillation counters alone; iii) all counters are located such that their surfaces are not directly exposed to the target; iv) to reject the hadron pairs, the Čerenkov counters are separated by magnets so that knock-on electrons from the pions interacting with gas radiators in the first pair of counters LC, RC are swept away by the magnet MA and do not enter the second pair of counters HL, HR. The low-energy knock-on electrons from HL, HR are rejected by shower counters.

The large number of Čerenkov counters and shower counters enables us to perform redundant checks on hadron rejection. Since each Čerenkov counter is 100% efficient on electrons and not efficient on hadrons, the observation that:

the yield of e^+e^- from 3 Čerenkov counters =

the yield of e^+e^- from 4 Čerenkov counters,

ensures that we are measuring pure e^+e^- pairs. The combined rejection is $>>10^8$.

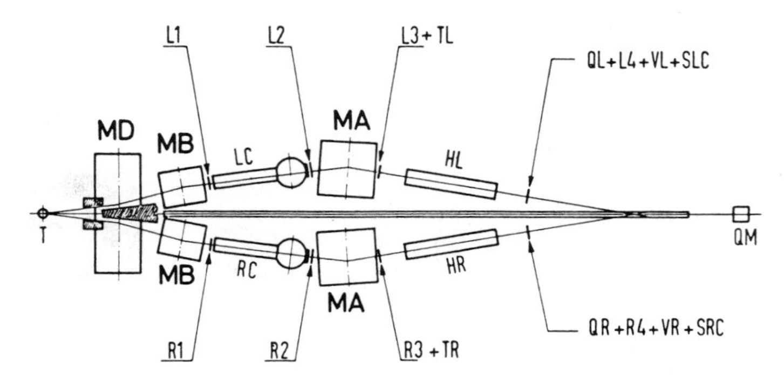

Fig. 1a. Plan view of the spectrometer. MD, MA, MB are dipole magnets; L1, ..., L4, and R1, ..., R4, are triggering counters; LC, RC, and HL, HR are large-aperture threshold Čerenkov counters; SLC, SRC are shower counters; and TL, QL, VL, and TR, QR, VR are hodoscopes. QM is a quantameter.

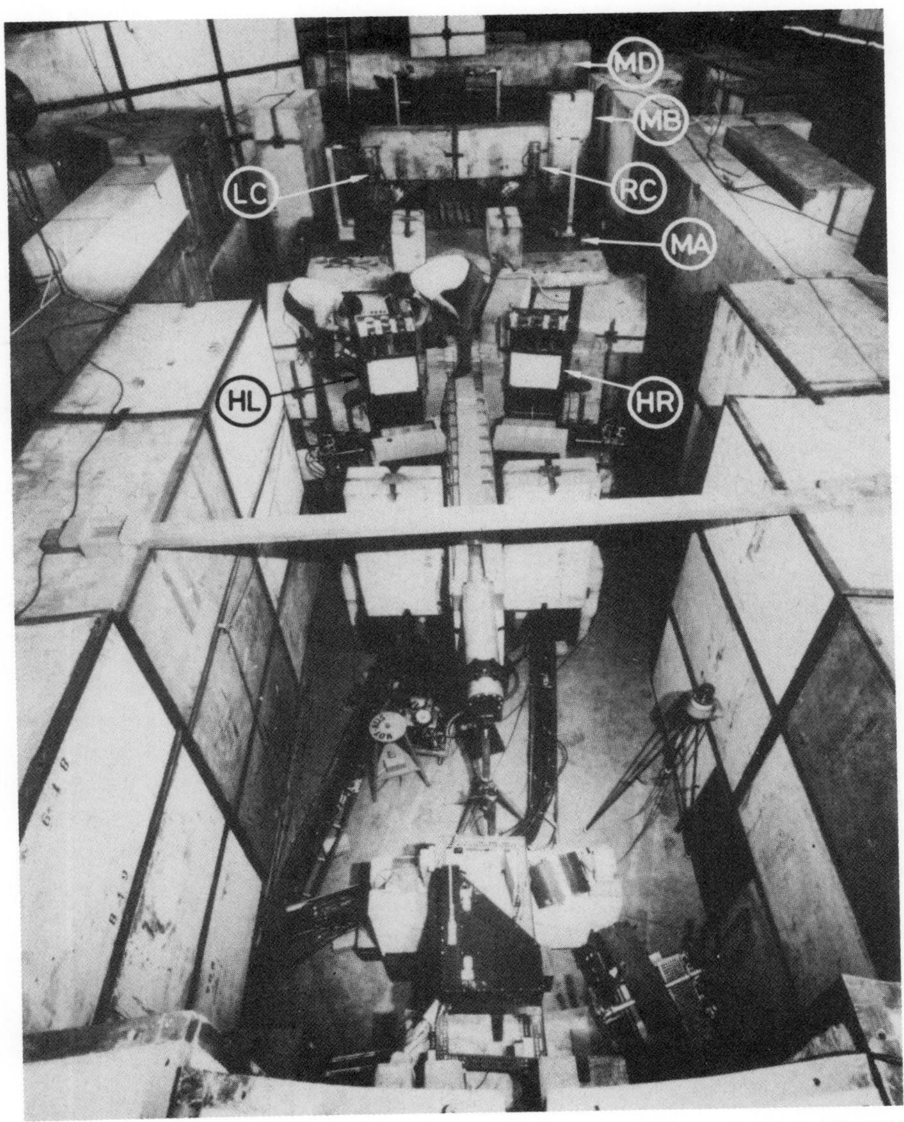

Fig. 1b. Over-all view of my first experiment at DESY. The position of LC, RC, HL, HR, MA, and MD are all marked. The physicist on the left is Dr. A. J. S. Smith; on the right is Dr. C. L. Jordan.

After we had finished this experiment, which showed that quantum electrodynamics correctly describes the pair production process to a distance of $\approx 10^{-14}$ cm, we tuned the spectrometer magnets so that the maximum pair mass acceptance is centred near $m \approx 750$ MeV. We observed a large increase in the e^+e^- yield and an apparent violation of QED. This deviation is caused by an enhancement of the strong interaction contribution to the e^+e^- yield where the incident photon produces a massive photon-like particle, the ρ meson, which decays into e^+e^- [6—8] with a decay probability of order α^2.

In order to show that this is indeed the case, we made another measurement at a larger e^+e^- opening angle and observed an even larger deviation from QED. This is to be expected since the QED process decreases faster than the strong interaction process when we increase the opening angle of the e^+e^- pair.

The observation of $\rho \to e^+ + e^-$ decay started a series of experiments by my group on this subject [9—12]. Basically the heavy photons ρ, ω, φ, are resonance states of $\pi^+\pi^-(\rho)$, $\pi^+\pi^-\pi^0(\omega)$, K^+K^- or $\pi^+\pi^-\pi^0(\varphi)$ with a rather short lifetime of typically $\approx 10^{-23}$—10^{-24} s. The widths of these particles are $\Gamma_\rho \approx \approx 100$ MeV, $\Gamma_\omega \sim 10$ MeV, and $\Gamma_\varphi \approx 5$ MeV. They are unique in that they all have quantum numbers J (spin) $= 1$, C (charge conjugation) $= -1$, P (parity) $= -1$. Thus they are exactly like an ordinary light-ray except for their heavy mass. The mass of ρ is $m_\rho \simeq 760$ MeV, and $m_\omega \simeq 783$ MeV; $m_\varphi \approx 1019.5$ MeV.

The production of heavy photons by photons on nucleon and nuclear targets shows that it is a diffraction process very much like the classical scattering of light from a black disk. The experiments on photoproduction of heavy photons and observation of their e^+e^- decay measure the coupling strength between each heavy photon and the photon. The interference between the e^+e^- final state from heavy photon decays and e^+e^- from QED measures the production amplitude of the heavy photon. The interference between these amplitudes can be viewed classically as a simple two-slit experiment, where in front of one of the slits we placed a thin piece of glass (corresponding to $\gamma \to \rho \to \gamma \to \to e^+e^-$) thus disturbing the interference pattern. The QED pairs alone would correspond to passing of light without the glass in front of the slit. The interference between $\rho(2\pi) \to e^+ + e^-$ and $\omega(3\pi) \to e^+ + e^-$ and the interference between $\rho(2\pi) \to 2\pi$ and $\omega(3\pi) \to 2\pi$ are measurements of strength of isospin non-conservation in electromagnetic interactions [13].

In the course of these experiments, since the width of ω is ~ 10 MeV and φ is ~ 5 MeV, we developed a detector with a mass resolution of ~ 5 MeV.

Some of the measurements have low event rates. In one particular experiment where we studied the e^+e^- mass spectra in the mass region above the ρ and ω mesons, the yield of e^+e^- pairs was about one event per day, with the full intensity of the accelerator. This implies that for about half a year the whole laboratory was working on this experiment alone. The rate of one event per day also implies that often there were no events for 2—3 days, and then on other days we had 2—3 events. It was during the course of this experiment that we developed the tradition of checking all voltages manually every 30 minutes, and calibrating the spectrometer by measuring the QED yields every 24 hours. To ensure that the detector was stable, we also established the practice of having physicists on shift, even when the accelerator was closed down for maintenance, and never switched off any power supplies. The net effect of this is that for many years our counting room has had a different grounding system from that of the rest of the laboratory. The Control Room for this series of experiments is shown in Fig. 2.

Some of the quantitative results from the above experiments may be explained if we assume that there are three kinds of fundamental building blocks

Fig. 2. Earlier Control Room at DESY. The three other people in the picture are Miss I. Schulz, Dr. U. Becker and Dr. M. Rohde. All have worked with me during the last 10 years.

in the world, known as quarks, which combine to form various elementary particles. The interactions between photons, heavy photons, and nuclear matter are results of interactions of the various quarks.

Sakurai [14] was the first to propose that the electromagnetic interaction of elementary particles may be viewed as through the heavy photon (vector meson) intermediate states.

2. NEW PARTICLES

After many years of work, we have learnt how to handle a high intensity beam of $\sim 10^{11}$ γ/s with a 2–3% duty cycle, at the same time using a detector that has a large mass acceptance, a good mass resolution of $\Delta M \approx 5$ MeV, and the ability to distinguish $\pi\pi$ from e^+e^- by a factor of $>>10^8$.

We can now ask a simple question: How many heavy photons exist? and what are their properties? It is inconceivable to me that there should be only three of them, and all with a mass around 1 GeV. To answer these questions, I started a series of discussions among members of the group on how to proceed. I finally decided to first perform a large-scale experiment at the 30 GeV proton accelerator at Brookhaven National Laboratory in 1971, to search for more heavy photons by detecting their e^+e^- decay modes up to a mass (m) of 5 GeV. Figure 3 shows the photocopy of one page of the proposal; it gives some of the reasons I presented, in the spring of 1972, for performing an e^+e^- experiment in a proton beam rather than in a photon beam, or at the DESY colliding beam accelerator then being constructed.

The best way to search for vector mesons is through production experiments of the type $p + p \to V^0 + X$, $V^0 \to e^+e^-$. The reasons are:

(a) The V^0 are produced via strong interactions, thus a high production cross section.

(b) One can use a high intensity, high duty cycle extracted beam.

(c) An e^+e^- enhancement limits the quantum number to 1^-, thus enabling us to avoid measurements of angular distribution of decay products.

Contrary to popular belief, the e^+e^- storage ring is not the best place to look for vector mesons. In the e^+e^- storage ring, the energy is well-defined. A systematic search for heavier mesons requires a continuous variation and monitoring of the energy of the two colliding beams—a difficult task requiring almost infinite machine time. Storage ring is best suited to perform detailed studies of vector meson parameters once they have been found.

Fig. 3. Page 4 of proposal 598 submitted to Brookhaven National Laboratory early in 1972 and approved in May of the same year, giving some of the reasons for performing this experiment in a slow extracted proton beam.

Historically, to my knowledge, the Zichichi Group was the first one to use hadron-hadron collisions to study e^+e^- yields from proton accelerators [15]. This group was the first to develop the earlier shower development method so as to greatly increase the e/π rejection [16]. In later years the Lederman Group made a study of the $\mu^+\mu^-$ yield from proton nuclei collisions [17]. Some of the early theoretical work was done by Preparata [18], Drell and Yan [19], and others.

Let me now go to the J-particle experiment [20—22].

I. To perform a high-sensitivity experiment, detecting narrow-width particles over a wide mass range, we make the following four observations.

i) Since the e^+e^- come from electromagnetic processes, at large mass m, the yield of e^+e^- is lower than that of hadron pairs ($\pi^+\pi^-$, K^+K^-, $\bar{p}p$, $K^+\bar{p}$, etc.) by a factor $<<10^{-6}$.

ii) Thus, to obtain sufficient e^+e^- rates, a detector must be able to stand a high flux of protons, typically of $10^{11}-10^{12}$ protons/s, and

iii) it must be able to reject hadron pairs by a factor of $>>10^8$.

iv) For a detector with finite acceptance, there is always the question of where is the best place to install it to look for new particles. *A priori* we do not know what to do. But we do know that in reactions where ordinary hadrons are produced, the yield is maximum when they are produced at rest in the centre-of-mass system [23]. If we further restrict ourselves to the 90° e^+e^- decay of new particles, then we quickly arrive at the conclusion that the decayed e^+ or e^- emerge at an angle of 14.6° in the laboratory system for an incident proton energy of 28.5 GeV, independent of the mass of the decaying particle.

II. Figure 4 shows the layout of the slow-extracted intense proton beam from the Alternating Gradient Synchrotron (AGS) at Brookhaven, during the period 1973—1974. Our experiment (No. 598) was located in a specially designed beam line (the A-line). To design a clean beam with small spot sizes, I remembered having a conversation with Dr. A. N. Diddens of CERN who had used a slow-extracted beam at the CERN Proton Synchrotron. He advised me to focus the beam with magnets alone without using collimators.

The incident beam of intensity up to 2×10^{12} protons per pulse was focused to a spot size of 3×6 mm². The position of the beam was monitored by a closed-circuit TV. The stability and the intensity of the beam were monitored by a secondary emission counter and six arrays of scintillation counter telescopes, located at an angle of 75° with respect to the beam, and buried behind 12 feet of concrete shielding. Daily calibrations were made of the secondary emission counter with the Al and C foils.

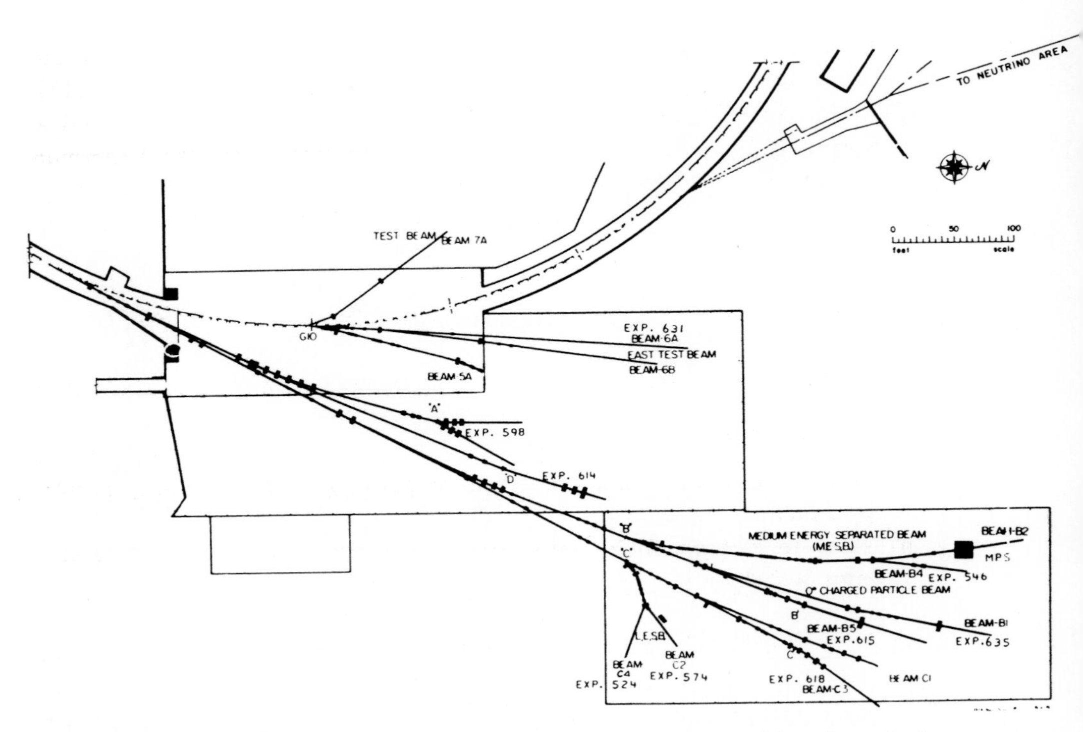

Fig. 4. The AGS East experimental area. The MIT experiment is No. 598 at the end of Station A. Experiment 614 is that of Prof. M. Schwartz (see Ref. 22).

III. From our early experience at DESY, we felt the best way to build an electron-pair detector that could handle high intensities, and at the same time have a large mass acceptance and a good mass resolution, is to design a large double-arm spectrometer and to locate most of the detectors behind the magnets so that they would not "view" the target directly. To simplify analysis and to obtain better mass resolution, we used the "p, θ independent" concept in which the magnets bend the particles vertically to measure their momentum, while the production angles are measured in the horizontal plane. Figures 5a and 5b show the plan and side views of the spectrometer and detectors.

The main features of the spectrometer are the following:

1) *The target:* The target consists of nine pieces of 1.78 mm thick beryllium, each separated by 7.5 cm so that particles produced in one piece and accepted by the spectrometer do not pass through the next piece. This arrangement also helps us to reject pairs of accidentals by requiring two tracks to come from the same origin.

2) *The magnet system:* The bending powers of the dipole magnets M_0, M_1, M_2 are such that none of the counters sees the target directly. The field of the magnets in their final location was measured with a three-dimensional Hall probe at a total of 10^5 points.

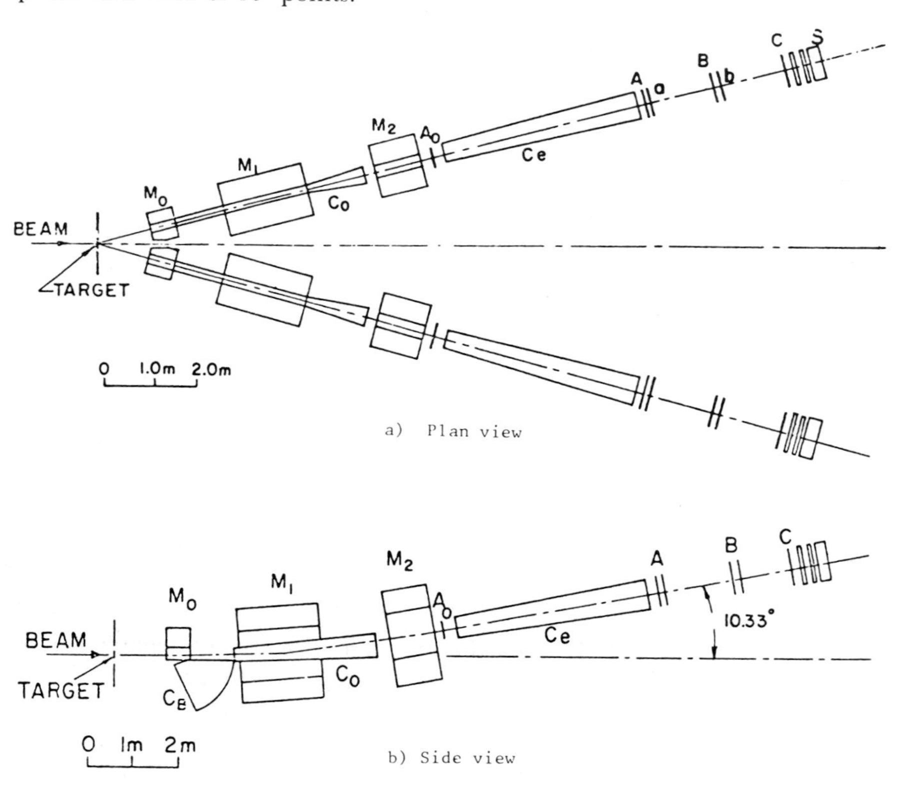

Fig. 5. Schematic diagram of the experimental set-up for the double-arm spectrometer used in our discovery of the J particle. M_0, M_1, and M_2 are dipole magnets; A_0, A, B, and C are 8000-wire proportional chambers; a and b are each 8×8 hodoscopes; S designates three banks of lead-glass and shower counters; C_B, C_0, and C_e are gas Čerenkov counters.

3) *The chambers:* A_0, A, B, and C are multiwire proportional chambers. They consist of more than 8000 very fine, 20 μm thick, gold-plated wires, 2 mm apart, each with its own amplifier and encoding system. The wire arrangement is shown in Fig. 6. The 11 planes all have different wire orientation. In each of the last three chambers the wires are rotated 60° with respect to each other, so that for a given hit, the numbers of wires add up to a constant—a useful feature for sorting out multitracks and rejecting soft neutrons and γ-rays which do not fire all planes. We developed special gas mixtures to operate the chambers at low voltage and high radiation environment. To help improve the timing resolution, two planes of thin (1.6 mm thick) hodoscopes (8×8) are situated behind each of the chambers A and B. These chambers are able to operate at a rate of $\sim$20 MHz and are also able to sort out as many as eight particles simultaneously in each arm.

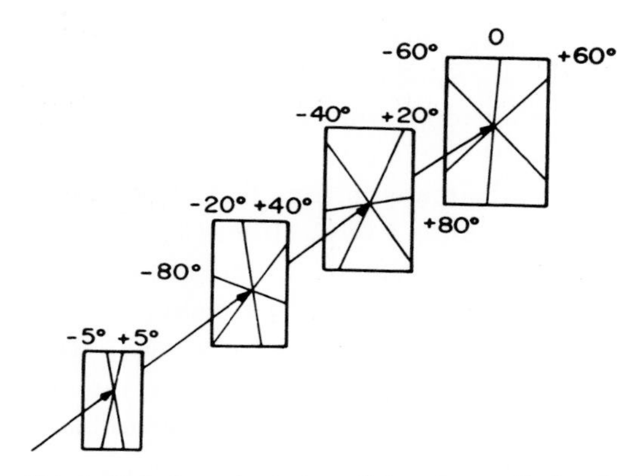

Fig. 6. Relative orientation of the planes of wires in the proportional chambers.

It is essential that all 8000 wires should function properly because to repair a single wire would involve removing close to a thousand tons of concrete.

These chambers and the magnets yield a mass resolution of ± 5 MeV and a mass acceptance of 2 GeV at each magnet current setting. The good mass resolution makes it possible to identify a very narrow resonance. The large mass acceptance is very important when searching over a large mass region for narrow resonances.

4) *Čerenkov counters and shower counters:* The Čerenkov counters marked C_0 and C_e together with the lead-glass and shower counters marked S, enable one to have a rejection against hadron pairs by a factor of $>>1 \times 10^8$.

The Čerenkov counter in the magnet (C_0, see Fig. 7a) has a large spherical mirror with a diameter of 1 m. This is followed by another Čerenkov counter behind the second magnet with an elliptical mirror of dimensions 1.5×1.0 m^2. The Čerenkov counters are filled with hydrogen gas so that the knock-on electrons are reduced to the minimum. As in our earlier DESY experiments, the separation of the two counters by strong magnetic fields ensures that the

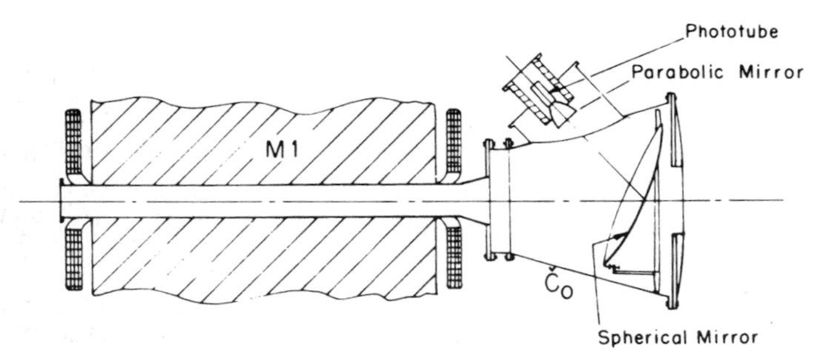

Fig. 7a. Plan view of the C_0 counter shown in its location in the experiment.

small number of knock-on electrons produced in the first counter is swept away and does not enter into the second counter.

To reduce multiple scattering and photon conversion, the material in the beam is reduced to a minimum. The front and rear windows of C_0 are 126 μm and 250 μm thick, respectively. To avoid large-angle Čerenkov light reflection, the mirrors of C_0 and C_e are made of 3 mm thick black lucite, aluminized on the forward (concave) surface only. The mirrors in the experiment were made at the Precision Optical Workshop at CERN. We measured the curvature of the mirrors with a laser gun, and out of the many mirrors that were made a total of 24 were used in this experiment (4 in C_0, 4 in C_e, 16 in C_B).

The counters are painted black inside so that only the Čerenkov light from electrons along the beam trajectory will be focused onto the photomultiplier cathode. Special high-gain, high-efficiency phototubes of the type RCA C31000M are used, so that when we fill the counter with He gas as radiator (where we expect, on the average, 2—3 photoelectrons) we are able to locate the single photoelectron peak (see Fig. 7b).

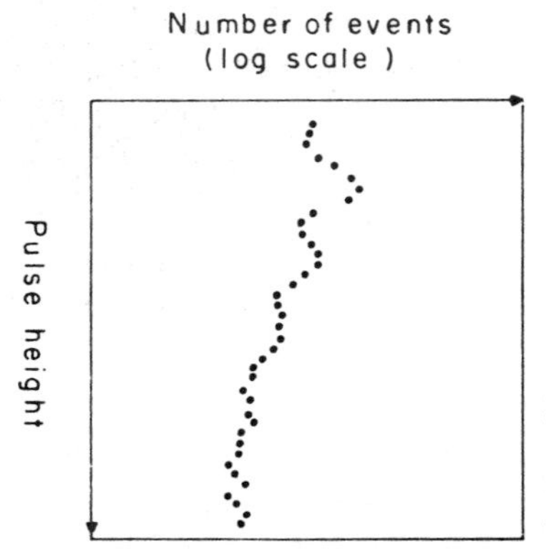

Fig. 7b. Pulse-height spectrum from the phototube (RCA C31000M) of the C_0 Čerenkov counter with He as radiator. Clearly visible are the one, two, and three photoelectron peaks.

The counter C_0 is very close to the target, which is a high-radiation-level area. To reduce random accidentals and dead-time, the excitation voltage on the photomultiplier has to be kept as low as possible. Yet we must still ensure that the counter is efficient. We have to avoid mistakingly setting the voltage so low that the counter is only efficient on an e^+e^- pair from $\pi^0 \to \gamma + e^+ + e^-$, which may enter the counter. When C_0 is filled with hydrogen gas, a single electron will yield about eight photoelectrons, a pair will yield about sixteen. The knowledge of the location of one photoelectron peak enables us to distinguish between these two cases. The counters are all calibrated in a test beam to make sure they are 100% efficient in the whole phase space.

At the end of each arm there are two orthogonal banks of lead-glass counters of three radiation lengths each, the first containing twelve elements, the second thirteen, followed by one horizontal bank of seven lead-lucite shower counters, each ten radiation lengths thick, to further reject hadrons from electrons. The subdividing of the lead-glass and lead-lucite counters into ~ 100 cells also enables us to identify the electron trajectory from spurious tracks.

Figure 8 shows an over-all view of the detector with the roof removed. Figure 9 shows the end section of one arm of the detector, showing part of the Čerenkov counter C_e, the proportional chambers, and counters.

Fig. 8. Over-all view of the detector.

Fig. 9. End view of one arm, showing part of the Čerenkov counter C_e, the chambers A, B, C, with part of the 8000 amplifiers X, cables Y, and hodoscopes Z. The lead-glass counter is at the end of chamber U.

5) *A pure electron beam for calibration:* To obtain a high rejection against hadron pairs and to ensure that the detectors are 100% efficient for electrons, we need to calibrate the detectors with a clean electron beam. In an electron accelerator such as DESY we can easily produce a clean electron beam with an energetic photon beam hitting a high-Z target thus creating 0° e^+e^- pairs. In a proton accelerator the best way to create a clean electron beam is to use the reaction $\pi^0 \to \gamma + e^+ + e^-$, tagging the e^+ in coincidence with the e^-. To accomplish this, the very directional Čerenkov counter C_B is placed close to the target and below a specially constructed magnet M_0 (Fig. 10a). This counter also is painted black inside; it is sensitive to electrons above 10 MeV/c and rejects pions below 2.7 GeV/c. The coincidence between C_B and C_0, C_e, the shower counter, and the hodoscopes, indicates the detection of an e^+e^- pair from the process $\pi^0 \to \gamma + e^+e + e^-$. A typical plot of the relative timing of this coincidence is shown in Fig. 10b. We can trigger on C_B and provide a pure electron beam to calibrate C_0, C_e, the lead-glass and shower counters.

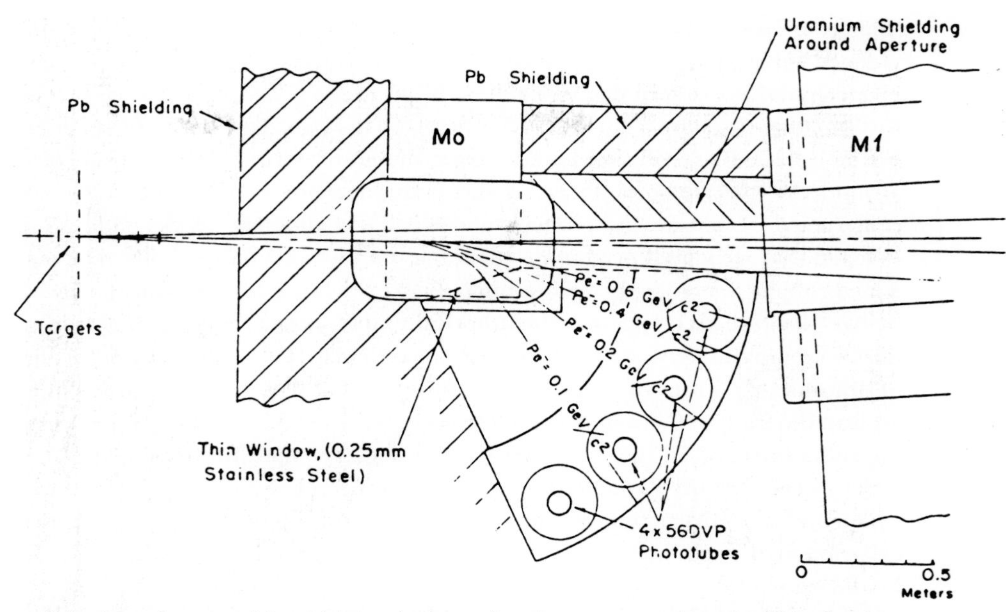

a. Side view of magnet M_0 which bends the various low-energy trajectories (P_e) of $e^\pm$ into C_B.

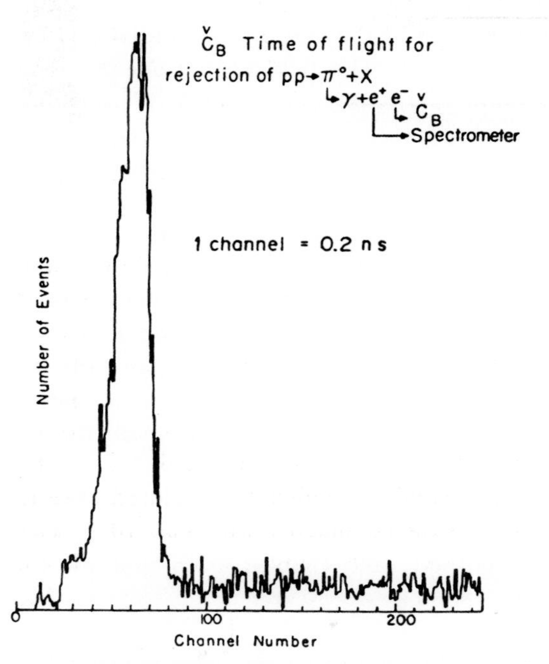

b. The relative timing between an electron pulse from C_B and a positron trigger from the main spectrometer arm or vice versa.

Fig. 10. Measurement of e^+e^- from $\pi^0 \to \gamma + e^+ + e^-$ decay.

This is another way of setting the voltage of the C_0 counters, since the coincidence between C_e and C_B will ensure that the counter is efficient for a single electron and not a zero degree pair.

6) *Shielding:* As shown in Fig. 8 the detector is large, and with 10^{12} protons incident on a 10% collision length target there are $\sim 10^{12}$ particles generated around the experimental area. To shield the detector and the physicists, we constructed scaled-down wooden models of the concrete blocks, and soon realized that we would need more shielding than was available at Brookhaven. This problem was solved by obtaining all the shielding blocks from the Cambridge Electron Accelerator, which had just closed down. The total shielding used is approximately a) 10,000 tons of concrete, b) 100 tons of lead, c) 5 tons of uranium, d) 5 tons of soap—placed on top of C_0, between M_1 and M_2, and around the front of C_e to stop soft neutrons. Even with this amount of shielding, the radiation level in the target area, one hour after the shutting down of the proton beam, is still 5 röntgen/hour, a most dangerous level.

During the construction of our spectrometers, and indeed during the entire experiment, I encountered much criticism. The problem was that in order to gain a good mass resolution it was necessary to build a spectrometer that was very expensive. One eminent physicist made the remark that this type of spectrometer is only good for looking for narrow resonances—and there are no narrow resonances. Nevertheless, since I usually do not have much confidence in theoretical arguments, we decided to proceed with our original design.

In April 1974, we finished the set-up of the experiment and started bringing an intense proton beam into the area. We soon found that the radiation level in our counting room was 0.2 röntgen/hour. This implied that our physicists would receive the maximum allowable yearly dose in 24 hours! We searched very hard, for a period of two to three weeks, looking for the reason, and became extremely worried whether we could proceed with the experiment at all.

One day, Dr. U. Becker, who has been working with me since 1966, was walking around with a Geiger counter when he suddenly noticed that most of the radiation was coming from one particular place in the mountains of shielding. Upon close investigation we found out that even though we had 10,000 tons of concrete shielding blocks, the most important region—the top of the beam stopper—was not shielded at all! After this correction, radiation levels went down to a safe level and we were able to proceed with the experiment.

From April to August, we did the routine tune-ups and found the detectors performing as designed. We were able to use 10^{12} protons per second. The small pair spectrometer also functioned properly and enabled us to calibrate the detector with a pure electron beam.

IV. Owing to its complexity, the detector required six physicists to operate it. Before taking data, approximately 100 hours were spent ensuring that all the detectors were close to 100% efficient. I list some examples:

i) The efficiency of the Čerenkov counters was measured over the whole phase space, and voltages set so that they were efficient everywhere. A typical result for C_e is shown in Fig. 11a.

ii) The voltages and the response of all the lead-glass and shower counters were calibrated to ensure that the response did not change with time.

iii) The efficiency of the hodoscopes at the far end, furthest away from the photomultiplier tube, was checked.

iv) The timing of the hodoscopes was also checked to ensure that signals from each counter generated by particles produced at the target arrived simultaneously. During the experiment, the time-of-flight of each of the hodoscopes and the Čerenkov counters, the pulse heights of the Čerenkov counters and of the lead-glass and shower counters, the single rates of all the counters together with the wire chamber signals, were recorded and continuously displayed on a storage/display scope.

v) To ensure that the proportional wire chambers were efficient over their whole area, a small test counter was placed behind the chambers at various positions over the chambers' area, and voltage excitation curves were made at those positions. A typical set of curves for all the planes is shown in Fig. 11b.

vi) To check the timing between the two arms, two tests were performed. Firstly, the test counter was physically moved from one arm to the other

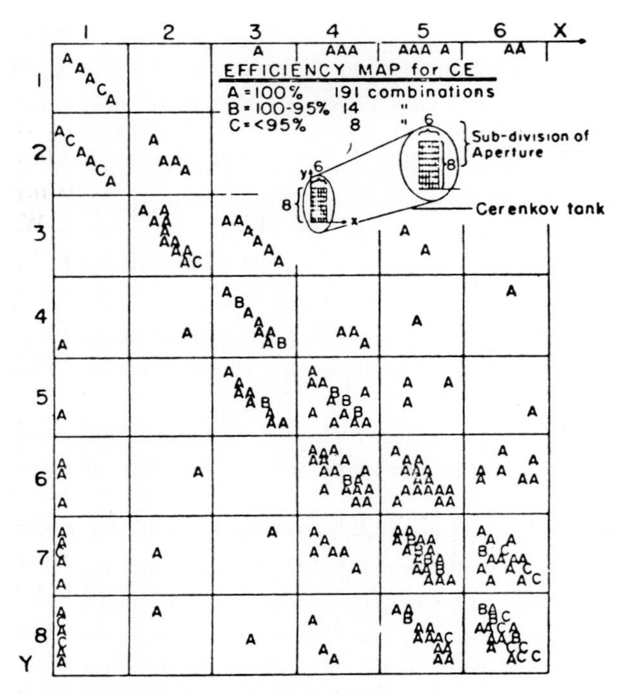

Fig. 11a. Mapping of the efficiency of the C_e counter over its whole phase space. The letters on the plot refer to efficiencies measured for trajectories between the corresponding points marked on the grid at each end of the counter.

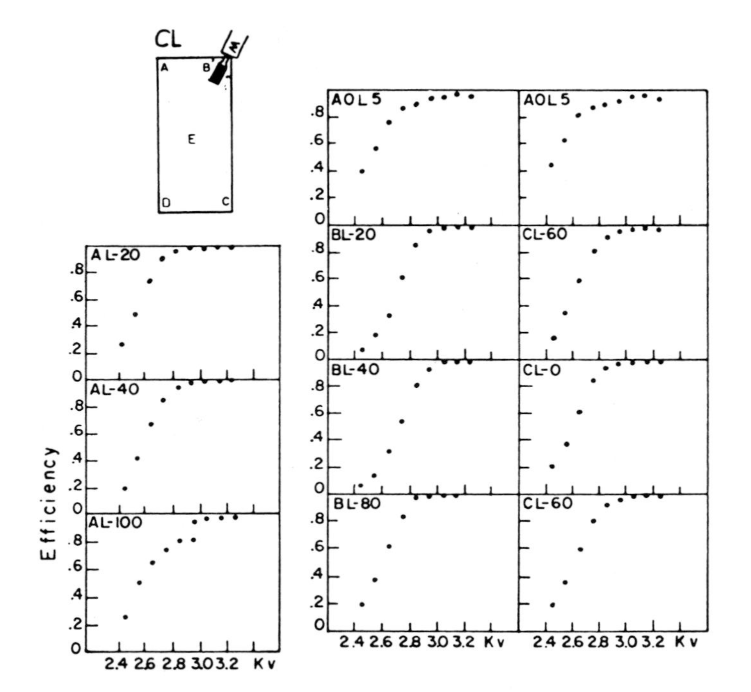

Fig. 11b. Efficiency of all the wire planes as a function of the applied voltage. The measurements were done by placing a small test counter W in various positions, marked A, B, C, D, E, in every chamber.

so that the relative timing could be compared. Secondly, the e^+e^- yield was measured at low mass, $m_{ee} < 2$ GeV/c^2, where there is an abundance of genuine e^+e^- pairs.

In the early summer of 1974 we took some data in the high mass region of 4—5 GeV. However, analysis of the data showed very few electron-positron pairs.

By the end of August we tuned the magnets to accept an effective mass of 2.5—4.0 GeV. Immediately we saw clean, real, electron pairs.

But most surprising of all is that most of the e^+e^- pairs peaked narrowly at 3.1 GeV (Fig. 12a). A more detailed analysis shows that the width is less than 5 MeV! (Fig. 12b).

Throughout the years, I have established certain practices in the group with regard to experimental checks on our data and on the data analysis. I list a few examples:

i) To make sure the peak we observed was a real effect and not due to instrumentation bias or read-out error of the computer, we took another set of data at a lower magnet current. This has the effect of moving the particles into different parts of the detector. The fact that the peak remained fixed at 3.1 GeV (Fig. 12a) showed right away that a real particle had been discovered.

ii) We used two completely different sets of programs to ensure that the analysis was correct. This means that two independent groups of physi-

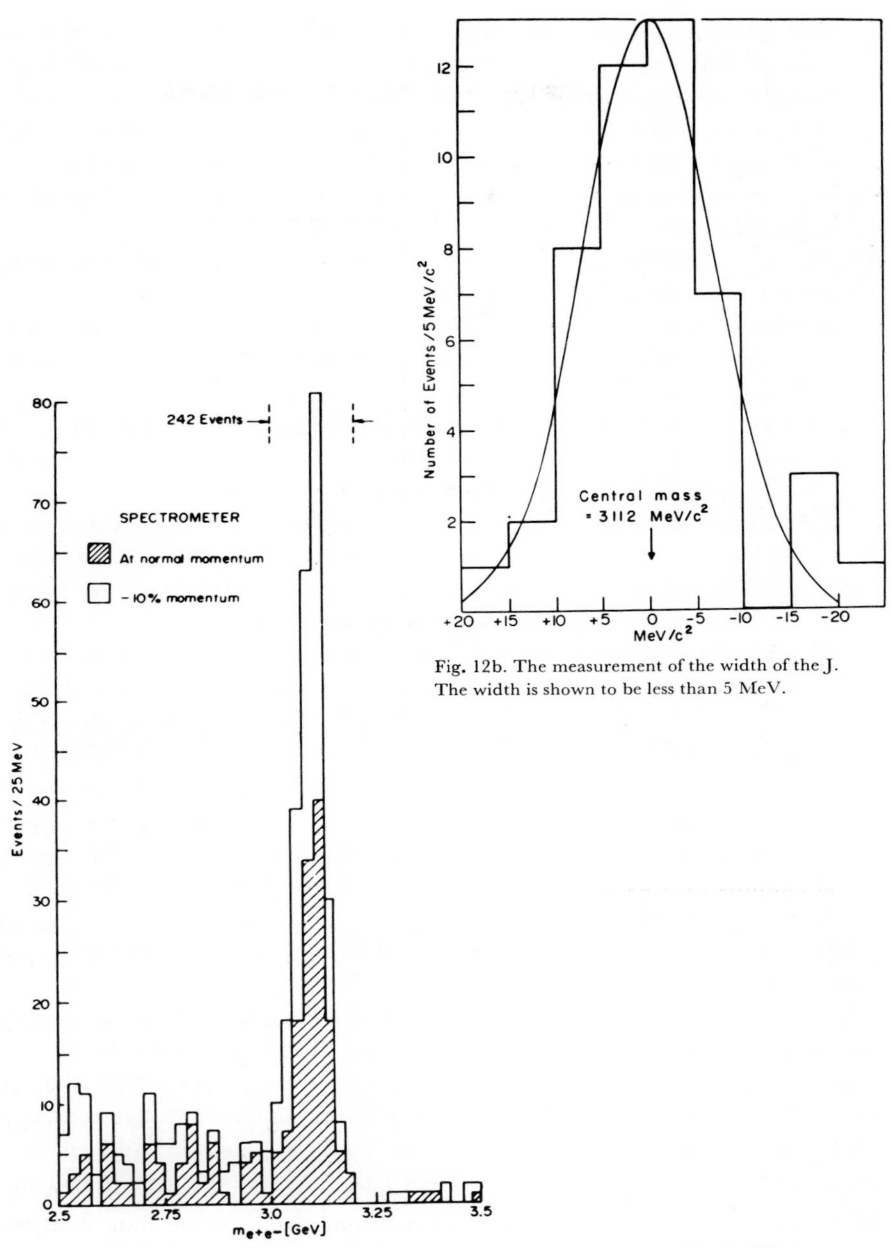

Fig. 12b. The measurement of the width of the J. The width is shown to be less than 5 MeV.

Fig. 12a. Mass spectrum for events in the mass range $2.5 < m_{ee} < 3.5$ GeV/c. The shaded events correspond to those taken at the normal magnet setting, while the unshaded ones correspond to the spectrometer magnet setting at -10% lower than normal value.

cists analysed the data, starting from the reduction of raw data tapes, to form their own data summary tapes, and then performed two sets of Monte Carlo acceptance calculations, two sets of event reconstruction, two sets of data corrections, and finally, two sets of results which must agree with each other. Although this procedure uses twice as much computer time, it provides greater confidence in our results after the two independent approaches have reached the same conclusions.

iii) To understand the nature of various second-order background corrections, we made the following special measurements:

a) To check the background from pile-up in the lead-glass and shower counters, different runs were made with different voltage settings on the counters. No effect was observed in the yield.

b) To check the background from scattering from the sides of the magnets, cuts were made in the data to reduce the effective aperture. No significant reduction in the yield was found.

c) To check the read-out system of the chambers and the triggering system of the hodoscopes, runs were made with a few planes of chambers deleted and with sections of the hodoscopes omitted from the trigger. No unexpected effect was observed on the yield.

d) Since the true event rate is proportional to incident beam intensity and the accidental backgrounds from the two arms are proportional to the square of the incident intensity, a sensitive way to check the size of the background is to run the experiment again with different intensities. This was done and the background contribution in the peak was found to be unnoticeable.

iv) To understand the nature of production properties of the new peak, we increased the target thickness by a factor of two. The yield increased by a factor of two, not by four.

These and many other checks convinced us that we had observed a real massive particle.

We discussed the name of the new particle for some time. Someone pointed out to me that the really exciting stable particles are designated by Roman characters—like the postulated W^0, the intermediate vector boson, the Z^0, etc.—whereas the "classical" particles have Greek designations like ρ, ω, etc. This, combined with the fact that our work in the last decade had been concentrated on the electromagnetic current $j_\mu(x)$, gave us the idea to call this particle the J particle.

V. I was considering announcing our results during the retirement ceremony for V. F. Weisskopf, who had helped us a great deal during the course of many of our experiments. This ceremony was to be held on 17 and 18 October 1974. I postponed the announcement for two reasons. First, there were speculations on high mass e^+e^- pair production from proton-proton collisions as coming from a two-step process: $p+N\rightarrow\pi+\ldots$, where the pion undergoes a second collision $\pi+N\rightarrow e^++e^-+\ldots$. This could be checked by a measurement based on target thickness. The yield from a two-step process would increase quadratically with target thickness, whereas for a one-step process the yield increases linearly. This was quickly done, as described in point (iv) above.

Most important, we realized that there were earlier Brookhaven measurements [24] of direct production of muons and pions in nucleon-nucleon collisions which gave the μ/π ratio as 10^{-4}, a mysterious ratio that seemed not to change from 2000 GeV at the ISR down to 30 GeV. This value was an order of magnitude larger than theoretically expected in terms of the three known vector mesons, ρ, ω, φ, which at that time were the only possible "intermediaries" between the strong and electromagnetic interactions. We then added the J meson to the three and found that the linear combination of the four vector mesons could not explain the μ^-/π^- ratio either. This I took as an indication that something exciting might be just around the corner, so I decided that we should make a direct measurement of this number. Since we could not measure the μ/π ratio with our spectrometer, we decided to look into the possibility of investigating the e^-/π^- ratio.

We began various test runs to understand the problems involved in doing the e/π experiment. The most important tests were runs of different e^- momenta as a function of incident proton intensities to check the single-arm backgrounds and the data-recording capability of the computer.

On Thursday, 7 November, we made a major change in the spectrometer (see Fig. 13) to start the new experiment to search for more particles. We began by measuring the mysterious e/π ourselves. We changed the electronic logic and the target, and reduced the incident proton beam intensity by almost two orders of magnitude. To identify the e^- background due to the decay of π^0 mesons, we inserted thin aluminium converters in front of the spectrometer to increase the $\gamma \to e^+ + e^-$ conversion. This, together with the C_B counter which measures the $\pi \to \gamma + e^+ + e^-$ directly, enabled us to control the major e^- background contribution.

We followed the e/π measurements with another change in the spectrometer by installing new high-pressure Čerenkov counters and systematically measuring hadron pairs (K^+K^-, $\pi^+\pi^-$, $\bar{p}p$, etc.) to find out how many other particles exist that do not decay into e^+e^- but into hadrons. But, after a long search, none was found.

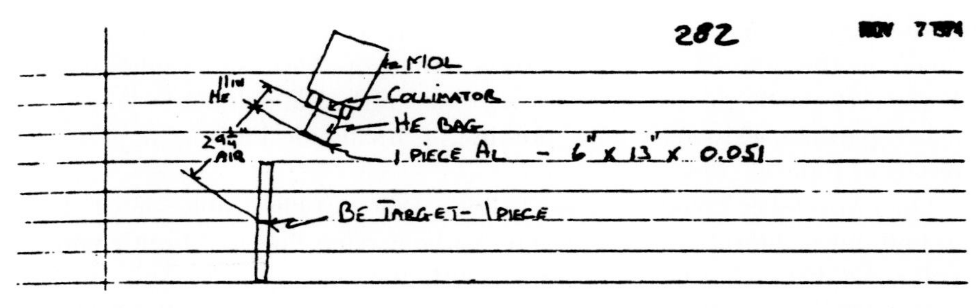

Fig. 13a. Aluminium foil arrangement in front of magnet M_0 in our new experiment to determine the e/π ratio. The converter was used to determine the electron background yield.

284 NOV 7 1974

1 φ ATTEN.

TRIGGER CO1.CE1.EFGH MISC att CHECK HV BEAM CONDITION INTENSITY 2×10^{10}

(SINGLE) ARM L OR R He BAGS OK? SPOT SIZE:

(e) P K μ π MOL

DOUBLE ARM MOR HEIGHT

ee pp̄ μμ K̄K ππ M2L ENERGY

TARGET: 1×Be M2R SPILL LENGTH

2 foils Al PIPE OTHER

SEM-Al VAC

SEM-Al HV

CHAMBER CONDITION CHERENKOV COUNTERS

CHAMBER	HV	OTHER		GAS	PRESSURE	OTHERS
AOL		ON	CBL			ON
AOR		ON	CBR			ON
AL			COL			
AR			COR			
BL			CEL			
BR			CER			
CL						
CR						

METHYAL TEMP: FILLED:

FLAGS 1 PIECES

(TARGET) POSITIONS center

Thickness 29 mil

MATERIAL

TAPE: DA 54 MAGNET SETTINGS PEOPLE ON SHIFTS

RUN# 221 MOL 60.00 Ting

DATE NOV 7 1974 MOR Kary

TIME START: 21 50 M1L 45.123 Wu

- STOP: 22 35 M1R 45.123 Chen

M2L 26.212 Leong

M2R 26.212

NOV 7 1974 MOMENTUM: Med P (−) POLARITY R−

COIL ON

COMMENTS

TSI SCALERS.

SEM-Al SEM

G1 G2 G3 G4

MA

MB MRD

MC REX MR

MD LEK MLD

ML

Fig. 13b. Data sheet for a typical run under the new experimental conditions. Blank spaces imply either data entered in the computer or conditions identical to the prior run. In this run the electrons pass through the right detector arm with a momentum of about 6 GeV. Two pieces of aluminium foil in front of the magnet M_0 serve as converters. [From the group's data book, pp. 282 and 284, 7 November 1974.]

In the meantime, since the end of October, M. Chen and U. Becker and others in the group had been insisting that we publish our results quickly. I was very much puzzled by the $\mu/\pi = 10^{-4}$ ratio and wanted to know how many particles existed. Under pressure, I finally decided to publish our results of J alone.

On 6 November I paid a visit to G. Trigg, Editor of Physical Review Letters, to find out if the rules for publication without refereeing had been changed. Following that visit, I wrote a simple draft in the style of our quantum electrodynamics paper of 1967 (Ref. 5). The paper emphasized only the discovery of J. and the checks we made on the data without mention of our future plans.

On 11 November we telephoned G. Bellettini, the Director of Frascati Laboratory, informing him of our results. At Frascati they started a search on 13 November, and called us back on 15 November to tell us excitedly that they had also seen the J signal and obtained a $\Gamma^2_{\mu\mu}/\Gamma_{\text{total}} = 0.8 \pm 0.2$ keV. Their first spectrum is shown in Fig. 14a. The Frascati Group were able to publish their results in the same issue of Physical Review Letters [25] as ours. Very shortly after, they made a more detailed study of J (Fig. 14b) and also established that its total width is only ~60 keV. (It lives ~1000 times longer than the ρ meson.) They have since made a systematic search for more particles at lower mass—but have found none [26].

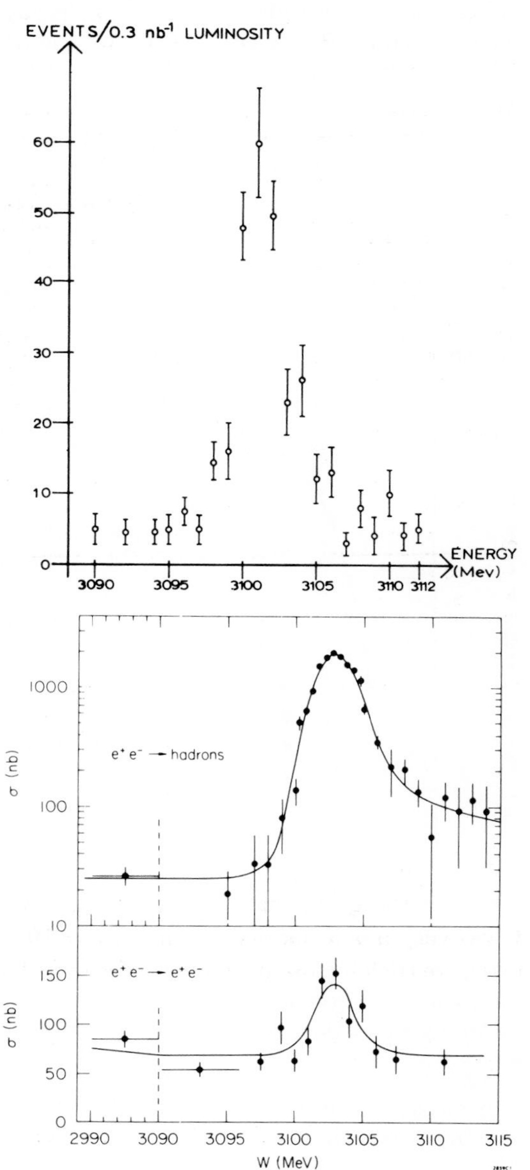

Fig. 14a. Result from one of the Frascati groups on J-particle production. The number of events per 0.3 nb^{-1} luminosity is plotted versus the total c.m. energy of the machine. (From Ref. 25.)

Fig. 14b. Excitation curves for the reactions $e^+ + e^- \to$ hadrons and $e^+ + e^- \to e^+ + e^-$.
The solid line represents the best fit to their data. (From Ref. 26.)

VI. Now, immediately after the discovery of J, because of its heavy mass and unusually long lifetime, there were many speculations as to the nature of this particle. Lee, Peoples, O'Halloran and collaborators [27] were able to photoproduce the J particle coherently from nuclear targets with an $\sim$100 GeV photon beam. They showed that the photoproduction of the J is very similar to ρ production and thus were the first to establish that J is a strongly interacting particle.

Pilcher, Smith and collaborators [28] have ingeniously used a large acceptance spectrometer to perform an accurate and systematic study of J production at energies $>$100 GeV. By using π beams as well as proton beams, and by measuring a wide range of mass and the momentum transfer dependence of $\mu\mu$ production, they were the first to state that the single muon yield which produced the mysterious $\mu/\pi = 10^{-4}$, which had puzzled me for a long time, comes mostly from the production of muon pairs. The J yield from the π meson seems to be much higher than from the proton.

In Fig. 15 are listed some of the relative yields of J production from various proton accelerators. It seems that I had chosen the most difficult place to discover the J.

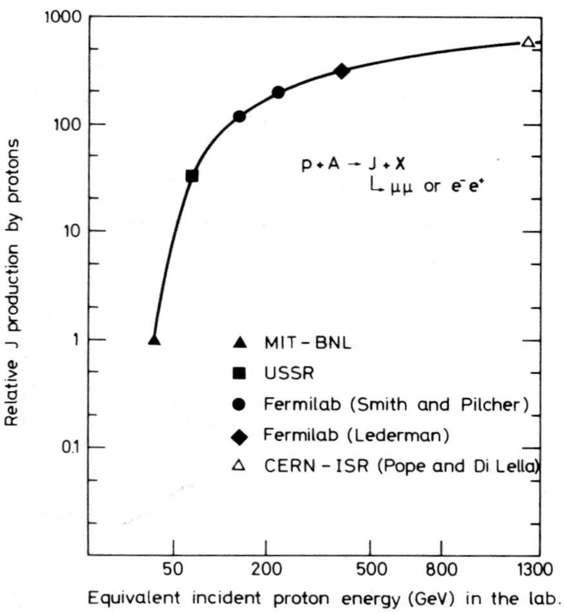

Fig. 15. Relative J production, at 90° in the centre of mass, as a function of the energy of the incident proton beam. For experiments using nuclear targets, a linear A-dependence has been used to obtain the yield on a nucleon. Refs: MIT-BNL: J. J. Aubert et al., Phys. Rev. Letters *33*, 1404 (1974); CERN-ISR: F. W. Büsser et al., Phys. Letters *56B*, 482 (1975); USSR: Yu. M. Antipov et al., Phys. Letters *60B*, 309 (1976); Lederman Group: H. D. Snyder et al., Phys. Rev. Letters *36*, 1415 (1976); Smith-Pilcher Group: K. J. Anderson et al., paper submitted to the 18th Internat. Conf. on High-Energy Physics, Tbilisi, USSR (1976).

3. SOME SUBSEQUENT DEVELOPMENTS

The discovery of the J has triggered off many new discoveries. Some of the most important experimental work was done at SLAC [29] and at DESY [30].

The latest results [31] from the 4π superconducting magnet detector, called "Pluto", measuring the $e^+ + e^- \to$ hadrons near the mass of ψ' (the sister state of J) first discovered at SLAC, are shown in Fig. 16a. The yield of ψ' (and of J) goes up by $>10^2$. It can be seen that an electron-positron storage ring is an ideal machine for studying these new particles. The same group has recently carried out a careful search for new particles at a higher mass region. Their accurate results, shown in Fig. 16b, confirm the indication by SLAC that there may be many more states in this high mass region.

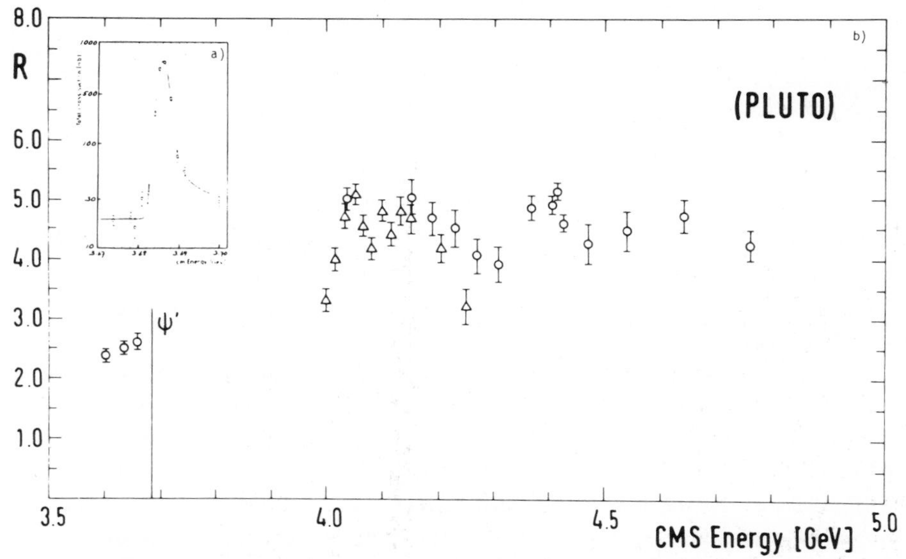

Fig. 16. a) Excitation curve for ψ'. b) Ratio R = ($e^+ + e^- \to$hadrons) over ($e^+ + e^- \to \mu^+ + + \mu^-$), measured by the DESY Pluto group. (Ref. 31.)

One of the most important discoveries after that of the J is the observation by the double-arm spectrometer (DASP) Group at DESY [32] of the chain reaction

$$
\begin{array}{l}
e^+ + e^- \to \psi' \\
\qquad\quad \lfloor\!\to P_c \to \gamma_1 \\
\qquad\qquad\quad \lfloor\!\to \gamma_2 + J \\
\qquad\qquad\qquad\quad \lfloor\!\to \mu + \mu.
\end{array}
$$

By tuning the storage ring so that the electron-positron energy reaches 3.7 GeV to produce the ψ', using the double-arm spectrometer to select the $J \to \mu^+ + \mu^-$ events and detecting both the γ_1 and γ_2 as well, they found that the two photons γ_1 and γ_2 are strongly correlated into two groups. The first group has $E_{\gamma_1} = 169 \pm 7$ MeV and $E_{\gamma_2} = 398 \pm 7$ MeV (or vice versa, since they did not

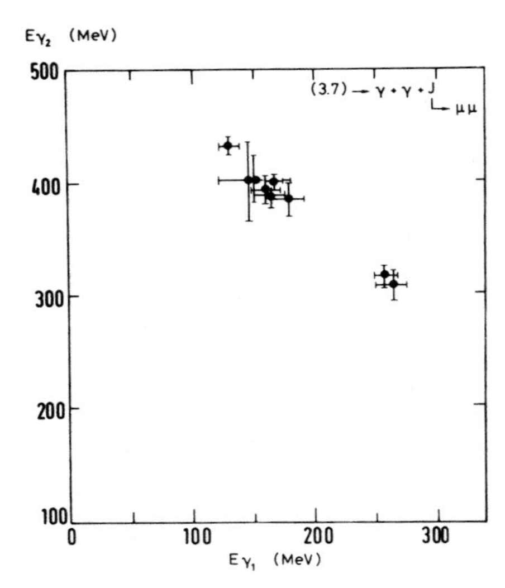

Fig. 17. Scatter plot of the two-photon energies for candidates for the decay $\psi' \to (J \to \mu^+ + \mu^-) + \gamma + \gamma$. (Ref. 32.)

determine which γ came first), and the second group has $E_{\gamma_1} = 263 \pm 8$ MeV and $E_{\gamma_2} = 315 \pm 8$ MeV. This correlation, called scatter plot, is shown in Fig. 17. The emission of monochromatic γ-rays indicates the existence of intermediate states with even-spin quantum number.

The narrow width of the J and the existence of the P_c and many other states, strongly suggests that the J may be a bound state of two new quarks. The existence of charmed quarks was first proposed by Bjorken and Glashow [33], and Glashow, Iliopoulos and Maiani [34], originally as a cure for certain difficulties in the weak interaction of hadrons. Indeed, the energy levels of the observed states are very similar to the positronium state discovered by Deutsch in 1951 [35].

Recently there are indications from experiments at BNL [36], from DESY [37, 38], from the Fermi Laboratory [39] and from SLAC [40] of the existence of further narrow states, indications which very much follow the general prediction of Glashow.

4. CONCLUSION

In conclusion, we can ask ourselves some further questions:

1) We know that the photon transforms itself into ρ, ω, and φ with a mass of about 1 GeV. It can transform into J and its various associated states with a mass of about 3—5 GeV. What happens when we go to higher and higher energies? It seems very unlikely that there should not be many more new series of photon-like particles.
2) The existence of J implies that we need at least four quarks to explain the phenomena observed so far. How many more quarks will we need if we find a new series of particles in higher energy regions?
3) If we need a large family of quarks, are they the real fundamental blocks of nature? Why has none of them been found?

REFERENCES

1. See for example: J. Needham, Science and civilization in China (Cambridge University Press, New York, 1962), Vol. 4.
2. G. Herzberg, Atomic spectra and atomic structure (Dover Reprint, New York, 1944). A. I. Akhiezer and V. B. Berestetskii, Quantum electrodynamics (translated by Oak Ridge Technical Information Service, ORNL, Tennessee, USA, 1957). S. D. Drell, Ann. Phys. *4*, 75 (1958).
3. S. J. Brodsky and S. C. C. Ting, Phys. Rev. *145*, 1018 (1966).
4. J. Bailey, K. Borer, F. Combley, H. Drumm, C. Eck, F. J. M. Farley, J. H. Field, W. Flegel, P. M. Hattersley, F. Krienen, F. Lange, G. Petrucci, E. Picasso, H. I. Pizer, O. Runolfsson, R. W. Williams and S. Wojcicki, Phys. Letters *55B*, 420 (1975). E. Picasso, private communication.
5. J. G. Asbury, W. K. Bertram, U. Becker, P. Joos, M. Rohde, A. J. S. Smith, S. Friedlander, C. Jordan and S. C. C. Ting, Phys. Rev. Letters *18*, 65 (1967). H. Alvensleben, U. Becker, W. K. Bertram, M. Binkley, K. Cohen, C. L. Jordan, T. M. Knasel, R. Marshall, D. J. Quinn, M. Rohde, G. H. Sanders and S. C. C. Ting, Phys. Rev. Letters *21*, 1501 (1968).
6. J. G. Asbury, U. Becker, W. K. Bertram, P. Joos, M. Rohde, A. J. S. Smith, C. L. Jordan and S. C. C. Ting, Phys. Rev. Letters *19*, 869 (1967).
7. For theoretical papers on leptonic decay of vector mesons, see:
M. Gell-Mann, D. Sharp and W. G. Wagner, Phys. Rev. Letters *8*, 261 (1962).
S. L. Glashow, Phys. Rev. Letters *7*, 469 (1961).
M. Gell-Mann and F. Zachariasen, Phys. Rev. *124*, 953 (1961).
Y. Nambu and J. J. Sakurai, Phys. Rev. Letters *8*, 79 (1962).
S. M. Berman and S. D. Drell, Phys. Rev. B *133*, 791 (1964).
N. M. Kroll, T. D. Lee and B. Zumino, Phys. Rev. *157*, 1376 (1967).
H. Joos, Phys. Letters *24B*, 103 (1967).
8. See also: J. K. de Pagter, J. I. Friedman, G. Glass, R. C. Chase, M. Gettner, E. von Goeler, R. Weinstein and A. M. Boyarski, Phys. Rev. Letters *16*, 35 (1966). A. Wehman, E. Engels, L. N. Hand, C. M. Hoffman, P. G. Innocenti, R. Wilson, W. A. Blanpied, D. J. Drickey and D. G. Stairs, Phys. Rev. Letters *18*, 929 (1967). B. D. Hyams, W. Koch, D. Pellett, D. Potter, L. von Lindern, E. Lorenz, G. Lütjens, U. Stierlin and P. Weilhammer, Phys. Letters *24B*, 634 (1967). M. N. Khachaturyan, M. A. Azimov, A. M. Baldin, A. S. Belousov, I. V. Chuvilo, R. Firkowski, J. Hladky, M. S. Khvastunov, J. Manca, A. T. Matyushin, V. T. Matyushin, G. A. Ososkov, L. N. Shtarkov and L. I. Zhuravleva, Phys. Letters *24B*, 349 (1967).
9. H. Alvensleben, U. Becker, W. K. Bertram, M. Chen, K. J. Cohen, T. M. Knasel, R. Marshall, D. J. Quinn, M. Rohde, G. H. Sanders, H. Schubel and S. C. C. Ting, Phys. Rev. Letters *24*, 786 (1970).
10. H. Alvensleben, U. Becker, M. Chen, K. J. Cohen, R. T. Edwards, T. M. Knasel, R. Marshall, D. J. Quinn, M. Rohde, G. H. Sanders, H. Schubel and S. C. C. Ting, Phys. Rev. Letters *25*, 1377 (1970).
11. H. Alvensleben, U. Becker, W. K. Bertram, M. Chen, K. J. Cohen, R. T. Edwards, T. M. Knasel, R. Marshall, D. J. Quinn, M. Rohde, G. H. Sanders, H. Schubel and S. C. C. Ting, Phys. Rev. Letters *25*, 1373 (1970).
12. H. Alvensleben, U. Becker, W. Busza, M. Chen, K. J. Cohen, R. T. Edwards, P. M. Mantsch, R. Marshall, T. Nash, M. Rohde, H. F. W. Sadrozinski, G. H. Sanders, H. Schubel, S. C. C. Ting and S. L. Wu, Phys. Rev. Letters *27*, 888 (1971).
13. For an excellent review of earlier work on ρ-ω interference, see G. Goldhaber, *in* Experimental meson spectroscopy (eds. C. Baltay and A. H. Rosenfeld) (Columbia Univ. Press, New York, 1970), p. 59. See also E. Gabathuler, same Proceedings, p. 645.

14. J. J. Sakurai, Ann. Phys. *11*, 1 (1960) and Nuovo Cimento *16*, 388 (1960). For an excellent review of earlier work, see H. Joos, Review talk at the Internat. Conf. on Elementary Particles, Heidelberg, 1967 (North-Holland Publ. Co., Amsterdam, 1968), p. 349.
15. M. Conversi, T. Massam, Th. Muller, M. A. Schneegans and A. Zichichi, Proc.12th Internat. Conf. on High-Energy Physics, Dubna, 1964 (Atomizdat, Moscow, 1966), p. 895, and T. Massam, Th. Muller and A. Zichichi, Nuovo Cimento *40*, 589 (1965).
16. T. Massam, Th. Muller and A. Zichichi, CERN 63-25 (1963). See also, T. Massam, Th. Muller, M. Schneegans and A. Zichichi, Nuovo Cimento *39*, 464 (1965).
17. J. H. Christenson, G. S. Hicks, L. M. Lederman, P. J. Limon, B. G. Pope and E. Zavattini, Phys. Rev. Letters *25*, 1523 (1970).
18. One of the first theoretical works on dileptonic production from pp collisions is that of: G. Altarelli, R. A. Brandt and G. Preparata, Phys. Rev. Letters *26*, 42 (1971).
19. S. D. Drell and T. M. Yan, Phys. Rev. Letters *25*, 316 (1970).
20. J. J. Aubert, U. Becker, P. J. Biggs, J. Burger, M. Chen, G. Everhart, P. Goldhagen, Y. Y. Lee, J. Leong, T. McCorriston, T. G. Rhoades, M. Rohde, S. C. C. Ting and S. L. Wu, Phys. Rev. Letters *33*, 1404 (1974).
21. J. J. Aubert, U. Becker, P. J. Biggs, J. Burger, M. Chen, G. Everhart, J. W. Glenn III, P. Goldhagen, Y. Y. Lee, J. Leong, P. Mantsch, T. McCorriston, T. G. Rhoades, M. Rohde, S. C. C. Ting and S. L. Wu, Nuclear Phys. *B89*, 1 (1975).
22. See, S. C. C. Ting, Discovery story *in* Adventures in Experimental Physics *5*, 115 (1976).
23. T. T. Wu, private communication. V. Blobel, H. Fesefeldt, H. Franz, B. Hellwig, W. Idschok, J. W. Lamsa, D. Mönkemeyer, H. F. Neumann, D. Roedel, W. Schrankel, B. Schwarz, F. Selonke and P. Söding, Phys. Letters *48B*, 73 (1974).
24. L. B. Leipuner, R. C. Larsen, L. W. Smith, R. K. Adair, H. Kasha, C. M. Ankenbrandt, R. J. Stefanski and P. J. Wanderer, Phys. Rev. Letters *34*, 103 (1975).
25. C. Bacci, R. Baldini Celio, M. Bernardini, G. Capon, R. Del Fabbro, M. Grilli, E. Iarocci, L. Jones, M. Locci, C. Mencuccini, G. P. Murtas, G. Penso, G. Salvini, M. Spano, M. Spinetti, B. Stella, V. Valente, B. Bartoli, D. Bisello, B. Esposito, F. Felicetti, P. Monacelli, M. Nigro, L. Paoluzi, I. Peruzzi, G. Piano Mortari, M. Piccolo, F. Ronga, F. Sebastiani, L. Trasatti, F. Vanoli, G. Barbarino, G. Barbiellini, C. Bemporad, R. Biancastelli, M. Calvetti, M. Castellano, F. Cevenini, V. Ćostantini, P. Lariccia, S. Patricelli, P. Parascandalo, E. Sassi, C. Spencer, L. Tortora, U. Troya and S. Vitale, Phys. Rev. Letters *33*, 1408 (1974).
26. C. Bemporad, Proc. Internat. Symposium on Lepton and Photon Interactions at High Energies, Stanford, 1975 (SLAC, Calif., USA, 1975), p. 113.
27. B. Knapp, W. Lee, P. Leung, S. D. Smith, A. Wijangco, K. Knauer, D. Yount, D. Nease, J. Bronstein, R. Coleman, L. Cormell, G. Gladding, M. Gormley, R. Messner, T. O'Halloran, J. Sarracino, A. Wattenberg, D. Wheeler, M. Binkley, J. R. Orr, J. Peoples and L. Read, Phys. Rev. Letters *34*, 1040 and 1044 (1975). The group has recently finished a series of experiments on J production with a neutron beam on nuclear targets, and has learned valuable information on the transmission properties of J. M. Binkley, I. Gaines, J. Peoples, B. Knapp, W. Lee, P. Leung, S. D. Smith, A. Wijangco, J. Knauer, J. Bronstein, R. Coleman, G. Gladding, M. Goodman, M. Gormley, R. Messner, T. O'Halloran, J. Sarracino and A. Wattenberg, Phys. Rev. Letters *37*, 571 and 574 (1976).
28. K. J. Anderson, G. G. Henry, K. T. McDonald, J. E. Pilcher, E. I. Rosenberg, J. G. Branson, G. H. Sanders, A. J. S. Smith and J. J. Thaler, Phys, Rev. Letters *36*, 237 (1976), and *37*, 799 and 803 (1976).
29. See, for example, review papers by: R. F. Schwitters, Proc. Internat. Symposium on Lepton and Photon Interactions at High Energies, Stanford, 1975 (SLAC, Calif., USA, 1975), p. 5. G. S. Abrams, same Proceedings, p. 25. G. J. Feldman, same Proceedings, p. 39. A. D. Liberman, same Proceedings, p. 55.

30. For a review of DESY work, see: Review paper by B. H. Wiik, Proc. Internat. Symposium on Lepton and Photon Interactions at High Energies, Stanford, 1975 (SLAC, Calif., USA, 1975), p. 69.
31. J. Burmester, L. Criegee, H. C. Dehne, K. Derikum, R. Devenish, J. D. Fox, G. Franke, G. Flügge, Ch. Gerke, G. Horlitz, Th. Kahl, G. Knies, M. Rössler, G. Wolff, R. Schmitz, T. N. Rangaswamy, U. Timm, H. Wahl, P. Waloschek, G. G. Winter, W. Zimmermann, V. Blobel, H. Jensing, B. Koppitz, E. Lohrmann, A. Bäcker, J. Bürger, C. Grupin, M. Rost, H. Meyer and K. Wacker, DESY preprint 76/53 (1976). Also, E. Lohrmann, private communication.
32. W. Braunschweig, H.-U. Martyn, H. G. Sander, D. Schmitz, W. Sturm, W. Wallraff, K. Berkelman, D. Cords, R. Felst, E. Gadermann, G. Grindhammer, H. Hultschig, P. Joos, W. Koch, U. Kötz, H. Krehbiel, D. Kreinick, J. Ludwig, K.-H. Mess, K. C. Moffeit, A. Petersen, G. Poelz, J. Ringel, K. Sauerberg, P. Schmüser, G. Vogel, B. H. Wiik, G. Wolf, G. Buschhorn, R. Kotthaus, U. E. Kruse, H. Lierl, H. Oberlack, R. Pretzl, M. Schliwa, S. Orito, T. Suda, Y. Totsuka and S. Yamada, Phys. Letters *57B*, 407 (1975).
33. B. J. Bjorken and S. L. Glashow, Phys. Letters *11*, 255 (1964). See also earlier paper by S. L. Glashow and M. Gell-Mann, Ann. Phys. *15*, 437 (1961).
34. S. L. Glashow, J. Iliopoulos and L. Maiani, Phys. Rev. D *2*, 1285 (1970).
35. M. Deutsch, Phys. Rev. *82*, 455 (1951).
36. E. Cazzoli, A. M. Cnops, P. L. Connolly, R. I. Louttit, M. J. Murgtagh, R. B. Palmer, N. P. Samios, T. T. Tso and H. H. Williams, Phys. Rev. Letters *34*, 1125 (1975).
37. W. Braunschweig, H.-U. Martyn, H. G. Sander, D. Schmitz, W. Sturm, W. Wallraff, D. Cords, R. Felst, R. Fries, E. Gadermann, B. Gittelman, H. Hultschig, P. Joos, W. Koch, U. Kötz, H. Krehbiel, D. Kreinick, W. A. McNeely, K. C. Moffeit, P. Petersen, O. Römer, R. Rüsch, B. H. Wiik, G. Wolf, G. Grindhammer, J. Ludwig, K. H. Mess, G. Poelz, J. Ringel, K. Sauerberg, P. Schmüser, W. De Boer, G. Buschhorn, B. Gunderson, R. Kotthaus, H. Lierl, H. Oberlack, M. Schliwa, S. Orito, T. Suda, Y. Totsuka and S. Yamada, Phys. Letters *63B*, 471 (1976).
38. J. Burmester, L. Criegee, H. C. Dehne, K. Derikum, R. Devenish, J. D. Fox, G. Franke, G. Flügge, Ch. Gerke, G. Horlitz, Th. Kahl, G. Knies, M. Rössler, G. Wolff, R. Schmitz, T. N. Rangaswamy, U. Timm, H. Wahl, P. Waloschek, G. G. Winter, W. Zimmermann, V. Blobel, H. Jensing, B. Koppitz, E. Lohrmann, A. Bäcker, J. Bürger, C. Grupin, M. Rost, H. Meyer and K. Wacker, DESY preprint 76/50 (1976), and private communication from B. Wiik.
39. B. Knapp, W. Lee, P. Leung, S. D. Smith, A. Wijangco, J. Knauer, D. Yount, J. Bronstein, R. Coleman, G. Gladding, M. Goodman, M. Gormley, R. Messner, T. O'Halloran, J. Sarracino, A. Wattenberg, M. Binkley, I. Gaines and J. Peoples, Phys. Rev. Letters *37*, 882 (1976).
40. G. Goldhaber, F. M. Pierre, G. S. Abrams, M. S. Alam, A. M. Boyarski, M. Breidenbach, W. C. Carithers, W. Chinowsky, S. C. Cooper, R. G. DeVoe, J. M. Dorfan, G. J. Feldman, C. E. Friedberg, D. Fryberger, G. Hanson, J. Jaros, A. D. Johnson, J. A. Kadyk, R. R. Larsen, D. Lüke, V. Lüth, H. L. Lynch, R. J. Madaras, C. C. Morehouse, H. K. Nguyen, J. M. Paterson, M. L. Perl, I. Peruzzi, M. Piccolo, T. P. Pun, P. Rapidis, B. Richter, B. Sadoulet, R. H. Schindler, R. F. Schwitters, J. Siegrist, W. Tanenbaum, G. H. Trilling, F. Vannucci, J. S. Whitaker and J. E. Wiss, Phys. Rev. Letters *37*, 255 (1976). I. Peruzzi, M. Piccolo, G. J. Feldman, H. K. Nguyen, J. E. Wiss, G. S. Abrams, M. S. Alam, A. M. Boyarski, M. Breidenbach, W. C. Carithers, W. Chinowsky, R. G. DeVoe, J. M. Dorfan, G. E. Fischer, C. E. Friedberg, D. Fryberger, G. Goldhaber, G. Hanson, J. A. Jaros, A. D. Johnson, J. A. Kadyk, R. R. Larsen, D. Lüke, V. Lüth, H. L. Lynch, R. J. Madaras, C. C. Morehouse, J. M. Paterson, M. L. Perl, F. M. Pierre, T. P. Pun, P. Rapidis, B. Richter, R. H. Schindler, R. F. Schwitters, J. Siegrist, W. Tanenbaum, G. H. Trilling, F. Vannucci and J. S. Whitaker, Phys. Rev. Letters *37*, 569 (1976).

Physics 1977

PHILIP W ANDERSON, NEVILL F MOTT and JOHN H VAN VLECK

for their fundamental theoretical investigations of the electronic structure of magnetic and disordered systems

THE NOBEL PRIZE FOR PHYSICS

Speech by Professor **PER-OLOV LÖWDIN** of the Royal Academy of Sciences
Translation from the Swedish text

Your Majesties, Your Royal Highnesses, Ladies and Gentlemen,

This year's Nobel Prize in physics is shared equally between Philip Anderson, Sir Nevill Mott and John Van Vleck for their fundamental contributions to the theory of the electronic structure of magnetic and disordered systems.

All matter consists of positive and negative electricity: partly heavy positive elementary particles gathered in atomic nuclei, partly light negative elementary particles—electrons—which move in wonderful patterns around the nuclei—always attracted to them but difficult to catch because of their own movement. It is this electron dance which is essentially responsible for the electric, magnetic, and chemical properties of matter.

The 1937 Nobel Prize winner in medicine, Albert Szent-Györgyi, has often compared the chemical process in living cells with a great drama played with the electrons as actors on a stage formed by the biomolecules—with the only difference that the scene as well as the actors may be a thousand billion times smaller than we are accustomed to from the Royal Opera. No scientist has seen the score of this musical of life itself, and no one will probably ever be able to see it in its entirety—only a few have been granted the privilege of seeing small fragments in the form of isolated ballets, often with a hero and sometimes with a ballerina.

In the crystal and ligand field theories developed by Van Vleck, there is always a metal atom playing the role of the hero in the drama. In many of the enzymes fundamental for the life of our body, there is often a metal atom in the active center which regulates the action. The haemoglobin in our red blood cells contains an iron atom which carries the oxygen molecule to its given place in the body—in the same way as the hero carries the ballerina on his strong arms. It is Van Vleck who has developed the basic theory for such processes, which are also of great importance in the chemistry of complex compounds, geology, and laser technology.

The electronic dance is of similar importance also in the solid bodies surrounding us—in the ladies' diamonds, in the every-day rock salt, or in the amorphous glasses. Such materials have characteristic electric and magnetic properties which depend on the motions of the electrons. In the same way as it is easier in an ordinary waltz to waltz forward than backward, there is in the electronic dance a specific spin-orbit coupling between the rotations of the electrons and their translational movements, which is of importance for the magnetic properties. Like the dancers in a ballet are constantly changing place, the electrons have also their own exchange and superexchange phenomena—their own characteristic "pas de deux". Both Van Vleck and Anderson have studied the local magnetic properties of matter, where the hero is a metal atom with strong personal magnetism whose special properties may vary strongly with the environment—a theory basic for the construction of dilute magnetic alloys. Here one dares perhaps to speak about a successful localization policy.

One of the greatest current problems of humanity is the so-called energy problem—it has been said that the modern society uses too much energy. According to the laws of physics, such a statement is quite absurd, since energy can neither be created nor destroyed. The whole thing is instead a problem of order—at the level of the elementary particles. What happens is that energy of higher order is transformed into energy of lower order, that mechanical and electric energy are changed into heat, that the motions of the elementary particles involved will be more and more disordered. It is the merit of Anderson to have shown that even the reverse may sometimes happen: that geometrically disordered materials, as for instance glass, have their own laws, and that the electronic dance in them may lead to localized states with a high form of order, which influence the properties of the material. Perfectly ordered systems are of great importance in electronics, but they are usually very expensive to produce, so disordered systems with similar properties are hence of essential importance.

In some of his work, Sir Nevill Mott has taken up these and similar ideas in order to study the electrical properties of materials and the transition between conductors, semi-conductors and insulators. In this connection, Mott has also investigated the importance of the interaction between the electrons —that the electrons indeed like to dance in pairs, but also that there is a mutual repulsion which sometimes causes them to guard their own domains and stop the hand-in-hand dance which is essential for the electronic conductivity of the material. The theory for Mott-transitions and Mott-Anderson transitions is today of fundamental importance for the understanding of certain materials and for the construction of new ones. Anderson and Mott have shown that properly controlled disorder may be technically as important as perfect order.

This year's Laureates in physics are all three giants within solid-state theory, and it is actually rather remarkable how small a portion of their total work has been considered in connection with this year's Nobel award. Even if these discoveries already now have shown their technical value, it is their fundamental contributions to the free basic research—to the human knowledge of the electronic structure of solids—which has primarily been awarded, with the understanding that it may be even more awarded, with the understanding that it may be even more practically important in the future. Through their work, Anderson, Mott, and Van Vleck have shown that the understanding of the electronic choreography is not only remarkably beautiful from the point of view of science but also of essential importance for the development of the technology of our every-day life.

I have the pleasure and the honour on behalf of the Academy to extend to you our warmest congratulations and I now invite you to receive your prizes from the hands of His Majesty the King.

John H Van Vleck

JOHN HASBROUCK VAN VLECK

I was born in Middletown, Connecticut, March 13, 1899 where my father and grandfather were respectively professors of mathematics and of astronomy at Wesleyan University. However, when I was seven years old father accepted a professorship at the University of Wisconsin, so I grew up in Madison, Wisconsin, where I attended the public schools, and graduated from the University of Wisconsin in 1920. As a sort of revolt against having two generations of academic forbears, I vowed as a child that I would not be a college professor, but after a semester of graduate work at Harvard, I outgrew my childish prejudices, and realized that the life work for which I was best qualified was that of a physicist, not of the experimental variety, but in an academic environment.

I have been lucky in a number of respects. Coming from an academic family, I had invaluable parental guidance or advice at various times. At Harvard I took most of my courses under Professor Bridgman or Professor Kemble. The latter's course on quantum theory fascinated me, so I decided to write my doctor's thesis under Kemble's supervision. He was the one person in America at that time qualified to direct purely theoretical research in quantum atomic physics. My doctor's thesis was the calculation of the binding energy of a certain model of the helium atom, which Kemble and Niels Bohr suggested independently and practically simultaneously, with Kramers making the corresponding calculation in Copenhagen. The results did not agree with experiment for the "old quantum theory" was not the real thing. However, when the true quantum mechanics was discovered by Heisenberg and others in 1926, my background in the old quantum theory and its correspondence principle was a great help in learning the new mechanics, particularly the matrix form which is especially useful in the theory of magnetism.

I was fortunate in being offered an assistant professorship at the University of Minnesota in 1923, a year after my Ph. D. at Harvard, with purely graduate courses to teach. This was an unusual move by that institution, as at that time, posts with this type of teaching were generally reserved for older men, and recent Ph. D.'s were traditionally handicapped by heavy loads of undergraduate teaching which left little time to think about research. Also it was at Minnesota that I met Abigail Pearson, a student there, whom I married June 10, 1927, and on Nobel Day, December 10, 1977 we had been married exactly 50 1/2 years!

I was also lucky in choosing the theory of magnetism as my principal research interest, as this is a field which has continued to be of interest over the years, with new ramifications continuing to make their appearance (magnetic resonance, relaxation, microwave devices, etc.). So often a particular field loses general interest after a span of time. My last paper dealing with magnetism was published fifty years after my first one.

Besides my work on magnetism, and the closely related subjects of ligand fields and of dielectrics, one of my interests has been molecular spectra. The theoretical problems associated with the fine structures therein appeared rather academic at the time, but recently have burgeoned in interest in connection with radioastronomical investigations, including notably those of the observatory at Gothenburg.

Degrees, positions, awards, etc.

A.B. University of Wisconsin, 1920

Ph. D., Harvard University, 1922 (instructor 1922—3)

Honorary D. Sc. or D. Honoris Causa, Wesleyan U., 1936; U. Wisconsin, 1947; Grenoble U., 1950; U. Maryland, 1955; Oxford U., 1958; U. Paris, 1960; Rockford College, 1961; U. Nancy, 1961; Harvard U., 1966; U. Chicago, 1968; U. Minnesota 1971.

On faculty, University of Minnesota, 1923—28; University of Wisconsin 1928—34 Harvard University 1934—69, emeritus 1969— (Dean of Engineering and Applied Physics 1951—57).

Lorentz (visiting) professor, Leiden, 1960; Eastman Professor, Oxford, 1961—62; Guggenheim Fellow, 1930.

Foreign member, Royal Swedish Academy, Uppsala Academy, Netherlands Academy, Academie des Sciences, Royal Society of London.

National Medal of Science, USA; Lorentz Medal (Netherlands); Cresson Medal (Franklin Institute); Michelson Prize of Case Institute of Technology; Langmuir Award in Chemical Physics; General Electric Foundation; Chevalier, Legion of Honor.

Member, National Academy of Sciences, American Academy of Arts and Sciences, American Philosophical Society, International Academy of Quantum Molecular Science; Honorary Member, French Physical Society; President, American Physical Society, 1952.

Professor Van Vleck died in 1980.

QUANTUM MECHANICS
THE KEY TO UNDERSTANDING MAGNETISM

Nobel Lecture, 8 December, 1977

by
J. H. VAN VLECK
Harvard University, Cambridge, Massachusetts, USA

The existence of magnetic materials has been known almost since prehistoric times, but only in the 20th century has it been understood how and why the magnetic susceptibility is influenced by chemical composition or crystallographic structure. In the 19th century the pioneer work of Oersted, Ampère, Faraday and Joseph Henry revealed the intimate connection between electricity and magnetism. Maxwell's classical field equations paved the way for the wireless telegraph and the radio. At the turn of the present century Zeeman and Lorentz received the second Nobel Prize in physics for respectively observing and explaining in terms of classical theory the so-called normal Zeeman effect. The other outstanding early attempt to understand magnetism at the atomic level was provided by the semi-empirical theories of Langevin and Weiss. To account for paramagnetism, Langevin (1) in 1905 assumed in a purely ad hoc fashion that an atomic or molecular magnet carried a permanent moment μ, whose spacial distribution was determined by the Boltzmann factor. It seems today almost incredible that this elegantly simple idea had not occurred earlier to some other physicist inasmuch as Boltzmann had developed his celebrated statistics over a quarter of a century earlier. With the Langevin model, the average magnetization resulting from N elementary magnetic dipoles of strength μ in a field H is given by the expression

$$M = \frac{N\mu\int\int \cos\theta e^{\mu H\cos\theta/kT} d\omega}{\int\int e^{\mu H\cos\theta/kT} d\omega} = NL\left(\frac{\mu H}{kT}\right), \text{ where } L(x) = \coth x - \frac{1}{x} \tag{1}$$

At ordinary temperatures and field strengths, the argument x of the Langevin function can be treated as small compared with unity. Then $L(x) = \frac{1}{3}x$, and Eq. (1) becomes

$$M = N\frac{\mu^2}{3kT}H \tag{2}$$

so that the magnetic susceptibility $\chi = \frac{M}{H}$ is inversely proportional to temperature, a relation observed experimentally for oxygen ten years earlier by Pierre Curie (2) and hence termed Curie's law.

To explain diamagnetism, Langevin took into account the Larmor precession of the electrons about the magnetic field, and the resulting formula for the diamagnetic susceptibility is

$$\chi = -\frac{Ne^2}{6mc^2} \sum_i (r_i)^2 \tag{3}$$

where $(r_i)^2$ is the mean square radius of an electron orbit, and the summation extends over all the electrons in the atom. The important thing about (3) is that, in substantial agreement with experiment, it gives a diamagnetic susceptibility independent of temperature, provided the size of the orbits does not change.

Two years later, in 1907, Pierre Weiss (3), another French physicist, took the effective field acting on the atom or molecule to be the applied field augmented by a mysterious internal or molecular field proportional to the intensity of magnetization. The argument of the Langevin function then becomes $\frac{\mu(H+qM)}{kT}$ rather than $\frac{\mu H}{kT}$, and in place of (2) one has

$$\chi = \frac{M}{H} = \frac{N\mu^2}{3k(T-T_c)} \text{ where } T_c = \frac{Nq\mu^2}{3k} \tag{4}$$

Since the right side of (4) becomes infinite for $T = T_c$, the Weiss model predicts the existence of a Curie point below which ferromagnetism sets in. This model also describes qualitatively quite well many ferromagnetic phenomena. Despite its many successes there was one insuperable difficulty from the standpoint of classical electrodynamics. Namely the coefficient q of the molecular field qM should be of the order $\frac{4\pi}{3}$ whereas it had to be of the order 10^3 to describe the observed values of T_c.

There was, moreover, an even worse difficulty. If one applies classical dynamics and statistical mechanics consistently, a very simple calculation, which can be made in only a few lines but I shall not reproduce it here, shows that the diamagnetic and paramagnetic contributions to the susceptibility exactly cancel. Thus there should be no magnetism at all. This appears to have been first pointed out by Niels Bohr (4) in his doctor's dissertation in 1911, perhaps the most deflationary publication of all time in physics. This may be one reason why Bohr broke with tradition and came forth with his remarkable theory of the hydrogen spectrum in 1913. That year can be regarded as the debut of what is called the old quantum theory of atomic structure, which utilized classical mechanics supplemented by quantum conditions. In particular it quantized angular momentum and hence the magnetic moment of the atom, as was verified experimentally in the molecular beam experiments of Stern and Gerlach (5). Hence there was no longer the statistical continuous distribution of values of the dipole moment which was essential to the proof of zero magnetism in classical theory. When Langevin assumed that the magnetic moment of the atom or molecule had a fixed value μ, he was quantizing the system without realizing it, just as in Molière's *Bourgeois Gentilhomme*, Monsieur Jourdain had been writing prose all his life, without appreciating it, and was overjoyed to discover he had been doing anything so elevated. Magnetism could be understood qualitatively in terms of incomplete shells of electron orbits, and a sentence of Bohr which I like to quote reads "In short an examination of the magnetic properties and colors of the long periods gives us a striking illustration of how a wound in the otherwise symmetrical inner structure of the atom is first created and then healed." However, with the passage of time it became increasingly clear that the old quantum theory could give quantitatively correct results for energy levels or spectral frequencies only in hydrogen. One historian of science has referred to the early 1920's as the crisis in quantum theory, but I would characterize this era as one of increasing disillusion and disappointment in contrast to the hopes which were so high in the years immediately following 1913.

The advent of quantum mechanics in 1926 furnished at last the real key to the quantitative understanding of magnetism, I need not elaborate on the miraculous coincidence of three developments, the discovery of the matrix form of quantum mechanics by Heisenberg and Born, the alternative but equivalent wave mechanical form by de Broglie and Schrödinger, and the introduction of electron spin by Uhlenbeck and Goudsmit. A quantum mechanics without spin and the Pauli exclusion principle would not have been enough — one wouldn't have been able to understand even the structure of the periodic table or most magnetic phenomena. Originally spin was a sort of appendage to the mathematical framework, but in Dirac 1928 synthesized everything in his remarkable four first order simultaneous equations. To stress the importance of the quantum mechanical revolution, I cannot do better than to quote an often-metnioned sentence from one of Dirac's early papers, which reads "*The general theory of quantum mechanics is now almost complete. The underlying physical laws necessary for the mathematical theory of a large part of physics and all of chemistry are thus completely known*".

With at last the key available for the proper analysis of what was going on inside the atom, it was natural that more than one physicist would try applying it to a particular problem. So it is not surprising that four different researchers independently calculated and reported in practically simultaneous publications (6) the susceptibility of a rotating diatomic molecule carrying a permanent dipole moment, which could be either electric or magnetic depending on whether one was interested in an electric or magnetic susceptibility. (I was one of the four. The others were Kronig, Manneback, and Miss Mensing working in collaboration with Pauli. The new mechanics happily restored the factor $\frac{1}{3}$ in the Langevin formula) (or the corresponding Debye expression in the electric case), as shown in Table I. Thus was ended the confusion of the old quantum theory, where half quanta worked better in band spectra even though whole integers were required with rational application of Bohr's 1913 ideas.

There are three common paramagnetic gases, viz. O_2, NO_2, and NO. I shall discuss NO first as its behavior is the most interesting of the three. In 1926 Robert Mulliken, who has a sixth sense for deducing molecular energy levels from band spectra, had decided that the ground state of the NO molecule was a $^2\Pi$ state, whose two components were separated by about 122 cm^{-1} but he wasn't sure whether the doublet was regular rather than inverted. I tried

Table 1. Value of C in Relation $\chi = CN\mu^2/kT$

Value of C	Form and Year of Theory
$\frac{1}{3}$	Classical, 1905
1.54	Whole quanta, 1921
4.57	Half Quanta, 1924
$\frac{1}{3}$	Quantum mechanics, 1926

calculating the susceptitility of NO on the basis of Mulliken's energy levels and found (7) that the observed susceptibility at room temperatures could be explained on the basis that the doublet was regular, i.e. the ${}^2\Pi_{1/2}$ component lower than the ${}^2\Pi_{3/2}$. I wasn't entirely convinced that the agreement was real rather than spurious, as molecular quantum mechanics was then in its infancy. If the theory was correct there should be deviations from Curie's law, and so measurements on the susceptibility as a function of temperature should be decisive. To my surprise, experiments to test this prediction were performed in 1929 at three different laboratories in different parts of the world, with each going to a lower temperature than the preceding (8). As shown in Fig. 1, the agreement with theory was gratifying. The ordinate in Fig. 1 is not the susceptibility itself, but rather the effective magneton number μ_{eff} defined by $\chi = \mathcal{N}\mu^2_{eff}\,\beta^2/3kT$, where β is the Bohr magneton $he/4\pi mc$. The non-constancy of μ_{eff} is a measure of the deviation from Curie's law.

My calculations on NO started me thinking on the general conditions under which Curie's law should be valid or non-valid. I noted the fact, often over-

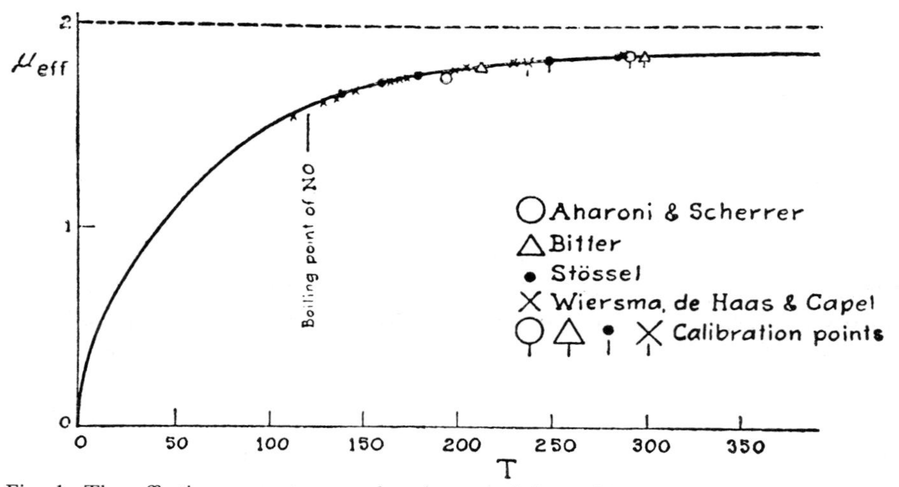

Fig. 1. The effective magneton number (measured in multiples of the Bohr unit $\beta = he/4\pi mc$) of nitric oxide as a function of temperature. Were Curies law valid, the curve would be a horizontal straight line.

looked in those early days, that to make a proper computation of the susceptibility even in weak fields, it is necessary to know the energy of the stationary states, or alternatively the partition function, to the second order in the field strength H., corresponding to including the second as well as first order Zeeman effect. If the energy of a stationary state is

$$E_i = E_i^{(0)} + E_i^{(1)}H + E_i^{(2)}H^2 + \dots$$

the correct formula for the susceptibility is

$$\chi = \frac{N}{\sum_i p_i} \sum_i \left(\frac{E_i^{(1)\,2}}{kT} - 2E_i^{(2)} \right) p_i \text{ with } p_i = \exp\left(\frac{-E_i^{(0)}}{kT} \right) \tag{5}$$

Perturbation theory tells us that

$$E_i^{(1)} = \langle i \mid \mu_H \mid i \rangle, \; E_i^{(2)} = \sum_j \frac{|\langle i \mid \mu_H \mid j \rangle|^2}{h\nu_{ij}} \; (j \neq i) \tag{6}$$

where $h\nu_{ij}$ is the energy interval $E_i^{(0)} - E_j^{(0)}$ spanned by the matrix element $\langle i|\mu_H|j\rangle$ of the magnetic moment in the direction of the field H. From (5) and (6) one derives (7) the results presented in Table II.

Table II. Behavior of the Susceptibility in Various Situations

(a) χ is proportional to $1/T$ if all $|h\nu_{ij}|$ are $<<kT$.
(b) χ is independent of T if all $|h\nu_{ij}|$ are $>>kT$.
(c) $\chi = A + B/T$ if all $|h\nu_{ij}|$ are either $>>kT$ or $<<kT$.
(d) no simple dependence of χ on T if $|h\nu_{ij}|$ is comparable with kT.

In connection with the above it is to be understood that the relevant $h\nu_{ij}$ are only those which relate to the energy intervals spanned by $\langle i|\mu|j\rangle$, which because of selection principles can often be less than the total spread in the populated energy levels.

From too cursory examination of Eq. (5) one might conclude that case (a) could never arise when there is a second order Zeeman effect, but this is not so. Since $h\nu_{ji} = -h\nu_{ij}$, $|\langle i|\mu_H|j\rangle|^2 = |\langle j|\mu_H|i\rangle|^2$ the various terms in (4) can be so paired as to involve a factor $(p_j - p_i)/h\nu_{ij})$ which is approximately $\frac{1}{2}(p_i + p_j)/kT$ if $|h\nu_{ij}|<<kT$. The fact that the factor $h\nu_{ij}$ has thereby disappeared shows that there is no catastrophe in the expression for the susceptibility even when the denominators in the expression (6) for the second order perturbed energy are very small.

The NO molecule, as we have seen, is an illustration of the situation (d). On the other hand, the O_2 and NO_2 molecules are examples of (a) and hence obey Curie's law. The oxygen molecule exhibits the same susceptibility as though its spin of unity ($S = 1$) were completely uncoupled from the molecule. Actually the spin is coupled to the molecule so that most of the Zeeman energy becomes of the second rather than first order, but this complication is immaterial as regards the susceptibility since the binding energy is only of the order 2 cm^{-1}, small compared to kT. The third common paramagnetic gas NO_2 should have a suceptibility corresponding to a free spin $\frac{1}{2}$, as it is an odd molecule. Existing data were in disagreement with this prediction when I made it, but new magnetic measurements made by Havens at Wisconsin at my suggestion removed this discrepancy (9).

In 1925 Hund (10) wrote a paper on the magnetic susceptibilities of rare earth compounds which was the crowning achievement of the empiricism of the old quantum theory. He utilized Landé's then phenomenological *g*-factor and the Hund rule that the state of lowest energy is that of maximum spin, and of maximum *L* compatible with this *S*. At the time this rule was an inspired conjecture, but today physicists justify it by examining nodes in the wave function. He thus obtained the formula

$$\chi = \frac{N\beta^2 J(J+1)g_J^2}{3kT}$$

for the susceptibility, and found that this expression agreed remarkably well with experiments for all the trivalent rare earths compounds except those containing Sm or Eu. In 1928 Laporte (11) pointed out that for these particular two ions, the multiplet structure was such that the interval separating the lowest multiplet component from the one next above it is not large compared to kT. So he summed Hund's expression for χ over the multiplet's various values of J weighted in accordance with the Boltzmann factor. Even so, he was not able to raise the susceptibility to the values found experimentally. When I read his paper it occurred to me that probably the cause for the discrepancy was that the second order energy had been omitted. So Miss Frank and I made the relevant calculations (12), and then there was agreement with experiment, as shown in Fig. 2. The reason that Hund was able to obtain agreement with experiment for other rare earths was that his

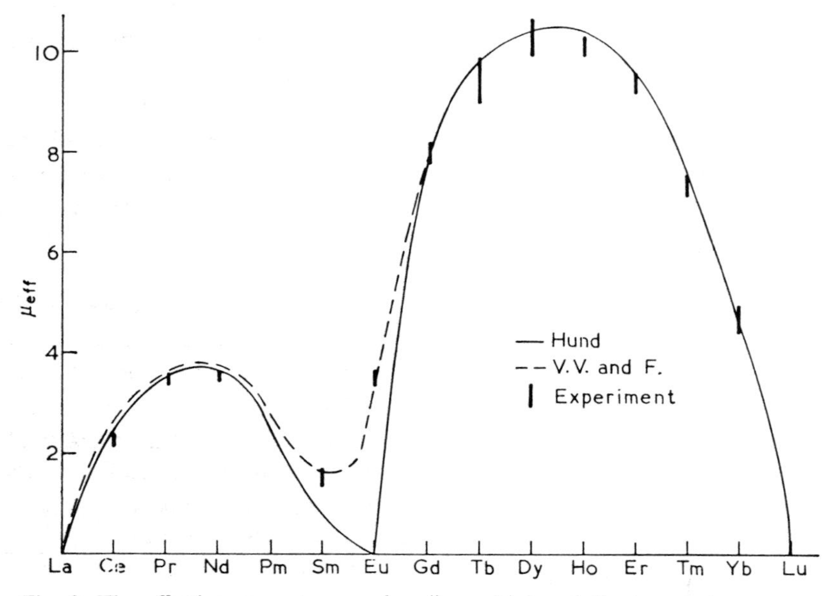

Fig. 2. The effective magneton number (in multiples of β) at room temperature for the sequence of trivalent ions in the configurations $4f^0$, 4f, $4f^2$, . . ., $4f^{14}$.

empirical expression for the first order energy was the same as the true quantum-mechanical one, and that the second order energy could be omitted without too much error. The latter was the case because the interval separating the lowest multiplet component from the next one above is large except for Sm^{3+}, Eu^{3+}, and the second order energy involves this interval in the denominator. Since Sm^{3+} and Eu^{3+}, unlike the other rare earth ions, correspond to case (d) of table II, deviations from Curie's law are to be expected for salts containing these ions. This was indeed confirmed by the limited amount of experimental data available at the time.

In 1930 and 1931 a great deal of my time went into writing my book on the Theory of Electric and Magnetic Susceptibilities, which appeared in 1932 (13). In this volume I aimed to include the major theoretical developments which had taken place up to the time of writing. Besides the things which I have already mentioned, there were other major developments in the theory of magnetism in the early days of quantum mechanics. Heisenberg (14) took the mystery out of the then twenty year old Weiss molecular field. He showed that it arose from exchange effects connecting the different magnetic atoms, which had the effect of introducing the needed strong coupling between the spins. Other notable theoretical developments prior to 1932 included Landau's paper (15) on the diamagnetism of free electrons, in which he showed that spinless free electrons had a small susceptibility of diamagnetic sign, in contrast to the zero result of classical mechanics. Pauli (18) showed that the spin moment of conduction electrons gives rise to only a small paramagnetic susceptibility practically independent of temperature. This paper was notable because it was the first application of Fermi-Dirac statistics to the solid state. If one used the Boltzmann statistics one would have a large susceptibility obeying Curie's law.

On the other hand, there were some important development which arrived just a little too late for me to include them in my volume. Néel's first paper on antiferromagnetism appeared in 1932, and in later years he introduced an important variant called ferri-magnetism, in which the anti-parallel dipoles are of unequal strength, so that they do not compensate and the resulting behavior can be ferromagnetic (17). There was also Peirls' (18) theoretical explanation of the de Haas-van-Alphen effect, and Bloch's 1932 paper (19) on the width of the boundaries (now called Bloch walls) separating the elementary domains in ferromagnetic materials. The corresponding domain structure was explained and elaborated by Landau and Lifschitz two years later (20).

In 1930 I held a Guggenheim fellowship for study and travel in Europe. I spent most of the time in Germany, but by far the most rewarding part of the trip scientifically was a walk which I took with Kramers along one of the canals near Utrecht. He told me about his own theorem (21) on degeneracy in molecules with an odd number of electrons and also of Bethe's long paper (22) concerned with the application of group theory to the determination of the quantum mechanical energy levels of atoms or ions exposed to a crystalline electric field, and in my book I referred to the role of the crystalline field only in a qualitative way, stressing the fact that it could largely suppress the orbital part of the magnetic moment in salts of the iron group. In the process of writing I did not have the time or energy to attempt quantitative numerical computations. I was most fortunate when, beginning in the fall of 1931 I had two post-doctoral students from England, namely William (now Lord) Penney, and Robert Schlapp. I suggested to these two men that they make calculations respectively on salts of the rare earth and of the iron group. The basic idea of the crystalline field potential is an extremely simple one, namely that the magnetic ion is exposed not just to the applied magnetic field but experiences in addition a static field which is regarded as an approximate representation of the forces exerted upon it by other atoms in the crystal. The form of the crystalline potential depends on the type of crystalline symmetry. For some of the most common types of symmetry the terms of lowest order in x, y, z are respectively

axial, tetragonal or hexagonal	$A(x^2 + y^2 - 2z^2)$	(7a)
rhombic	$Ax^2 + By^2 - (A + B)z^2$	(7b)
cubic	$D(x^4 + y^4 + z^4 - \frac{3}{5} r^4)$	(7c)

If the potential satisfies Laplace's equation, the factors A, B, D are constants, but because of charge overlap they can be functions of the radius.

The $4f$ electrons responsible for the magnetism of the rare earths are sequestered in the interior of the atom, and so experience only a small crystalline field. The general formalism which I developed in 1927 and which is displayed in table II shows that it is a good approximation to treat the atom as free provided the decomposition of the energy levels caused by the crystalline field is small compared to kT. This condition is fulfilled fairly well for the rare earths at room temperatures, and explains the success of Hund's theory. At low temperatures inclusion of the crystalline potential is usually imperative, and so Penney utilized it to interpret the existing experimental data mainly by Cabrera and by Becquerel. Fig. 3 is taken from the original paper of Penney and Schlapp (23). The ordinate is the reciprocal of the susceptibility. Hence for Nd^{3+} one expects it to approach zero as $T \to 0$ inasmuch as Nd^{3+} is an ion with an odd number of electrons, and even at $T = 0$ there is still the Kramers degeneracy which implies a first order Zeeman effect and a $1/T$ term in the susceptibility. On the other hand for the even ion Pr^{3+} a sufficiently asymmetrical field should completely lift the degeneracy (case (b) of Table II) and the susceptibility should remain finite as one approaches $T = 0$. This difference is strikingly exhibited in the two sides of Fig. 3.

When applied to the iron group the results of crystal field theory are particularly striking and form the basis of much of what may be called modern magnetochemistry. The crystalline potential is much larger than for the rare earths and is so powerful that it quenches a large part of the orbital part of the magnetic moment even at room temperatures. Schlapp found that the magnetic behavior in the iron group required a large crystalline field of nearly (but usually not entirely) cubic symmetry.

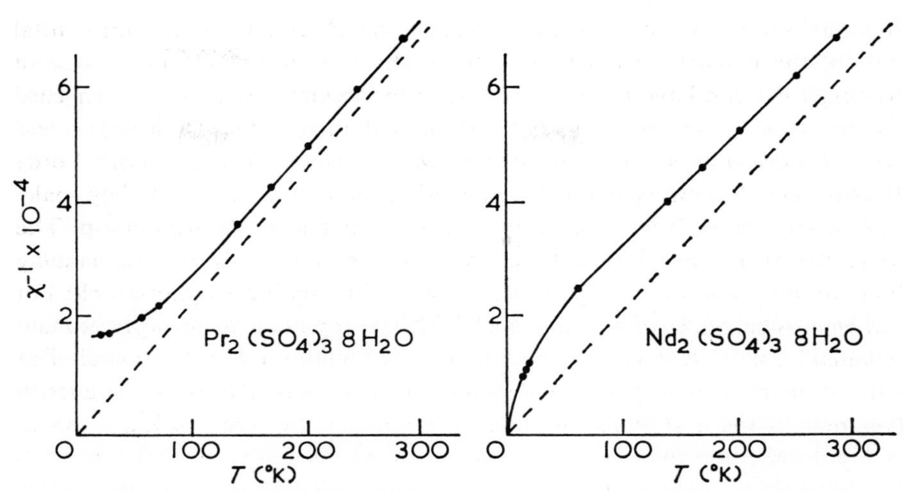

Fig. 3. The reciprocal of the susceptibility as a function of temperature, for two rare earth compounds containing respectively an even and odd number of electrons.

Each time I read the paper of Schapp and Penney (24) I am impressed with how it contains all the essential ingredients of modern crystalline field theory, although there have been changes in the best quantitative estimate of D in (7c). For instance it accounted for the fact that most nickel salts are nearly isotropic magnetically and follow Curie's law down to quite low temperatures, whereas the corresponding cobalt salts are highly anisotropic and deviate greatly from Curie's law. However, for a while we thought that there was a difficulty and inconsistency. Let us focus attention on the ions in F states; e.g. Ni^{++}, Co^{++}. In a nearly cubic field an F state will decompose in the fashion shown in Figure 4. If a non-degenerate level is deepest, as in Figure 4, then the orbital moment is completely quenched, and there should be almost complete isotropy. On the other hand, if Figure 4 is upside down, and if the components a, b, c of the ground level do not coincide because of deviations

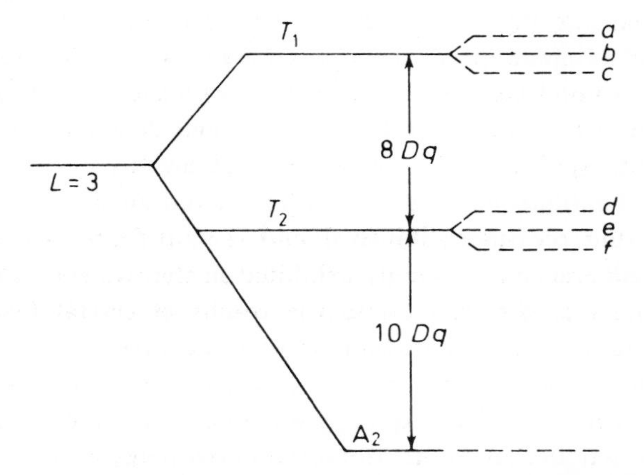

Fig. 4. Orbital energies of an F state in a nearly cubic field The decompositions (a-b-c) and (d-e-f) ensue only because of deviations from cubic symmetry. The quantity Dq is connected with the constant D of (7c) by the relation $Dq = 2\,D\langle r^4\rangle/105$.

from cubic symmetry, and so have different Boltzmann factors, the anisotropy will be considerable. The very different behavior of nickel and cobalt can thus be explained if it supposed that Figure 4 is rightside up for Ni^{++} but is upside down in Co^{++}. The calculations of Schlapp then worked fine. However, this seemed to us for a while a thoroughly dishonest procedure, as it appeared to require a change in the sign of D

Then one day it dawned on me that a simple calculation based on the invariance of the trace shows that the splitting pattern does indeed invert in going from nickel to cobalt even though the constant D is nearly the same.

The article (25) in which I published this result is my favorite of the various papers I've written as it involved only a rather simple calculation, and yet it gave consistency and rationality to the apparently irregular variations in magnetic behavior from ion to ion.

The iron group salts I have discussed are of the 6-coordinated type, e.g. $Co(NH_4)_2(SO_4)_2\,6H_2O$. A simple electrostatic calculation made by Gorter (26) shows that the constant D in (7c) should change sign when the coordination is 4 rather than 6 fold and then Fig. 4 should be upright in Co^{++} and inverted in Ni^{++}. Krishnan and Mookherji (27) in 1937 verified experimentally this theoretical prediction. They prepared some tetracoordinated cobalt compounds, which are a beautiful cobalt blue in color and found that they indeed show very much less anisotropy than do the pink six-coordinated ones.

In 1935 I published a paper (28) in which I amplified and generalized in two respects the primitive crystal field theory employed a few years previously by Penney, Schlapp, and myself. In the first place I showed that Bethe's grouping of energy levels according to symmetry type was still valid even if one allowed the electrons in the unclosed shells to wander away sometimes from the central paramagnetic ion and take a look at the diamagnetic atoms clustered around it. In more technical language, the wave function of the electron has mixed into it small terms which correspond to such excursions. This generalization corresponds to the use of molecular rather than atomic orbitals. Following Ballhausen (29) it is convenient to designate this more general model as ligand rather than crystal field theory, as chemists sometimes refer to the neighboring atoms clustering about the central ion as ligands. The use of ligand in distinction from crystal field theory can also be characterized as making allowance for incipient covalence.

The other modification I made of the conventional theory was to note that under certain conditions, the levels may be split so much by the crystalline field as to break down the Hund rule that the deepest state is that of maximum multiplicity permitted by the Pauli principle. This situation is shown schematically in Fig. 5, which is drawn for the configuration d^6. According to the Hund rule the deepest state is ^5D ($S = 2$) and this necessitates all but one of the five Stark components being singly inhabited, as in the left side of Fig. 5. It is obvious that the energy in the crystalline or ligand field is lower if the three deepest Stark components are doubly populated, with antiparallel spin because of the exclusion principle. However, then the resultant spin is

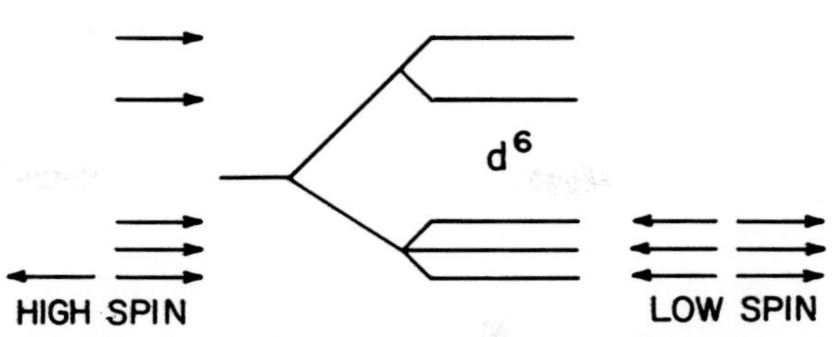

Fig. 5. The central diagram of the figure shows the decomposition of a single 3d level in a field of mainly cubic symmetry. The arrows indicate how the different crystalline field components are filled in case the ion contains six 3d electrons, and also the direction of alignment of each spin. The situation in the left side of the figure represents conformity to the Hund rule, while the right exemplifies what happens when minimization of the energy in the crystalline field is so important as to break down this rule.

only 0, the Russell-Saunders coupling is broken up, and the part of the energy not associated with the crystalline field is raised. The two cases represented by the two sides of Fig. 5 are sometimes referred to as the high and low spin cases. When the susceptibility of a compound is found to conform to the low rather than high spin situation, this is something of interest to chemists. It shows that the inter-atomic bonding is strong, since it is large enough to break down the Hund rule. Beginning with Pauling and Coryell (30) in 1936, this magnetic criterion has even been used to study the chemical behavior of iron in blood. For example, the ferro-haemoglobin ion exhibits high and low spin values 2 and 0 in the presence of H_2O and O_2 molecules respectively. I should by all means mention that prior to my own paper Pauling (31) also stressed the role of covalency effects in magnetism, and the fact that sometimes the low rather than high spin case may be realized. However, in my opinion the method of electron pair bonds which he employed is less flexible and realistic without some modification than is that of molecular orbitals which I used.

On 1937 Jahn and Teller (32) established a remarkable theorem that when in a crystal there is a degeneracy or coincidence of levels for reasons of symmetry, the ligands experience forces which distort the crystalline arrangement, thereby lowering both the symmetry and the energy.

I realized that the Jahn-Teller effect might have an important effect on magnetic susceptibilities, and in 1939 I published a paper on this subject (33). The energetic effect of Jahn-Teller distortions, is very similar to that of molecular vibrations. Consequently I was able to make the calculations which I performed do double duty using them also in connection with the theory of paramagnetic relaxation caused by spin-lattice coupling. The work I have discussed so far all has related mainly to static susceptibilities but when I visited Leiden in 1938, Gorter (34) aroused my interest in the behavior of the susceptibility at radio frequencies and related problems in relaxation. In a landmark pioneer paper written in 1932 Waller (35) showed that there could be a transfer of energy between the magnetic and phonon systems because of the modulation of the dipolar energy by the lattice vibrations, and a little later Heitler and Teller, Fierz, and Kronig (36) showed that there could be a similar relaxation effect, usually of larger magnitude, because of the vibrational modulations of the energy associated with the crystalline potential. I made a more detailed explicit calculation (37) of the numerical values of the relaxation times to be expected for titanium, chronium and ferric ions. On the whole the agreement with experiment was rather miserable. In an attempt to explain away part of the disgreement, I suggested in another paper (38) that there might be what is usually called a phonon bottleneck. The point is that because of the conservation of energy only a portion of all the phonons, those in a narrow frequency range, can exchange energy with the spin or magnetic system. Because of their limited heat capacity, these phonons are easily saturated and brought to the same temperature as the spin system, except insofar as they exchange energy by anharmonic processes coupling them to other oscillators, or transport the excess energy to a surrounding bath that serves as a thermostat. Consequently the relaxation process may be considerably slower than one would calculate otherwise.

This brings me up to the years of world war II, during which very little was done in the way of pure research. Even before the war, the number of physicists interested in magnetism was limited, both because at that time there were few theoretical physicists in the world, and because there were many different fields in which quantum mechanics could be applied. So I seldom ran into problems of duplicating the work of other physicists, except for the calculations with the rotating dipole I mentioned near the beginning of my talk, and some duplication with Kronig on paramagnetic relaxation. As an example of the rather relaxed rate of development I might mention that while the first successful experiments on adiabatic demagnetization were made by Giauque (39) at California in 1933, the first attempt to interpret these experiments in the light of crystal field theory was not until Hebb and Purcell (40) published an article in 1937 which was essentially a term paper in my course in magnetism which had only two students. Shortly after the war, the whole tempo of research in magnetism changed abruptly. The development of radar in the war created apparatus and instruments for microwave spectroscopy, permitting exploration of a spectral low frequency spectral region previously practically untouched. Also infrared and optical spectroscopy of solids was pursued much more vigorously, with improved apparatus. On the theoretical side, crystalline and ligand field calculations were made in various centers, notably in Japan, going into much more detail and lengthly computation than in the work of my group at Wisconsin in the 1930's.

For the rare earths the pre-war period may be described as the era of the rare earth sulphate octohydrates, as the meager magnetic measurements at that time were mainly on these compounds. These materials are particularly annoying as they have a very complicated crystal structure, with eight rare earth ions in the unit cell. However, the x-ray analysis (41) that yielded this disconcerting information had not been made at the time of Penney and Schlapp's work, and so they obtained the theoretical curve shown in Fig. 3, by making faute de mieux the simplifying assumption that the local crystalline field had cubic symmetry, and was the same for all the paramagnetic ions.

Undoubtedly the local potential is more complicated. Even today there have been few attempts to revaluate the crystalline field parameters for sulphate octohydrates, both because of theoretical complexity and the paucity of new experimental data. The most comprehensive crystalline field analysis for rare earth salts in modern times is on the ethyl sulphates ($Re(C_2H_5SO_4)_3\ 9H_2O$), which have only one ion in the unit cell and are magnetically dilute. One important result is that the higher order harmonics in the series development of the crystalline potential are much more important than one thought in the early days. These ethyl sulphates have hexagonal symmetry. Were only second order terms important, the crystalline potentia would be of the simple type (7a), but actually there are also important terms involving fourth and sixth order harmonics, including those of the type $(x \pm iy)^6$. One sometimes worries how meaningful and reliable are the crystalline field parameters deduced from spectroscopic data, but very comforting magnetic measurements have been published by Cooke and collaborators (42). They measured the susceptibility both parallel and perpendicular to the hexagonal axis, and as shown in Fig. 6 found that the experimental results agreed exceedingly well with the theoretical curve calculated with the spectroscopically determined (43) crystalline field parameters.

One of the spectacular developments associated with spectroscopy of the solid state was the first optical laser constructed by Maiman (44) in 1960. By a sheer coincidence it involved transitions between the same ruby energy levels that were interpreted in terms of crystal field theory by Finkelstein and myself (45) in 1940. Cynics can well claim that our theoretical labelling of the energy levels was no more germane to the successful instrumentation of a laser than the prior naming of a star was to astrophysical studies thereof. Still it may be true that any theoretical understanding of the nature and

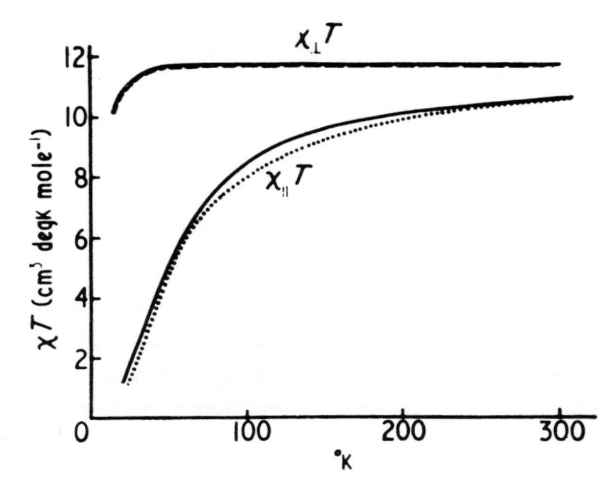

Fig. 6. The product of susceptibility times temperature for erbium ethyl sulphate as a function of temperature for directions parallel and perpendicular to the hexaconal axis. The broken curves represent experimental measurements of the susceptibility by Cooke, Lazemby and Leark, (42) the solid curves are calculated theoretically with the crystalline field parameters of Erath. (43)

relaxation rates of the different energy levels in solids may help the experimentalists a little.

Particularly gratifying to me were the improved determinations of spin-lattice relaxation times made at various laboratories (46). These confirmed the reality of the bottleneck effect. They also verified the proportionality of the relaxation time in a certain temperature range to T^{-9} which I had predicted for salts with Kramers degeneracy and of sufficient magnetic dilution that there is no bottleneck.

The year 1946 brought about the discovery of nuclear magnetic resonance independently by Purcell, Torrey and Pound, and by Bloch, Hansen and Packard (47). I need not tell you how enormously important the field of nuclear magnetism has become both for its basic scientific interest and its surprising technological applications. The nuclear magnetic resonance spectrometer has become a standard tool for any laboratory concerned with analytical chemistry, completely usurping the role of the Bunsen burner in earlier days. Measurements of transferred hyperfine-structure give a quantitative measure of incipient covalence in molecular orbital or ligand field theory. Little of my own research has been concerned with nuclear magnetism, but in 1948 Purcell asked me if I could explain theoretically the size of the line widths he and Pake (48) were observing in the resonance of the F nucleus in CaF_2. It occurred to me that this could be done by applying the method of moments that Waller (35) developed in 1932. The predicted magnitude of the mean square line breadth and its dependence on direction agreed on the whole very well with experiment. The only difference in this calculation (49) of the mean square dipolar broadening as compared with that originally performed by Waller is that he was concerned with the width in a weak magnetic field, whereas in the experiments by Pake and Purcell the dipolar energy is small compared to the Zeeman energy, and this necessitates the truncation of the Hamiltonian function, i.e., the omission of certain terms. A year previously I had also used Waller's method of moments in connection with explaining some apparently anomalous line shapes in some of the Leiden experiments on paramagnetic dispersion. Gorter was a visiting Professor at Harvard in 1947, and one morning we came to the laboratory and discovered that we had both overnight come to the conclusion that the explanation is to be found in an effect now generally known as exchange narrowing. Gorter had reached this conclusion on the basis of an intuitive picture, that the spin waves associated with exchange spoiled the coherence of the dipolar coupling, analogous to the motional narrowing discussed by Bloembergen, Purcell, and Pound in connection with nuclear magnetic resonance in liquids (50). On the other hand I used a more mathematical approach, showing that exchange enhanced the fourth but not the second moment, thereby narrowing the line. The result was a joint paper by Gorter and myself (51).

So far I have not said much about ferromagnetism, partly because more of my own work has been in paramagnetism, but mainly because most ferromagnetic metals are very complicated since they are conductors. Over the years there have been arguments *ad infinitum* as to which is the best model to use, each researcher often pushing his own views with the ardor of a religious zealot (52). Heisenberg's original model (14) was one in which the spins responsible for the ferromagnetism did not wander from atom to atom, whereas in the band picture developed by Stoner (53) the electrons carrying a free spin can wander freely through the metal without any correlation in their relative positions, as the exchange effects are approximated by an uncorrelated molecular field. Undoubtedly the truth is between the two extremes, and I have always favoured as a first approximation a sort of compromise model, which may be called that of minimum polarity (54). In nickel for instance, this model there is continual interchange of electrons between the configurations d^{10} and d^9 but no admixture of d^8, d^7 etc. as then the correlation energy is increased.

Neutron diffraction is a very powerful new tool for disclosing how atomic magnets are arranged relative to each other. It has led to the surprising and spectacular discovery that in certain materials notably rare earth metals, the elementary magnets are arranged in a spiral conical or wavy fashion, rather than pointing all in the same direction within an elementary domain (55). They can be ferromagnetic in one temperature region and antiferromagnetic in another. This weird kind of magnetism is sometimes called helical magnetism. Most rare earth metals belong to this category and the mathematical interpretation of the experimental results is complicated and difficult despite the fact that the 4*f* electrons participate but little in electrical conductivity, unlike the 3*d* electrons in iron or nickel. I have not been involved in any of the theoretical analysis except for a point connected with the magnetic anisotropy. When I attended the conference on quantum chemistry sponsored by Professor Lowdin in Florida in 1971, Bozorth presented some measurements

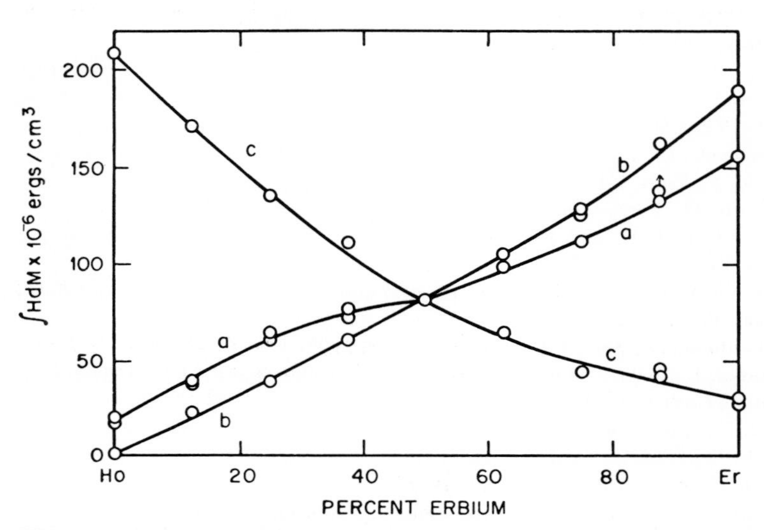

Fig. 7. The energy of magnetization for various amounts of Ho relative to Er. The three curves are for three different directions and would coincide were there is no magnetic anisotropy. The latter is measured by the differences between the ordinates of the three curves.

of the ferromagnetic anisotropy of Ho-Er alloys. He found that the anisotropy of pure holmium was approximately the negative of that of erbium, and vanished when there was an equal amount of Ho and Er, as shown on Fig. 7. It finally occurred to me that precisely the same property of spherical harmonics that explained the inversion of Fig. 4 in passing from Co^{++} to Ni^{++} also explained (56) the inversion of the anisotropy of Ho as compared to Er, with the obvious corollary that the Ho and Er contributions should cancel each other out for a 50% mixture. So sometimes primitive theory can still be useful, but in general a higher degree of mathematical sophistication is required as time progresses, and as more and more exotic magnetic phenomena are discovered by the experimentalists. This you will learn from the addresses by Anderson and Sir Neville Mott but one can still say that quantum mechanics is the key to understanding magnetism. When one enters the first room with this key there are unexpected rooms beyond, but it is always the master key that unlocks each door.

REFERENCES

1. Langevin, P. J. de Physique *4,* 678 (1905), Annales de Chimie et Physique, *5,* 70 (1905).
2. Curie, P., Ann. Chim. Phys. *5,* 289 (1895), Oeuvres, p. 232.
3. Weiss, P., J. de Physique *6,* 667 (1907).
4. Bohr, N., Dissertation, 1911; reprinted in vol I of his collected works. The vanishing of the susceptibility in classical statistics was also proved independently with a slightly different method by Miss J. H. van Leeuwen, Dissertation, Leiden 1919, J. de Physique *2,* 361 (1921); the two methods of proof are summarized in p. 94 of ref. 13.
5. Gerlach, W. and Stern, O., Zeit. f. Physik *9,* 349 (1922).
6. Mensing, L. and Pauli, W., Phys. Zeit. *27,* 509 (1926); R. de L. Kronig, Proc. Nat. Acad. Sci. *12,* 488 (1926), C. Manneback Phys. Zeits. *27* 563 (1926), J. H. Van Vleck, Nature, *118,* 226 (1926).
7. Van Vleck, J., Nature, May 7, 1927. Phys. Rev. *29,* 727 (1927); *31,* 587 (1928).
8. Bitter, F., Proc. Nat. Acad. *15* 632 (1929); Aharoni and Scherrer Zeits. f. Physik *58,* 749 (1929); Wiersma, de Haas and Capel, Leiden Communications 212b.
9. Havens, G. G., Phys. Rev. *43,* 992 (1932).
10. Hund, F., Zeits. f. Physik *33,* 855 (1925).
11. Laporte, O., Zeits. f. Physik *47,* 761 (1928).
12. Van Vleck, J. H. and Frank, A., Phys. Rev. *34,* 1494 and 1625 (1929) Frank, A., Phys. Rev. *39,* 119 (1932); 48, 765 (1935).
13. Van Vleck, J. H., the Theory of Electric and Magnetic Susceptibilities, Oxford University Press, 1932.
14. Heisenberg, W., Zeits. f. Physik *49,* 619 (1928).
15. Landau, L., Zeits. f. Physik *64,* 629 (1930, E. Teller, ibid 67, 311 (1931).
16. Pauli, W., Zeits. f. Physik *41,* 81 (1927).
17. Néel, L., Ann. de Physique *17,* 64 (1932), *5,* 256 (1936).
18. Peirls, R., Zeits. f. Physik *81,* 186 (1933).
19. Bloch, F., Zeits. f. Physik *74,* 295 (1932).
20. Landau, L. and Lifshitz, E., Phys. Zeits. d. Sowjetsunion *8,* 153 (1935).
21. Kramers, H. A., Proc. Amsterdam Acad. *33,* 959 (1930) or collected works, p. 522.
22. Bethe, H., Ann. der Physik *3,* 133 (1929).
23. Penney, W. G. and Schlapp, R., Phys. Rev. *41,* 194 (1932).
24. Schlapp, R. and Penney, W. G., Phys. Rev. *42,* 666, (1932).

25. Van Vleck, J. H., Phys. Rev. *41*, 208 (1932).
26. Gorter, C. J., Phys. Rev. *42*, 487 (1932).
27. Krishnan, K. S. and Mookherji, A., Phys. Rev. *51*, 428 and 774 (1937).
28. Van Vleck, J. H., J. Chem. Physics *3*, 807 (1925).
29. Ballhausen, C. J., *Introduction to Ligand Field Theory*, (McGraw-Hill, 1962). This volume is recommended as an excellent survey of the subject, including comparison with experiment.
30. Pauling, L. and Coryell, C. D., Proc. Nat. Acad. *22*, 159 and 210 (1936).
31. Pauling, L., J. Amer. Chem. Soc. *53*, 1367 (1931).
32. Jahn, H. A. and Teller, E., Proc. Roy. Soc. *161*, 220, (1937).
33. Van Vleck, J. H., J. Chem. Phys. *7*, 61 and 72 (1939)
34. Gorter, C. J., Physica *3*, 503 (1936) and other later papers in that journal, also his book "Paramagnetic Relaxation" (Elsevier, 1947).
35. Waller, I., Zeits. f. Physik, *79*, 370 (1932).
36. Heitler, W. and Teller, E., Proc. Roy. Soc. *155*, 629 (1936); Fierz, M., Physica *5*, 433 (1938), R. de L. Kronig, ibid *6*, 33 (1939).
37. Van Vleck, J. H., Phys. Rev. *57*, 426 and 1052 (1940).
38. Van Vleck, J. H., Phys. Rev. *59*, 724 and 730 (1940).
39. Giauque, W. F. and Mac Dougall, D. P., Phys. Rev. *43*, 768 (1933) *47*, 885 (1935); F. Simon p. 763 Nature (1935).
40. Purcell, E. M. and Hebb, M. H., J. Chem. Physics, *5* 338 (1937).
41. Zachariasen, W. H., J. Chem. Phys. *3*, 197 (1935).
42. Cooke, A. H., Lazenby, R. and Leask, M. J., Proc. Phys. Soc. London *85*, 767 (1965).
43. Erath, E. H., J. Chem. Phys. *34*, 1985 (1961).
44. Maiman, T. H., Nature 187, 493 (1960).
45. Finkelstein, R. and Van Vleck, J. H., J. Chem. Physics *8*, 790 (1940); Van Vleck, J. H., ibid, *8*, 787 (1940). These papers relate to chrome alum whereas Maiman used chromium embedded in Al_2O_3, but the spectroscopic properties of the chromium ion are similar in the two cases, cf. p. 238 of Ballhausen, ref. 29.
46. See especially Scott, P. L. and Jeffries, C. D., Phys. Rev. *127*, 32 (1962), Ruby, R. R., Benoit, H. and Jeffries, C. D., ibid *127*, 51 (1962).
47. Purcell, E. M., Torrey, H. C. and Pound, R. V., Phys. Rev. *69*, 37 (1946); Bloch, F., Hansen, W. W. and Packard, M., ibid *69*, 127 (1946).
48. Pake, J. E. and Purcell, E. M., Phys. Rev. *74*, 1184 and *75*, 534 (1948); see also ref. 50.
49. Van Vleck, J. H., Phys. Rev. *74*, 1168 (1948).
50. Bloembergen, N., Purcell, E. M. and Pound, R. V., Phys. Rev. *73*, 679 (1948).
51. Gorter, C. J. and Van Vleck, J. H., Phys. Rev. *72*, 1128 (1947); also ref. 49.
52. For a very complete review of all the different models and their limitations see Herring, C., Vol. IV of *Magnetism* (edited by Rado, J. E. and Suhl, H.) Academic Press, 1966.
53. Stoner, E. C., Phil. Mag. *21*, 145 (1936); Proc. Roy. Soc. A *165*, 372 (1938); A. *169*, 339 (1939).
54. Van Vleck, J. H., Rev. Mod. Phys. *25*, 220 (1953), also pages 475—484 of *Quantum Theory of Atoms, Molecules and the Solid State* (edited by P. Lowdin) Academic Press, 1966.
55. For an excellent discussion of the theory of the magnetic ordering in rare earth metals see the chapter by R. J. Elliott in Vol IIa of *Magnetism* (Edited by Rado, J. E. and Suhl, H), Academic Press, 1966.
56. Bojorth, R. M., Clark, A. E., and Van Vleck, J. H., Intern. J. Magnetism, 2, 19 (1972).

Philip W. Anderson

PHILIP W. ANDERSON

My father, Harry Warren Anderson, was a professor of plant pathology at the University of Illinois in Urbana, where I was brought up from 1923 to 1940. Although raised on the farm—my grandfather was an unsuccessful fundamentalist preacher turned farmer—my father and his brother both became professors. My mother's father was a professor of mathematics at my father's college, Wabash, in Crawfordsville, Indiana, and her brother was a Rhodes Scholar, later a professor of English, also at Wabash College; on both sides my family were secure but impecunious Midwestern academics. At Illinois my parents belonged to a group of warm, settled friends, whose life centered on the outdoors and in particular on the "Saturday Hikers", and my happiest hours as a child and adolescent were spent hiking, canoeing, vacationing, picnicking, and singing around the campfire with this group. They were unusually politically conscious for that place and time, and we lived with a strong sense of frustration and foreboding at the events in Europe and Asia. My political interests were later strengthened by the excesses in the name of "security" and "loyalty" of the "McCarthy" years, to the extent that I have never accepted work on classified matters and have from time to time worked for liberal causes and against the Vietnam war.

Among my parents' friends were a number of physicists (such as Wheeler Loomis and Gerald Almy) who encouraged what interest in physics I showed. An important impression was my father's one Sabbatical year, spent in England and Europe in 1937. I read voraciously, but among the few intellectual challenges I remember at school was a first-rate mathematics teacher at the University High School, Miles Hartley, and I went to college intending to major in mathematics. I was one of several students sent to Harvard from Uni High in those years on the new full-support National Scholarships. The first months at Harvard were more than challenging, as I came to the realization that the humanities could be genuinely interesting, and, in fact, given the weaknesses of my background, very difficult. Nonetheless in time I relaxed and enjoyed the experience of Harvard, and was in the end pleasantly surprised to come out with a good record.

In those wartime years (1940—43) we were urged to concentrate in the immediately applicable subject of "Electronic Physics" and I was then bundled off to the Naval Research Laboratory to build antennas (1943—45). (It may be remembered that such war work was advisable for those of us who wore glasses, the "services" at that time being convinced that otherwise we would be best utilized as infantry.) This work left me with a lasting admiration for Western Electric equipment and Bell engineers, and for the competence of my former physics (not electronics) professors at Harvard; after the war, I went back to learn what the latter could teach me.

Graduate school (1945—49) consisted of excellent courses; a delightful group of friends, including for instance Dave Robinson and Tom Lehrer, centered around bridge, puzzles, and singing; a happy decision that Schwinger and Q.E.D. would lead only to standing in the long line outside Schwinger's office, whereas van Vleck, whom I already knew from undergraduate school and a wartime incident, seemed to have time to think about what I might do; meeting and marrying one summer the niece of old family friends, Joyce Gothwaite, and therefore settling down to work on my problem. Further motivation was provided by the birth of a daughter, Susan. When I did settle down, I rather suddenly came to realize that the sophisticated mathematical techniques of modern quantum field theory which I was learning in advanced courses from Schwinger and Furry were really genuinely useful in the experimental problem of spectral line broadening in the new radio-frequency spectra, just then being exploited because of wartime electronics advances. Although I didn't know it, across the world—in England with Fröhlich and Peierls, in Princeton with Bohm and later Pines, and in Russia with Bogoliubov and especially Landau—the new subject of many-body physics was being born from similar marriages of maturing mathematical techniques with new experimental problems.

In spite of a number of contretemps, with the help of Van and of an understanding recruiter, Deming Lewis, who seemed to be the only person who believed me when I said I *had* solved my problem and wanted to do something else, I got to Bell Laboratories to work with the constellation of theorists who were then there: Bill Shockley, John Bardeen, Charles Kittel, Conyers Herring, Gregory Wannier, Larry Walker, John Richardson, and later others. Kittel in particular fostered my interest in linebroadening problems and introduced Wannier and me to antiferromagnetism, while Wannier taught me many fundamental techniques, and Herring put me in touch with the ideas of Landau and Mott and kept us all abreast of the literature in general. I learned crystallography and solid state physics from Bill Shockley, Alan Holden, and Betty Wood. And I learned most of all the Bell mode of close experiment-theory teamwork—at first with Jack Galt, Bill Yager, Bernd Matthias, and Walter Merz.

Much of the rest is a matter of record. One important experience was Ryogo Kubo's convincing the Japanese in 1952 that they should invite as their first Fulbright scholar in physics an unknown 28-year-old. This Sabbatical was postponed to 1953, the year of the Kyoto International Theoretical Physics Conference, which was dominated by Mott as the president of IUPAP, and was my first meeting with many other friends of later years. Lecturing has never come easily to me, but I gave, as best I could, lectures on magnetism and a seminar on linebroadening which included Kubo, Toru Moriya, Kei Yosida, Jun Kanamori, among other wellknown Japanese solid staters. I acquired an admiration for Japanese culture, art, and architecture, and learned of the existence of the game of GO, which I still play.

Another milestone for me was a year at the Cavendish Laboratory and Churchill College (1961—62), which was not at Oxford because Brian Pippard promised me that I could lecture and that the lectures would be attended. Mott kept asking me what my 1958 paper meant, and there were a lot of discussions centered around broken symmetry and some ideas of Brian Josephson, who attended my lectures.

When he left Princeton for Illinois in 1959, David Pines bequeathed me a French student named Pierre Morel; Morel and I worked in 1959—61 on some unconventional ideas on anisotropic superfluidity I had, which became related to He_3 by discussions with Keith Brueckner; later we worked on solving the Eliashberg equations for superconductivity. Some of these ideas came to fruition working with a young experimentalist, John Rowell, on my return to Bell: we discovered the Josephson effect and worked on "phonon bumps".

In 1967 Nevill Mott managed what must have been a most difficult arrangement to steer through the Cambridge system: a permanent "Visiting Professorship" for two terms out of three at the Cavendish. This arrangement would have been totally impossible without the self-effacing and unsparing cooperation of Volker Heine who joined with me in leading the "TCM Group" (Theory of Condensed Matter) for eight productive and exciting years, spiced with warm encounters with students, visitors and associates from literally the four corners of the earth. One of our brainchildren is a still viable Science and Society course. Through the good offices of John Adkins, Jesus College gave me a Fellowship for this period. A souvenir of those years is a small cottage on the cliffs of Cornwall, where Joyce and I spend a spring month every year, hiking and seeing friends. After eight years the sense of being tourists in each of two cultures, with no really satisfactory role in either, led us reluctantly to return to the United States, and in 1975 the job at Cambridge was replaced with a half-time appointment at Princeton.

The years since the Nobel prize have been productive ones for me. For instance, in 1978, shortly after receiving the prize in part for localization theory, I was one of the "Gang of Four" (with Elihu Abrahams, T.V. Ramakrishnan, and Don Licciardello) who revitalized that theory by developing a scaling theory which made it into a quantitative experimental science with precise predictions as a function of magnetic field, interactions, dimensionality, etc.; a major branch of science continues to flow from the consequences of this work. (Most recently, "photon localization" has been in th news.)

In 1975 S. F. (now Sir Sam) Edwards and I wrote down the "replica" theory of the phenomenon I had earlier named "spin glass", followed up in '77 by a paper of D. J. Thouless, my student Richard Palmer, and myself. A brilliant further breakthrough by G. Toulouse and G. Parisi led to a full solution of the problem, which turned out to entail a new form of statistical mechanics of wide applicability in fields as far apart as computer science, protein folding, neural networks, and evolutionary modelling, to all of which directions my students and/or I contributed. The field of quantum valence fluctuations was another older interest which became much more active during this period, partly as a consequence of my own efforts.

Finally, in early 1987 the news of the new "high-T_c" cuprate superconductors galvanized the world of many-body quantum physics, and led many of us to reexamine older ideas and dig for new ones. Putting together a cocktail of older ideas of my own (the "RVB" singlet pair fluid state) and of many others, mixed with brand new insights, I have been able to arrive at an account of most of the wide variety of unexpected anomalies observed in these materials. The theory involves a new state of matter (the two-dimensional "Luttinger liquid") and a quite new mechanism for electron pairing ("deconfinement"). Experimental confirmations of the predictions of this theory are appearing regularly.

The prize seemed to change my professional life very little. Management chores at AT&T Bell Labs continued and culminated in an informal arrangement as consultant for the new Vice President of Research, Arno Penzias, during the first two years of his tenure, which coincided with the first difficult years of "divestiture" for the AT&T company. I thereupon gratefully retired in 1984 from Bell and am now full-time Joseph Henry Professor of Physics at Princeton. I served a 5-year stint as Chairman of the Board of the Aspen Center for Physics, retiring 3 years ago, and for 4 years was on the Council and Executive Committee of the American Physical Society. Since 1986 or so I have been deeply involved (though officially I am merely a co-vice-chairman) with a new, interdisciplinary institution, the Sante Fe Institute, dedicated to emerging scientific syntheses, especially those involving the sciences of complexity. Two other Nobelists are involved: Murray Gell-Mann, who is our science board chairman and an eloquent spokesperson for our ideas and ideals; and Ken Arrow, with whom I co-chaired the workshops founding our interdisciplinary study of the bases of economic theory. My own work in spin glass and its consequences has formed some of the intellectual basis for these interests.

The Nobel prize gives one the opportunity to take public stands. I happened to be in a position to be caught up in the campaign against "Star Wars" very early (summer '83) and wrote, spoke and testified repeatedly, with my finest moment a debate with Secretary George Schultz in the Princeton Alumni Weekly, reprinted in *Le Monde* in 1987. I have also testified repeatedly and published some articles in favor of Small Science.

Some further honors after the Nobel prize of which I am particularly conscious were the National Medal of Science; an ScD from my father's, mother's, sister's and wife's Alma Mater, the University of Illinois; foreign membership in the Royal Society, the Accademia Lincei, and the Japan Academy; and honorary fellowship of Jesus College, Cambridge.

We have kept our cottage on the cliffs of Cornwall, and our custom of seeing English and other friends in April there. We abandoned our much loved house, designed by Joyce, in New Vernon near Bell Labs for another of her good designs on some brushy acres with a view across the Hopewell

Valley near Princeton. Susan is established as a painter in Boston of, at the moment, primarily scenes of Martha's Vineyard, and teaching some drawing classes at MIT. A prize of which I am, vicariously, enormously proud is the designation as Northeast U.S. Tree Farmers of the Year earned by my sister and her husband of New Milford, Pa in 1990.

VITA

P. W. ANDERSON

Birth Date: December 13, 1923, Indianapolis, Indiana. B.S. summa cum laude Harvard, '43; MA '47, PhD '49, MA (Cantab) '67. M. 1947, Joyce Gothwaite, d. Susan Osborne b. 1948. Naval Research Lab. 1943—5 (Chief Spec. X, USNR 1944—5) Bell Tel Labs. 1949—53, 1954—; Memb. Tech. Staff to 1959, Dept. Head 1959—61, MTS 1961—74, Asst. Dir. 1974—76, Consult. Dir. 1977—. Fulbright Lecturer, Tokyo, 1953—4; Visiting Fellow, Churchill College, Cambridge, 1961—2; O.E. Buckley Prize A.P.S. 1964; Am. Acad. A&S 1966, N.A.S. 1967; Visiting Prof. Theor. Phys. U. of Cambridge 1967—75, Fellow Jesus College 1969—75; Prof. of Physics, Princeton U. 1975—; Dannie Heinemann Prize, Gottingen Acad. Sci. 1975.

LOCAL MOMENTS AND LOCALIZED STATES

Nobel Lecture, 8 December, 1977
by
PHILIP W. ANDERSON
Bell Telephone Laboratories, Inc, Murray Hill, New Jersey, and Princeton University, Princeton, New Jersey, USA

I was cited for work both in the field of magnetism and in that of disordered systems, and I would like to describe here one development in each field which was specifically mentioned in that citation. The two theories I will discuss differed sharply in some ways. The theory of local moments in metals was, in a sense, easy: it was the condensation into a simple mathematical model of ideas which were very much in the air at the time, and it had rapid and permanent acceptance because of its timeliness and its relative simplicity. What mathematical difficulty it contained has been almost fully cleared up within the past few years.

Localization was a different matter: very few believed it at the time, and even fewer saw its importance; among those who failed to fully understand it at first was certainly its author. It has yet to receive adequate mathematical treatment, and one has to resort to the indignity of numerical simulations to settle even the simplest questions about it. Only now, and through primarily Sir Nevill Mott's efforts, is it beginning to gain general acceptance.

Yet these two finally successful brainchildren have also much in common: first, they flew in the face of the overwhelming ascendancy at the time of the band theory of solids, in emphasizing *locality*: how a magnetic moment, or an eigenstate, could be permanently pinned down in a given region. It is this fascination with the local and with the failures, not successes, of band theory, which the three of us here seem to have in common. Second, the two ideas were born in response to a clear experimental signal which contradicted the assumptions of the time; third, they intertwine my work with that of my two great colleagues with whom I have been jointly honored; and fourth, both subjects are still extremely active in 1977.

I. The "Anderson Model": Local Moments in Metals

To see the source of the essential elements of the model I set up for local moments in metals, it will help to present the historical framework. Just two years before, I had written a paper on "superexchange" (1) discussing the source and the interactions of the moments in insulating magnetic crystals such as MnO, $CuSO_4 \cdot 5H_2O$, etc. I had described these substances as what we should now call "Mott insulators" on the insulating side of the Mott transition, which unfortunately Sir Nevill says he will not describe. Briefly, following a suggestion of Peierls, he developed the idea that these magnetic insulating salts were so because to create an ionized electronic excitation would require an additional excitation energy U, the energy necessary to change the configurations of two distant atoms from d^n+d^n to $d^{n-1}+d^{n+1}$. This energy U

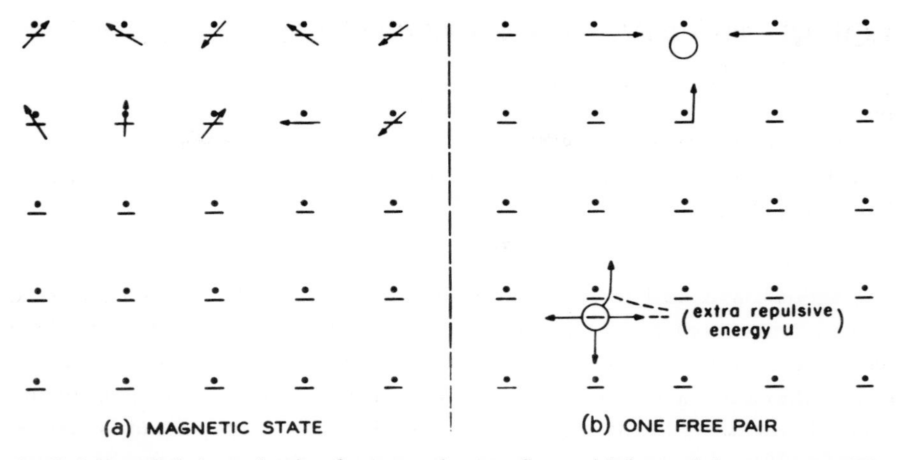

Fig. 1. Mott—Peierls mechanism for magnetic state. State with free pair has extra repulsive energy "U" of two electrons on same site.

is essentially the Coulomb repulsive energy between two electrons on the same site, and can be quite large (see Fig. 1). To describe such a situation, I set up a model Hamiltonian (now called the "Hubbard" Hamiltonian).

$$H = \sum_{i,j,\sigma} b_{ij}\, \overset{+}{c_{i\sigma}}\, c_{j\sigma} + \sum_i U n_{i\uparrow} n_{i\downarrow} \tag{1}$$

Here b_{ij} represents the amplitude for the electron to "hop" from site to site—such hops as shown in Fig. 1, right half—and U the repulsion energy between two opposite spin electrons on the same site (parallel, of course, being excluded). With (1)—appropriately generalized—it was possible to understand the predominantly antiferromagnetic interactions of the spins in these Mott insulators, which include the ancient "lodestone" or magnetite, as well as the technically important garnets and ferrites. These interactions are caused by the virtual hopping of electrons from a site to its neighbor and return, which is only

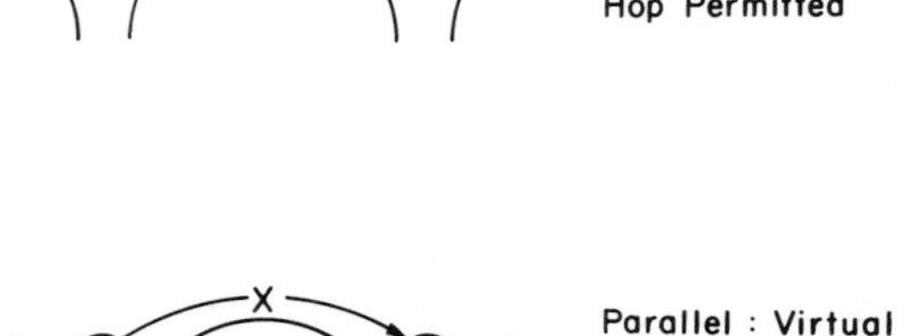

Fig. 2. Virtual hopping as the origin of superexchange.

possible for antiferromagnetism, where the requisite orbital is empty. From simple perturbation theory, using this idea,

$$\mathcal{J}_{ij} = \frac{-2b^2{}_{ij}}{U} \tag{2}$$

where b represents the tendency of electrons to hop from site to site and form a band. (The provenance of (2) is made obvious in Fig. 2.) In fact, I showed later in detail (2) how to explain the known empirical rules describing such interactions, and how to estimate parameters b and U from empirical data.

The implications for magnetism in metals—as opposed to insulators--of this on-site Coulomb interaction U were first suggested by Van Vleck and elaborated in Hurwitz' thesis (3) during the war, and later in a seminal paper which I heard in 1951, published in 1953 (4). Also, very influential for me was a small conference on magnetism in metals convened at Brasenose College, Oxford, September 1959, by the Oxford-Harwell group, where I presented some very qualitative ideas on how magnetism in the iron group might come about. More important was my first exposure to Friedel's and Blandin's ideas on resonant or virtual states (5, 6) at that conference. The essence of Friedel's ideas were 1) that impurities in metals were often best described not by atomic orbitals but by scattering phase shifts for the band electrons, which would in many cases be of resonant form; 2) that spins in the case of magnetic impurities might be described by spin-dependent scattering phase shifts.

Matthias and Suhl, at Bell, were at that time much involved in experiments and theory on the effect of magnetic scatterers on superconductivity (7). For many rare earth atoms, the decrease in T_c due to adding magnetic impurities is clear and very steep; (see Fig. 3a), and even steeper for most transition metal impurities. For instance, Fe at the 10^{-5} level completely wipes out superconductivity in Mo. But in many other cases, e.g., Fe in Ti, a nominally magnetic atom had no effect, or raised T_c (as in Fig. 3b). A systematic study of the occurrence of moments was carried out by Clogston *et al* (8). As yet, no real thought (except see Ref. (6)) had been given to what a magnetic moment in a metal *meant*: the extensive investigations of Owen *at al* (9) and of Zimmermann (10), for instance, on Mn in Cu, and the Yosida calculation (11), essentially postulated a local atomic spin given by God and called S, connected to the free electrons by an empirical exchange integral $\mathcal{J}$; precisely what we now call the "Kondo Hamiltonian":

$$H = \sum_{k\sigma} \varepsilon_k \, n_{k\sigma} + \mathcal{J} \, \vec{S} \cdot \vec{s} \tag{3}$$

$$\text{where } s = \sum c^{+}_{k\sigma} \, \vec{\sigma}_{\sigma\sigma'} \, c_{k'\sigma'}$$

is the local spin density of free electrons at the impurity.

The "Anderson model" (12) is the simplest one which provides an electronic mechanism for the existence of such a moment. We insert the vital on-site exchange term U, and we characterize the impurity atom by an additional orbital φ_d, with occupancy $n_{d\sigma}$ and creation operator $c^{+}{}_{d\sigma}$, over and above

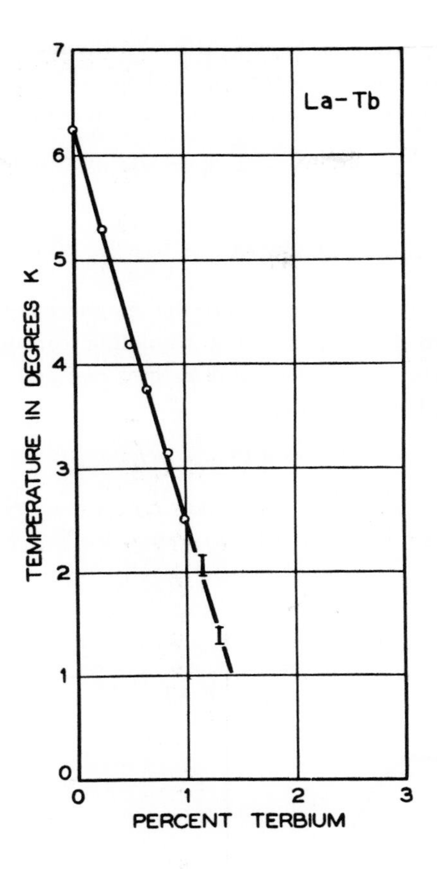

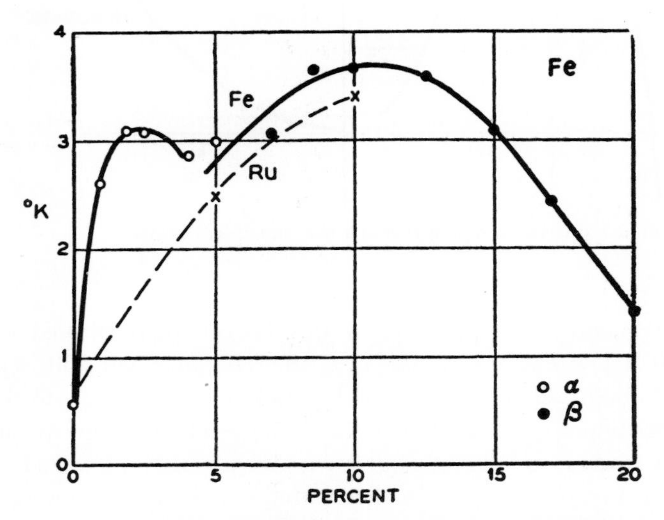

Superconducting transition temperatures of iron or ruthenium solid solutions in titanium.

Fig. 3. Effect of magnetic impurities on T_c of a superconductor (a); when nonmagnetic T_c goes up (b).

the free electron states near the Fermi surface of the metal (the obvious over-completeness problem is no real difficulty, as I showed later (13). The physics should be clear by reference to Fig. 4. The Hamiltonian is

$$H = \sum_{k\sigma} \varepsilon_k n_{k\sigma} + U n_{d\uparrow} n_{d\downarrow} + E_d (n_{d\uparrow} + n_{d\downarrow}) \qquad (4)$$

$$+ \sum_{k\sigma} V_{dk} (c^{+}_{d\sigma} c_{k\sigma} + cc)$$

where in addition to free electrons and the magnetic term U, we have a d-to-k tunneling term V_{dk} representing tunneling through the centrifugal barrier which converts the local orbital φ_d into one of Friedel's resonances. The resonance would have a width

$$\Delta = \pi < V^2_{dk} > \varrho(E_d) \qquad (5)$$

and in the absence of U would be centered at E_d, the energy of the d resonance (if the density of states ϱ is sufficiently constant—see Fig. 4 again).

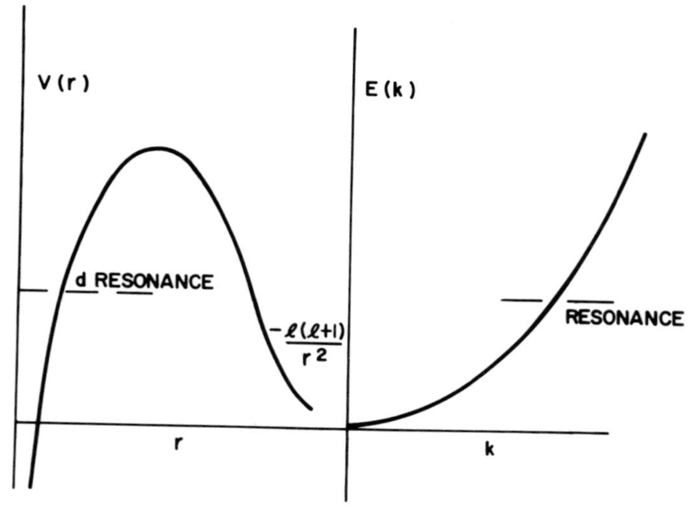

Fig. 4. d-Resonance due to tunneling through the centrifugal barrier.

A simple Hartree-Fock solution of this Hamiltonian showed that if E_d is somewhat below E_F, and if $\Delta/U < \pi$, the resonance will split as shown in Fig. 5 (from the original paper). One has two resonances; one for each sign of spin, a mostly occupied one below the Fermi level and a mostly empty one above. This leads to a pair of equivalent magnetically polarized solutions, one for each direction of spin. In these solutions, the local state φ_d is mixed into scattered free-electron states: there are no local bound electronic states, but there *is* a local moment. Again, in Hartree-Fock theory, the magnetic region is shown in Fig. 6. The parameters could be estimated from chemical data or from first principles, and it was very reasonable that Mn or Fe in polyelectronic metals should be non-magnetic as was observed, but magnetic in, for instance, Cu.

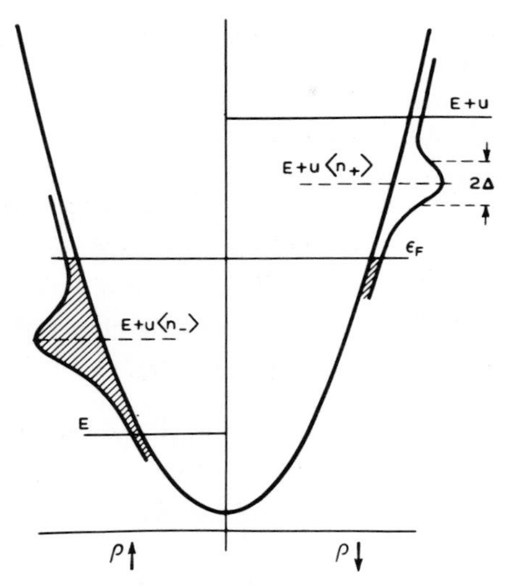

Fig. 5. Spin-split energy levels in the magnetic case.

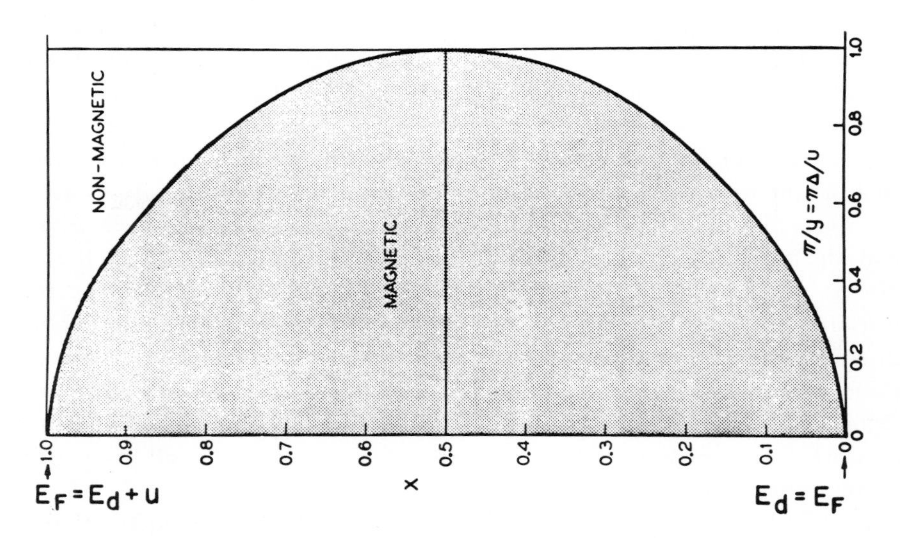

Fig. 6. Magnetic region of parameter space in the "Anderson Model".

This seems and is a delightfully simple explanation of a simple effect. The mathematics is shamelessly elaborated (or simplified) from nuclear physics (Friedel's improvements on Wigner's theory of resonances) and similar things occur in nuclear physics called "analog resonances". Nonetheless, it has led to an extraordinary and still active ramification of interesting physics.

Before discussing some of these branchings, let me say a bit about the model's simplicity, which is to an extent more apparent than real. The art of model-building is the exclusion of real but irrelevant parts of the problem, and entails hazards for the builder and the reader. The builder may leave out something genuinely relevant; the reader, armed with too sophisticated an experimental probe or too accurate a computation, may take literally a schematized model whose main aim is to be a demonstration of possibility. In this case, I have left out (1) the crystal structure and in fact the atomic nature of the background metal, which is mostly irrelevant indeed. (2) The degeneracy of the *d* level, which leads to some important physics explored in an Appendix of the paper and later and much better by Caroli and Blandin (14). In the Appendix I showed that if the resonance was sufficiently broad compared to other internal interactions of the electrons in the *d* orbitals, the different *d* orbitals would be equally occupied as is usually observed for transition metal impurities; in the opposite case the orbital degrees of freedom will be "unquenched", as is almost always the case for rare earth atoms. (3) Left out are all correlation effects except U; this relies on the basic "Fermi liquid" idea that metallic electrons behave as if free, but detaches all parameters from their values calculated naively: they are *renormalized*, not "bare" parameters. This is the biggest trap for the unwary, and relies heavily on certain fundamental ideas of Friedel on scattering phase shifts and Landau on Fermi liquids. I have also left out a number of real possibilities some of which we will soon explore.

One of my strongest stylistic prejudices in science is that many of the facts Nature confronts us with are so implausible given the simplicities of non-relativistic quantum mechanics and statistical mechanics, that the mere demonstration of a reasonable mechanism leaves no doubt of the correct explanation. This is so especially if it also correctly predicts unexpected facts such as the correlation of the existence of moment with low density of states, the quenching of orbital moment for all *d*-level impurities as just described, and the reversed free-electron exchange polarization which we shall soon discuss. Very often such a simplified model throws more light on the real workings of nature than any number of "ab initio" calculations of individual situations, which even where correct often contain so much detail as to conceal rather than reveal reality. It can be a disadvantage rather than an advantage to be able to compute or to measure too accurately, since often what one measures or computes is irrelevant in terms of mechanism. After all, the perfect computation simply reproduces Nature, does not explain her.

To return to the question of further developments from the model: I should like to have had space to lead you along several of them. Unfortunately, I shall not, and instead, I shall show you a Table of the main lines, and then follow one far enough to show you an equation and a picture from the recent literature.

The one of these lines I would like to take time to follow out a bit is the "model" aspect I. This started as a very physical question: what is the sign and magnitude of the spin-free electron interaction? Already in '59 *before* the model appeared, I made at the Oxford Discussion a notorious bet of one pound with (now Sir) Walter Marshall that the free-electron polarization caused by the spins in metals would be negative, for much the same reason as in

Table 1: Ramifications of the Anderson Model

I. AM as an exact field-theoretic model—see text:
- a) AM = Kondo; Anderson, Clogston, Wolff, Schrieffer
- b) Fundamental difficulties of both: Alexander, Schrieffer, Kondo, Suhl, Nagaoka, Abrikosov
- c) Solution of Kondo: PWA, Yuval, Hamann, Yosida, Wilson, Nozieres, etc.
- d) Solution of AM: Hamann, Wilson, Krishna-Murthy, Wilkins, Haldane, Yoshida, etc.

II. "Microcosmic" view of magnetism in metals; interacting AM's and rules for alloy exchange interactions, Alexander, PWA (15), Moriya (16)

III. Applications to Other Systems
- a) Adatoms and molecules on surfaces, Grimley (17), Newns (18), etc.
- b) Magnetic impurities in semiconductors, Haldane (19)
- c) With screening + phonons, $-U$: mixed valence, surface centers, etc., Haldane (20). The sky seems to be the limit.

superexchange: the occupied spin state *below* the Fermi level is repulsive, that above is attractive because it can be occupied by the free electrons of the same spin. Clogston and I published this for the Anderson model (21). This was formalized by Peter Wolff, and published later with Schrieffer (22), into a perturbative equivalence of "Kondo" and "Anderson" models with the exchange integral $\mathcal{J}$ of (3) being

$$\mathcal{J} = \frac{2\Delta}{\pi}\left(\frac{1}{E_d} - \frac{1}{E_d+U}\right) \tag{6}$$

$$(\Delta = \pi <|V_{dk}|^2> \varrho(E_F))$$

Soon, however, it came to be realized that neither Kondo nor Anderson models behaved reasonably at low temperatures (Kondo (23), Suhl (24), Schrieffer (25), etc.), but exhibited nasty divergences at low temperatures which seemed to signal disappearance of the local moment. The best physical description of what happens (for a more extensive review for nonspecialists perhaps my series of papers in Comments on Solid State Physics will suffice) is that at high temperatures or on high energy (short time) scales, the Hartree-Fock theory given above is correct, and there is a free spin. But as the energy scale is lowered, the effective antiferromagnetic coupling between this spin and the free-electron gas "bootstraps" itself up to a very large value, eventually becoming strong enough to bind an antiparallel electron to it and become non-magnetic. This is a very precise analog of the process of continuous "confinement" of the color degrees of freedom of modern quark theories (26) and is a delightful example of the continuing flow of ideas and techniques back and forth between many-body physics and quantum field theory.

In the past few years extensive investigations via renormalization group theory (which, in a nearly modern form, was first applied to this problem (27)) have led to the essential solution of this "Kondo problem". A very succinct way of describing that solution is the computation of the scaling of the susceptibility as a function of temperature by Wilson (28) (Fig. 7). For comparison, and to show the remarkable precision of the Schrieffer-Wolff transformation,

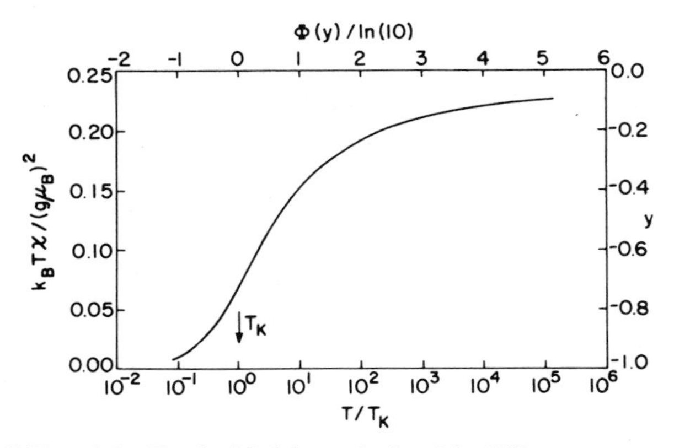

Fig. 7. Susceptibility of the Kondo Model as calculated by Wilson.

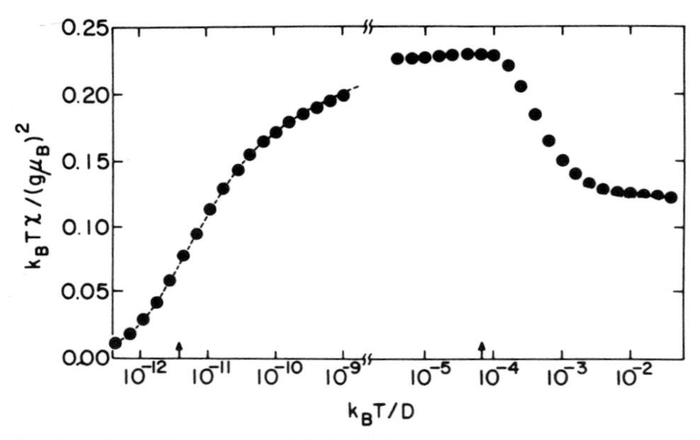

Fig. 8. Susceptibility of the "Anderson Model" showing equivalence to Kondo. (from Krishna-Murthy et al (29))

we give as the last figure of this subject Krishna-Murthys' corresponding calculation (29) for the Anderson model (Fig. 8) and one equation: Haldane's precise equivalencing of the parameters of the two models, from his thesis (20):

$$T_K = \frac{1}{2\pi}\left(\frac{2\Delta U}{\pi}\right)^{1/2} \exp\left[\frac{E_d(E_d+U)}{2\Delta U}\right] \tag{7}$$

which may be used to find the properties of one model from the other: e.g.,

$$X(t\rightarrow 0) = \cdot\frac{103}{T_K} \text{ etc.}$$

I am indebted to a London Times article about Idi Amin for learning that in Swahili "Kondoism" means "robbery with violence." This is not a bad description of this mathematical wilderness of models; H. Suhl has been heard to say that no Hamiltonian so incredibly simple has ever previously done such violence to the literature and to national science budgets.

II. The Origins of Localization Theory

In early 1956, a new theoretical department was organized at the Bell Laboratories, primarily by P. A. Wolff, C. Herring and myself. Our charter was unusual in an industrial laboratory at the time: we were to operate in an academic mode, with postdoctoral fellows, informal and democratic leadership, and with an active visitor program, and that first summer we were fortunate in having a large group of visitors of whom two of those germane to this story were David Pines and Elihu Abrahams.*

The three of us took as our subject magnetic relaxation effects in the beautiful series of paramagnetic resonance experiments on donors in Si begun by Bob Fletcher and then being carried on by George Feher. Feher was studying (primarily) paramagnetic resonance at liquid He temperatures of the system of donor impurities (e.g., P, As, etc.) in very pure Si, in the concentration range 10^{15}—10^{18} impurities/cc encompassing the point of "impurity band" formation around 6×10^{17}. At such temperatures most of the donors were neutral (except those emptied by compensating "acceptor" impurities such as B, Al or Ga), having four valences occupied by bonds, leaving a hydrogenic orbital for the last electron which, because of dielectric screening and effective mass, has an effective Bohr radius of order 20 Å (Fig. 9). The free spin of

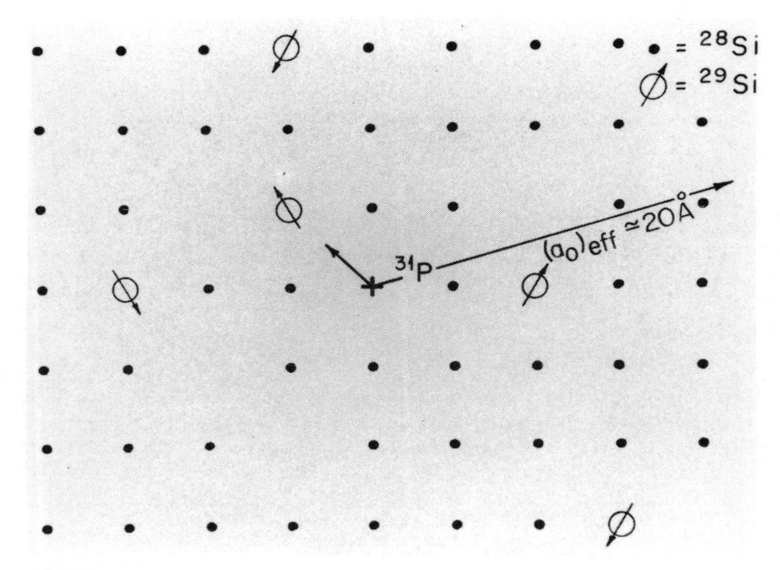

Fig. 9. Donor wave functions in Si and Si_{29} nuclei: schematic.

* It may be of interest to note that theorists permanently or temporarily employed at Bell Labs that summer were at least the following: a) (permanent or semipermanent) P. W. Anderson, C. Herring, M. Lax, H. W. Lewis, G. H. Wannier, P. A. Wolff, J. C. Phillips; b) (temporary) E. Abrahams, K. Huang, J. M. Luttinger, W. Kohn, D. Pines, J. R. Schrieffer, P. Nozieres; c) (permanent but not in theory group): L. R. Walker, H. Suhl, W. Shockley.

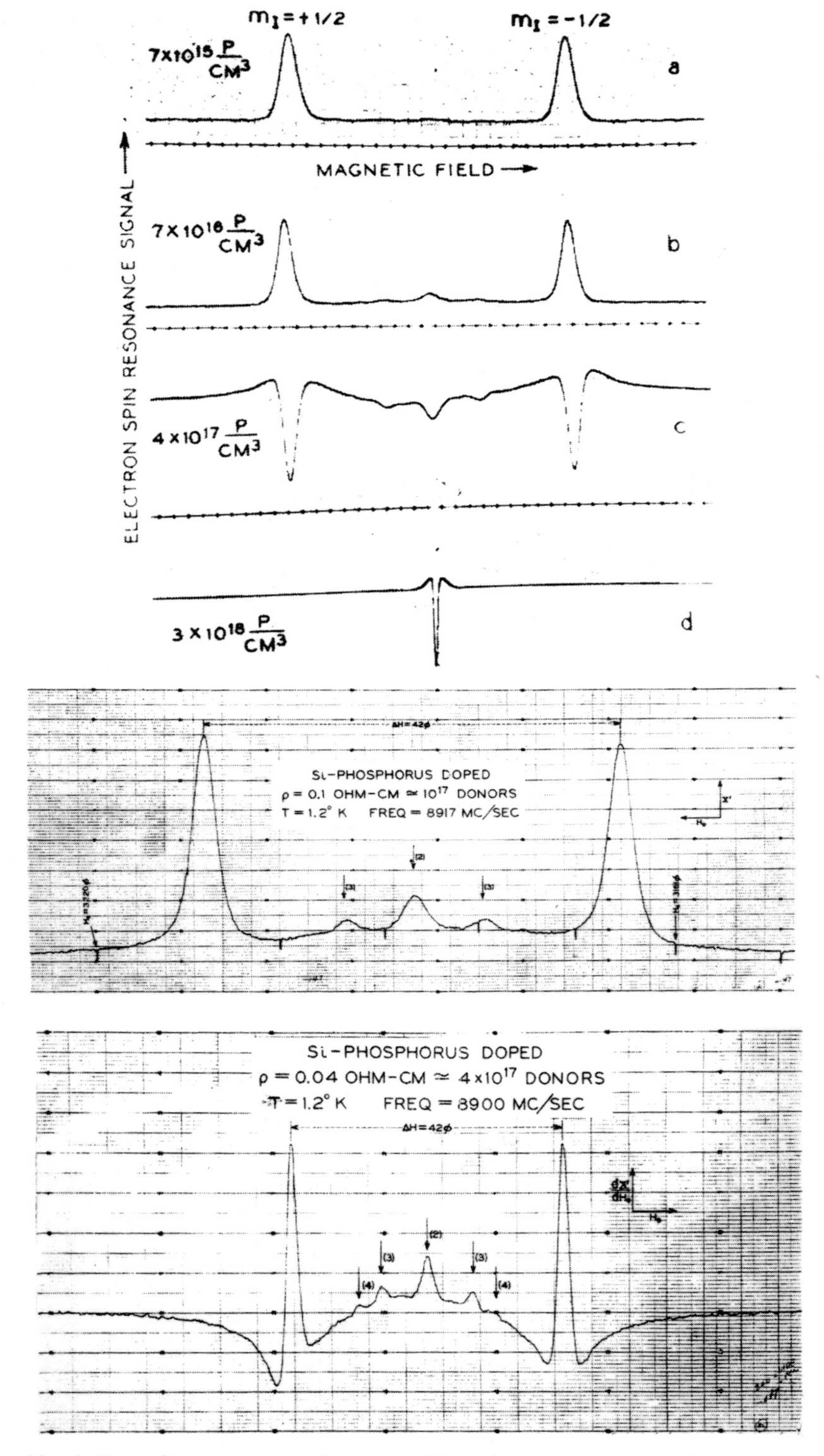

Fig. 10. a) Hyperfine structures of donor EPR at increasing donor (P) concentrations the Mott—Anderson metal-insulator transition. b) Example of well-developed cluster lines.

this extra donor orbital has a hyperfine interaction with the donor nucleus (^{31}P or As, for instance) leading to the clean hfs (30) shown in Fig. 10. In addition, isotopic substitution proved that most of the residual breadth of the lines is also caused by hfs interactions, of the very extended electronic orbital with the random atmosphere of $\sim 5\ \%$ of Si_{29} nuclei in natural Si, and for reasonably low donor densities of $\sim 10^{16}$/cc the actual spin-spin and spin-lattice relaxation times were many seconds. That is, the lines were "inhomogeneously broadened", so that many very detailed experimental techniques were available. Feher and Fletcher (31) had already probed what we would now call the Mott-Anderson transition in these materials (Fig. 10a). As the concentration was raised, first lines with fractional hfs appeared, signifying clusters of 2, 3, 4, or more spins in which the exchange integrals between donors overweighed the hf splitting and the electron spins saw fractionally each of the donor nuclei in the cluster. (A good example is shown in Fig. 10b.) Finally, at $\sim 6 \times 10^{17}$, came a sudden transition to a homogeneously broadened free-electron line: the electrons went into an "impurity band" at that point. Pines, Bardeen and Slichter (33) had developed a theory of spin-lattice relaxation for donors, and it was our naive expectation that we would soon learn how to apply this to Feher's results. In fact, no theoretical discussion of the relaxation phenomena observed by Feher was ever forthcoming, only a description of the experiments (34). What the three of us soon realized was that we were confronted with a most complex situation little of which we understood. In particular, we could not understand at all the mere fact of the extremely sharp and well-defined "spin-packets" evinced by such experiments as "hole digging" and later the beautiful "ENDOR" effect (32, 34). (In the ENDOR experiment Feher would select a spin packet by saturating the line at a specific frequency ("digging a hole", Fig. 11a) and monitor the nmr frequencies of ^{29}Si nuclei in contact with packet spins by exciting with the appropriate radio frequency and watching the desaturation of the packet (Fig. 11b). In this figure, the many seconds recovery time after passing the ENDOR line is actually an underestimate of the packet T_2 because the system is driven.) Thus every individual P electron had its own frequency and kept it for seconds or minutes at a time.

We assumed from the start the basic ideas of Mott with regard to actual electron motion: that since there were few compensating acceptors, Coulomb repulsion kept most of the donors singly-occupied leaving us with the paramagnetic spin system we observed. W. Kohn seems to have suggested that even the empty donors would be pinned down by staying close to their compensating negatively charged acceptors because of Coulomb attraction (see Fig. 12). Thus there was little actual electron motion, and we noticed only some speeding up of the relaxation times as we approached what now would be called the "Mott-Anderson" transition. Stretching our gullibility a bit, we could believe that nothing spectacular was *necessarily* required to prevent mobility of the actual charged electron excitations. (It was, however, at this time that I suggested to Geballe the study of dielectric relaxation in these materials to probe this motion, which led to the discovery of the now well-

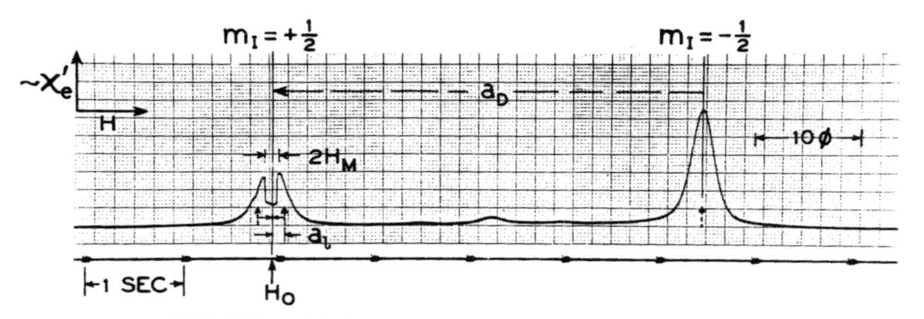

DEMONSTRATION OF THE INHOMOGENEOUS BROADENING IN P-DOPED SILICON (ρ = 0.25 Ω CM, T = 1.2° K, H_0 = 3000 ϕ) IN THE ENDOR TECHNIQUE SPIN PACKETS ARE FLIPPED AS INDICATED BY ARROWS

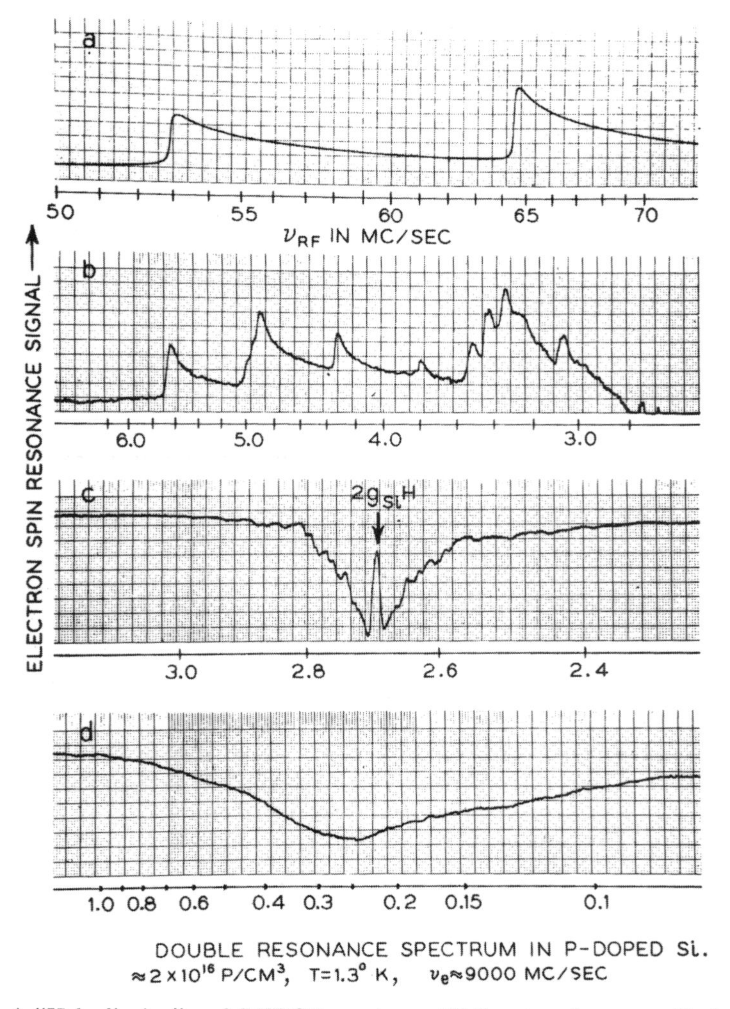

DOUBLE RESONANCE SPECTRUM IN P-DOPED Si.
≈2×10¹⁶ P/CM³, T=1.3° K, ν_e≈9000 MC/SEC

Fig. 11. a) "Hole-digging" and ENDOR spectrum. While saturating a specific frequency (11a) an rf signal of variable frequency is applied (11b). Note slow refilling of hole (exponential recoveries) in Fig. 11b: sweep time is several minutes. This is the Si^{29} spin-lattice relaxation, enhanced by the rf power applied.

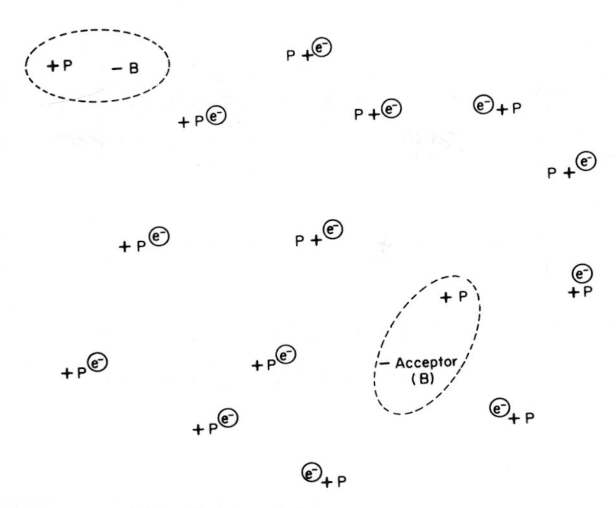

Fig. 12. Hypothetical binding of charged donors to acceptors.

known Pollak-Geballe "$\omega^{0.8}$" conductivity (34). I felt that the absence of conduction in the impurity band was also a serious question, in this as in many other systems.)

No arguments using Coulomb interaction saved us from a second dilemma: the absence of spin diffusion. Bloembergen, in 1949 (36), had proposed the idea of spin diffusion in nuclear spin systems, which has since had much experimental verification. His idea was that the dipolar interactions caused mutual precessions which, in the high temperature paramagnetic state of a spin system, could by diffusion equilibrate the spin temperature in space, thereby giving a means—for example—for nuclear spins to relax by diffusing to the neighborhood of an electronic spin impurity. To calculate the process he used a simple estimate from the Golden Rule plus random walk theory.

Portis (37), in 1953, introduced the idea of random "inhomogeneous broadening" where complete equilibration within a spectral line is impeded, and instead one speaks of "spin packets" of spins having a definite resonance frequency within the line (Fig. 13). (Such packets are spatially random, of

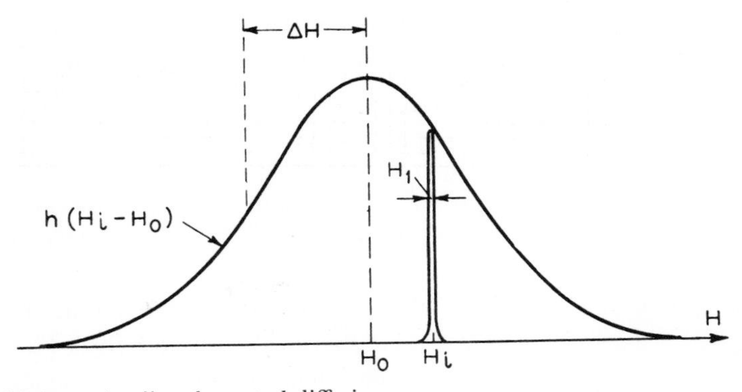

Fig. 13. "Spin packets" and spectral diffusion.

course; in macroscopically inhomogeneous systems the same phenomena had been seen much earlier.) Portis estimated that if the interaction of neighboring spins was J_{ij}, the lifetime of a spin packet (38) should be of order

$$\hbar/\tau \simeq \pi|J_{ij}|^2 \times \frac{Z}{W} \tag{8}$$

W being the width of the line and Z the number of neighbors: this is apparently obvious by the Golden Rule. But when Elihu Abrahams estimated J_{ij} for our system, he found that according to (8), τ should have varied from .1 to 10^{-6} s, whereas Feher's spin packets stayed saturated for 10—100 s in a typical ENDOR experiment. His estimates were checked by the "cluster" phenomenon of Fig. 10.

I find in my notes a reference on 6/20/56 to a discussion with Pines where I suggested an "All or Nothing" theorem to explain this. Later, on 10/31/56, comes an optimistically claimed "proof" of "Anderson's Theorem", much like an unsophisticated version of my final paper which even so is hardly a "proof"; such does not yet really exist. I also seem to have spoken to an uninterested audience at the Seattle International Theoretical Physics Symposium. But the actual work was not completed until shortly before I talked about it to much the same group of residents and visitors on July 10th and 17th, 1957. By that time, I had clearly been a nuisance to everyone with "my" theorem: Peter Wolff had given me a short course in perturbation theory, Conyers Herring had found useful preprints from Broadbent and Hammersley on the new subject of percolation theory, Larry Walker had made a suggestion and Gregory Wannier posed a vital question, etc. But my recollection is that, on the whole, the attitude was one of humoring me.

Let me now give you the basics of the argument I then presented (39) but in much more modern terminology (the mathematics is the same, essentially). I don't think this is the only or final way to do it; a discussion which is more useful in many ways, for instance, can be based on Mott's idea of minimum metallic conductivity as used by Thouless and co-workers and as he will touch upon; but I think this way brings out the essential nature of this surprising nonergodic behavior most clearly. I apologize for this brief excursion into mathematics, but please be assured that I include the least amount possible.

The first problem was to create a model which contained only essentials. This was simple enough: a linearized, random "tight-binding" model of non-interacting particles:

$$H = \sum_i E_i n_i + \sum_{ij} V_{ij} \overset{+}{c_i} c_j \tag{9}$$

in which the "hopping" integrals V_{ij} were taken to be nonrandom functions of r_{ij} (the sites i can sit on a lattice if we like) but E_i was chosen from a random probability distribution of width W (Fig. 14). The objects c_i could be harmonic oscillator (phonon) coodinates, electron operators, or spinors for which $V_{ij} \simeq J_{ij}$ and we neglect the $J_{ij} S_i^z S_i^z$ interactions of the spin flips. The essential thing is that (9) leads to the linear equation of motion

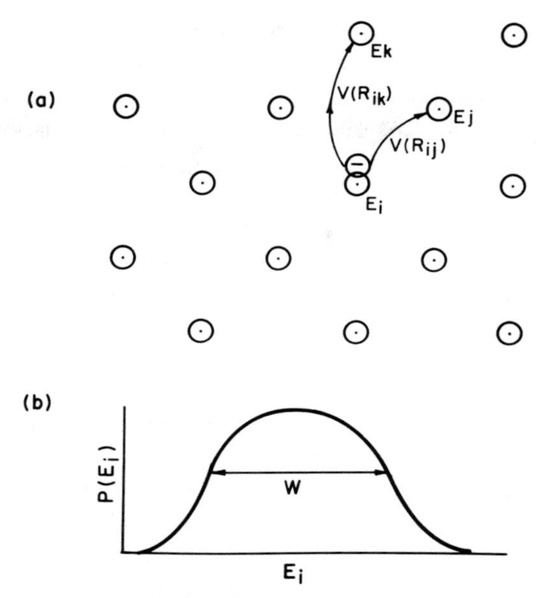

Fig. 14. Model for diffusion in a random lattice
a) sites and hopping integrals
b) probability distribution of E_i

$$i\hbar \dot{c}_i^{+} = E_i c_i^{+} + \sum_i V_{ij}\, c_j^{+} \tag{10}$$

If W is zero and all E_i the same (say 0), (10) describes a band of Bloch states of width about $Z\ \overline{V}_{ji}$. For $W \ll \overline{V}_{ij} = V$, the theories of transport recently developed by Van Hove and Luttinger (40) clearly would describe resistive impurity scattering of free waves (say, electrons, for simplicity). If, on the other hand, $W \gg V$, that would describe our system of local hf fields large compared to J_{ij}; or of random Coulomb and strain energies large compared to the hopping integrals for the electrons from donor to donor.

What is clearly called for is to use W as a perturbation in the one case, and V_{ij} in the other; but what is not so obvious is that the behavior of perturbation theory is absolutely different in the two cases. For definiteness, let us talk in terms of the "resolvent" or "Greenian" operator which describes all the exact wavefunctions φ_a and their energies E_a:

$$G = \frac{1}{\mathrm{E}-\mathrm{H}};\ \text{i.e. } G(r, r') = \sum_a \varphi_a(r)\, \frac{1}{E-E_a}\, \varphi_a(r') \tag{11}$$

where the φ_a and E_a are the exact eigenfunctions of the Hamiltonian (9). In the conventional, "transport" case, we start our perturbation theory with plane-wave-like states

$$\varphi_k^0 = \frac{1}{\sqrt{N}} \sum_i e^{ik\cdot R_i}\, \varphi_i$$

with energy

$$E_k = \sum_{(i-j)} V_{ij} \cos k \cdot (R_i - R_j)$$

which we assume are only weakly perturbed by the scattering caused by randomly fluctuating E_i's. The E_k's are a continuum in the limit of a large system and we take advantage of this to rearrange perturbation theory and get

$$G_{kk} = \frac{1}{E - E_k - \sum(k, E)} \tag{12}$$

where $\sum$, the "self-energy", is itself a perturbation series (Fig. 15a)

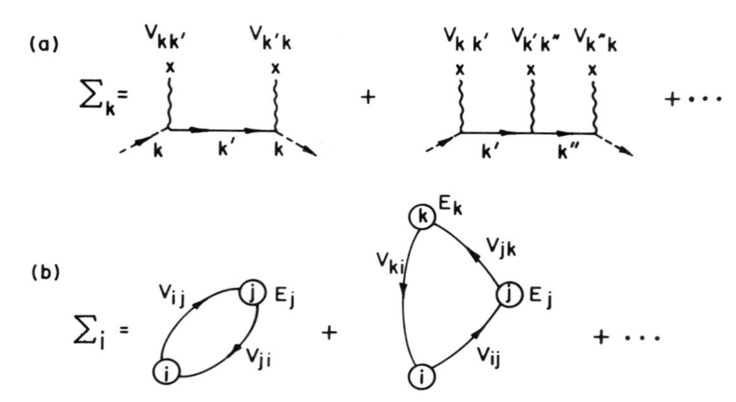

Fig. 15. a) Self-energy diagrams in conventional "propagator" theory.
b) Self-energy diagrams in "locator" theory.

$$\sum = \sum_{k' \neq k} (V_{kk'})^2 \frac{1}{E - E_{k'}} + \sum_{k'', k' \neq k} \frac{V_{kk''} V_{k''k'} V_{k'k}}{(E - E_{k'})(E - E_{k''})} + \ldots \tag{13}$$

which, since E_k is a continuum, has a finite imaginary part as E approaches the real axis

$$\lim_{\mathrm{Im}E \to \pm 0} \mathrm{Im} \sum = \pm \pi \int |V_{kk'}|^2 \delta(E_{k'} - E) + \ldots \tag{14}$$

Note that $V_{kk'}$ in this case comes from the width "W" not V_{ij}.)

This equation means that E_k has a finite width in energy, and ImG, the density of states, is a finite, continuous function of E (Fig. 16).

$$\lim_{s \to \pm 0} G_k (E + \mathrm{i}s) = \frac{1}{E - E_k - R_e \sum \mp \mathrm{i}\Delta(E)} \tag{15}$$

G has a genuine *cut* on the real axis, and there is a continuum of energy states at every site, of every energy in the band: the states are what we now call "extended". That is, the definition (11) of G basically tells us

$$\mathrm{Im}G(i, i; E) = \pi \sum_a |\varphi_a(r_i)|^2 \delta(E - E_a). \tag{16}$$

Transforming (15) to find G_{ii}, we find that the $|\varphi_a(r_i)|^2$ are each infinitesimal

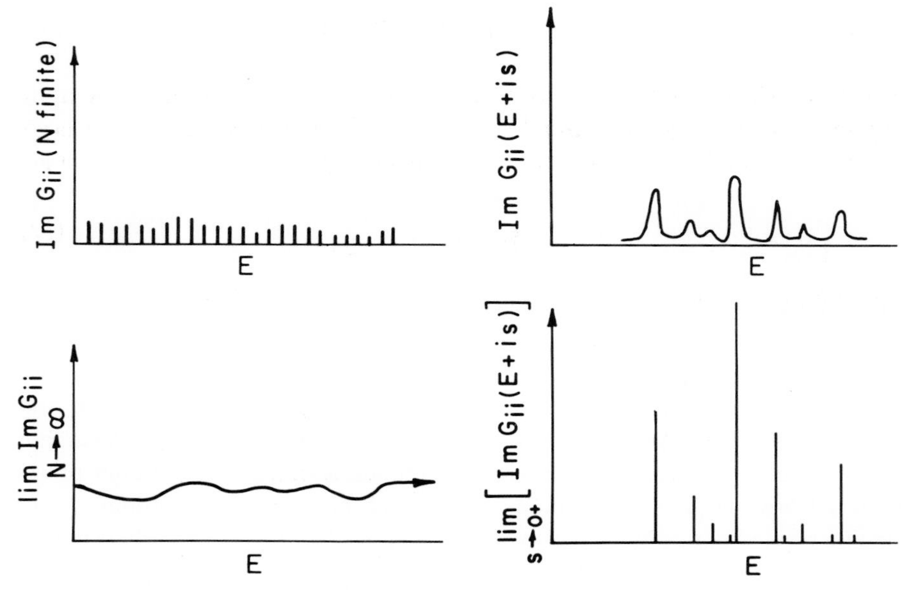

Fig. 16. a) Im G_{ii} in extended case
b) Im G_{ii} in localized case

of order $\frac{1}{\sqrt{N}}$, forming in the limit $N \to 0$ a true continuum of states of every energy at site i. Of course, there are sum rules stating that every state is somewhere and that no states get lost:

$$\sum_i |\varphi_a(i)|^2 = \sum_a |\varphi_a(i)|^2 = 1 \tag{17}$$

and these are satisfied by $\varphi_a(i) \sim (\sqrt{N})^{-1}$, where N is the total number both of a's and i's.

My contribution was just to show that this is not the only possible case, other than just an empty band of energies, or a set of discrete states as one may have near a single attractive potential like a hydrogen atom. What I showed is that one may have a continuum in *energy* but not in *space*. This is immediately made plausible just by doing perturbation theory in the opposite order.

In this case one takes E_i as the big term, and the starting eigenfunctions and eigen-energies are just

$$\varphi_i^0 = \varphi_i,\ E_i^0 = E_i \tag{18}$$

and V_{ij} is the perturbation. In this case, (which Larry Walker suggested I call "cisport") we use a "locator" instead of a "propagator" series, for the "locator" G_{ii} not the "propagator" G_{kk}:

$$G_{ii}(E+\mathrm{i}s) = \frac{1}{E+\mathrm{i}s-E_i-\sum_i(E+\mathrm{i}s)} \tag{19}$$

where now the self-energy $\sum$ is a superficially similar series to (13) (Fig. 15b)

$$\Sigma_i = \sum_{j \neq i} \frac{(V_{ij})^2}{E-E_j} + \sum \frac{V_{ij}\, V_{jk}\, V_{ki}}{(E-E_j)(E-E_k)} + \ldots \tag{20}$$

If at this point we make one tiny mistake, we immediately arrive back at Portis' answer (8): namely if we *average* in any way, we get

$$\text{Ave}\left\{ \lim_{s \to 0+} (\text{Im} \sum_{j,\, k \neq i} (E+is)) \right\} \simeq \frac{< V_{ij}^2 >}{W} \tag{21}$$

But there is a very important fundamental truth about random systems we must always keep in mind: *no real atom is an average atom, nor is an experiment ever done on an ensemble of samples.* What we really need to know is the *probability distribution* of $\text{Im}\sum$, *not* its average, because it's only each specific instance we are intèrested in. I would like to emphasize that this is the important, and deeply new, step taken here: the willingness to deal with *distributions*, not *averages*. Most of the recent progress in the fundamental physics of amorphous materials involves this same kind of step, which implies that a random system is to be treated not as just a dirty regular one, but in a fundamentally different way.

Having taken this point of view, it is sufficient to study only the first term of (20), it turns out. Let us first pick a finite s, and then take the limit as $s \to$ o. With a finite s,

$$\text{Im}\left(\frac{\Sigma}{s}\right) = \sum_j \frac{|V_{ij}|^2}{(E-E_j)^2+s^2}$$

The condition that E_j appear as a peak of $\text{Im}\frac{\Sigma}{s}$ is that E_j be within s of E, and that $V_{ij} > s$. To assess the probability that V_{ij} is large enough, use the physically realistic assumption of exponential wavefunctions:

$$V(R) = V_0\, \mathrm{e}^{-R/R_0}.$$

In the energy interval of size s, there will be ns/W energies E_j per unit volume ($\mathcal{N}$ is the site density per unit volume), while $V > s$ implies

$$V > s : R < R_0\, ln\, \frac{s}{V_0}$$

and the probability that both $V > s$ and $E-E_j < s$ is

$$P(V > s,\, |E-E_j| < s) = \mathcal{N} \cdot \frac{4\pi R_0^3}{3} \left(\frac{s}{W}\right) \left(ln\, \frac{s}{V_0}\right)^3$$

$$P(s \to \text{o}) = 0$$

It is easy to formalize this: one may show that the probability distribution of $\text{Im}\sum$ is essentially

$$\lim_{s \to 0} P\left(\text{Im}\left(\frac{\Sigma}{s}\right) = X\right) \text{d}X = \frac{\text{d}X}{X^{3/2}}\, \mathrm{e}^{-\frac{s}{X}} \tag{22}$$

which indeed has a divergent average as it should, but is finite nonetheless, so that $\text{Im}\sum \propto s$ and there is *not* a finite cut at the real axis.

When we stop and think about what this means, it turns out to be very simple. It is just that we satisfy the sum rules (17) *not* by each $\varphi_a(i)$ being infinitesimal, but by a discrete series of finite values: the biggest φ_a^i is of order 1, the next of order 1/2, etc., etc., (see Fig. 16b). Thus, $\text{Im}G_{ii}$ is a sum of a discrete infinite series of δ-functions with convergent coefficients. This is the *localized* case.

That is more than enough mathematics, and is all that we will need. The rest boils down simply to the question of when this lowest-order treatment is justified, and how it breaks down.

The bulk of the original paper was concerned with how to deal with the higher terms of the series and show that they don't change things qualitatively: what they do, actually, is just to *renormalize* V_{ij} and the E_j's so that even if V_{ij} is short-range initially, it becomes effectively exponential; and, of course, the V_{ij}'s broaden the spectrum. If this is the case, one then realizes that the extended case can only occur because of a breakdown of perturbation theory. This comes about as the higher terms of perturbation theory "renormalize" $V(R_j)$ and stretch it out to longer and longer range, so that the exponentially localized function become less so and finally one reaches a "mobility edge" or "Anderson transition".

Here we begin to tie in to some of the ideas which Professor Mott will describe. First, it is evident that the self-energy series is a function of E—i.e., of where we are on the real energy axis—so it will cease to converge first at one particular energy E, the "mobility edge." For a given model, it is reasonable—in fact usual—to have the localized case for some energies, the extended one for others, separated by a "mobility edge". The significance of this fact was realized by Mott.

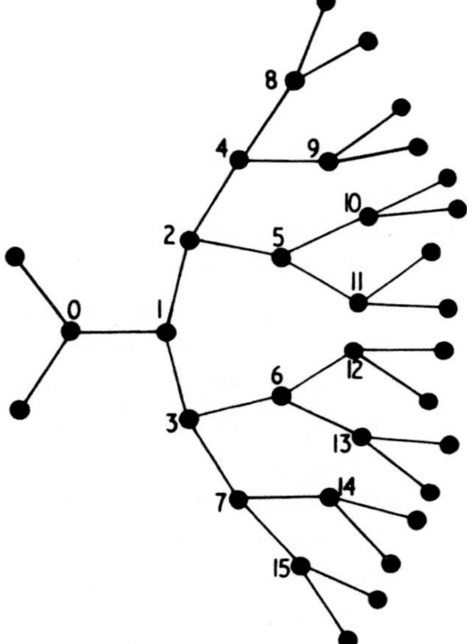

Fig. 17. "Cayley Tree" on which localization theory is exact.

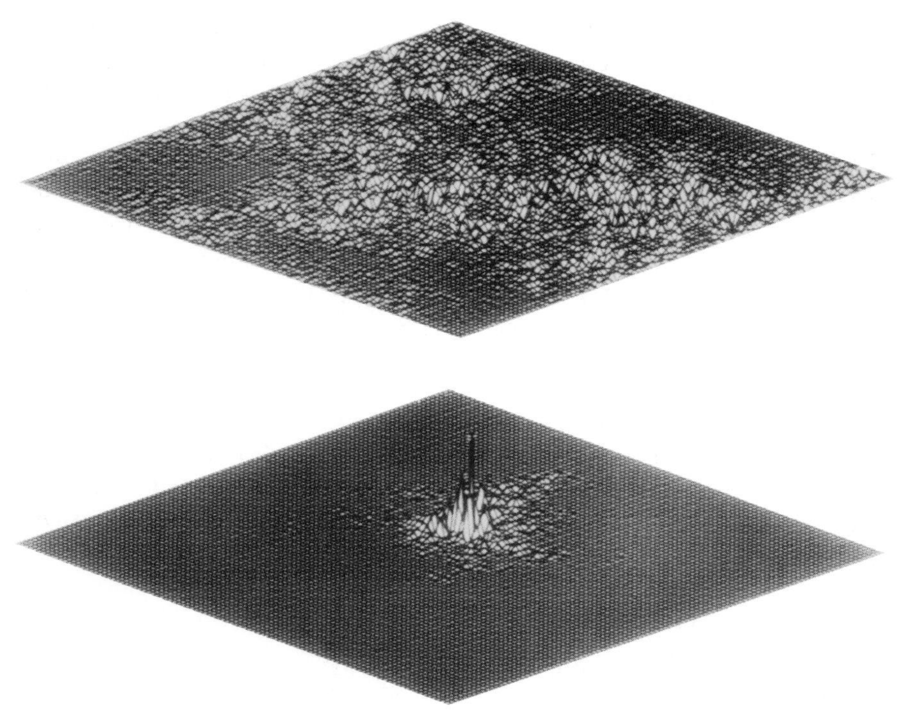

Fig. 18. Computer demonstration of localization (courtesy of Yoshino and Okazaki).
a) W/V = 5.5
b) W/V = 8.0

The actual calculation of this divergence or "Anderson transition" was carried out by me using conservative approximations in the original paper, but it was only much later realized (41) that that calculation was *exact* on a "Cayley tree" or Bethe Lattice (Fig. 17). Much earlier, Borland (42), and Mott and Twose (43) had shown that localization always occurs in one dimension (also a Cayley tree case, with $K = 1$). Since it is easy to convince oneself that the Cayley tree is a lattice of infinite dimensionality d (though finite neighbor number) it is likely that delocalization first occurs at some lower critical dimensionality d_c, which we now suspect to be 2, from Thouless' scaling theory (44). This dimensionality argument (or equivalent ones of Thouless) first put to rest my earliest worry that my diagram approximations were inexact: in fact, they *under*estimate localization, rather than otherwise. A second reason why I felt discouraged in the early days was that I couldn't fathom how to reinsert interactions, and was afraid they, too, would delocalize. The realization that, of course, the Mott insulator localizes without randomness, because of interactions, was my liberation on this: one can see easily that the Mott and Anderson effects supplement, not destroy, each other, as I noted in some remarks on the "Fermi Glass" (45) which more or less marked my re-entry into this problem. The present excitement of the field for me is that I feel a theory of localization with interactions is beginning to appear, in work within my group as well as what Professor Mott will describe. It is remarkable that in almost all cases interactions play a vital role, yet many results are not changed too seriously by them.

I will close, then, and leave the story to be completed by Professor Mott. I would like, however, to add two things: first, a set of figures of a beautiful computer simulation by Yoshino and Okazaki (46), which should convince the most skeptical that localization does occur. The change in W between these two figures is a factor 1.5, which changed the amplitudes of a typical wave function as you see, from extended to extraordinarily well localized. (see Fig. 18).

Finally, you will have noted that we have gone to extraordinary lengths just to make our magnetic moments—in the one case—or our electrons—in the other—stay in one place. This is a situation which was foreshadowed in the works of an eminent 19th century mathematician named Dodson, as shown in the last figure (Fig. 19). "Now here, you see, it takes all the running *you* can do, to keep in the same place."

"Well, in *our* country," said Alice, still panting a little, "you'd generally get to somewhere else—if you ran very fast for a long time as we've been doing."

"A slow sort of country!" said the Queen. "Now, *here*, you see, it takes all the running *you* can do, to keep in the same place.

Fig. 19. Efforts to avoid localization (Dodson).

REFERENCES

1. Anderson, P. W., Phys. Rev. *115*, 2 (1959).
2. Anderson, P. W., Solid State Phys. (F. Seitz and Turnbull, D., eds.) *14*, 99 (1963).
3. Hurwitz, H., unpublished doctoral thesis, Harvard, 1941.
4. van Vleck, J. H., Rev. Mod. Phys. *25*, 220 (1953).
5. Friedel, J., Can. J. Phys. *34*, 1190 (1956); Suppl. Nuovo Cim. VII, 287 (1958).
6. Blandin, A. and Friedel, J., J. Phys. Radium. *20*, 160 (1959).
7. Suhl, H. and Matthias, B. T., Phys. Rev. *114*, 977 (1959).
8. Clogston, A. M., Matthias, B. T., Peter, M., Williams, H. J., Corenzwit, F. and Sherwood, R. C., Phys. Rev. *125*, 541 (1962).
9. Owen, J., Browne, M., Knight, W. D. and Kittel, C., Phys. Rev. *102*, 1501 (1956).
10. Crane, L. T. and Zimmerman, J. E., Phys. Rev. *123*, 113 (1961).
11. Yosida, K., Phys. Rev. *106*, 893 (1957).
12. Anderson, P. W., Phys. Rev. *124*, 41 (1961).
13. Anderson, P. W. and McMillan, W. L., Proc. Varenna School of Physics XXXVII (1966), p. 50, Academic Press, N.Y. 1967.
14. Caroli, B. and Blandin, A., J. Phys. Chem. Solids *27*, 503 (1966).
15. Alexander, S. and Anderson, P. W., Phys. Rev. *133*, A1594 (1964).
16. Moriya, T., Proc. Varenna School XXXVII (1966), p. 206, Academic Press, N.Y. 1967.
17. Grimley, T. B., Proc. Phys. Soc. (Lond.) *90*, 757; *92*, 776 (1967). See also A. J. Bennett and L. M. Falicov, Phys. Rev. *151*, 512 (1966).
18. Newns, D. M., Phys. Rev. *178*, 1123 (1969).
19. Haldane, F. D. M. and Anderson, P. W., Phys. Rev. *B13*, 2553 (1976).
20. Haldane, F. D. M., Thesis (Cambridge 1977).
21. Anderson, P. W. and Clogston, A. M., Bull. Amer. Phys. Soc. *6*, 124 (1961).
22. Schrieffer, J. R. and Wolff, P. A., Phys. Rev. *149*, 491 (1966).
23. Kondo, J., Prog. Theor. Phys. *32*, 37 (1964).
24. Suhl, H., Phys. Rev. *138*, A515 (1965); Abrikosov, A. A., Physics *2*, 21 (1965).
25. Schrieffer, J. R. and Mattis, D. C., Phys. Rev. *140*, A1412 (1965).
26. Anderson, P. W., Proc. Conf. on Gauge Theories, pg. 311, Northeastern Univ., Arnowitt and Nath, eds., MIT Press, 1976.
27. Anderson, P. W., Yuval, G. and Hamann, D. R., Phys. Rev. *B1*, 4464 (1970).
28. Wilson, K. G., Revs. Mod. Phys. *47*, 773 (1975).
29. Krishna-Murthy, H. R., Wilkins, J. W., Wilson, K. G., Phys. Rev. Lett. *35*, 1101 (1975).
30. Fletcher, R. C., Yager, W. A., Holden, A. N., Read, W. T., Merritt, F. R., Phys. Rev. *94*, 1392 (1954).
31. Feher, G., Fletcher, R. C. and Gere, E. A., Phys. Rev. *100*, 1784 (1955).
32. Feher, G., Phys. Rev. *114*, 1219 (1959).
33. Pines, D., Bardeen, J. and Slichter, C. P., Phys. Rev. *106*, 489 (1957). Also see E. Abrahams, Phys. Rev. *107*, 491 (1957).
34. Feher, G. and Gere, E. A., Phys. Rev. *114*, 1245 (1959).
35. Pollak, M. and Geballe, T. H., Phys. Rev. *122*, 1742 (1961).
36. Bloembergen, N. Physica *15*, 386 (1949).
37. Portis, A. M., Phys. Rev. *91*, 1071 (1953).
38. Portis, A. M., Phys. Rev. *104*, 584 (1956).
39. Anderson, P. W., Phys. Rev. *109*, 1492 (1958).
40. For example, Kohn, W. and Luttinger, J. M., Phys. Rev. *108*, 590 (1957).
41. Anderson, P. W., Abou-Chacra, R. and Thouless, D. J., J. Phys. C. *6*, 1734 (1973).
42. Borland, R. E., Proc. Roy. Soc. A *274*, 529—45 (1963).
43. Mott, N. F. and Twose, W. D., Adv. Phys. *10*, 107—63 (1961).
44. Licciardello, D. C. and Thouless, D. J., Phys. Rev. Lett. *35*, 1475—8 (1975).
45. Anderson, P. W., Comments in Solid State Physics *2*, 193 (1970).
46. Yoshino, S. and Okazaki, M., J. Phys. Soc. Japan *43*, 415—423 (1977).

Nevill Mott

NEVILL FRANCIS MOTT

Nevill Francis Mott was born in Leeds, U.K., on September 30th, 1905. His parents, Charles Francis Mott and Lilian Mary (née) Reynolds, met when working under J. J. Thomson in the Cavendish Laboratory; his great grandfather was Sir John Richardson, the arctic explorer. He was educated at Clifton College, Bristol and St. John's College, Cambridge, where he studied mathematics and theoretical physics. He started research in Cambridge under R. H. Fowler, in Copenhagen under Niels Bohr and in Göttingen under Max Born, and spent a year as a lecturer at Manchester with W. L. Bragg before accepting a lectureship at Cambridge. Here he worked on collision theory and nuclear problems in Rutherford's laboratory. In 1933 he went to the chair of theoretical physics at Bristol, and under the influence of H. W. Skinner and H. Jones turned to the properties of metals and semiconductors. Work during his Bristol period before the war included a theory of transition metals, of rectification, hardness of alloys (with Nabarro) and of the photographic latent image (with Gurney). After a period of military research in London during the war, he became head of the Bristol physics department, publishing papers on low-temperature oxidation (with Cabrera) and the metal-insulator transition.

In 1954 he was appointed Cavendish Professor of Physics, a post which he held till 1971, serving on numerous government and university committees. The research for which he was awarded the Nobel prize began about 1965. Some of his main books are "The Theory of Atomic Collisions" (with H. S. W. Massey), "Electronic Processes in Ionic Crystals" (with R. W. Gurney) and "Electronic Processes in Non-Crystalline Materials" (with E. A. Davis).

Outside research in physics he has taken a leading part in the reform of science education in the United Kingdom and is still active on committees about educational problems. He was chairman of a Pugwash meeting in Cambridge in 1965. He was chairman of the board and is now president of Taylor & Francis Ltd., scientific publishers since 1798. He was Master of his Cambridge college (Gonville and Caius) from 1959–66. He was President of the International Union of Physics from 1951 to 1957, and holds more than twenty honorary degrees, including Doctor of Technology at Linköping.

In 1930 he married Ruth Eleanor Horder. They have two daughters and three grandchildren, Emma, Edmund and Cecily Crampin.

For the last ten years he has lived in a village, Aspley Guise, next door to his son-in-law and family. During this period he has written an autobiography "A Life in Science" (Taylor and Francis) and edited a book with several authors on a religion-science interface "Can Scientists Believe?" (James and James, London), together with many scientific papers, mainly in the last 3 years on high-temperature superconductors.

ELECTRONS IN GLASS

Nobel Lecture, 8 December, 1977

by
NEVILL MOTT
Cavendish Laboratory, Cambridge, England

The manufacture of glass, along with the forming of metals, is an art that goes back to prehistoric times. It always seems to me remarkable that our first understanding of the ductility of metals in terms of atomic movements came *after* the discovery of the neutron. Geoffrey Taylor (1) was the great name here, and Nabarro (2) and I first tried to explain why metallic alloys are hard. The years that passed before anyone tried to get a theoretical understanding of electrons in glass surprises me even more. After all, the striking fact about glass is that it is transparent, and that one does not have to use particularly pure materials to make it so. But, in terms of modern solid state physics, what does "transparent" mean? It means that, in the energy spectrum of the electrons in the material, there is a gap of forbidden energies between the occupied states (the valence band) and the empty states (the conduction band); light quanta corresponding to a visible wave-length do not have the energy needed to make electrons jump across it. This gap is quite a sophisticated concept, entirely dependent on quantum mechanics, and introduced for solids in the 1930's by the pioneering work of Bloch, Peierls and A. H. Wilson. The theory was based on the assumption that the material was crystalline. The gap, in most treatments, was closely related to Bragg reflection of the electron waves by the crystal lattice and the mathematical analysis was based on the assumption of a perfect crystal. Glass, and amorphous materials generally, do not give a sharp Bragg reflection; it is curious, therefore, that no one much earlier than my coworkers and I (3) in Cambridge less than ten years ago seems to have asked the question "how can glass be transparent?".

Actually our curiosity was stimulated by the investigation of the Leningrad school under Kolomiets (4) from 1950 onwards of electrical rather than the optical properties of the glassy semiconductors. These are black glasses, containing arsenic, tellurium and other elements, and for them the band-gap lies in the infra-red. The gap is sufficiently small to ensure that at room temperature an electron can be excited across it. The Leningrad experiments showed, it seems to me, that the concepts of a conduction and a valence band could be applied to glasses, and, more remarkably, that the gap, and hence the conductivity, did not depend sensitively on composition. This is related to the fact that oxide glasses are normally transparent and can only be coloured, as in medieval stained glass, by the addition of transition metal atoms, where an inner shell produces its own absorption spectrum, depending little on the surroundings. These properties of glass are in sharp contrast with the behaviour of crystals, where the whole of silicon technology depends on the fact that if, for instance, phosphorus with its five electrons is added, four form bonds but the fifth is very loosely bound. The discovery of this property of glasses certainly makes Kolomiets one of the fathers of the branch of science that I am describing, as were others in Eastern European countries, notably Grigorovici in Bucurest and Tauc in Prague. The explanation in chemical terms (5) of this property seems to be that in a glass each atom will have the right number of neighbours to enable *all* electrons to be taken up in bonds. There are important exceptions to this, mainly for deposited films, which I will come to, but in most glasses cooled from the melt it seems to be true.

This being so, what is the nature of the "conduction band" in amorphous materials? Is there necessarily a "tail" of states extending through the gap, as assumed in an early and important paper by Cohen, Fritzsche and Ovshinsky (6)? The fact that most glasses are transparent makes this unlikely. Clues came from another Leningrad idea due to Ioffe and Regel (7), namely that the mean free path cannot be shorter than the electron wavelength, and from the vastly important paper published by Anderson (8) in 1958, "Absence of diffusion in certain random lattices", described in his Nobel lecture this year.

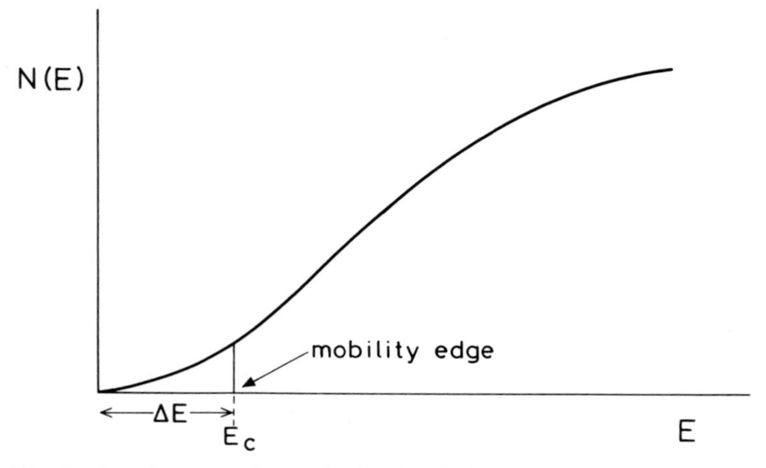

Fig. 1. The density of states in the conduction band of a non-crystalline material, showing the mobility edge E_c separated by an energy ΔE from the band edge.

We now understand that in any non-crystalline system the lowest states in the conduction band are "localized", that is to say traps, and that on the energy scale there is a continuous range of such localized states leading from the bottom of the band up to a critical energy (9) E_c, called the mobility edge (6), where states become non-localized or extended. This is illustrated in fig. 1, which shows the density of states. There is an extensive literature calculating the position of the mobility edge with various simple models (10), but it has not yet proved possible to do this for a "continuous random network" such as that postulated for SiO_2, As_2Se_3, amorphous Si or any amorphous material where the co-ordination number remains the same as in the crystal. This problem is going to be quite a challenge for the theoreticians—but up till now we depend on experiments for the answer, particularly those in which electrons are injected into a non-crystalline material and their drift mobilities measured. What one expects is that at low temperatures charge transport is by "hopping" from one localized state to another, a process involving interaction with phonons and with only a small activation energy, while at high temperatures current is carried by electrons excited to the mobility edge, the mobility behaving as $\mu_0 \exp(-\Delta E/kT)$. With this model the drift mobility, conductivity, and thermopower are illustrated in fig. 2 and (following a theory due to Friedman (11)) the Hall mobility can also be calculated. Owing to the brilliant work of Spear, Le Comber (12) and co-workers it is clear that this is just what happens in at least one material, silicon deposited from SiH_4 in a glow-discharge. As regards other materials, there is good evidence (13) that "holes" in arsenic telluride behave the same way, though there are other

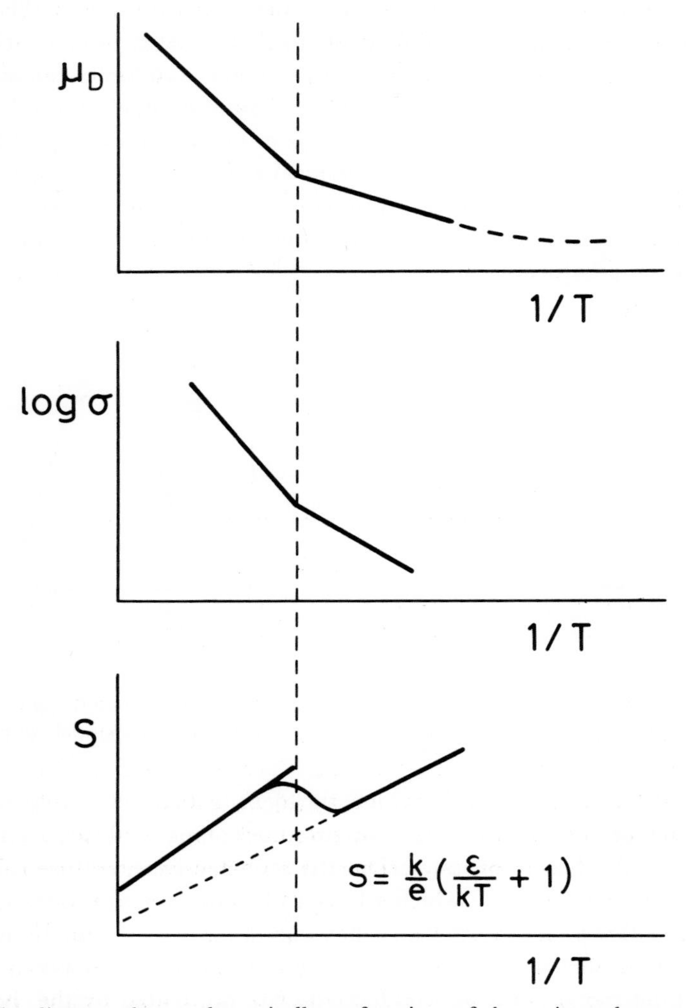

Fig. 2. The diagram shows schematically as functions of the reciprocal temperature the drift mobility μ_D, the conductivity σ and thermopower S of a material where the conduction band is as in fig. 1. ε is equal to $E_c - E_F$.

interpretations (14). But in other non-crystalline materials, notably for electrons in liquid rare gases (15), vitreous silicon dioxide (16) and some others there is no evidence for a mobility edge at all, the drift mobility *decreasing* with increasing temperature. In some materials, then, the range of localized states (ΔE in fig. 1) must be smaller than kT at room temperature. We await theoretical predictions of when this should be so.

For semiconductors, then, the data are rather scanty and we may ask how strong is the evidence for the existence of localized states and for a mobility edge generally for electrons in disordered systems? Apart from glow-discharge deposited silicon, far and away the strongest evidence, in my view, comes from systems of the type which Anderson has called "Fermi glasses". Here one must go back to the model of a metal introduced in the very early days of quantum mechanics by Sommerfeld. Electron states in a crystalline metal are occupied up to a limiting Fermi energy E_F, as in fig. 3. The density of states at the Fermi level, which I denote by $N(E_F)$, determines the electronic specific heat and the Pauli paramagnetism. These statements remain true if the medium is non-crystalline, or if there is a random field of any kind as in an alloy; but in this case states at the bottom of the band, or possibly right through it, are localized. They may be localized at the Fermi energy. If so, we call the system a Fermi glass. Although the specific heat and Pauli magnetism behave as in a metal, the conductivity does not; it tends to zero with decreasing temperature.

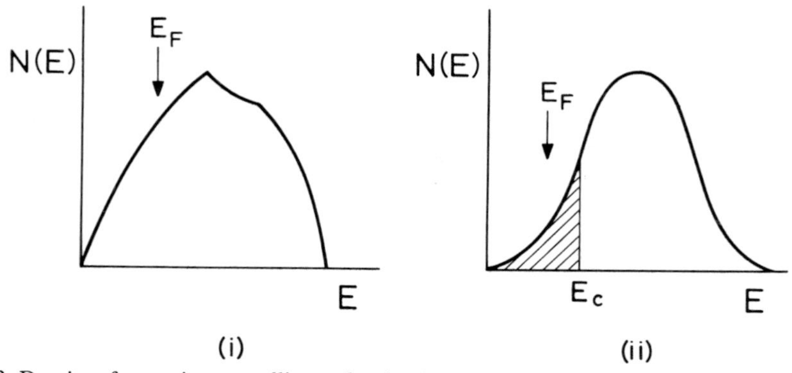

Fig. 3. Density of states in a metallic conduction band, with states occupied up to a limiting Fermi energy E_F. (i) is for a crystal, (ii) for an amorphous or liquid material, with localized states shaded and a mobility edge at E_c.

Let us examine a system in which the density of electrons or degree of disorder can be varied, either by changing the composition or in some other way. Thus if the Fermi energy crosses the mobility edge, a "metal-insulator transition" occurs, of a kind which I have called an Anderson transition (17). I will now examine the electrical behaviour of such a system. If the Fermi energy E_F lies well above any mobility edge, we expect the behaviour familiar in most liquid metals, and the conductivity can be treated by the theory put forward by Ziman (18) in 1961—one of the first successful approaches to conduction in non-crystalline materials, which showed that such problems were capable of exact treatment and encouraged the rest of us to try our hands. Ziman's theory is a "weak scattering" theory, the mean free path (L) being large compared with the distance between atoms (a). As one increases the strength of the scattering, one reaches the Ioffe-Regel condition (in this case $L \sim a$), and the conductivity is then about

$$\frac{1}{3} e^2/\hbar a \sim 3000 \ \Omega^{-1} \ \mathrm{cm}^{-1}$$

if $a \sim 3$ Å. If the disorder gets stronger and stronger, Anderson localization sets in. The conductivity just before it occurs is then

$$\text{const } e^2/\hbar a,$$

where the constant depends on the Anderson localization criterion, and is probably in the range 0.1—0.025. I have called this quantity the "minimum metallic conductivity (9, 19) and denoted it by $\sigma_{\min}$. For $a \sim 3$ Å it is in the range 250—1000 Ω^{-1} cm^{-1}, though in systems for which a is larger, such as impurity bands, it is smaller. I have maintained for several years that if the conductivity is finite in the limit of low temperatures, it cannot be less than this. This really does seem to be the case, and there is quite strong evidence for it, some of which I will describe. But the proposal proved very controversial (20), and only recently due to the numerical work of Licciardello and Thouless (21), and other analytical work is it carrying conviction among most theorists.

Now let me ask what happens when the Fermi energy lies below the mobility edge, so that states at the Fermi energy are localized, and the material is what I called a "Fermi glass". There are two mechanisms of conduction; at high temperatures electrons are excited to the mobility edge, so that

$$\sigma = \sigma_{\min} \exp\left\{-(E_c - E_F)/kT\right\}, \tag{1}$$

and at low temperatures conduction is by thermally activated hopping from one level to another. In 1969 I was able to show[5] that the latter process should give a conductivity following the law

$$\sigma = A \exp(-B/T^{\frac{1}{4}}) \tag{2}$$

with B depending on the radial extension of the wave functions and the density of states. In two dimensions $T^{\frac{1}{4}}$ becomes $T^{\frac{1}{3}}$. There has been quite a literature on this (22), following my elementary proof, and perhaps the effect of correlation is not yet perfectly understood, but I am convinced (23) that $T^{\frac{1}{4}}$ behaviour is *always* to be expected in the limit of low temperatures.

It follows, then, that for a system in which one can vary the number of electrons, the plot of resistivity against $1/T$ will be as in fig. 4. If there is a high density of electrons, and E_F lies above E_c, the conductivity should be nearly independent of temperature. As the density of electrons is lowered, the Fermi energy falls till it reaches E_c, and then $\sigma = \sigma_{\min}$. If the density falls still further, states are localized giving conduction by the two mechanisms of (1) and (2) at high and low temperatures respectively.

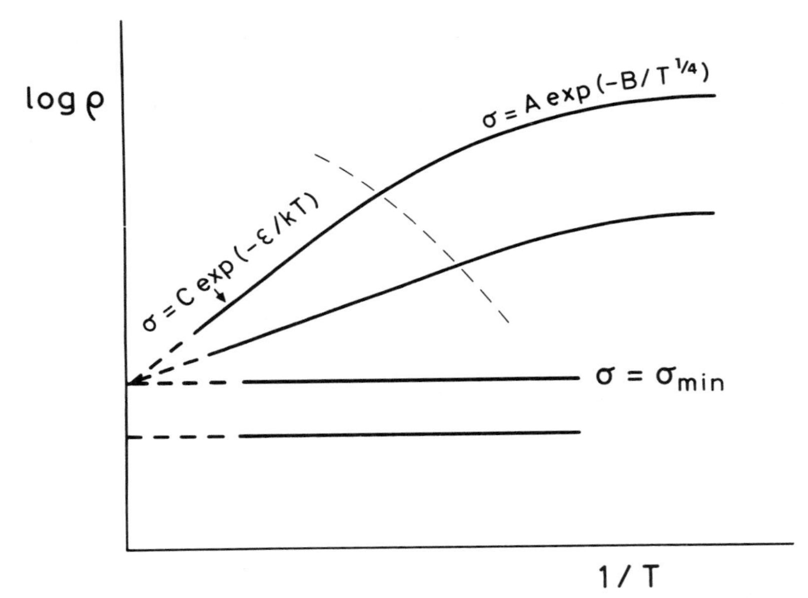

Fig. 4. Plot of log(resistivity) against 1/T for a system in which the density of electrons can be altered so that ε ($= E_c - E_F$) changes sign, giving a metal-insulator transition of Anderson type.

As regards the systems to which this concept can be applied, there are many. One is the alloy $La_{1-x}Sr_xVO_3$, which I owe to my colleagues (24) in Professor Hagenmuller's laboratory at Bordeaux. In these, a vanadium *d* band contains a number of electrons which varies with *x*, and thus with composition. But the simplest system is the MOSFET (metal-oxide-silicon-field-effect-transistor) illustrated in fig. 5. In this, *two dimensional* conduction takes place in an inversion layer at the Si—SiO_2 interface, the "band bending" being illustrated

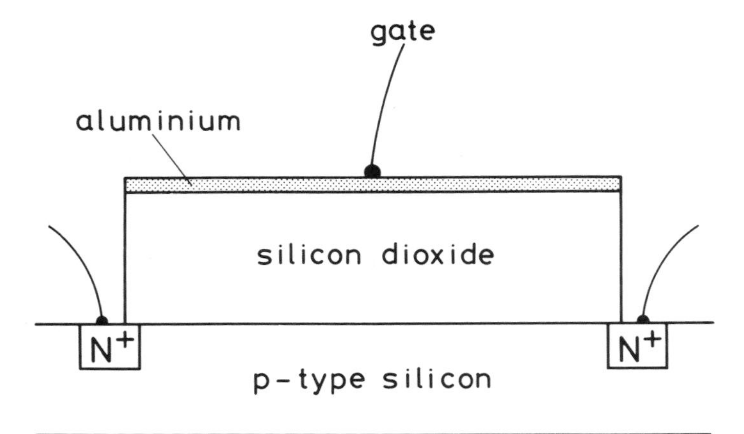

Fig. 5. A MOSFET device, for demonstration of two-dimensional conduction along the interface between the p-type Si and SiO_2.

in fig. 6. The electron gas in the inversion layers is degenerate at helium temperatures, and the beauty of the system is that the density of electrons can be varied simply by changing the gate voltage. Disorder arises because the oxide contains random charges—capable of being controlled by the technology. The investigations of Pepper and co-workers (25, 15) showed behaviour confirming the pattern of fig. 4 in every detail, and reasonable values of $\sigma_{\min}$ (expected to be 0.1 $e^2/\hbar$ in two dimensions).

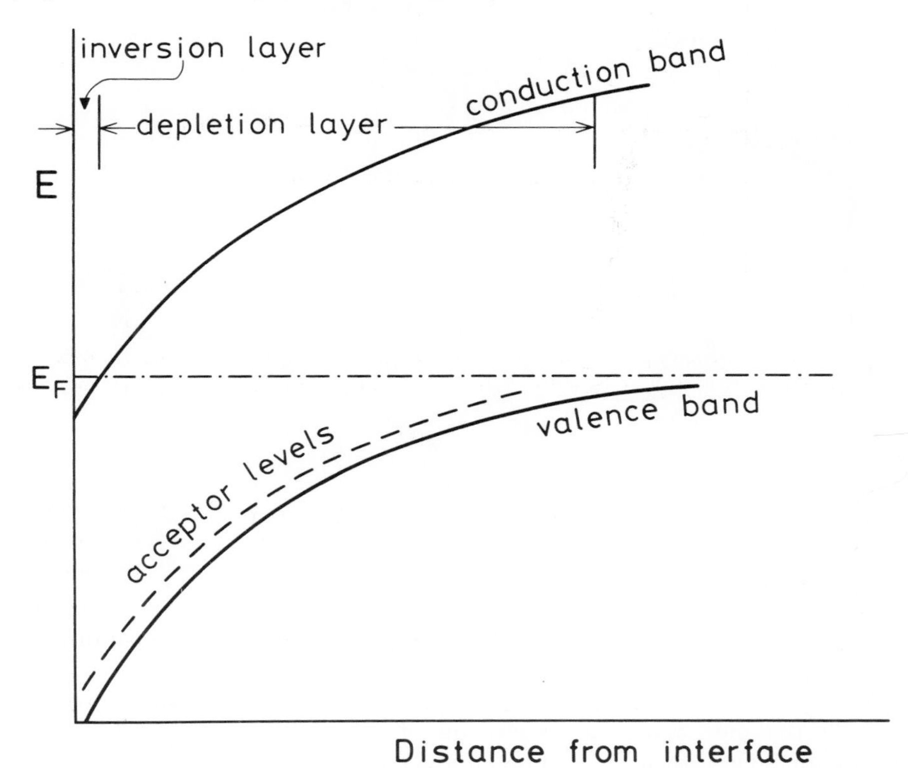

Fig. 6. Application of a field to the surface of a p-type semiconductor inducing an n-type surface layer.

$T^{\frac{1}{4}}$ behaviour occurs also in many amorphous semiconductors, such as Si and Ge, and indeed was first observed in amorphous silicon by Whalley (26) and $T^{\frac{1}{3}}$ in thin films by Knotek, Pollak et al (44). The Marburg group under Professor Stuke (27) has investigated this phenomenon and its relation to electron spin resonance in detail. The idea here is that many amorphous materials contain "deep levels" due to defects such as dangling bonds; a photograph (fig. 7) is included to show what is meant. Some of these may be charged and some not; if so, the density of states at the Fermi level is finite, and electrons hopping from one of these levels to another can occur, giving a conductivity following eqn (2).

Now I would like to finish the scientific part of this lecture by mentioning two new things and two old ones.

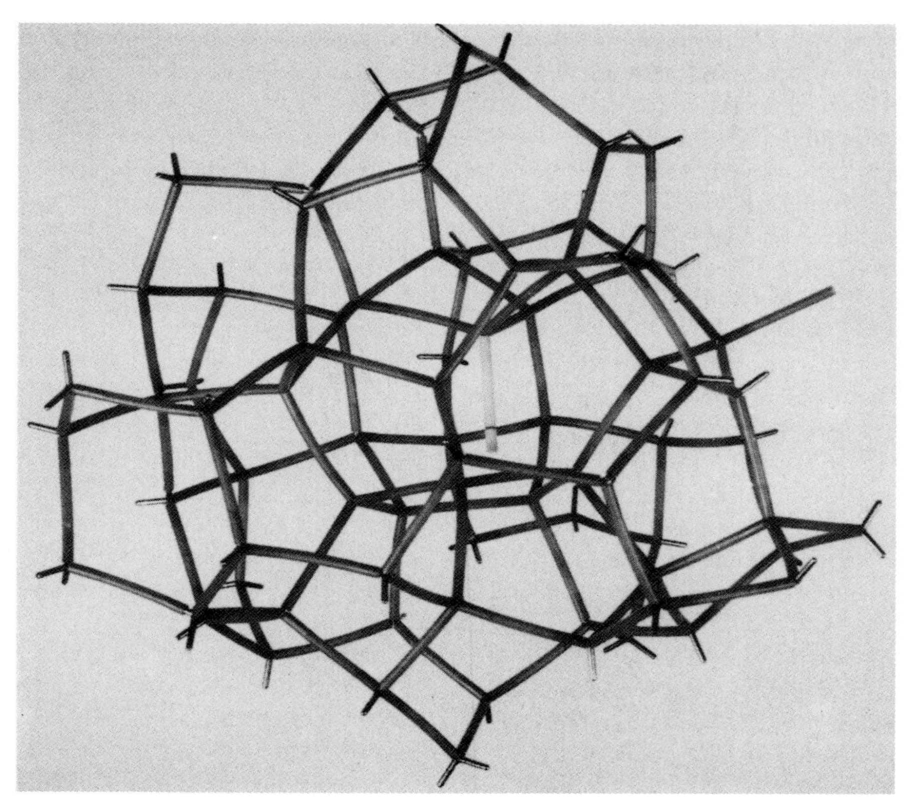

Fig. 7. A "dangling bond" in a continuous random network with fourfold co-ordination (courtesy of Dr E. A. Davis).

One of the new things is the important discovery by Spear and co-workers (28, 29) that one *can* dope deposited films of silicon, for instance by depositing PH_3 with SiH_4. Much of the phosphorus seems to go in with three nearest neighbours, so that there are no loosely bound electrons, but sufficient take up fourfold co-ordination so as to give donors. These lose their electrons to states in the gap, but the Fermi energy can be shifted very near to the conduction or the valence bands. It is thus possible to make comparatively cheap p-n junctions, with important implications for the economics of solar cells.

The other new thing is the introduction of the "negative Hubbard U" by Anderson (30), and the application of the idea to specific defects by Street and Mott (31), and by Mott, Davis and Street (32), with subsequent development by Kastner, Adler and Fritzsche (33). It is here supposed by the latter authors that there is a real difference in glasses between defects and fluctuations in density, each making their specific contribution to the entropy (34).

We think the model is applicable to materials in which the top of the valence band consists of lone pair orbitals (35), for instance in selenium p orbitals that do not take part in a bond. If so, we believe that "dangling bonds" as shown in fig. 5 will *either* contain two electrons or none, and thus show no free spin and be positively or negatively charged. The repulsive energy (the "Hubbard U") due to two electrons on one site is compensated because the positive centre can form a strong bond if it moves towards another selenium, which is thus threefold co-ordinated. The positive and negative centres thus formed have been called by Kastner et al "valence alternating pairs". The important point that these authors show is that one can form a pair without breaking a bond, while a neutral centre (dangling bond) costs much more energy to form it. The evidence that there are charged centres in these materials comes mainly from the experimental work of Street, Searle and Austin (36) on photoluminescence. We now think that the model is capable of explaining a great many of the properties of chalcogenide glasses, and perhaps of oxide glasses too. In particular, it shows how the Fermi energy can be pinned without introducing free spins, it seems capable of giving an explanation of dielectric loss and it provides traps which limit the drift mobility. I feel that this work, particularly as formulated by Anderson, is another example of the Kolomiets principle, that glasses cannot be doped; they form complete bonds whenever they can, even if the cost is negative and positive centres.

I said I would end by talking about two old things. One of course is the use of amorphous selenium for office copying by the Xerox company—a multi-billion dollar industry developed, as is so often the case, before anybody had tried to make theories of the processes involved. When the subject became fashionable all over the world, we found of course that the Xerox scientists knew a great deal about it; and their recent contributions, particularly on dispersive transport (37), are of the highest importance.

The other comparatively "old" thing is the threshold switch invented by S. R. Ovshinsky (38). This in its simplest form consists of a deposited film of a chalcogenide glass about one micron thick, with a molybdenum or carbon electrode on each side. Such a system switches into a highly conducting state as the potential across it is increased, switching off again when the current through it drops below a certain value (fig. 8). The claims made for this device generated a considerable amount of controversy, it being suggested that a thermal instability was involved and that similar phenomenon had been observed many years ago. I do not think this is so, and proposed (39) in 1969, soon after the phenomenon was brought to my notice, that the phenomenon is an example of double injection, holes coming in at one electrode and electrons attheother. This is still my opinion. Experimental work, notably by Petersen and Adler (40) and by Henisch (41), make it practically certain that the conducting channel is not hot enough appreciably to affect the conductivity. The work of Petersen and Adler shows that in the on-state the current flows in a channel in which the density of electrons and holes and the current density do not depend on the total current; as the current increases, the channel simply gets wider, and can be much thicker than the thickness of the film. My own belief (42) is that the channel has strong similarities to the electron-hole droplets in crystalline germanium, that even at room temperature

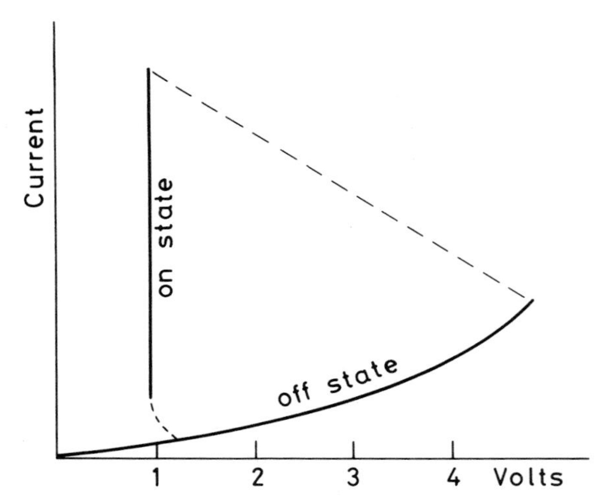

Fig. 8. Current voltage curve of a threshold switch, consisting of a thin chalcogenide film between two electrodes.

one has to do with a *degenerate* plasma of electrons and holes, and that the density of carriers is such that the Fermi energies of both gases lie above the respective mobility edges; only thus can the observed mobilities ($\sim$ 1 cm^2/V s) be explained. But we are still far from a full understanding of the behaviour of this fascinating device.

Finally, since I think that mine is the first Nobel prize to be awarded wholly for work on amorphous materials, I would like to say that I hope this will give a certain status to a new, expanding and at times controversial subject. The credit for the prize must certainly be shared with people with whom I've talked and corresponded all over the world. I myself am neither an experimentalist nor a real mathematician; my theory stops at Schrödinger equation. What I've done in this subject is to look at all the evidence, do calculations on the back of an envelope and say to the theoretician, "if you apply your techniques to this problem, this is how it will come out" and to the experimentalists just the same thing. This is what I did for $T^{\frac{1}{4}}$ hopping and the minimum metallic conductivity. But without these others on both sides of the fence I would have got nowhere. My thanks are due particularly to my close collaborator Ted Davis, joint author of our book on the subject (43), to Walter Spear and Mike Pepper in the U.K., to Josef Stuke in Marburg, to Karl Berggren in Sweden, to Hiroshi Kamimura in Japan, to Mike Pollak, Hellmut Fritzsche and to many others in the United States and of course to Phil Anderson.

REFERENCES

1. Taylor, G. I., Proc. Roy. Soc. A *145*, 362 (1934).
2. Mott, N. F. and Nabarro, F. R. N., Proc. Phys. Soc. *52*, 86 (1940); Report on Strength of Solids (The Physical Society, London) p. 1.
3. See papers in Amorphous and Liquid Semiconductors (3rd International Conference), ed. N. F. Mott, North Holland, 1970.
4. For a review, see B. T. Kolomiets, Phys. Stat. Solidi *7*, 359 (1964).
5. Mott, N. F., Phil. Mag., *19*, 835 (1969).
6. Cohen, M. H., Fritzsche, H. and Ovshinsky, S. R., Phys. Rev. Lett., *22*, 1065 (1969).
7. Ioffe, A. F. and Regel, A. R., Prog. Semicond., *4*, 237 (1960).
8. Anderson, P. W., Phys. Rev., *109*, 1492 (1958).
9. Mott, N. F., Adv. Phys., *16*, 49 (1967).
10. Edwards, J. T. and Thouless, D. J., J. Phys. C (Solid St. Phys.) *5*, 807 (1972).
11. Friedman, L., J. Non-Cryst. Solids, *6*, 329 (1971).
12. See review by W. E. Spear, Adv. Phys., *23*, 523 (1974).
13. Nagels, P., Callaerts, R. and Denayer, M., Amorphous and Liquid Semiconductors (5th Int. Conf.), ed. J. Stuke and W. Brenig, Taylor & Francis, London, p. 867. (1974).
14. Emin, D., Seager, C. H. and Quinn, R. K., Phys. Rev. Lett., *28*, 813 (1972).
15. Miller, L. S., Howe, S. and Spear, W. E., Phys. Rev., *166*, 871 (1968).
16. Hughes, R. C., Phys. Rev. Lett., *30*, 1333 (1973); Appl. Phys. Lett., *26*, 436 (1975); Phys. Rev. B, *15*, 2012 (1977).
17. Mott, N. F., Pepper, M., Pollitt, S., Wallis, R. H. and Adkins, C. J., Proc. R. Soc. A, *345*, 169 (1975).
18. Ziman, J. M., Phil. Mag., *6*, 1013 (1961).
19. Mott, N. F., Phil. Mag., *26*, 1015, (1972).
20. Cohen, M. H. and Jortner, J., Phys. Rev. Lett., *30*, 699 (1973).
21. Licciardello, D. C. and Thouless, D. J., J. Phys. C: Solid St. Phys., *8*, 4157, (1957).
22. Shklovskii, B. I. and Efros, A. L., Soviet Phys. JETP, *33*, 468 (1971); V. Ambegaokar, B. I. Halperin and J. S. Langer, Phys. Rev. B, *4*, 2612 (1971); M. Pollak, J. Non-Cryst. Solids, *11*, 1 (1972).
23. Mott, N. F., Phil. Mag., *34*, 643 (1976).
24. Dougier, P. and Casalot, A. J., Solid St. Chem., *2*, 396 (1970).
25. Pepper, M., Pollitt, S., Adkins, C. J. and Oakley, R. E., Phys. Lett. A, *47*, 71 (1974).
26. Walley, P. A., Thin Solid Films, *2*, 327 (1968).
27. Stuke, J., 6th Int. Conference on Amorphous and Liquid Semiconductors, Leningrad, p. 193, ed. B. T. Kolomiets, Nauka.
28. Spear, W. E. and Le Comber, P. G., Solid St. Commun., *17*, 1193 (1975).
29. Spear, W. E., Adv. Phys. *6*, 811 (1977).
30. Anderson, P. W., Phys. Rev. Lett., *34*, 953 (1975).
31. Street, R. A. and Mott, N. F., Phys. Rev. Lett., *35*, 1293 (1975).
32. Mott, N. F., Davis, E. A. and Street, R. A., Phil. Mag., *32*, 961 (1975).
33. Kastner, M., Adler, D. and Fritzsche, H., Phys. Rev. Lett., *37*, 1504 (1976).
34. Bell, R. J. and Dean, P., Physics and Chemistry of Glasses, 9, 125 (1968).
35. Kastner, M., Phys. Rev. Lett., *28*, 355 (1972).
36. For a review, see R. A. Street, Adv. Phys., *25*, 397 (1976).
37. Scher, H. and Montroll, E. W., Phys. Rev. B, *12*, 2455 (1975).
38. Ovshinsky, S. R., Phys. Rev. Lett., *21*, 1450 (1968).
39. Mott, N. F., Contemporary Physics, *10*, 125 (1969).
40. Petersen, K. E. and Adler, D., J. Appl. Phys., *47*, 256 (1976).
41. Henisch, H. K. and Pryor, R. W., Solid-St. Electronics, *14*, 765 (1971).
42. Adler, D., Henisch, H. K. and Mott, N. F., Rev. Mod. Phys. 50 (in press) 1978.
43. Mott, N. F. and Davis, E. A., Electronic Processes in Non-Crystalline Meterials, Oxford University Press, 1st ed. 1971, 2nd ed. 1978.
44. Knotek, M. L., Pollak, M., Donovan, T. M. and Kurzman, H., Phys. Rev. Lett., 30, 853 (1973).

Physics 1978

PETER LEONIDOVITCH KAPITZA

for his basic inventions and discoveries in the area of low-temperature physics

ARNO A PENZIAS and ROBERT W WILSON

for their discovery of cosmic microwave background radiation

THE NOBEL PRIZE FOR PHYSICS

Speech by Professor LAMEK HULTHÉN of the Royal Academy of Sciences. Translation from the Swedish text

Your Majesties, Your Royal Highnesses, Ladies and Gentlemen,

This year's prize is shared between Peter Leonidovitj Kapitza, Moscow, "for his basic inventions and discoveries in the area of low-temperature physics" and Arno A. Penzias and Robert W. Wilson, Holmdel, New Jersey, USA, "for their discovery of cosmic microwave background radiation".

By low temperatures we mean temperatures just above the absolute zero, $-273°C$, where all heat motion ceases and no gases can exist. It is handy to count degrees from this zero point: "degrees Kelvin" (after the British physicist Lord Kelvin) E.g. 3 K (K = Kelvin) means the same as $-270°C$.

Seventy years ago the Dutch physicist Kamerlingh-Onnes succeeded in liquefying helium, starting a development that revealed many new and unexpected phenomena. In 1911 he discovered *superconductivity* in mercury: the electric resistance disappeared completely at about 4 K. 1913 Kamerlingh-Onnes received the Nobel prize in physics for his discoveries, and his laboratory in Leiden ranked for many years as the Mekka of low temperature physics, to which also many Swedish scholars went on pilgrimage.

In the late twenties the Leiden workers got a worthy competitor in the young Russian Kapitza, then working with Rutherford in Cambridge, England. His achievements made such an impression that a special institute was created for him: the Royal Society Mond Laboratory (named after the donor Mond), where he stayed until 1934. Foremost among his works from this period stands an ingenious device for liquefying helium in large quantities—a pre-requisite for the great progress made in low temperature physics during the last quarter-century.

Back in his native country Kapitza had to build up a new institute from scratch. Nevertheless, in 1938 he surprised the physics community by the discovery of the *superfluidity* of helium, implying that the internal friction (viscosity) of the fluid disappears below 2.2 K (the so-called lambda-point of helium). The same discovery was made independently by Allen and Misener at the Mond Laboratory. Later Kapitza has pursued these investigations in a brilliant way, at the same time guiding and inspiring younger collaborators, among whom we remember the late Lev Landau, recipient of the physics prize 1962 "for his pioneering theories for condensed matter, especially liquid helium". Among Kapitza's accomplishments we should also mention the method he developed for producing very strong magnetic fields.

Kapitza stands out as one of the greatest experimenters of our time, in his domain the uncontested pioneer, leader and master.

We now move from the Institute of Physical Problems, Moscow, to Bell Telephone Laboratories, Holmdel, New Jersey, USA. Here Karl Jansky, in the beginning of the thirties, built a large movable aerial to investigate sources of radio noise and discovered that some of the noise was due to radio waves coming from the Milky Way. This was the beginning of radio astronomy that has taken such an astounding development after the second World War—as an illustration let me recall the discovery of the pulsars, honoured with the physics prize 1974.

In the early 1960ies a station was set up in Holmdel to communicate with the satellites Echo and Telstar. The equipment, including a steerable horn antenna, made it a very sensitive receiver for microwaves, i.e. radio waves of a few cm wavelength. Later radio astronomers Arno Penzias and Robert Wilson got the chance to adapt the instrument for observing radio noise e.g. from the Milky Way. They chose a wave length c. 7 cm where the cosmic contribution was supposed to be insignificant. The task of eliminating various sources of errors and noise turned out to be very difficult and time-consuming, but by and by it became clear that they had found a background radiation, equally strong in all directions, independent of time of the day and the year, so it could not come from the sun or our Galaxy. The strength of the radiation corresponded to what technicians call an antenna temperature of 3 K.

Continued investigations have confirmed that this background radiation varies with wave length in the way prescribed by wellknown laws for a space, kept at the temperature 3 K. Our Italian colleagues call it "la luce fredda"—the cold light.

But where does the cold light come from? A possible explanation was given by Princeton physicists Dicke, Peebles, Roll and Wilkinson and published together with the report of Penzias and Wilson. It leans on a cosmological theory, developed about 30 years ago by the Russian born physicist George Gamow and his collaborators Alpher and Herman. Starting from the fact that the universe is now expanding uniformly, they concluded that it must have been very compact about 15 billion years ago and ventured to assume that the universe was born in a huge explosion—the "Big Bang". The temperature must then have been fabulous: 10 billion degrees, perhaps more. At such temperatures lighter chemical elements can be formed from existing elementary particles, and a tremendous amount of radiation of all wave lengths is released. In the ensuing expansion of the universe, the temperature of the radiation rapidly goes down. Alpher and Herman estimated that this radiation would still be left with a temperature around 5 K. At that time, however, it was considered out of the question, that such a radiation would ever be possible to observe. For this and other reasons the predictions were forgotten.

Have Penzias and Wilson discovered "the cold light from the birth of the universe"? It is possible—this much is certain that their exceptional perseverance and skill in the experiments led them to a discovery, after which cosmology is a science, open to verification by experiment and observation.

Piotr Kapitsa, Arno Penzias, Robert Wilson, In accordance with our tradition I have given a brief account in Swedish of the achievements, for which you share this year's Nobel prize in Physics. It is my privilege and pleasure to congratulate you on behalf of the Royal Swedish Academy of Sciences and ask you to receive your prizes from the hands of His Majesty the King!

PJOTR LEONIDOVICH KAPITZA

Pjotr Leonidovich Kapitza was born in Kronstadt, near Leningrad, on the 9th July 1894, son of Leonid Petrovich Kapitza, military engineer, and Olga Ieronimovna née Stebnitskaia, working in high education and folklore research.

Kapitza began his scientific career in A.F. Ioffe's section of the Electromechanics Department of the Petrograd Polytechnical Institute, completing his studies in 1918. Here, jointly with N.N. Semenov, he proposed a method for determining the magnetic moment of an atom interacting with an inhomogeneous magnetic field. This method was later used in the celebrated Stern-Gerlach experiments.

At the suggestion of A.F. Ioffe in 1921 Kapitza came to the Cavendish Laboratory to work with Rutherford. In 1923 he made the first experiment in which a cloud chamber was placed in a strong magnetic field, and observed the bending of alfa-particle paths. In 1924 he developed methods for obtaining very strong magnetic fields and produced fields up to 320 kilogauss in a volume of 2 cm^3. In 1928 he discovered the linear dependence of resistivity on magnetic field for various metals placed in very strong magnetic fields. In his last years in Cambridge Kapitza turned to low temperature research. He began with a critical analysis of the methods that existed at the time for obtaining low temperatures and developed a new and original apparatus for the liquefaction of helium based on the adiabatic principle (1934).

Kapitza was a Clerk Maxwell Student of Cambridge University (1923–1926), Assistant Director of Magnetic Research at Cavendish Laboratory (1924–1932), Messel Research Professor of the Royal Society (1930–1934), Director of the Royal Society Mond Laboratory (1930–1934). With R.H. Fowler he was the founder editor of the International Series of Monographs on Physics (Oxford, Clarendon Press).

In 1934 he returned to Moscow where he organized the Institute for Physical Problems at which he continued his research on strong magnetic fields, low temperature physics and cryogenics.

In 1939 he developed a new method for liquefaction of air with a low-pressure cycle using a special high-efficiency expansion turbine. In low temperature physics, Kapitza began a series of experiments to study the properties of liquid helium that led to discovery of the superfluidity of helium in 1937 and in a series of papers investigated this new state of matter.

During the World War II Kapitza was engaged in applied research on the production and use of oxygen that was produced using his low pressure expansion turbines, and organized and headed the Departement of Oxygen Industry attached to the USSR Council of Ministers.

Late in the 1940's Kapitza turned his attention to a totally new range of physical problems. He invented high power microwave generators – planotron and nigotron (1950–1955) and discovered a new kind of continuous high pressure plasma discharge with electron temperatures over a million K.

Kapitza is director of the Institute for Physical Problems. Since 1957 he is a member of the Presidium of the USSR Academy of Sciences. He was one of the founders of the Moscow Physico-Technical Institute (MFTI), and is now head of the department of low temperature physics and cryogenics of MFTI and chairman of the Coordination Council of this teaching Institute. He is the editor-in-chief of the Journal of Experimental and Theoretical Physics and member of the Soviet National Committee of the Pugwash movement of scientists for peace and disarmament.

He was married in 1927 to Anna Alekseevna Krylova, daughter of Academician A.N. Krylov. They have two sons, Sergei and Andrei.

Honorary degrees

D.Phys.-Math.Sc., USSR Academy of Sciences, 1928; D.Sc., Algiers University, 1944; Sorbonne, 1945; D.Ph., Oslo University, 1946; D.Sc., Jagellonian University, 1964; Technische Universität Dresden, 1964; Charles University, 1965; Columbia University, 1969; Wroclaw Technical University, 1972; Delhi University, 1972; Université de Lausanne, 1973; D.Ph., Turku University, 1977.

Honorary memberships

Member of the USSR Academy of Sciences, 1939 (corresponding member – 1929); Fellow of the Royal Society, London, 1929; French Physical Society, 1931; Institute of Physics, England, 1934; International Academy of Astronautics, 1964; Honorary Member of the Moscow Society of Naturalists, 1935; the Institute of Metals, England, 1943; the Franklin Institute, 1944; Trinity College Cambridge, 1925; New York Academy of Sciences, 1946; Indian Academy of Sciences, 1947; the Royal Irish Academy, 1948; National Institute of Sciences of India, 1957; German Academy of Naturalists "Leopoldina", 1958; International Academy of the History of Science, 1971; Tata Institute of Fundamental Research, Bombay, India, 1977. Foreign Member of Royal Danish Academy of Sciences and Letters, 1946; National Academy of Sciences, USA, 1946; Indian National Sciences Academy, 1956; Polish Academy of Sciences, 1962; Royal Swedish Academy of Sciences, 1966; American Academy of Arts and Sciences, 1968; Royal Netherlands Academy of Sciences, 1969; Serbian Academy of Sciences and Arts, 1971; Finnish Academy of Arts and Sciences, 1974. Honorary Fellow of Churchill College Cambridge, 1974.

Awards

Medal of the Liége University, 1934; Faraday Medal of the Institute of Electrical Engineers, 1942; Franklin Medal of the Franklin Institute, 1944; Sir Devaprasad Sarbadhikary Gold Medal of the Calcutta University, 1955; Kothenius Gold Medal of the German Academy of Naturalists "Leopoldina", 1959; Frédéric Joliot-Curie Silver Medal of the World Peace Committee, 1959; Lomonosov Gold Medal of the USSR Academy of Sciences, 1959; Great Gold Medal of the USSR Exhibition of Economic Achievements, 1962; Medal for Merits in Science and to Mankind of the Czechoslovak Academy of Sciences, 1964; International Niels Bohr Medal of Dansk Ingeniørvorening, 1964; Rutherford Medal of the Institute of Physics and

Physical Society, England, 1966; Golden Kamerlingh Onnes Medal of the Netherlands Society of Refrigeration, 1968; Copernic Memorial Medal of the Polish Academy of Sciences, 1974.

USSR State Prize – 1941, 1943; Simon Memorial Award of the Institute of Physics and Physical Society, England, 1973; Rutherford Memorial Lecture, Royal Society of London; Bernal Memorial Lecture, Royal Society of London, 1976.

Order of Lenin – 1943, 1944, 1945, 1964, 1971, 1974; Hero of Socialist Labour, 1945, 1974; Order of the Red Banner of Labour, 1954;
Order of the Jugoslav Banner with Ribbon, 1967.

Publications

Collected Papers of P.L. Kapitza, 3 vol., Pergamon Press, Oxford, 1964–1967;
High Power Microwave Electronics, Pergamon Press, 1964.
Experiment. Theory. Practice. "Nauka", Moscow, 1977.
Le livre du problème de physique, CEDIC, Paris, 1977.

Professor Kapitza died in 1984.

PLASMA AND THE CONTROLLED THERMONUCLEAR REACTION

Nobel Lecture, 8 December, 1978

by

P. L. KAPITZA
Institute for Physical Problems of the Academy of Sciences, Moscow, USSR

The choice of the theme for my Nobel lecture presents some difficulty for me. Usually the lecture is connected with work recognized by the prize. In my case the prize was awarded for work in low temperature physics, at temperatures of liquid helium, a few degrees above absolute zero. It so happened that I left this field some 30 years ago, although at the Institute under my directorship low temperature research is still being done. Personally I am now studying plasma phenomena at those very high temperatures that are necessary for the thermonuclear reaction to take place. This research has led to interesting results and has opened new possibilities, and I think that as a subject for the lecture this is of more interest than my past low temperature work. For it is said, "les extrêmes se touchent".

It is also well recognized that at present the controlled thermonuclear reaction is the process for producing energy that can effectively resolve the approaching global energy crisis, resulting from the depletion of fossil fuels used now as our principal energy source.

It is also well known that intensive research on fusion is done in many countries and is connected with fundamental studies of high temperature plasmas. The very possibility of fusion is well beyond doubt, for it takes place in the explosion of hydrogen bombs. We also have a detailed theoretical understanding of nuclear fusion reactions that is in agreement with experiments. But in spite of the great effort and large sums spent up to now, it is impossible to conduct the process of fusion as to make it a useful source of energy. This certainly is a cause for some bewilderment.

One could expect that during the decades of experimental and theoretical plasma work in studying the conditions for fusion we would have reached a sufficient understanding of the various facts that hinder us from setting up a controlled thermonuclear reaction. It could be expected that we should have discovered and revealed the main difficulties that bar our progress. In this lecture I hope to clarify, what are these difficulties, and what are the chances that these difficulties will be resolved. I will also try to explain the divergence of opinions of different scientists on the practical possibilities for obtaining useful thermonuclear energy.

Before embarking into this subject I would like to speak on the practical importance of obtaining energy from nuclear sources.

The reality of the approaching global energy crisis is connected with the unavoidable lack of raw materials: gas, oil, coal. This is now generally appreciated. It is also known that the GNP (gross national product) that determines the wellbeing of people is proportional to the expenditure of energy. Energy resources depletion will inevitably lead to general impoverishment.

Two possible ways out of the approaching energy crisis are discussed. The first, maybe the more attractive, is to extensively use the inexhaustable sources of energy: hydroelectric power, the power of wind, solar energy, geothermal energy. The second way is to use nuclear energy discovered by man less than a hundred years ago. At present heavy element fission power is already cheaper than energy from some nonexhaustable sources.

It is well known that the main fuel in these reactors is uranium. It has been shown that as used at present there is enough uranium for only a hundred years. If in the future uranium will be more fully used in breeder reactors, it will last 50 times longer, for a few thousand years. Many consider that uranium dissolved in sea water may also be efficiently used for cheap energy production. Thus it may seem that the processes now used in modern nuclear reactions may resolve the approaching energy crisis. But there are important reasons against using uranium as a source of energy. These arguments are mainly connected with security.

In the first place, the use of uranium leads to accumulation of longlived radioactive wastes and the problem of safely storing a growing amount of these waste materials. This is a problem that at present has not been definitely solved.

In the second place, in a large energy-producing nuclear power plant a vast amount of radioactive material is accumulated, so that in a hypothetical accident, the dispersion of this material might lead to a catastrophe comparable in scale to that of Hiroshima.

I think that eventually modern technology will resolve these two dangers. But there is still a third hazard, even more grave. This is the danger, that the construction of great numbers of nuclear power stations will inevitably lead to such a huge amount of radioactive material disseminated around the world, that an efficient control on its proper uses will be practically impossible. In the long run not only a small country, but even a wealthy man or a large industrial organization will be able to build its own atomic bomb. There is at present no secret of the bomb. The necessary amount of plutonium, especially if breeders are to be widely built, will be readily available. Thus recently in India a small bomb was built and exploded. With the present system of international organizations there is nobody with sufficient authority that could execute the necessary control of the peaceful use of uranium as a source of energy. Moreover, it is not clear now how such an organization could be set up. This is the main reason why it is most important to obtain energy by the third way, through the process of thermonuclear fusion.

It is common knowledge, that this process will not lead to generation of large amounts of radioactive wastes and thus to a dangerous accumulation of radioactive material, and mainly it does not open any chances for a feasible nuclear explosion. This is the main reason why the solution of the scientific and technical problems involved in controlled thermo-nuclear fusion is considered of prime importance by many physicists.

The conditions for the thermonuclear reaction for energy production are well known and firmly established. There are two reactions of importance: the D + D and D + T process. The first one is the reaction between two nuclei of deuterium. The second occurs in the interaction of deuterium with tritium. In both cases fast neutrons are emitted, whose energy may be used. As a small amount of deuterium is present in water and is easy to extract, an abundant source of fuel is available. Free tritium practically does not exist in nature and tritium has to be produced, as it is usually done, through the interaction of neutrons with lithium.

The thermonuclear reaction is to take place in a high temperature plasma. So as to practically use the energy of neutrons the production of energy has to be greater than the power used to sustain the high plasma temperature.Thus the energy, obtained from the neutrons, has to be much greater than the bremsstrahlung radiation of the electron gas in the plasma. Calculations show that for useful energy production for the D + D reaction the necessary ion temperature is 10 times greater than in the case of the D + T reaction. Although the D + T reaction works at a lower temperature, it is hampered by the necessity to burn lithium, whose amount in nature is limited. Moreover, it seems that the use of lithium greatly complicates the design of the reactor. Calculations show that for obtaining useful energy the temperature of ions in a plasma for the D + D reaction should be about 10^9K and for the D + T reaction about 10^8K.

From research in plasma and nuclear physics it is thus known that for practical energy generation purposes, the technical problem of realising a controlled thermonuclear reaction is reduced to obtaining a plasma ion temperature at least 10^8K with a density $10^{13}-10^{14}$ cm^{-3}. It is obvious that the containment of a plasma in this state by any ordinary vessel cannot be done, as there is no material that can withstand the necessary high temperature.

A number of methods for the containment of plasma and its thermal isolation have been suggested.

The most original and promising method was the "Tokamak" proposed in the Soviet Union and under development for more than a decade /(1) page 15/. The principle of its operation can be seen from the design shown in fig. 1. The plasma is confined by a magnetic field, generated in a toroidal solenoid. The plasma has the form of a ring of a radius R and a cross section of the radius a, placed in the coil. The plasma has a pressure of a few atmospheres. As it expands in the magnetic field, currents are excited that retard this expansion. The plasma is surrounded by a vacuum insulation. This is necessary to sustain the sufficiently high temperatures at which thermonuclear reactions take place. It is obvious that this method of confinement is limited in time. Calculations show that due to the low thermal capacity of the plasma, the energy for initial plasma heating, even in cases when the plasma exists for a few seconds, will be small as com-

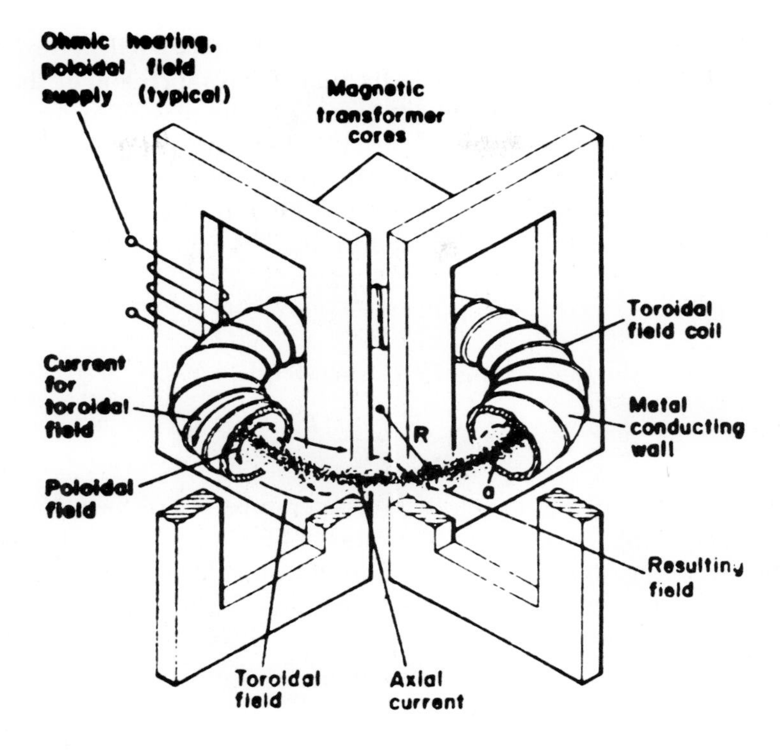

Fig. 1. Main features of a Tokamak

pared with the thermonuclear energy. Thus a reactor of this type may effectively work only in a pulsed mode. The Tokamak is started as a betatron: by discharging condensors through the coils of the transformer yoke. In practice plasma confinement by this method is not simple. In the first place there are difficulties in stabilising the plasma ring in the magnetic field. With the growth of the cross section radius a and moreover of the torus radius R, the ring loses its proper form and becomes unstable. This difficulty may be circumvented by choosing the appropriate ratio of R to a, and by properly designing the magnetic field, although at present the time for plasma confinement is only a small fraction of a second. It is assumed that with scaling the Tokamak up this time will be proportional to the square of the size of the machine.

But the main difficulty is due to reasons, not fully appreciated in the beginning. For the thermonuclear reaction one has to heat the D and T ions. The main difficulty in passing heat to them is due to the fact that the plasma is heated by an electric field. In this case all the energy is transferred to the electrons and is only slowly transferred to the ions because of their large mass as compared to the mass of the electrons. At higher temperatures this heat transfer gets even less efficient. In the Tokamak the plasma is heated by the betatron current induced through the condensor discharge. Thus, all the energy for heating the plasma is confined to the electrons and is transferred to the ions by collisions. To heat the ions to the desired temperatures the necessary time Δ t is much longer than the time during which we may maintain heating of the plasma by an electric current. The calculations that are usually done are complicated, as attempts were made to do them as exactly as possible, and so they lose in clarity. It is easy to estimate the lower time limit in which the ion heating may be made by the following simple formula /(2) page 24 expression 14/

$$\Delta t > -2.5 \cdot 10^2 \frac{f}{\Lambda} \frac{T_e^{3/2}}{n} ln\,(1 - \frac{T_i}{T_e}).$$

We assume that during heating the plasma density n,

$$n = \frac{7.3 \cdot 10^{21} P}{T_e},$$

the pressure P (atm) and the electron temperature T_e are constant.

The coefficient f is equal to the ratio of the ion mass to that of the proton, Λ is the well known logarithmic factor /2 in (4)/, T_i—the ion temperature. For modern Tokamaks, operating with the D + T reaction and at plasma temperatures $T_i = 5.10^8$ and $n = 3.10^{13} cm^{-3}$ (with an initial electron temperature $T_e = 10^9 K$) the time necessary to heat the ions to nuclear process temperatures is more than 22 seconds, at least two orders of magnitude more than confinement times in the modern Tokamaks. The plasma confinement time may be made greater only by building a larger machine, as it seems that the time Δ t is proportional to the square of the size. From this formula it also follows that the time Δ t for the D + D reaction is greater by another two orders of magnitude and then $\Delta t \sim 2.10^3$ sec. The difficulties with the time for heating the ions is now fully recognized, although one cannot see how to shorten this time and how a Tokamak may work if, before the plasma ions have been heated, all the betatron energy from the condensors will be fully radiated by the electrons. That is why in the current Tokamak projects extraneous energy sources are envisaged that are greater than the energy of the betatron process, used only for initially firing the plasma.

Extra energy must be transferred to the ions by a more efficient way than Coulomb scattering of electrons on ions. There are two possible processes for this. The first /(1) page 20/ already used, consists in injecting into the plasma ring atoms of deuterium or tritium, already accelerated to temperatures necessary for the thermonuclear reaction. The second process of heating is through exciting radial Alfvèn magnetoacoustic waves in the external magnetic field by the circulating high frequency current. It is known /(3)/ that the energy dissipated by magnetoacoustic waves is directly passed into the ions and the transmitted power is sufficient to heat the ions and sustain their temperature for a sufficiently long time. Thus the problem of heating the ions may be solved, although the mode of operation of Tokamak will be more complicated than at first suggested. The design of the Tokamak becomes more complicated and its efficiency diminishes.

In all nuclear reactors the power generated is proportional to the volume of the active zone and the losses are proportional to its surface. Therefore the efficiency of nuclear reactors is greater for larger sizes and there exists a critical size for a nuclear reactor after which it may generate useful power. The practically necessary dimension is determined not by scientists but by the engineers who design the machine in general with proper choice of all the auxiliaries and the technology necessary for energy production. The following development is to a great measure determined by the talent and inventive ability of the design engineers. That is why the critical size of the Tokamak will be mainly determined by the proposed designs. Personally I think that the existing published design solutions lead us to a critical size for Tokamaks that make them unfeasible. But certainly life does show that the ingenuity of man has no limits and therefore one cannot be sure that a practically useful critcial size of Tokamaks may not be reached in the future.

One must note that although the main difficulty for obtaining a thermonuclear reaction in Tokamaks is the heating of deuterium and tritium ions, there is a difficulty of still another kind that does not have a well defined solution. In a Tokamak, for example, the plasma attracts and absorbs impurities extracted from the walls of the container. These impurities greatly lower the reaction rate. The plasma emits neutral atoms that hit and erode the wall. Moreover, the extraction of energy from neutrons also complicates the design of the Tokamak and leads to a larger critical size. Will we be able to bring the critical dimension of the Tokamak to a practically possible size? Even if it will eventually happen, of course we have no means to say when it will happen. Now we may only state that there are no theoretical reasons why in a Tokamak controlled thermonuclear reactions are not feasible, but the possiblity to release useful energy is as yet beyond the scale of our current practice.

Among other approaches to controlled thermonuclear fusion serious considerations should be given to pulsed methods without magnetic confinement /(1) page 33/. The idea is to heat a D + T pellet about 1 mm in diameter in a short time so as it will not have time to fly apart. For this very high pressures are necessary, that ensure intensive heat transfer between ions and electrons. It is assumed that in this way the thermonuclear reaction in a D + T pellet may fully take place. For this it is necessary to have a very powerful source of focussed laser light that should heat the pellet from all sides simultaneously in about a nanosecond. This heating is a complicated process, but using modern computers one may calculate all necessary conditions. If we illuminate a pellet by a well focussed laser beam, this may lead to a surplus of thermonuclear energy. But when one considers this process in detail, it is not clear how one can possibly resolve the technical and engineering difficulties. How, for instance, can one ensure uniform and simultaneous illumination and how can one usefully exploit the neutron energy?

In this case one may also say that the basic theoretical idea is sound, but the consequent engineering development with current technology is beyond our reach. Once again one cannot completely exclude a solution to this problem, although the design for laser implosion seems to me even less probable than the pulsed magnetic methods like the Tokamak.

The third approach to a thermonuclear reactor is based on continuously heating the plasma. Up to now this method has been developed only at our Institute. Our work was described 9 years ago /(4). Since then this type of reactor has been studied in detail, and now we see the main difficulties which we have to encounter. I will describe here in general terms what are the problems demanding a scientific solution.

As distinct from Tokamaks and the laser implosion method for producing conditions for the thermonuclear process, our method was not specially invented, but while developing a high power CW microwave generator accidentally we discovered a hot plasma phenomenon. We constructed an efficient microwave generator operating at 20 cm wave length with a power of a few hundred kW. This generator was called the "Nigotron" and its principles are described in (5) where full details of its construction with operating characteristics are given. In the process of its development beginning in 1950, during tests of our early model, high power microwave radiation was passed through a quartz sphere, filled with helium at 10 cm Hg pressure. We observed a luminiscent discharge with well defined boundaries. The phenomenon was observed only for a few seconds, as the quartz sphere in one place melted through.

These observations led us to the suggestion that the ball lightening may be due to high frequency waves, produced by a thunderstorm cloud after the conventional lightening discharge. Thus the necessary energy is produced for sustaining the extensive luminosity, observed in a ball lightening. This hypothesis was published in 1955 (7). After some years we were in a position to resume our experiments. In March 1958 in a spherical resonator filled with helium at atmospheric pressure under resonance conditions with intense H_{01} oscillations we obtained a free gas discharge, oval in form. This discharge was formed in the region of the maximum of the electric field and slowly moved following the circular lines of force.

We started to study this type of discharges where the plasma was not in direct contact with the walls of the resonator. We assume that this plasma may be at a high temperature. During a number of years we studied this interesting phenomenon in various gases and at different pressures, up to some tens of atmospheres at different power levels, reaching tens of kW. We also studied the effect of a magnetic field reaching 2,5 T in our experiments. This work is described in detail (4). A sketch of our setup is shown in fig. 2.

The plasma discharge has a cord-like form 10 cm long, equal to half the wavelength. Intense microwave oscillations E_{01} are excited in a cylindrical resonator (1). The cord of the discharge is situated at the maximum of the electric field and its stability along the longitudinal axis was due to the high

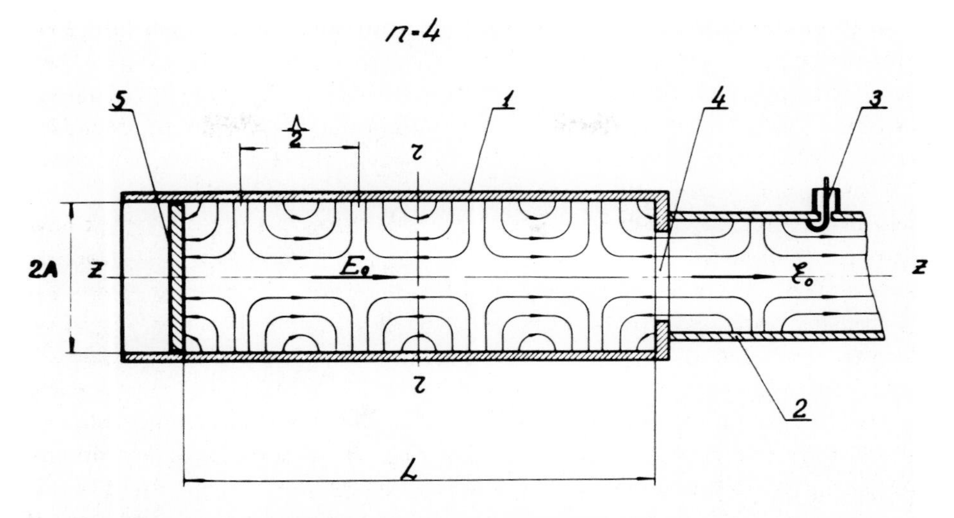

Fig. 2. Structure of the HF field in a resonator for E_{01} oscillations

frequency electric field. In a radial direction the stability was provided by rotating the gas. The discharge in hydrogen or deuterium was of great interest. At low powers the discharge did not have a well defined boundary and its luminosity was diffuse. At higher power the luminosity was greater and the diameter of the discharge increased. Inside the discharge a well defined filamentary cord-like nucleus was observed. In our initial experiments the power dissipated in the discharge was up to 15 kW and the pressure reached 25 atm. The higher the pressure, the more stable was the discharge with a well defined shape. A photograph of the discharge is shown in fig. 3. By measuring the conductivity of the plasma and by using passive and active spectral diagnostics we could firmly establish that the central part of the discharge had a very high temperature – more than a million K. So at the boundary of the plasma cord in the space of a few millimeters we had a discontinuity of temperature more than a million K. This meant that at its surface there was a layer of very high heat isolation. At first some doubt was expressed about the existence of such a layer. Various methods of plasma diagnostics were used, but they all and always confirmed the high temperature – more than a million K. Later we found out how it is possible to explain the physical nature of this temperature jump. It is easy to show that at these high temperatures electrons scattered at the boundary and freely diffusing into the surrounding gas will carry away a power of hundreds of kW. The lack of such a thermal flux may be explained by assuming the existence of electrons reflected without losses at the boundary of a double layer. The occurrence of a similar phenomenon is well known as such a layer exists in hot plasmas surrounded by dielectric walls, say, of glass or ceramics.

It is well known that in these conditions even at high pressures the electrons may have a temperature of many ten thousands of K and not

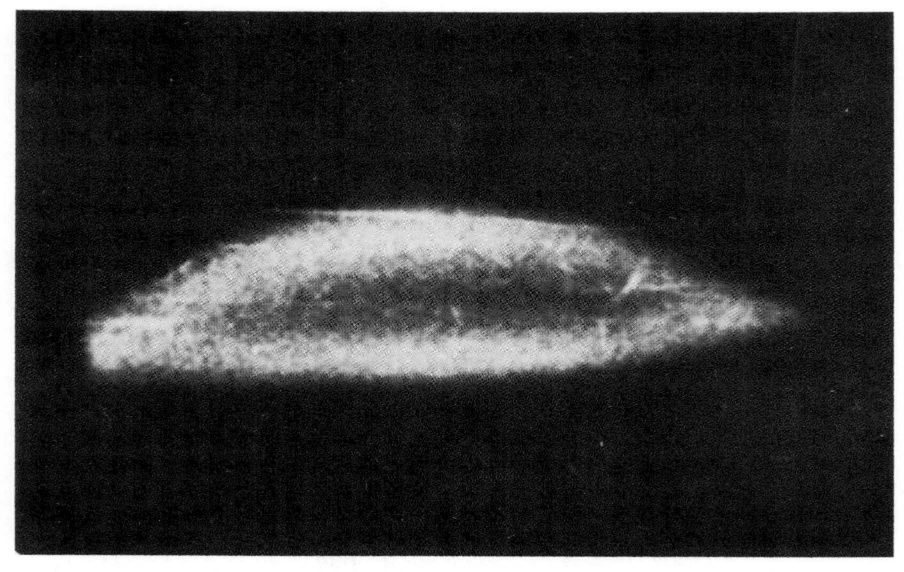

Fig. 3. Photograph of a cord discharge in deuterium with an admixture of 5 % argon at high power P = 14, 7 kW and high pressure p = 3,32 atm. Length of the discharge ~ 10 cm. The left edge of the discharge is blocked by the window. Oscillations of E_{01} type (1969)

markedly heat the walls. This phenomenon is well explained by the existence of a double layer on the dielectric surface. The mechanism leading to its formation is simple. When the electron hits the surface, due to its greater mobility it penetrates the dielectric to a greater depth than the ions and leads to the formation of an electric double layer, the electric field of which is so directed that it elastically reflects the hot electrons. The low electron heat conductance at the surface of plasmas is widely used in gas discharge lamps and the method of plasma heat insulation was first suggested by Langmuir. We assume that at a sufficiently high pressure a similar mechanism of heat insulation may take place in our hot plasma. The existance of a double layer in the plasma on the boundary of the cord discharge as a discontinuity in density was experimentally observed by us. This mechanism for a temperature discontinuity may obviously exist only if the ion temperature is much lower than the electron temperature and not much above the temperature at which the plasma is noticibly ionised. But this is only necessary at the boundary of the discharge. In the central part of the discharge the ion temperature may reach high values. As we will see futher, the difference in temperatures inside the core and at the surface is determined by the value of the thermal flux and the heat conductivity of the ion gas. Usually the heat conductivity is high, but in a strong magnetic field the transverse heat conductivity may become very small. Thus we may expect that in a strong magnetic field the ion temperature in the core will not differ from the electron temperature and may be sufficiently high to obtain in a deuterium or tritium plasma a thermonucle-

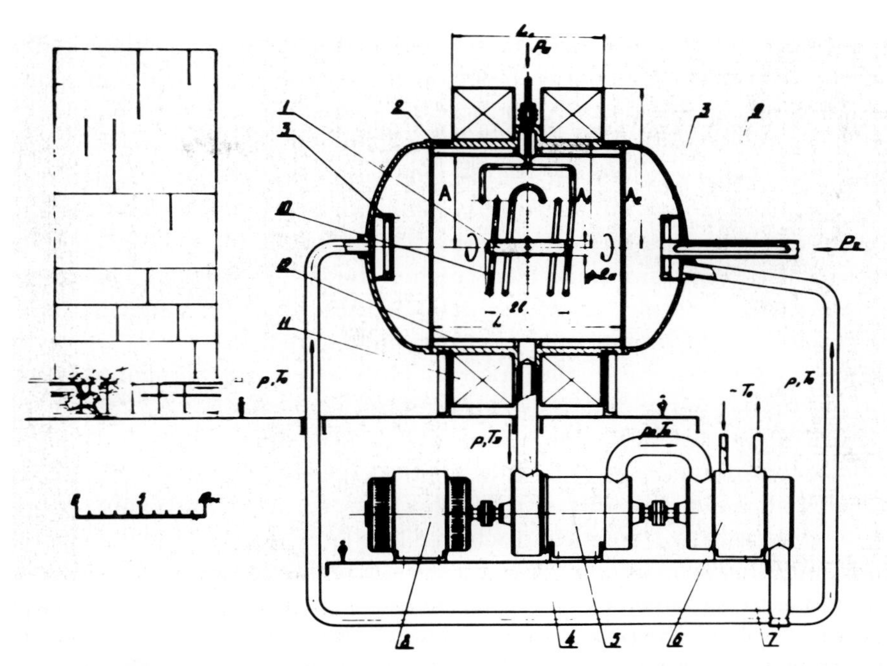

Fig. 4. Drawing of the construction of a thermonuclear reactor operating on a closed cycle. 1 – cord discharge, 2 – cylindrical container of the reactor, 3 – inclined nozzles, 4 – pipe connecting the container of the reactor with the gas turbine, 5 – gas turbine, 6 – isothermal compressor, 7 – cooling water, 8 – generator, 9 – coaxial waveguide, 10 – coil for the alternating magnetic field, 11 – solenoid, 12 – copper wall of the resonator, L – length of the resonator, L_1 – length of the solenoid, P_a – power of magnetoacoustic oscillations, P_r – high-frequency power, A – radius of the resonator, A_1 – internal radius of the winding, A_2 – external radius of the winding, 21 – length of the cord discharge, 2a – diameter of the cord discharge, h – distance between the wall of the container and the resonator.

ar reaction. This is the basis for designing a thermonuclear reactor to produce useful energy, and this has been worked out (8). The general outlay and the description of the reactor are shown in fig. 4.

The cord discharge (1) takes place in a confining vessel and resonator (2). The deuterium pressure is 30 atm, the magnetic field 1 T, produced by an ordinary solenoid. The design shows how the neutron energy is used. The gas heated by the neutrons passes through a gas turbine (5) where it adiabatically expands. Next it passes through a turbocompressor (6) and is isothermally compressed. The excess power is consumed in the generator (8). The cord discharge is heated by a high frequency field as it is done in cylindrical resonators (see fig. 2). The difference is in the coil surrounding the discharge and used to excite magnetoacoustic waves so as to raise the plasma ion temperature/(4) page 1003/. This design and pertinent calculations were published in 1970/(8) page 200/so as to demonstrate the expected parameters of our thermonuclear reactor, working with our plasma cord.

During the past time we have considerably increased our understanding of the processes in the plasma. We have mainly improved the microwave diagnostics and it is now possible to measure with 5 % accuracy the radial density distribution, its dependence on the magnetic field, pressure and supplied microwave power. The necessary stability conditions have been established. All this has allowed us to raise the microwave power by many times and in this way increase the electron temperature up to 50 million K. If we could establish temperature equilibrium between the electrons and ions in this case even without the extra heating of the plasma by magneto-acoustic oscillations, we could have reached the D + T reaction. The design of the reactor is simpler and its size is smaller. In this case the thermonuclear reactor would be not only easier to build but the neutron energy is easier to convert to mechanical power. Thus we escape the main difficulties on the way to building pulsed thermonuclear reactors.

But still we have also some unresolved difficulties which merit most serious consideration, because they might make the whole problem unsolvable. The main difficulty is the following. Now we can obtain in our installation a high frequency discharge at a pressure of 25 atmoshperes and continuously maintain the electrons at a temperature of 50 million K, and going to a greater size of our discharge even more. At present the size is limited only by the power conveyed to it. Thus we have permanently an electron gas with a record high temperature, even higher than the electron temperature inside the Sun. The main problem is to heat the ions to the same temperature, for although the electron gas interacts with the ions in the entire volume of the discharge, it is not easy to raise this temperature in such a way.

The temperature equalisation proceeds in two steps. In the first step the energy is passed from the electrons to the ions. This is simply due to the collisions of electrons with ions, and in this case it is obvious that the heat transfer will be proportional to the volume. The next stage is the transfer of energy from the ion gas to the surrounding media. This flux will be proportional to the surface of the plasma cord. At a given thermal conductivity of the ion gas the temperature will increase for larger sizes of the cross section o the plasma cord. Thus at a certain heat conductivity there will be a critical size for the diameter of the plasma cord, when the ion temperature will reach a value close to that of the electrons and the required D + D or D + T reaction can take place. If we know the heat conductivity of the plasma, then it is easy to calculate the critical dimension. If, for example, we make this calculation for ordinary ion plasma in the absence of a magnetic field, when the heat conductivity is determined by the mean free path, we will find that the plasma must have an unrealizably large size of many km. One can lower this cross section only by decreasing the heat conductivity of the ion gas by placing it in a magnetic field as it is done in the reactor shown in fig. 4. The heat conductivity of an ion gas in a magnetic field is markedly decreased and it is determined not by the mean free path but by the radius of Larmor orbits the size of which is inversely proportional to the magnetic field. The thermal conductivity of ion gas in a magnetic field is easy to calculate.

It is thus seen that the critical diameter of the cord is inversely proportional to the magnetic field and at a field of a few tesla the diameter of the cord to get thermal neutrons will be 5–10 cm, that can readily be provided for. For this we need a plasma installation considerably greater than the one in which we at present study the nature of the electron gas in the plasma. In the conditions of our laboratory this installation is quite feasible and is now under construction.

It may be shown that the thermonuclear reactor we have described makes it possible to obtain conditions not only for the D + T reaction but also for D + D, if it were not for yet another factor that could eventually make the whole process unfeasible.

We determined the heat conductivity of the ion gas by considering the mean free path of the ion, assuming it to be equal to the Larmor orbit radius, having not taken into account the effect of convection fluxes of heat in a gas. It is well known that even in ordinary gases the convection heat transfer is much larger then the heat conduction due to molecular collisions. It is also known that unfortunately it is virtually impossible to calculate theoretically the heat transfer by convective currents even for the simple case of random turbulent motion in an ordinary gas. In this case we usually can, by dimensional considerations, estimate the thermal conductivity in a similar case and then generalize it for a special case, determining the necessary coefficients empirically. In the case of plasma the process depends on many more parameters and the problem of determining the convectional thermal conductivity is even more complicated than in an ordinary gas. But theoretically we may estimate, which factors have most influence on the rate of convection. To sustain convection one must supply energy. In a gas this energy is drawn from the kinetic energy of flow and leads to loss of heat.

In a quiescent plasma there is no such source of energy. But in an ionized plasma there may be another source of energy that will excite convection. This source is connected with temperature gradients and some of the thermal energy flux could produce convection. Quantitatively this process is described by internal stresses and was first studied by Maxwell (9). Maxwell had shown that internal stresses are proportional to the square of viscosity and derivative of the temperature gradient. In an ordinary gas they are so small that up to now they have not yet been experimentally observed. This is because the viscosity, which is proportional to the mean free path, at normal pressures equals to $\sim 10^{-5}$ cm and so at low temperature gradients, the stresses are small.

In the plasma the mean free path of electrons and ions is of the order of cm and the temperature gradients are high. In this case the internal stresses following Maxwell's formula are 10 orders of magnitude greater than in a gas and we may expect both convection currents and turbulence. The presence of a magnetic field certainly can have effect on this phenomenon, and with additional effect of an electric field on convection it makes even a rough theoretical approach to estimating the magnitude of convection very unreliable. In this case there is only one alternative: to study these processes experimentally and this is what we are now doing.

In any case convectional thermal conductivity will lower the heating of ions and will lead to a greater critical cross section for the thermonuclear plasma cord. Correspondingly the size of the reactors for useful energy production will be greater.

If this size will be out of our practical reach, then we should consider methods to decrease convectional heat transfer. This may be done by creating on the boundary of the plasma a layer without turbulence, as it happens in fluids where we have the Prandtl boundary layer. This possibility has been theoretically considered /(4) page 1002/.

In conclusion we may say that the pulsed method used in Tokamaks can now be fully worked out theoretically, but the construction of a thermonuclear reactor, based on this method, leads to a large and complicated machine. On the other hand, our thermonuclear reactor is simple in construction, but its practical means of realisation and size depend on convection heat transfer processes, that cannot be treated purely theoretically.

The main attraction in scientific work is that it leads to problems, the solution of which it is impossible to foresee, and that is why for scientists research on controlled thermonuclear reactions is so fascinating.

LITERATURE

1. Ribe, F. L., Rev. of Modern Physics, 47, 7, 1975.
2. Kapitza, P. L., JETP Lett., *22* (1), 9, 1975.
3. Kapitza, P. L., Pitaevskii L. P., Sov. Phys. –JETP, *40* (4), 701, 1975.
4. Kapitza, P. L., Soviet Phys. –JETP, *30,* (6), 973, 1970.
5. Kapitza, P. L., High-Power Microwave Elecronics, Pergamon Press, Oxford, 1964.
6. Капица, П. Л., Филимонов, С. И., Капица, С. П., Сборник «Электроника больших мощностой,№ 6, «Наука», стр. 7, 1969.
7. Kapitza, P. L., Collected papers, vol. 2, 776, Pergamon Press, Oxford, 1965.
8. Kapitza, P. L., Sov. Phys. –JETP, *31,* (2), 199, 1970.
9. Maxwell, J. C., Phil. Trans. R. S., 170, 231, 1879.

The English translation from the Russian original text is authorized by the laureate.

Arno A Penzias

ARNO A. PENZIAS

I was born in Munich, Germany, in 1933. I spent the first six years of my life comfortably, as an adored child in a closely-knit middle-class family. Even when my family was rounded up for deportation to Poland it didn't occur to me that anything could happen to us. All I remember is a long train trip and scrambling up and down three tiers of narrow beds attached to the walls of a very large room. After some days of back and forth we were returned to Munich. All the grown-ups were happy and relieved, but I began to realize that there were bad things that my parents couldn't completely control, something to do with being Jewish. I learned that everything would be fine if we could only get to "America".

One night, shortly after my sixth birthday, my parents put their two boys on a train for England; we each had a suitcase with our initials painted on it and a bag of candy. They told me to be sure and take care of my younger brother. I remember telling him, "jetzt sind wir allein" as the train pulled out.

My mother received her exit permit a few weeks before the war broke out and joined us in England. My father had arrived in England almost as soon as the two of us, but we didn't see him because he was interned in a camp for alien men. The only other noteworthy event in the six or so months we spent in England awaiting passage to America occurred when I found that I could read my school books.

We sailed for America toward the end of December 1939 on the Cunard liner Georgic using tickets that my father had foresightedly bought in Germany a year and half earlier. The ship provided party hats and balloons for the Christmas and New Year's parties, as well as lots of lifeboat drills. The grey three-inch gun on the aft deck was a great attraction for us boys.

We arrived in New York in January of 1940. My brother and I started school and my parents looked for work. Soon we became "supers" (superintendents of an apartment building). Our basement apartment was rent free and it meant that our family would have a much-needed second income without my mother having to leave us alone at home. As we got older and things got better, we left our "super" job and my mother got a sewing job in a coat factory; my father's increasing wood-working skills helped him land a job in the carpentry shop of the Metropolitan Museum of Art. As the pressures on him eased, he later found time to hold office in a fraternal insurance company as well as to serve as the president of the local organization of his labor union.

It was taken for granted that I would go to college, studying science, presumably chemistry, the only science we knew much about. "College" meant City College of New York, a municipally supported institution then beginning its second century of moving the children of New York's immigrant poor into the American middle class. I discovered physics in my freshman year and switched my "major" from chemical engineering. Graduation, marriage and two years in the U.S. Army Signal Corps, saw me applying to Columbia University in the Fall of 1956. My army experience helped me get a research assistantship in the Columbia Radiation Laboratory, then heavily involved in microwave physics, under I. I. Rabi, P. Kusch and C. H. Townes. After a painful, but largely successful struggle with courses and qualifying exams, I began my thesis work under Professor Townes. I was given the task of building a maser amplifier in a radio-astronomy experiment of my choosing; the equipment-building went better than the observations.

In 1961, with my thesis complete, I went in search of a temporary job at Bell Laboratories, Holmdel, New Jersey. Their unique facilities made it an ideal place to finish the observations I had begun during my thesis work. "Why not take a permanent job? You can always quit," was the advice of Rudi Kompfner, then Director of the Radio Research Laboratory. I took his advice, and have remained here ever since.

Since the large horn antenna I had planned to use for radio-astronomy was still engaged in the ECHO satellite project for which it was originally constructed, I looked for something interesting to do with a smaller fixed antenna. The project I hit upon was a search for line emission from the then still undetected interstellar OH molecule. While the first detection of this molecule was made by another group, I learned quite a bit from the experience. In order to make some reasonable estimate of the excitation of the molecule, I adopted the formalism outlined by George Field in his study of atomic hydrogen. To make sure that I had it right, I took my calculation to him for checking. One of the factors in the calculation was the radiation temperature of space at the line wavelength, 18-cm. I used 2 K, a somewhat larger value than he had used earlier, because I knew that at least two measurements at Bell Laboratories had indications of a sky noise temperature in excess of this amount, and because I had noticed in Hertzberg's Diatomic Molecule book that interstellar CN was known to be excited to this temperature. The results of the calculation were used and forgotten. It was not until Dr. Field reminded me of them in December of 1966, that I had any recollection of the earlier connection. So much for the straight-line view of the progress of science!

The successful detection of OH at MIT made me look for a larger antenna. At the invitation of A. E. Lilley, I took key parts of my equipment to the Harvard College Observatory and spent several months participating in various OH observations. In the meantime, the horn antenna was pressed into service for another satellite project. A new Bell System satellite, TELSTAR, was due to be launched in mid-1962. While the primary earth station at Andover, Maine, was more-or-less on schedule, it was feared that the European partners in the project would not be ready at launch time, leaving Andover with no one to talk to. As it turned out, fitting the Holmdel horn with a 7-cm receiver for TELSTAR proved unnecessary; the Europeans were ready at launch time. This left the Holmdel horn and its beautiful new ultra low-noise 7-cm traveling wave maser available for radio astronomy. This stroke of good fortune came at just the right moment. A second radio astronomer, Robert Wilson, came from Caltech on a job interview, was hired, and set to work early in 1963.

In putting our radio astronomy receiving system together we were anxious to make sure that the quality of the components we added were worthy of the superb properties of the horn antenna and maser that we had been given. We began a series of radio astronomical observations. They were selected to make the best use of the careful calibration and extreme sensitivity of our system. Among these projects was a measurement of the radiation intensity from our galaxy at high latitudes which resulted in the discovery of the cosmic microwave background radiation, described in Wilson's lecture.

When our 7-cm program was accomplished, we converted the antenna to 21-cm observations including another microwave background measurement as well as galactic and intergalactic atomic hydrogen studies. As time went on, the amount of front line work that we could do became increasingly restricted. Much larger radio telescopes existed and they were being fitted with low-noise parametric amplifiers whose sensitivity began to approach that of our maser system. As a result we began looking for other things to do. An investigation of the cosmic abundance of deuterium was clearly an important problem. However nature had put the deuterium atomic line in an all but inaccessible portion of the long wavelength radio spectrum. I remember saying, to Bob Dicke, something to the effect that I didn't relish giving three years of my professional life to the measurement of the atomic deuterium line. He immediately replied, "Finding deuterium is worth three years". Fortunately, a better approach to the measurement of deuterium in space soon became available to me.

Up through the late 1960's the portion of the radio spectrum shortward of 1-cm wavelength was not yet available for line radio astronomy owing to equipment limitations. At Bell Laboratories, however, many of the key components required for such work had been developed for communications research purposes. With Keith Jefferts, a Bell Labs atomic physicist, Wilson and I assembled a millimeter-wave receiver which we carried to a precision radio telescope built by the National Radio Astronomy Observatory at Kitt Peak, Arizona, early in 1970. This new technique enabled us to discover and study a number of interstellar molecular species. Millimeter-wave spectral studies have proved to be a particularly fruitful area for radio astronomy, and are the subject of active and growing interest, involving a large number of scientists around the world. The most personally satisfying portion of this work for me was the discovery in 1973 of a deuterated molecular species, DCN. Subsequent investigations enabled us to trace the distribution of deuterium in the galaxy. This work provided us with evidence for the cosmological origin of this important substance, which earned the nickname "Arno's white whale" during this period.

From the first, I made it my business to engage in the communications work at Bell Labs in addition to my astronomical research. It seemed only reasonable to contribute to the pool of technology from which I was drawing. Similarly, Bell Labs has always been a contributor to, as well as a user of, the store of basic knowledge, as evidenced by their hiring of a radio astronomer in the first place.

As time went on, the applied portion of my efforts included administrative responsibilities. In 1972 I became the Head of the Radio Physics Research Department upon the retirement of A. B. Crawford, the brilliant engineer who built the horn antenna Wilson and I used in our discovery. In 1976, I became the Director of the Radio Research Laboratory, an organization of some sixty people engaged in a wide variety of research activitites principally related to the understanding of radio and its communication applications.

Early in 1979, my managerial responsibilities increased once again when I was asked to assume responsibility for Bell Labs' Communications Sciences Research Division. While I continued the personal research which traced the effects of nuclear processing in the Galaxy through the study of interstellar isotopes, pressure from other interests curtailed my entry into a new area — the nature and distribution of molecular clouds in interstellar space. Instead, I barely managed to introduce this subject to two of my graduate students who explored it in their PhD theses.

Then, toward the end of 1981, an unexpected event imposed an abrupt end to my career as a research scientist, when AT&T and the US Department of Justice decided to settle their anti-trust suit by breaking up the Bell System. In the process, I received yet another promotion — this time to Vice-President of Research — at a moment when two-thirds of the traditional research funding base moved off with the newly-divested local telephone companies.

As a result, I found myself facing several issues at once: What sort of research organization did the new AT&T require? How to create this new organization without destroying the world's premier industrial research laboratory in the process? Would the people in this large and tradition bound organization accept and support the changes needed to adapt to new economic and technological imperatives? Needless to say, such matters kept me quite busy.

In retrospect, the research organization which emerged from the decade following the Bell System's breakup deploys a far richer set of capabilities than its predecessor. In particular, our work features a growing software component, even as we strive to improve our hardware capabilities in areas such as lightwave and electronics. The marketplace upheaval brought forth by increased competition has helped speed the pace of technological revolution, and forced change upon the institutions all industrialized national, Bell Labs included. While change is rarely comfortable, I am happy to say that we not only survived but also grew in the process.

Except for two or three papers on interstellar isotopes, my tenure as Bell Labs' Vice-President of Research brought my personal research in astrophysics to an end. In its place, I have developed an interest in the principles which underlie the creation and effective use of technology in our society, and eventually found time to write a book on the subject *Ideas and Information,* published by W. W. Norton in 1989. In essence, the book depicts computers as a wonderful tool for human beings but a dreadful role model. In other words, if you don't want to be replaced by a machine, don't act like one. The warm reception this book received in the US, and the ten other countries which published it in various translations has given me much satisfaction.

I have also been a visiting member of the Astrophysical Sciences Department at Princeton University from 1972 to 1982. My occasional lecturing and research supervision were more than amply repaid by stimulating professional and personal relationships with faculty members and students.

Finally, most important of all is the love and support of my family, my wife Anne, our children and grandchildren.

B.Sc., City College of New York, 1954
M.A., Columbia University, 1958
Ph. D., Columbia University, 1962

Docteur Honoris Causa, Paris Observatory, 1976
Henry Draper Medal, National Academy of Sciences, 1977
Herschel Medal, Royal Astronomical Society, 1977

Member, National Academy of Sciences
Fellow, American Academy of Arts and Sciences
Fellow, American Physical Society
Member, International Astronomical Union
Member, International Union of Radio Science.

THE ORIGIN OF ELEMENTS

Nobel Lecture, 8 December, 1978

by

ARNO A. PENZIAS
Bell Laboratories, Holmdel, N. J. USA

Throughout most of recorded history, matter was thought to be composed of various combinations of four basic elements; earth, air, fire and water. Modern science has replaced this list with a considerably longer one; the known chemical elements now number well over one hundred. Most of these, the oxygen we breathe, the iron in our blood, the uranium in our reactors, were formed during the fiery lifetimes and explosive deaths of stars in the heavens around us. A few of the elements were formed before the stars even existed, during the birth of the universe itself.

The story of how the modern understanding of the origin of the chemical elements was acquired is the subject of this review. A good place to begin is with Lavoisier who, in 1789, published the first scientific list of the elements. Five of the twenty or so elements in Lavoisier's list were due to the work of Carl Wilhelm Scheele of Köping. (He was rewarded with a pension by the same Academy to whom the present talk is adressed, more than a century before Alfred Nobel entrusted another task of scientific recognition to it.) Toward the end of the last century the systematic compilation of the elements into Mendeleev's periodic table carried with it the seeds of hope for a systematic understanding of the nature of the elements and how they came to be.

The full scientific understanding of the origin of the elements requires a description of their build-up from their common component parts (e. g., protons and neutrons) under conditions known to exist, or to have existed, in some accessible place. Thus, the quest for this understanding began with nuclear physics. Once plausible build-up processes were identified and the conditions they required were determined, the search for appropriate sites for the nuclear reactions followed. Although this search was begun in earnest in the nineteen thirties, it was only toward the end of the nineteen sixties that the full outlines of a satisfactory theoretical framework emerged. In the broad outlines of the relevant scientific thought during this period one can discern an ebb and flow between two views. In the first, the elements were thought to have been made in the stars of our galaxy and thrust back out into space to provide the raw material for, among other things, new suns, planets and the rock beneath our feet. In the second view, a hot soup of nuclear particles was supposed to have been cooked into the existing elements before the stars were formed. This pre-stellar state was generally associated with an early hot condensed stage of the expanding universe.

Historically, the first quantitative formulations of element build-up were attempted in the nineteen thirties; they were found to require conditions then thought to be unavailable in stars. As a consequence, attention turned in the 1940's to consideration of a pre-stellar state as the site of element formation. This effort was not successful in achieving its stated goal, and in the 1950's interest again turned to element formation in stars. By then the existence of a wide range of stellar conditions which had been excluded in earlier views had become accepted. Finally, the 1960's saw a reawakened interest in the idea of a pre-stellar state at the same time that decisive observational support was given to the "Big Bang" universe by the discovery of cosmic microwave background radiation and its identification as the relict radiation of the initial fireball.

Given the benefit of hindsight, it is clear that the process of understanding was severely impeded by limitations imposed by the narrow range of temperature and pressure then thought to be available for the process of nuclear build-up in stars. The theory of stellar interiors based upon classical thermodynamics (Eddington, 1926) seemed able to explain the state of the then known stars in terms of conditions not vastly different from those in our sun. The much higher temperatures and pressures suggested by the nuclear physics of element formation were thought to be possible only under conditions of irreversible collapse (i. e. the theory lacked mechanisms for withstanding the tremendous gravitational forces involved); hence no material produced under those conditions could have found its way back into the interstellar medium and ordinary stars. The arguments and mechanisms required to depict the formation of heavy elements and their ejection into space are subtle ones. In describing them, S. Chandrasekhar wrote, ". . . one must have faith in drawing the consequences of the existence of the white dwarf limit. But that faith was lacking in the thirties and forties for reasons set out in my (to be published) article 'Why are the Stars as they are?." Thus, our story of a forty-year-long journey begins with the absence of sufficient faith.

The nuclear physics picture of element formation in an astrophysical setting was the subject of von Weizsäcker's "Über Elementumwandlungen im Innern der Sterne" (1937, 1938). (Interested readers can find a guide to earlier literature in Alpher and Herman's 1950 review.) The central feature of von Weizsäcker's work is a "build-up hypothesis" of neutrons and intervening β -decays; the direct build-up from protons would be blocked by the Coulomb repulsion of the positively charged nuclei of the heavier elements. Quantitative predictions that follow from this hypothesis can be obtained from the general features of empirical abundance-stability data through use of thermodynamic equilibrium relations like those used in the study of chemical reactions.

Consider the reversible exothermic reaction of two elements A and B combining to form a stable compound AB with an energy of formation ΔE, i. e.,

$$A+B \rightarrow AB+\Delta E. \tag{1}$$

Using square brackets to indicate concentration, we can compute relative abundances at thermal equilibrium from the relation

$$\frac{[A]x[B]}{[AB]} \propto exp\ (-\Delta E/kT). \tag{2}$$

where k is Boltzmann's constant.

The stable isotopes of the lighter elements have approximately equal numbers of neutrons and protons (fig. 1). The sequential addition of neutrons to a nucleus, ^{16}O say, results in heavier isotopes of the same element, ^{17}O and then ^{18}O in this case, until the imbalance of neutrons and protons is large enough to make the nucleus unstable. (^{19}O β − decays to ^{19}F in ~ 29 seconds.) A measure of the stability of an isotope is the increment in binding energy due to the last particle added. In the case of ^{17}O, for example, we have for this increment,

$$\Delta E(17) = [M(16)+M(n)-M(17)]c^2, \tag{3}$$

where M(16), M(n) and M(17) are the masses of ^{16}O, a neutron and ^{17}O, respectively, and c^2 is the square of the speed of light. In our example, the mass of ^{17}O is 17.004533 A.M.U., that of the neutron is 1.008986 and that of ^{16}O is 16.00000. Substituting in eqn (3) we find the binding energy increment to be .004453 A.M.U. or 6.7×10^{-6}ergs. We can get some idea of the temperatures involved in the addition of a neutron to ^{16}O from the use of relation (2). Because of the exponential nature of this relation, we can

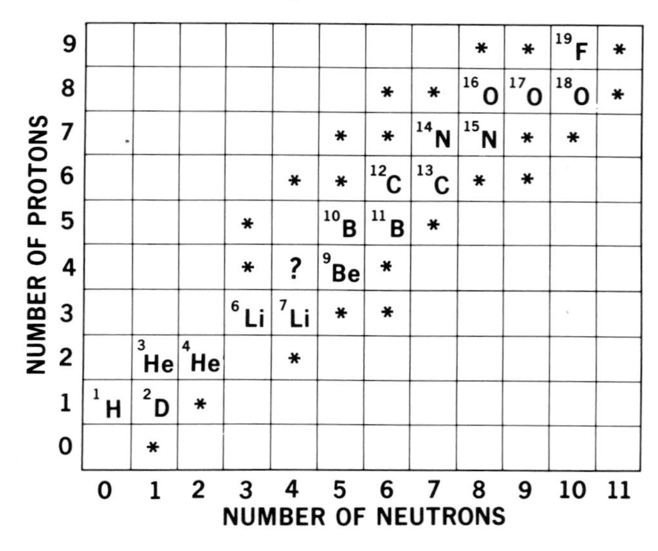

Fig. 1 *The Elements, Hydrogen Through Flourine.* The stable nuclei are plotted as a function of the number of protons and neutrons they contain. Radioactive combinations are indicated by an asterisk, an empty box indicates that the corresponding combination of protons and neutrons doesn't exist. (Note that both mass-5 boxes are empty.) The question mark indicates ^{8}Be; it can exist under special conditions as a metastable combination of two ^{4}He nuclei, thus providing the key stepping-stone in the transformation of three ^{4}He's into ^{12}C.

expect ΔE and kT to be of comparable magnitude for a wide range of relative isotopic abundances. Thus, from the approximation,

$$\Delta E \approx kT$$

we find that 6.7×10^{-6}ergs corresponds to a temperature of 5×10^{10}K.

Following earlier workers, von Weizsäcker applied the above relations to the relative abundance of the isotopes of a given element having three stable isotopes, (^{16}O, ^{17}O and ^{18}O for example) in a state of equilibrium established by thermal contact with a bath of neutrons at temperature T. If $[^{16}O]$, $[^{17}O]$, $[^{18}O]$ and $[n]$ are the concentrations of the two oxygen nuclei and the neutrons respectively, we may use the relations (2) and (3) to write

$$\frac{[^{16}O][n]}{[^{17}O]} \propto exp\ (\Delta E(17)/kT)$$

as well as

$$\frac{[^{17}O][n]}{[^{18}O]} \propto exp\ (\Delta E(18)/kT)$$

Thus the relative abundances of the three isotopes yield a pair of expressions involving the neutron density and temperature which permit the separate determination of these two quantities from the oxygen abundance data alone. (The abundances of several hundred stable nuclei -fig. 2 – had been determined from terrestrial samples supplemented by stellar spectra and meteorites.)

Using this three-isotope method, Chandrasekhar and Henrich (1942) obtained thermal equilibrium neutron densities and temperatures for five elements. Not surprisingly, in view of previous work, each element required a different temperature and neutron density. While the range of the temperature values was relatively small, between 2.9×10^9 for neon and 12.9×10^9 for silicon, the neutron densities ranged from $\sim 10^{31}\text{cm}^{-3}$ for silicon to $\sim 10^{19}\text{cm}^{-3}$ for sulphur, some twelve orders of magnitude! The high values of the temperatures and pressures derived as well as their lack of element-to-element consistency shows the shortcomings of this thermal equilibrium picture of stellar element formation.

Another problem with this neutron build-up picture was the simultaneous requirement of very rapid neutron capture in the formation of elements such as uranium and thorium, and very slow neutron capture for the formation of others. The "slow" elements require the capture sequence of neutrons to be *slow* enough to permit intervening β-decays, while others require *rapid* sequential neutron capture in order to permit their formation from a series of short-lived nuclei. The elements formed by these slow and rapid processes correspond, respectively, to the s and r peaks of fig. 2 [A concise early discussion of this problem is presented in the final chapter of Chandrasekhar's 1939 text.]

Another approach to the element formation problem provided an enormous contribution to understanding the nuclear physics of stars. In a

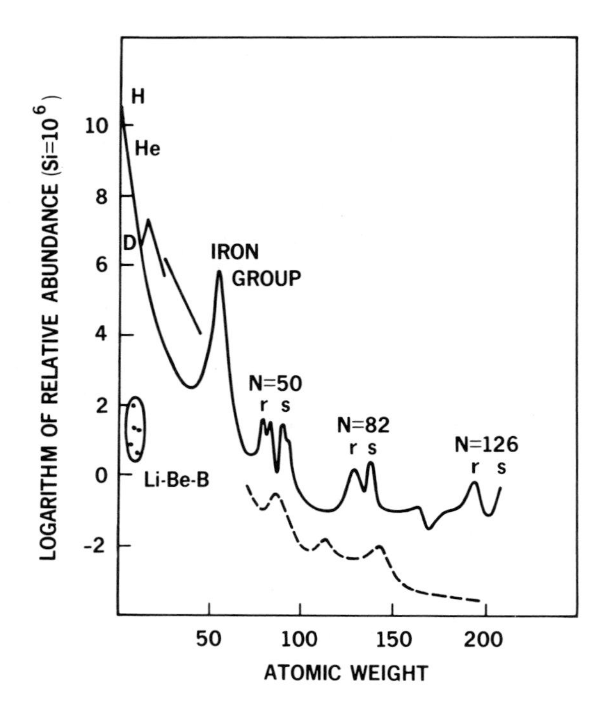

Fig. 2 *Relative Abundances of the Elements:* Smoothed curves representing the abundances of various groups of elements, after Burbidge et. al, 1957, who presented a total of eight processes to fit this data (See Clayton 1968 for a more modern treatment.) Lithium, Beryllium and Boron (circled) are not formed in the build-up process which goes from helium to carbon. The small amounts of these elements found in nature are fragments from the break-up of heavier elements.

beautiful paper entitled, "Energy Production in Stars", Bethe (1939) considered the individual nuclear reactions of the light nuclei, from hydrogen through oxygen. This paper established the role of the fusing of hydrogen into helium by two processes and demonstrated their quantitative agreement with observations. In the first process, protons combine to form a deuteron which is then transformed into ^{4}He by the further capture of protons. In the second, carbon and nitrogen are used as catalysts, viz

$$^{12}C+H = {}^{13}N+\gamma,\ {}^{13}N = {}^{13}C+e^+$$
$$^{13}C+H = {}^{14}N+\gamma$$
$$^{14}N+H = {}^{15}O+\gamma,\ {}^{15}O = {}^{15}N+e^+$$
$$^{15}N+H = {}^{12}C+{}^{4}He$$

(The notation and format are taken from the cited reference.)

As to the build-up of the heavier elements, however, no stable build-up process beyond the mass-4 nucleus had been found; a mass-4 nucleus cannot be combined with any other nucleus to form a heavier nucleus. In particular, no stable mass-5 nucleus exists, so the addition of a neutron or proton to ^{4}He doesn't work. Bethe wrote, "The progress of nuclear physics in the past few years makes it possible to decide rather definitely which processes can and which cannot occur in the interior of stars ... under present conditions, no elements heavier than helium can be built up to any appreciable extent". In an attempt to bypass the mass-4 barrier, Bethe considered, and correctly rejected, the direct formation of ^{12}C from the simultaneous collision of three helium nuclei. He also noted that the formation of ^{8}Be from two helium nuclei was prevented by the fact that this nucleus was known to be unstable, having a negative binding energy of "between 40 and 100 keV". This energy difference corresponds to a temperature of some 10^9K, again to be compared with the $\sim 2\times 10^7$K which was then thought to be the allowed stellar temperature. It was not realized at that time that it is possible to form ^{8}Be from ^{4}He at a sufficiently high ^{4}He density and temperature and so bypass the mass-4 barrier. So it was that recognition of the crucial role of ^{8}Be in the build-up of the elements had to await the acceptance, in the early 1950's, of a new understanding of the physics of stellar interiors.

In the intervening decade, therefore, attention was diverted toward processes which could have occurred before the formation of the stars, namely a hot dense state associated with the birth of the universe. The formalism associated with the birth of the universe had been laid out by Friedman (1922), Lemaitre (1927) and Einstein and deSitter (1932). The applicability of this formalism to the real world was established by the beautiful simplicity of Hubble's (1929) powerful result that the observed velocities of the "extragalactic nebulae" [i. e. the galaxies which make up the universe] were proportional to their distances from the observer. In its simplest form, the most distant galaxy is moving away at the fastest rate and the nearest at the slowest. This is exactly what one would expect if all the galaxies had begun their flight from a common origin and, at a common starting time, had been given their start in a tremendous explosion.

Not widely popular among respectable scientists of the time, this idea of an expanding universe was taken up in the 1940's in part because the theories of the stellar origin of the elements had failed in the 1930's. (The expanding universe picture was generally ignored again in the 1950's when the wide variety of stellar phenomena became understood. It was only in the 1960's that a more balanced view emerged, but that comes later in our story.) The title of Chandrasekhar and Henrich's 1942 paper "An Attempt to Interpret the Relative Abundances of the Elements and Their Isotopes" reflects the tentative and unsatisfactory nature of the state of understanding at that time. The paper begins, "It is now generally agreed that the chemical elements cannot be synthesized under conditions *now believed* (emphasis added) to exist in stellar interiors." As an alternative, the authors suggested that the expansion and cooling of the early universe might be a possible site for the processes. In this view, each of the elements had its abundance "frozen out" at an appropriate stage of the expansion of the hot ($\geqslant 10^9$K), dense ($\geqslant 10^6$gr/cm^3) universe.

As was shown by George Gamow (1946), however, the formation of elements in the early universe could not have occurred through these equilibrium processes. He accomplished this demonstration by a straightforward calculation of the time scales involved. (The interested general reader can find more on this and related points in the mathematical appendices of S. Weinberg's (1977) delightful book "The First Three Minutes".)

Consider a point mass m located on the surface of an expanding sphere with mass density ϱ. The energy E of the mass with respect to the center of the sphere is a fixed quantity, the sum of its kinetic and potential energies (the latter is a negative quantity), viz

$$E = const = \frac{mv^2}{2} - \frac{Gm(4\pi\rho R^3/3)}{R} \tag{5}$$

where G is the constant of gravitation, ϱ, the density, R, the radius of the sphere, and v, the outward velocity of the point mass, are all functions of time. Since $4\pi\varrho\ R^3/3$, the mass within the sphere, is not an increasing function of R, the far right-hand term must become arbitrarily large for sufficiently small values of R(t), i.e., at early times in the expansion. Under this "early time approximation" both right-hand terms must become very large because the difference between them is fixed. Thus we can regard the two terms as essentially equal at early times and, upon simple rearrangement, obtain

$$\frac{R(t)^2}{v(t)^2} \approx \frac{3}{8\pi\rho(t)G} \tag{6}$$

Now R/v is a characteristic time scale for the expansion: it is the reciprocal of Hubble's constant and is referred to as the Hubble age in cosmology. (Hubble's "constant" is constant in the spatial sense; it varies in time.) Putting numerical values in (6), we have

$$Age \approx \sqrt{\frac{10^6}{\rho}}\ sec. \tag{7}$$

where ϱ is expressed in gr/cm^3. Thus, as Gamow pointed out, a neutron density of $10^{30} cm^{-3}$ (about $10^6 gr/cm^3$) would exist for less than one second in the early universe. Since the β-decays necessary to establish the appropriate equalities between protons and neutrons are typically measured in minutes, it is clear that the time period needed to establish equilibrium with neutrons at the high densities required simply was not available in the early expanding universe.

This demonstration set the stage for the consideration of nonequilibrium processes. Fortunately, two timely developments for the undertaking of such a study had just occurred. The first was the publication of the values of neutron capture cross-sections in the open literature after the end of World War II. The second was a bright graduate student in need of a thesis topic. Lifshitz (1946) solved the problem that Gamow's student,

R. A. Alpher, had originally selected for a thesis topic, one having to do with turbulence and galaxy formation in the early universe. As a result, Alpher soon set to work on a new topic, the nonequilibrium formation of the elements by neutron capture. Since not all cross-sections were available, Alpher fitted a smooth curve through the published points, and used this curve for his calculations. The results of Alpher's calculation were introduced to the scientific world in a brief letter whose list of authors makes it part of the folklore of physics (Alpher, Bethe and Gamow 1948).

At this point the trail divides. Two different paths of investigation must be followed before they merge again into final results. We proceed to follow one of them with the understanding that we must return here later to follow the other.

In presenting his thesis results Alpher initiated a series of interactions between scientists which led to a succession of results very different from what he might have expected. First, Enrico Fermi, present at a seminar given by Alpher, soon raised an important objection: The straight line interpolation of capture cross-sections leads to a serious error in the case of the light nuclei. The neutron capture cross-section of a mass-4 particle is known to be essentially zero, whereas Alpher's curve was fitted to the average cross-sections of the nearby nuclei, which are quite large. Fermi had his student Turkevitch redo Alpher's calculations using explicit measured values for the cross-sections. Fermi and Turkevitch's results, never published separately but merely sent directly to Alpher, showed what Gamow and his co-workers knew and admitted privately, that their mechanism could produce nothing heavier than mass-4 from neutrons alone.

Second, Fermi pressed his friend Martin Schwartzschild for observational evidence of the formation of the heavy elements in stars. Together with his wife Barbara, Schwartzschild amply fulfilled this request. In one of the classic papers of observational astronomy (Schwartzschild and Schwartzschild 1950) they measured the faint spectra of two groups of stars of the same stellar type, F dwarfs, stars with long uneventful lifetimes. A separation into two groups, Population I and Population II, was done on the basis of velocity. This distinction, due to Baade, makes use of the fact that interstellar gas is almost totally confined to the galactic plane because vertical (i.e., perpendicular to the plane) gas motions are quickly damped out by cloud-to-cloud collisions. Thus, new stars born from this gas are to be found in the plane, without appreciable vertical motion. (These stars, which are easier to find, were found first and hence are called Population I.) Old stars, formed before the formation of the galactic disc retain the high velocities of the gas from which they were formed because dissipative encounters between stars are negligibly rare. Consequently, older (Pop II) stars can be distinguished by their higher velocities. The Schwartzschilds' comparison of the spectra of the two populations provided a clear answer: the younger Population I stars had the greater abundance of iron and other metals, thus revealing the enrichment of the interstellar medium between the times that the older and younger stars were formed.

This unmistakable evidence of metal production by stars during the lifetime of the galaxy removed the need for a pre-stellar mechanism for element formation. Only the path around the mass-4 barrier for element build-up in stars still had to be found. This was the third and final step.

Martin Schwartzschild presented this challenge to a young nuclear physicist, Ed Salpeter. Salpeter set to work, having a much wider range of accepted stellar conditions to work with than did Bethe in his earlier investigation. He soon found (Salpeter 1952) that ^{8}Be, unstable though it is, can be present in the hot dense cores of red giant stars in sufficient quantities to provide a convenient stepping stone for the formation of ^{12}C through the addition of a helium-4 nucleus.

With both observational support and the theoretical path around the mass-4 barrier, the triumph of stellar element formation now seemed complete. Fred Hoyle dismissed all pre-stellar theories of element build-up as "requiring a state of the universe for which we have no evidence" (Burbidge, et. al. 1957). So much for Alpher and Gamow's theory! "If the curve is simple the explanation must be simple" Gamow (1950) had said. But the curve of elemental abundances is not a simple one (Fig. 2). Burbidge et. al. presented no less than seven separate processes to account for the data, and left room for more under an eighth heading to fill in the few remaining gaps of their picture.

Ironically, it was Fred Hoyle himself who found a gap that could not be filled in the stellar picture, a gap in the best-understood process of them all, the formation of helium from hydrogen. Although the burning of hydrogen into helium provides the sun and the other stars with their energy and with building blocks for the formation of the heavier elements, Hoyle concluded that about ninety percent of the helium found in stars must have been made before the birth of the galaxy. The basis for this conclusion was an energy argument: the total amount of energy released by the formation of all the observed helium is some ten times greater than the energy radiated by the galaxies since their formation. Thus, "it is difficult to suppose that all the helium has been produced in ordinary stars" (Hoyle and Taylor 1964). Instead, attention was turned to helium formation in the early stages of an expanding universe, reviving work begun by George Gamow some sixteen years earlier. As indicated above, our description of Gamow's work was deferred in order to first follow the progress of the stellar picture of element build-up. We can now follow the second path.

Despite the problems inherent in Alphers treatment, (see, e.g., Alpher and Herman 1950), it provided the basis for a statement of profound simplicity and great power (Gamow 1948). Although wrong in almost every detail, Gamow's new insight pointed the way for others to follow. He noted that nuclear build-up cannot take place in the hottest, most condensed, state of the early universe because thermal photons at high temperatures $\geqslant 10^{10}$K are energetic enough to break up bound particle groups. Only when the temperature has cooled to $\sim 10^{9}$K, can nuclear reactions begin. Any build-up, however, must be completed during the few hundred seconds before all the free neutrons decay into protons. Gamow considered a cylinder (Fig. 3) swept out by a neutron with a 10^9K thermal velocity during its lifetime. The cross-section of the cylinder was the capture cross-section for deuteron formation. If there was to have been appreciable element build-up in the early universe, Gamow reasoned, some fraction, say one half, of the initial neutrons had to have collided with protons to form deuterons before they had time to decay. Thus, half of Gamow's sample cylinders should contain a proton. This statement determines the number of protons per unit volume. From this result, the mass of the proton, and his estimate of the fraction of matter that was in the form of protons (roughly one half), Gamow obtained the mass density of matter in the universe at 10^9K, about 10^{-6}gm/cm^3.

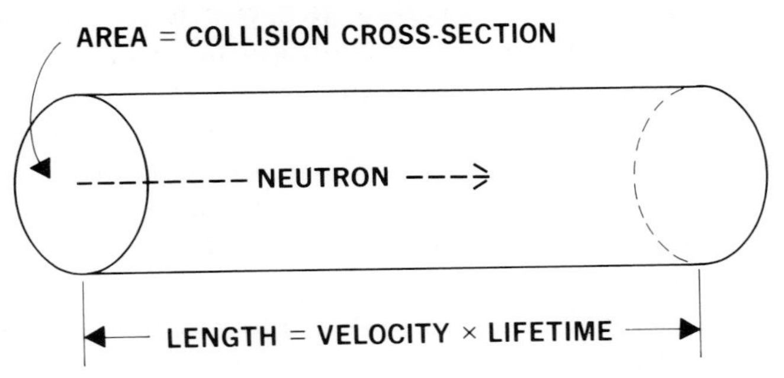

Fig. 3 *Gamows Sample Cylinder:* The volume swept out by a neutron in the early universe. The length of the cylinder is the product of the neutron's thermal velocity (at 10^9K) and its decay time. The cross-sectional area is the neutron-proton collision cross-section for deuteron formation. The fraction of neutrons forming deuterons is equal to the probability that the cylinder contains a proton.

Gamow then noted that the mass density of radiation at 10^9K (i.e., its energy density divided by c^2) was about 10gr/cm^3, as compared with only 10^{-6}gr/cm^3 for matter. This makes radiation the dominant component in the entropy of the early universe, permitting it to cool during the expansion as if the matter were not present. In that case, the temperature varies inversely with the radius of the expanding volume element (Tolman 1934, Peebles 1971) i.e.,

$$T \propto R^{-1}. \tag{8}$$

Now since ϱ, the density of matter, varies inversely as the cube of the radius, we can replace (8) with

$$T \infty \sqrt[3]{\rho}. \tag{9}$$

or

$$\frac{T_1}{T_2} = \sqrt[3]{\frac{\rho_1}{\rho_2}}\,.$$

This neat relation between temperature and matter density holds as long as radiation remains the dominant component. When the temperature drops below $\sim 3\times 10^3$K, the matter is too cool to remain ionized, and once it becomes neutral it is essentially transparent to the radiation. The radiation is then no longer coupled to the matter, it is free to expand forever in untroubled isolation, and eqn (9) continues to apply.

Gamow was only interested in tracing the radiation to the epoch when the matter becomes neutral and decouples from the radiation. From that point on, the matter has only its own thermal energy to support itself against gravitational collapse, so it fragments and condenses to form galaxies. Gamow used eqn (9) to find the density of matter at 3×10^3K and the Jeans criterion to determine the size of the collapsing fragments. Thus he was able to obtain a relation for the mass of galaxies containing only fundamental constants and the single assumption that half the initial neutrons collided to form deuterons. This was quite a trick, even for him!

Gamow's paper inspired his former student, Alpher and his collaborator Robert Herman to do the calculations more rigorously (Alpher and Herman 1949). Most importantly they replaced the "early-time" approximation Gamow used with a more exact formulation and traced the temperature of the relict primordial radiation to the present epoch. Taking the present matter density of the universe to be 10^{-30}gm/cm^3, they concluded that the present energy density of the relict radiation should correspond to a temperature of a few degrees Kelvin. Although mention of this prediction persisted in Gamow's popular writing, it was only repeated explicitly in a few of their subsequent scientific works. As for detection, they appear to have considered the radiation to manifest itself primarily as an increased energy density (Alpher and Herman 1949, pg. 1093). This contribution to the total energy flux incident upon the earth would be masked by cosmic rays and integrated starlight, both of which have comparable energy densities. The view that the effects of three components of approximately equal additive energies could not be separated may be found in a letter by Gamow written in 1948 to Alpher (unpublished, and kindly provided to me by R. A. Alpher from his files). "The space temperature of about 5° K is explained by the present radiation of stars (C-cycles). The only thing we can tell is that the residual temperature from the original heat of the Universe *is not higher* than 5° K." They do not seem to have recognized that the unique spectral characteristics of the relict radiation would set it apart from the other effects.

The first published recognition of the relict radiation as a detectable microwave phenomen appeared in a brief paper entitled "Mean Density of Radiation in the Metagalaxy and Certain Problems in Relativistic Cosmology", by A. G. Doroshkevich and I. D. Novikov (1964a) in the spring of 1964. Although the English translation (1964b) appeared later the same year in the widely circulated "Soviet Physics-Doklady", it appears to have escaped the notice of the other workers in this field. This remarkable paper not only points out the spectrum of the relict radiation as a blackbody microwave phenomenon, but also explicity focuses upon the Bell Laboratories twenty-foot horn reflector at Crawford Hill as the best available instrument for its detection! Having found the appropriate reference (Ohm 1961), they misread its results and concluded that the radiation predicted by the "Gamow Theory" was contradicted by the reported measurement.

Ohm's paper is an engineering report on a low-noise microwave receiving system. The reported noise of this system contained a residul excess of almost exactly three degrees! Ohm had measured a total system noise temperature of some 22K including the contribution of the receiver, the antenna, the atmosphere and the sky beyond. Separate measurements of each of the components of this noise temperature, except the sky beyond the atmosphere, totalled $\sim$ 19K. (From an analysis of his measurement errors, Ohm concluded that both sets of measurements, the total and the sum of individual contributions, could be consistent with an intermediate value). The atmospheric contribution was measured by moving the antenna in elevation and fitting the change in system temperature to a cosecant relation, a standard procedure which is described by Wilson (1978). To avoid confusion with other quantities, the atmospheric contribution thus derived was denoted T_{sky}, the "sky temperature". Ohm's value of 2.3K for this quantity was in good agreement with atmospheric attenuation theory. The background contribution due to the relict radiation has no elevation dependence and cannot be detected by this technique. Perhaps due to the unfortunate name, Doroshkevitch and Novikov regarded T_{sky} as containing the background radiation and therefore leading to a null result. The disappointment is reflected in Section IV of Zeldovitch's concurrent (1965) review.

The year 1964 also marked a reawakened interest in the "Gamow Theory" by Hoyle and Taylor (1964) as well as the first unambiguous detection of the relict radiation. The rough outlines of Gamow's initial treatment had long since been refined by the work of others. For example, it was pointed out by Hayashi (1950) that the assumption of an initial neutron material was incorrect. The radiation field at $T > 10^9$K generates electron- positron pairs which serve to maintain quasi-thermal equilibrium between neutrons and protons (see also Chandrasekhar and Henrich, 1942, who made the same point). Alpher, Follin and Herman (1953) incorporated this process into their rigorous treatment of the problem. Their work benefited from the availability of what was, by the standard of those days, a powerful electronic computer which permitted them to include the dynamic effects of expansion and cooling upon collisional and photo-disintegrated processes. Their results, which have not been substantially altered by subsequent work, are chiefly marked by (1) conversion of some 15% of the matter into helium, with the exact amount dependent only slightly upon the density at $T \approx 10^9 K$ and (2) production of deuterium whose surviving abundance is sensitively dependent upon the initial temperature/density relation. The same ground was covered in Hoyle and

Taylor's 1964 paper, which cited Alpher, Follin and Herman's paper and noted the agreement with the earlier results. Neither paper made any mention of surviving relict radiation.

Shortly thereafter, P. J. E. Peebles treated the same subject for a different reason. R.H. Dicke had, with P.G. Roll and D.T. Wilkenson, set out to measure the background brightness of the sky at microwave wavelengths. At his suggestion, Peebles began an investigation of the cosmological constraints that might be imposed by the results of such a measurement. Peebles' paper, which was submitted to the Physical Review and circulated in preprint form in March of 1965. This paper paralleled the above light element production picture and included Hoyle and Taylor (1964) among its references. In addition, it explicitly delineated the surviving relict radiation as a detectable microwave phenomenon. At about the same time, microwave background radiation was detected at Bell Laboratories and its extragalactic origin established. No combination of the then known sources of radio emission could account for it. Receipt of a copy of Peebles' preprint solved the problem raised be this unexplained phenomenon. Eddington tells us: "Never fully trust an observational result until you have at least one theory to explain it. "The theory and observation were then brougt together in a pair of papers (Dicke et al, 1965, Penzias and Wilson 1965) which led to decisive support for evolutionary cosmology and further renewal of interest in its observational consequences.

The existence of the relict radiation established the validity of the expanding universe picture with its cosmological production of the light elements, deuterium, helium-3 and helium-4 during the hot early stages of the expansion. The build-up of the heavier elements occurs at a much later stage, after the stars have formed. In stars, the cosmologically produced helium-4, together with additional amounts of helium produced by the stars themselves, is converted (via beryllium-8) into carbon-12 from which the heavier elements are then built. The stellar process described by Burbidge et al (1957) have been supplemented and, in some cases, replaced by processes whose existence was established trough later work, of which explosive nucleosynthesis is the most significant one. (See Clayton 1968 for a review.) Much of the build-up of the heavier elements goes on in a few violent minutes during the life of massive stars in which their outer shells are thrown outward in supernova explosions. This mechanism accounts both for the formation of the heavy elements as well as for their introduction into interstellar space. Thus, the total picture seems close to complete but puzzling gaps remain, such as the absence of solar neutrinos (Bahcall and Davis, 1976). One thing is clear however, observational cosmology is now a respectable and flourishing science.

Acknowledgment

My first thanks must go to the members of the Academy for the great honor they have bestowed upon me. The work which resulted in the occasion for this talk is described in an accompanying paper by my friend and colleague Robert W. Wilson. I am profoundly grateful for his unfailing help throughout our fifteen years of partnership.

The preparation of the talk upon which this manuscript is based owes much to many people. Conversations with R. A. Alpher, John Bahcall, S. Chandrasekhar and Martin Schwartzschild were particularly helpful. I am also grateful to A. B. Crawford, R. H. Dicke, G. B. Field, R. Kompfner, P. J. E. Peebles, D. Sciama, P. Thaddeus and S. Weinberg for earlier help given personally and through their published work.

REFERENCES

1. Alpher, R. A., Bethe, H. A., and Gamow, G., 1949, Phys. Rev. *73*. 803.
2. Alpher, R. A. and Herman, R. C., 1949, Phys. Rev. *75*, 1089
3. Alpher, R. A. and Herman, R. C., 1950, Rev. Mod. Phys. *22*. 153.
4. Alpher, R. A., Follin, J. W. and Herman, R. C., 1953, Phys. Rev. *92*, 1347.
5. Bethe, H. A., 1939, Phys. Rev. *55*, 434.
6. Burbidge, E. M., Burbidge, G. R., Fowler, W. A., and Hoyle, F., 1957, Rev. Mod. Phys. *29*, 547.
7. Chandrasekhar, S., 1939, *An Introduction to the Study of Stellar Structure*, (University of Chicago).
8. Chandrasekhar, S. and Henrich, L. R., 1942, Ap. J. *95*, 228.
9. Clayton, D. D., 1968 *Principles of Stellar Evolution and Nucleosynthesis* (McGraw-Hill).
10. Dicke, R. H., Peebles, P. J. E., Roll, P. G. and Wilkinson, D. T., 1965, Ap. J. *142*, 414.
11. Doroshkevich, A. G. and Novikov, I. D., 1964a Dokl. Akad. Navk. SSR *154*, 809.
12. Doroshkevich, A. G. and Novikov, I. D., 1964b Sov. Phys-Dokl. *9*. 111.
13. Einstein, A. and deSitter, W., 1932 Proc. Nat. Acad. Sci. *18*, 312.
14. Eddington, A. S., 1926, *The Internal Constitution of the Stars*, (Cambridge University Press).
15. Fowler, R. H., 1926, M.N.R.A. *87*, 114.
16. Friedmann, A., 1922, Zeits. Fur Physik *10*, 377.
17. Gamow, G., 1946, Phys. Rev. *70*, 572.
18. Gamow, G., 1948, Nature *162*, 680.
19. Gamow, G., 1950, Physics Today *3*, No. 8, 16
20. Hayashi, C., 1950, Prog. Theo. Phys. (Japan) *5*, 224.
21. Hoyle, F. and Taylor, R. J., 1964, Nature *203*, 1108.
22. Hubble, E. P., 1929, Proc. N.A.S. *15*, 168.
23. Lemaitre, G., 1927, Ann. Soc. Sci. Brux. *A47*, 49.
24. Lifschitz, E., 1946, J. Phys. USSR *10*, 116.
25. Ohm. E. A., 1961, Bell Syst. Tech. J. *40*, 1065.
26. Peebles, P. J. E., 1971, Physical Cosmology (Princeton University Press).
27. Penzias, A. A. and Wilson, R. W., 1965, Ap. J. *142*, 419.
28. Salpeter, E. E., 1952, Ap. J. *115*, 326.
29. Schwartzchild, B. and Schwartzschild, M. 1950, Ap. J. *112*, 248.
30. Tolman, R. C., 1934, *Relativity Thermodynamics and Cosmology*, (Clarendon Press, Oxford).
31. von Weizsäcker, C. F., 1937, Physik. Zeits. *38*, 176.
32. von Weizsäcker, C. F., 1938, Physik. Zeits. *39*, 633.
33. Weinberg, S., The First Three Minutes, (Basic Books).
34. Wilson, R. W., 1978 Nobel Lecture
35. Zeldovich, 1965, Advances in Astr. and Ap. *3*. 241.

ROBERT W. WILSON

My grandparents moved to Texas from the South after the U.S. Civil War and settled on small farms in the Dallas-Ft. Worth area. Both families emphasized education as the way to improve their children's lives and both my parents managed to graduate from college. After receiving an M.A. in chemistry from Rice University, my father worked for an oil well service company in Houston. I was born on January, 10, 1936. Two sisters followed, three and seven years later.

I attended public school in Houston. I took piano lessons for several years, and in high school, I played trombone in the marching band. I remember especially enjoying two seasonal activities: ice skating with the Houston Figure Skating Club in the winter and visiting an aunt and uncle's farm in west Texas in the summer.

During my pre-college years I went on many trips with my father into the oil fields to visit their operations. On Saturday mornings I often went with him to visit the company shop. I puttered around the machine, electronics, and automobile shops while he carried on his business. Both of my parents are inveterate do-it-yourselfers, almost no task being beneath their dignity or beyond their ingenuity. Having picked up a keen interest in electronics from my father, I used to fix radios and later television sets for fun and spending money. I built my own hi-fi set and enjoyed helping friends with their amateur radio transmitters, but lost interest as soon as they worked.

My high school career was undistinguished except for math and science. However, having barely been admitted to Rice University, I found that I enjoyed the courses and the elation of success and graduated with honors in physics. I did a senior thesis with C. F. Squire building a regulator for a magnet for use in low-temperature physics. Following that I had a summer job with Exxon and obtained my first patent. It covered the high-voltage pulse generator for a pulsed neutron source in a down-hole well-logging tool.

Following Rice, I went to Caltech for a Ph.D in physics, without any strong idea of what I wanted to do for a thesis topic. For the first year I lived in the Athenaeum (faculty club) where I became acquainted with a small group of graduate students and visiting faculty members, with whom I often dined and went on weekend outings. When the end of my second quarter approached, I needed a trial research project. David Dewhirst, a Cambridge astronomer and one of the Athenaeum group, suggested that I see John Bolton and Gordon Stanley about radio astronomy. The situation seemed perfect for me. John had come to Caltech to build the Owens

Valley Radio Observatory, and the heavy construction was finished. Radio astronomy offered a nice mixture of electronics and physics.

My introduction to radio astronomy was, however, delayed for a summer. I returned to Houston to court and marry Elizabeth Rhoads Sawin, whose spirit and varied interests have added much to my happiness during our twenty-year marriage.

The following year I took my first astronomy courses and went to the observatory during school breaks. That summer John Bolton asked me to join him in observing some of the bright regions on a radio map of the Milky Way which had been made by Westerhaut. By the end of the summer, this project had expanded to making a complete map of that part of the Milky Way which was visible to us. When it was time to measure our chart records and start drawing contour maps from the data, John set up a drawing board in his office, and worked with me on the project. This was typical of John. Whatever the project, whether digging a hole, surveying, laying cables, observing, or reducing data, John would work along with the others. His interest in our map-making and the location of the drawing board kept me at the map-making task instead of designing the next piece of equipment, which would have been my natural inclination.

Our first son, Philip, was born during my fourth year at Caltech. He had many trips to the Owens Valley Radio Observatory, the first at the age of two weeks. He and Betsy were readily accepted at the observatory.

My thesis project was to have been hydrogen-line interferometry, but when the first plans for a local oscillator system didn't work out, I used the galactic survey as the basis for my thesis. John Bolton returned to Australia before I completed my Ph.D. Maarten Schmidt, who had previously done galactic research and was currently working on quasars, saw me through the last months of thesis work. I remained at Caltech for an additional year as a postdoctoral fellow to finish several projects in which I was involved.

The project of setting up and running the Owens Valley Radio Observatory was very much a community effort. At one time or another I worked with all of the staff and other students and learned from all of them. My collaborations with V. Radhakrishram and B. G. Clark were especially fruitful. I also had the opportunity to meet many of the world's astronomers who visited Caltech.

In 1961, H. E. D. Scovil at Bell Labs offered to help us make a pair of traveling-wave maser amplifiers for the interferometer. V. Radhakrishran got the job of going to Bell Labs to make our masers. I had wanted to go, but had not yet completed my degree work. I worked with Rad on that project, though, and developed a good feeling toward Bell Labs which was later a strong influence on my decision to take a job there.

I joined Bell Laboratories at Crawford Hill in 1963 as part of A. B. Crawford's Radio Research department in R. Kompfner's laboratory. I started working with the only other radio astronomer, Arno Penzias, who had been there about two years. Our early radio astronomy projects are described in my Nobel lecture.

With the creation of Comsat by U. S. Congress, Bell System satellite efforts and related space research were reduced. In 1965 Arno and I were told that the radio astronomy effort could only be supported at the level of one full-time staff member, even though Art Crawford and Rudi Kompfner strongly supported our astronomical research. Arno and I agreed that having two half-time radio astronomers was a better solution to our problem than having one full-time one, so we started taking on other projects. The first one was a joint project—a propagation experiment on a terrestrial path using a 10.6μ carbon dioxide laser as a source. Following that, I did two applied radio astronomy projects. For the first, I designed a device we called the Sun Tracker. It automatically pointed to the sun while it was up every day and measured the attenuation of the sun's cm-wave radiation in the earth's atmosphere. Since, as we expected, the attenuation was large for too much of the time for a practical satellite system, I next set up three fixed-pointed radiometers at spaced locations to check on the feasibility of working around heavy rains.

In 1969 Arno suggested that we start doing millimeter wave astronomy. We could take the low noise millimeter-wave receivers which had been developed at Crawford Hill by C. A. Burrus and W. M. Sharpless for a waveguide communication system and make an astronomical receiver with them. We planned to use it at the National Radio Astronomy Observatory's new 36-foot radio telescope at Kitt Peak in Arizona. Our observations began in 1969 with a continuum receiver. The next year, K. B. Jefferts joined us, and with much help from C. A. Burrus at Crawford Hill and S. Weinreb at NRAO we made a spectral line receiver at 100–120 GHz. We were excited to discover unexpectedly large amounts of carbon monoxide in a molecular cloud behind the Orion Nebula. We quickly found that CO is widely distributed in our galaxy and so abundant that the rare isotopic species $^{13}C^{16}O$ and $^{12}C^{18}O$ were readily measurable. We soon observed a number of other simple molecules. Our major efforts were directed toward isotope ratios as a probe of nucleogenesis and understanding the structure of molecular clouds.

In 1972, S. J. Buchsbaum, who was our new executive director, revived an earlier proposal and suggested that we build a millimeter-wave facility at Crawford Hill. It was to be used partly for radio astronomy, and partly to monitor the beacons on the Comstar satellites which AT&T was planning to put up. I was project director for the design and construction of the antenna and was responsible for the equipment and programming necessary to make it a leading millimeter-radio telescope. The winter of 1977–78 was our first good observing season with the 7-meter antenna and I am looking forward to several more years of millimeter wave astronomy with it.

We still live in the house in Holmdel which we bought when I first came to Bell Laboratories. Our two younger children were born here, Suzanne in 1963, and Randal in 1967. We have come to enjoy the eastern woodlands and I now look forward to skiing and outdoor ice skating with my family and associates in the winter. I spend many evenings reading or continuing the day's work, but I also enjoy playing the piano, jogging, and traveling with the family.

B.A. 1957 Rice University "with honors in Physics".
Ph.D. 1962 California Institute of Technology.
Married 1958 Elizabeth Rhoads Sawin.
Employment:
Caltech, Research Fellow 1962–1963.
Bell Laboratories 1963–
Member of Technical Staff 1963–1976.
Head Radio Physics Research Department 1976–
Adjunct Professor, State University of New York (SUNY) 1978–
Member:
American Astronomical Society
International Astronomical Union
American Physical Society
International Union of Radio Sciences
American Academy of Arts and Sciences
Honors:
Phi Beta Kappa
Sigma Xi
Henry Draper Award 1977
Herschel Medal 1977

THE COSMIC MICROWAVE BACKGROUND RADIATION

Nobel Lecture, 8 December, 1978
by
ROBERT W. WILSON
Bell Laboratories
Holmdel, N.J. U.S.A.

I. INTRODUCTION

Radio Astronomy has added greatly to our understanding of the structure and dynamics of the universe. The cosmic microwave background radiation, considered a relic of the explosion at the beginning of the universe some 18 billion years ago, is one of the most powerful aids in determining these features of the universe. This paper is about the discovery of the cosmic microwave background radiation. It starts with a section on radio astronomical measuring techniques. This is followed by the history of the detection of the background radiation, its identification, and finally by a summary of our present knowledge of its properties.

II. RADIO ASTRONOMICAL METHODS

A radio telescope pointing at the sky receives radiation not only from space, but also from other sources including the ground, the earth's atmosphere, and the components of the radio telescope itself. The 20-foot horn-reflector antenna at Bell Laboratories (Fig. 1) which was used to discover the cosmic microwave background radiation was particularly suited to distinguish this weak, uniform radiation from other, much stronger sources. In order to understand this measurement it is necessary to discuss the design and operation of a radio telescope, especially its two major components, the antenna and the radiometer[1].

a. Antennas

An antenna collects radiation from a desired direction incident upon an area, called its collecting area, and focuses it on a receiver. An antenna is normally designed to maximize its response in the direction in which it is pointed and minimize its response in other directions.

The 20-foot horn-reflector shown in Fig. 1 was built by A. B. Crawford and his associates[2] in 1960 to be used with an ultra low-noise communications receiver for signals bounced from the Echo satellite. It consists of a large expanding waveguide, or horn, with an off-axis section parabolic reflector at the end. The focus of the paraboloid is located at the apex of the horn, so that a plane wave traveling along the axis of the paraboloid is focused into the receiver, or radiometer, at the apex of the horn. Its design emphasizes the rejection of radiation from the ground. It is easy to see

Fig. 1 The 20 foot horn-reflector which was used to discover the Cosmic Microwave Background Radiation.

from the figure that in this configuration the receiver is well shielded from the ground by the horn.

A measurement of the sensitivity of a small hornreflector antenna to radiation coming from different directions is shown in Fig. 2. The circle marked isotropic antenna is the sensitivity of a fictitious antenna which receives equally from all directions. If such an isotropic lossless antenna were put in an open field, half of the sensitivity would be to radiation from the earth and half from the sky. In the case of the hornreflector, sensitivity in the back or ground direction is less than 1/3000 of the isotropic antenna. The isotropic antenna on a perfectly radiating earth at 300 K and with a cold sky at 0° K would pick up 300 K from the earth over half of its response and nothing over the other half, resulting in an equivalent antenna temperature of 150 K. The horn-reflector, in contrast, would pick up less than .05 K from the ground.

This sensitivity pattern is sufficient to determine the performance of an ideal, lossless antenna since such an antenna would contribute no radiation of its own. Just as a curved mirror can focus hot rays from the sun and burn a piece of paper without becoming hot itself, a radio telescope can focus the cold sky onto a radio receiver without adding radiation of its own.

b. Radiometers

A radiometer is a device for measuring the intensity of radiation. A microwave radiometer consists of a filter to select a desired band of

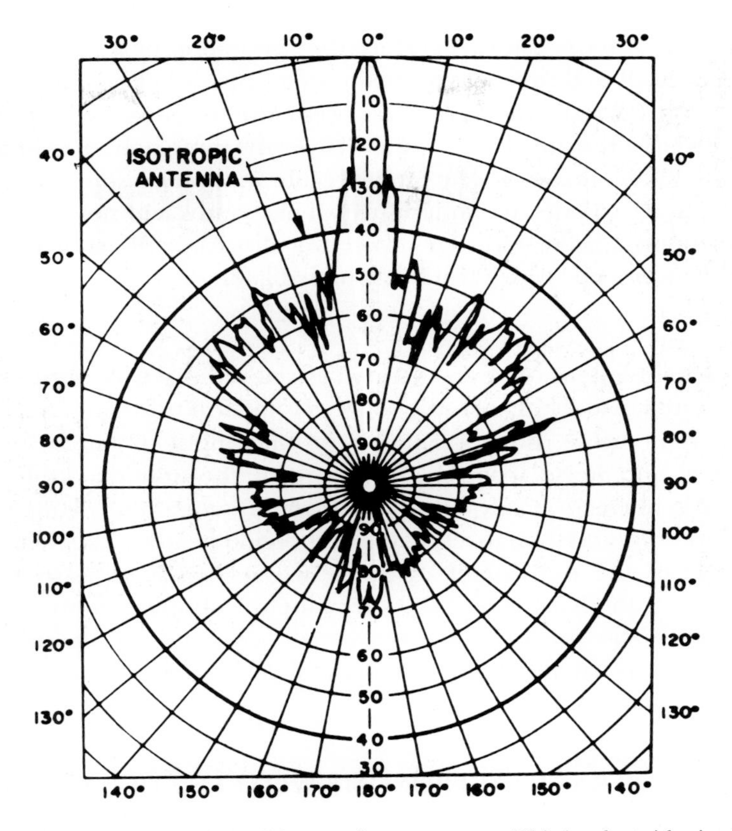

Fig. 2 Sensitivity pattern of a small horn-reflector antenna. This is a logarithmic plot of the collecting area of the antenna as a function of angle from the center of the main beam. Each circle below the level of the main beam represent a factor of ten reduction in sensitivity. In the back direction around 180 the sensitivity is consistently within the circle marked 70, corresponding to a factor of 10^{-7} below the sensitivity at 0.

frequencies followed by a detector which produces an output voltage proportional to its input power. Practical detectors are usually not sensitive enough for the low power levels received by radio telescopes, however, so amplification is normally used ahead of the detector to increase the signal level. The noise in the first stage of this amplifier combined with that from the transmission line which connects it to the antenna (input source) produce an output from the detector even with no input power from the antenna. A fundamental limit to the sensitivity of a radiometer is the fluctuation in the power level of this noise.

During the late 1950's, H. E. D. Scovil and his associates at Bell Laboratories, Murray Hill were building the world's lowest-noise microwave amplifiers, ruby travelling-wave masers[3]. These amplifiers were cooled to 4.2 K or less by liquid helium and contribute a correspondingly small amount of noise to the system. A radiometer incorporating these amplifiers can therefore be very sensitive.

Astronomical radio sources produce random, thermal noise very much like that from a hot resistor, therefore the calibration of a radiometer is usually expressed in terms of a thermal system. Instead of giving the noise power which the radiometer receives from the antenna, we quote the temperature of a resistor which would deliver the same noise power to the radiometer. (Radiometers often contain calibration noise sources consisting of a resistor at a known temperature.) This "equivalent noise temperature" is proportional to received power for all except the shorter wavelength measurements, which will be discussed later.

c. *Observations*

To measure the intensity of an extraterrestrial radio source with a radio telescope, one must distinguish the source from local noise sources—noise from the radiometer, noise from the ground, noise from the earth's atmosphere, and noise from the structure of the antenna itself. This distinction is normally made by pointing the antenna alternately to the source of interest and then to a background region nearby. The difference in response of the radiometer to these two regions is measured, thus subtracting out the local noise. To determine the absolute intensity of an astronomical radio source, it is necessary to calibrate the antenna and radiometer or, as usually done, to observe a calibration source of known intensity.

III. PLANS FOR RADIO ASTRONOMY WITH THE 20-FOOT HORN-REFLECTOR

In 1963, when the 20-foot horn-reflector was no longer needed for satellite work, Arno Penzias and I started preparing it for use in radio astronomy. One might ask why we were interested in starting our radio astronomy careers at Bell Labs using an antenna with a collecting area of only 25 square meters when much larger radio telescopes were available elsewhere. Indeed, we were delighted to have the 20-foot horn-reflector because it had special features that we hoped to exploit. Its sensitivity, or collecting area, could be accurately calculated and in addition it could be measured using a transmitter located less than 1 km away. With this data, it could be used with a calibrated radiometer to make primary measurements of the intensities of several extraterrestrial radio sources. These sources could then be used as secondary standards by other observatories. In addition, we would be able to understand all sources of antenna noise, for example the amount of radiation received from the earth, so that background regions could be measured absolutely. Traveling-wave maser amplifiers were available for use with the 20-foot horn-reflector, which meant that for large diameter sources (those subtending angles larger than the antenna beamwidth), this would be the world's most sensitive radio telescope.

My interest in the background measuring ability of the 20-foot horn-reflector resulted from my doctoral thesis work with J. G. Bolton at

Caltech. We made a map of the 31 cm radiation from the Milky Way and studied the discrete sources and the diffuse gas within it. In mapping the Milky Way we pointed the antenna to the west side of it and used the earth's rotation to scan the antenna across it. This kept constant all the local noise, including radiation that the antenna picked up from the earth. I used the regions on either side of the Milky Way (where the brightness was constant) as the zero reference. Since we are inside the Galaxy, it is impossible to point completely away from it. Our mapping plan was adequate for that project, but the unknown zero level was not very satisfying. Previous low frequency measurements had indicated that there is a large, radio-emitting halo around our galaxy which I could not measure by that technique. The 20-foot horn-reflector, however, was an ideal instrument for measuring this weak halo radiation at shorter wavelengths. One of my intentions when I came to Bell Labs was to make such a measurement.

In 1963, a maser at 7.35 cm wavelength[3] was installed on the 20-foot horn-reflector. Before we could begin doing astronomical measurements, however, we had to do two things: 1) build a good radiometer incorporating the 7.35 cm maser amplifier, and; 2) finish the accurate measurement of the collecting-area (sensitivity) of the 20-foot horn-reflector which D. C. Hogg had begun. Among our astronomical projects for 7 cm were absolute intensity measurements of several traditional astronomical calibration sources and a series of sweeps of the Milky Way to extend my thesis work. In the course of this work we planned to check out our capability of measuring the halo radiation of our Galaxy away from the Milky Way. Existing low frequency measurements indicated that the brightness temperature of the halo would be less than 0.1 K at 7 cm. Thus, a background measurement at 7 cm should produce a null result and would be a good check of our measuring ability.

After completing this program of measurements at 7 cm, we planned to build a similar radiometer at 21 cm. At that wavelength the galactic halo should be bright enough for detection, and we would also observe the 21 cm line of neutral hydrogen atoms. In addition, we planned a number of hydrogen-line projects including an extension of the measurements of Arno's thesis, a search for hydrogen in clusters of galaxies.

At the time we were building the 7-cm radiometer John Bolton visited us and we related our plans and asked for his comments. He immediately selected the most difficult one as the most important: the 21 cm background measurement. First, however, we had to complete the observations at 7 cm.

IV. RADIOMETER SYSTEM

We wanted to make accurate measurements of antenna temperatures. To do this we planned to use the radiometer to compare the antenna to a reference source, in this case, a radiator in liquid helium. I built a switch which would connect the maser amplifier either to the antenna or to Arno's helium-cooled reference noise source[5] (cold load). This would allow an accurate comparison of the equivalent temperature of the antenna to that of the cold load, since the noise from the rest of the radiometer would be constant during switching. A diagram of this calibration system[6] is shown in Figure 3 and its operation is described below.

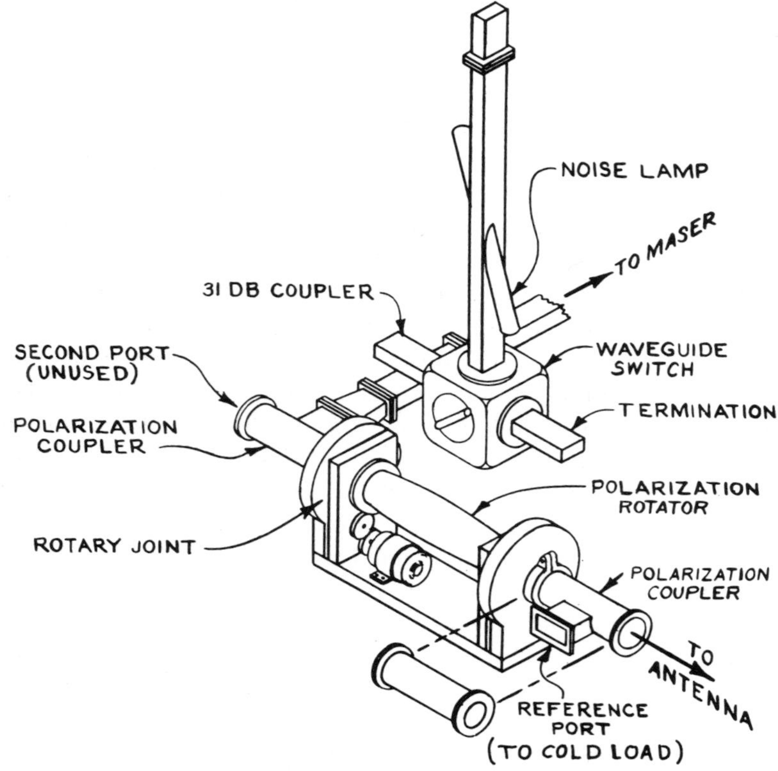

Fig. 3 The switching and calibration system of our 7.35 cm radiometer. The reference port was normally connected to the helium cooled reference source through a noise adding attenuator.

a. *Switch*

The switch for comparing the cold load to the antenna consists of the two polarization couplers and the polarization rotator shown in Fig. 3. This type of switch had been used by D. H. Ring in several radiometers at Holmdel. It had the advantage of stability, low loss, and small reflections. The circular waveguide coming from the antenna contains the two orthogonal modes of polarization received by the antenna. The first polarization coupler reflected one mode of linear polarization back to the antenna and substituted the signal from the cold load for it in the waveguide going to the rotator. The second polarization coupler took one of the two modes of linear polarization coming from the polarization rotator and coupled it to the rectangular (single-mode) waveguide going to the maser. The polarization rotator is the microwave equivalent of a half-wave plate in optics. It is a piece of circular waveguide which has been squeezed in the middle so that the phase shifts for waves traveling through it in its two principal planes of linear polarization differ by 180 degrees. By mechanically rotating it, the polarization of the signals passing through it can be rotated. Thus either the antenna or cold load could be connected to the maser.

This type of switch is not inherently symmetric, but has very low loss and is stable so that its asymmetry of .05 K was accurately measured and corrected for.

b. *Reference Noise Source*

A drawing of the liquid-helium cooled reference noise source is shown in Figure 4. It consists of a 122 cm piece of 90 percent-copper brass waveguide connecting a carefully matched microwave absorber in liquid He to a room-temperature flange at the top. Small holes allow liquid helium to fill the bottom section of waveguide so that the absorber temperature could be known, while a mylar window at a 30° angle keeps the liquid out of the rest of the waveguide and makes a low-reflection microwave transition between the two sections of waveguide. Most of the remaining parts are for the cryogenics. The gas baffles make a counter-flow heat exchanger between the waveguide and the helium gas which has boiled off, greatly extending the time of operation on a charge of liquid helium. Twenty liters of liquid helium cooled the cold load and provided about twenty hours of operation.

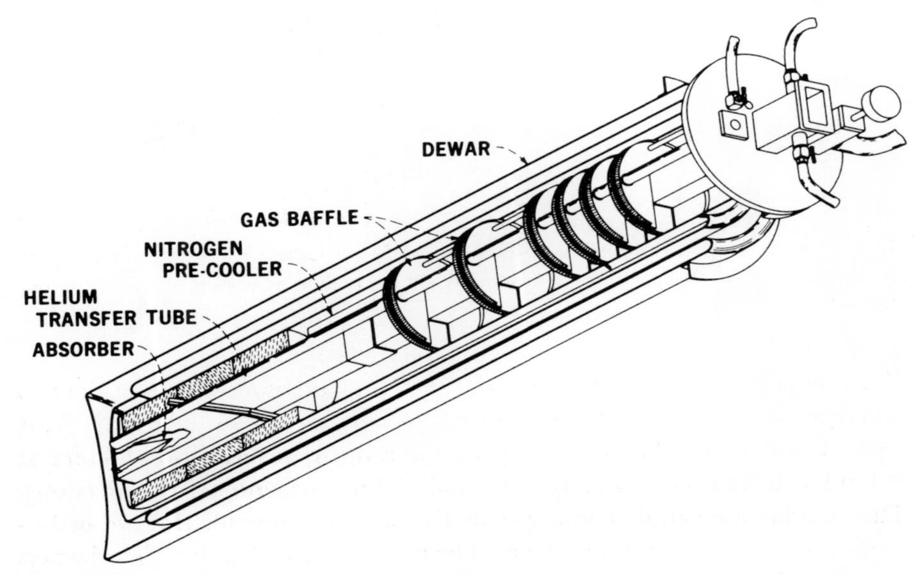

Fig. 4 The Helium Cooled Reference Noise Source.

Above the level of the liquid helium, the waveguide walls were warmer than 4.2 K. Any radiation due to the loss in this part of the waveguide would raise the effective temperature of the noise source above 4.2 K and must be accounted for. To do so we monitored the temperature distribution along the waveguide with a series of diode thermometers and calculated the contribution of each section of the waveguide to the equivalent temperature of the reference source. When first cooled down, the calculated total temperature of the reference noise source was about 5 K, and after several hours when the liquid helium level was lower, it increased to 6 K. As a check of this calibration procedure, we compared the antenna temperature (assumed constant) to our reference noise source during this period, and found consistency to within 0.1 K.

c. *Scale Calibration*

A variable attenuator normally connected the cold load to the reference port of the radiometer. This device was at room temperature so noise could be added to the cold load port of the switch by increasing its attenuation. It was calibrated over a range of 0.11 dB which corresponds to 7.4 K of added noise.

Also shown in Fig. 3 is a noise lamp (and its directional coupler) which was used as a secondary standard for our temperature scale.

d. *Radiometer Backend*

Signals leaving the maser amplifier needed to be further amplified before detection so that their intensity could be measured accurately. The remainder of our radiometer consisted of a down converter to 70 MHz followed by I. F. amplifiers, a precision variable attenuator and a diode detector. The output of the diode detector was amplified and went to a chart recorder.

Fig. 5 Our 7.35 cm radiometer installed in the cab of the 20 foot horn-reflector.

e. *Equipment Performance*

Our radiometer equipment installed in the cab of the 20-foot horn-reflector is shown in Fig. 5. The flange at the far right is part of the antenna and rotates in elevation angle with it. It was part of a double-choke joint which allowed the rest of the equipment to be fixed in the cab while the antenna rotated. The noise contribution of the choke joint could be measured by clamping it shut and was found to be negligible. We regularly measured the reflection coefficient of the major components of this system and kept it below 0.03 percent, except for the maser whose reflection could not be reduced below 1 percent. Since all ports of our waveguide system were terminated at a low temperature, these reflections resulted in negligible errors.

V. PRIOR OBSERVATIONS

The first horn-reflector-travelling-wave maser system had been put together by DeGrasse, Hogg, Ohm, and Scovil in 1959[7] to demostrate the feasibility of a low-noise, satellite-earth station at 5.31 cm. Even though they achieved the lowest total system noise temperature to date, 18.5 K, they had expected to do better. Fig. 6 shows their system with the noise temperature they assigned to each component. As we have seen in Section IIa,

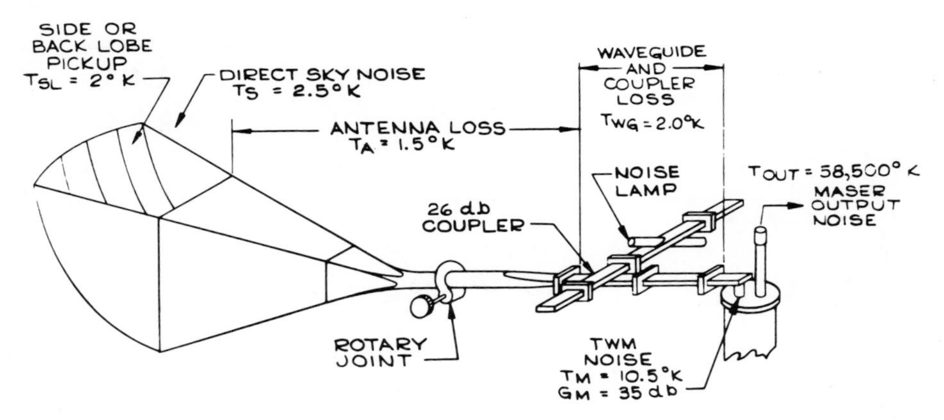

Fig. 6 A diagram of the low noise receiver used by deGrasse, Hogg, Ohm and Scovil to show that very low noise earth stations are possible. Each component is labeled with its contribution to the system noise.

the 2 K they assigned to antenna backlobe pickup is too high. In addition, direct measurements of the noise temperature of the maser gave a value about a degree colder than shown here. Thus their system was about 3 K hotter than one might expect. The component labeled T_s in Fig. 6 is the radiation of the earth's atmosphere when their antenna was aimed straight up. It was measured by a method first reported by R. H. Dicke[8]. (It is interesting that Dicke also reports an upper limit of 20 K for the cosmic microwave background radiation in this paper – the first such report.) If the antenna temperature is measured as a function of the angle above the horizon at which it is pointing, the radiation of the atmosphere is at a minimum when the antenna is directed straight up. It increases as the antenna points toward the horizon, since the total line of sight through the atmosphere increases. Figure 7 is a chart recording Arno Penzias and I

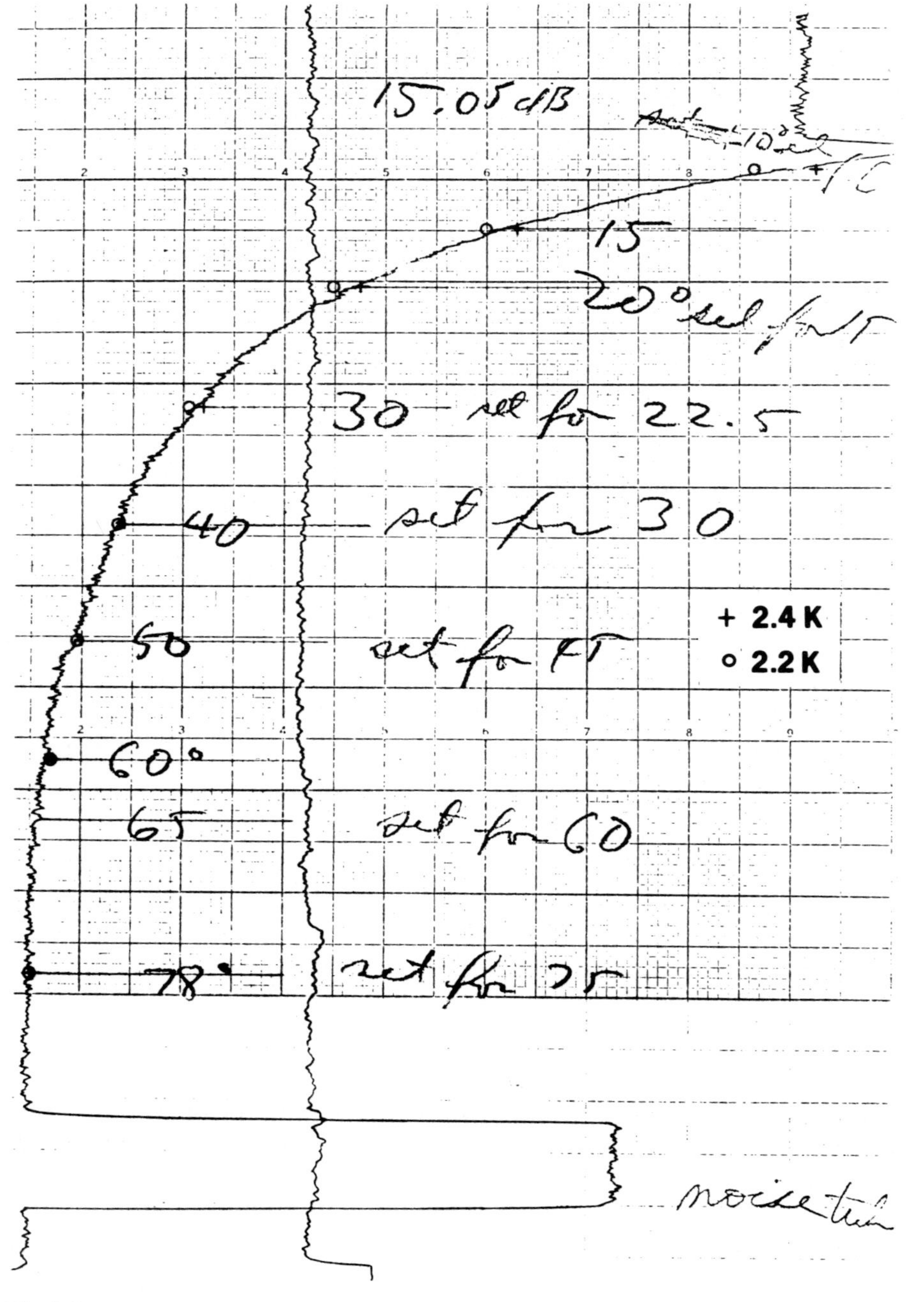

Fig. 7 A measurement of atmospheric noise at 7.35 cm wavelength with theoretical fits to the data for 2.2 and 2.4K Zenith atmospheric radiation.

made with the 20-foot horn-reflector scanning from almost the Zenith down to 10° above the horizon. The circles and crosses are the expected change based on a standard model of the earth's atmosphere for 2.2 and 2.4 K Zenith contribution. The fit between theory and data is obviously good leaving little chance that there might be an error in our value for atmospheric radiation.

Fig. 8 is taken from the paper in which E. A. Ohm[9] described the receiver on the 20-foot horn reflector which was used to receive signals bounced from the Echo satellite. He found that its system temperature was 3.3 K higher than expected from summing the contributions of the components. As in the previous 5.3 cm work, this excess temperature was smaller

TABLE II — SOURCES OF SYSTEM TEMPERATURE

Source	Temperature
Sky (at zenith)	2.30 ± 0.20°K
Horn antenna	2.00 ± 1.00°K
Waveguide (counter-clockwise channel)	7.00 ± 0.65°K
Maser assembly	7.00 ± 1.00°K
Converter	0.60 ± 0.15°K
Predicted total system temperature	18.90 ± 3.00°K

the temperature was found to vary a few degrees from day to day, but the lowest temperature was consistently 22.2 ± 2.2°K. By realistically assuming that all sources were then contributing their fair share (as is also tacitly assumed in Table II) it is possible to improve the over-all accuracy. The actual system temperature must be in the overlap region of the measured results and the total results of Table II, namely between 20 and 21.9°K. The most likely minimum system temperature was therefore

$$T_{system} = 21 \pm 1°K.^*$$

The inference from this result is that the "+" temperature possibilities of Table II must predominate.

Fig. 8 An excerpt from E. A. Ohm's article on the Echo receiver showing that his system temperature was 3.3K higher than predicted.

than the experimental errors, so not much attention was paid to it. In order to determine the unambiguous presence of an excess source of radiation of about 3 K, a more accurate measurement technique was required. This was achieved in the subsequent measurements by means of a switch and reference noise source combination which communications systems do not have.

VI. OUR OBSERVATIONS

Fig. 9 is a reproduction of the first record we have of the operation of our system. At the bottom is a list of diode thermometer voltages from which we could determine the cold load's equivalent temperature. The recorder trace has power (or temperature) increasing to the right. The middle part of this trace is with the maser switched to the cold load with various settings of the noise adding attenuator. A change of 0.1 dB corresponds to a temperature change of 6.6 K, so the peak-to-peak noise on the trace amounts to less than 0.2 K. At the top of the chart the maser is switched to

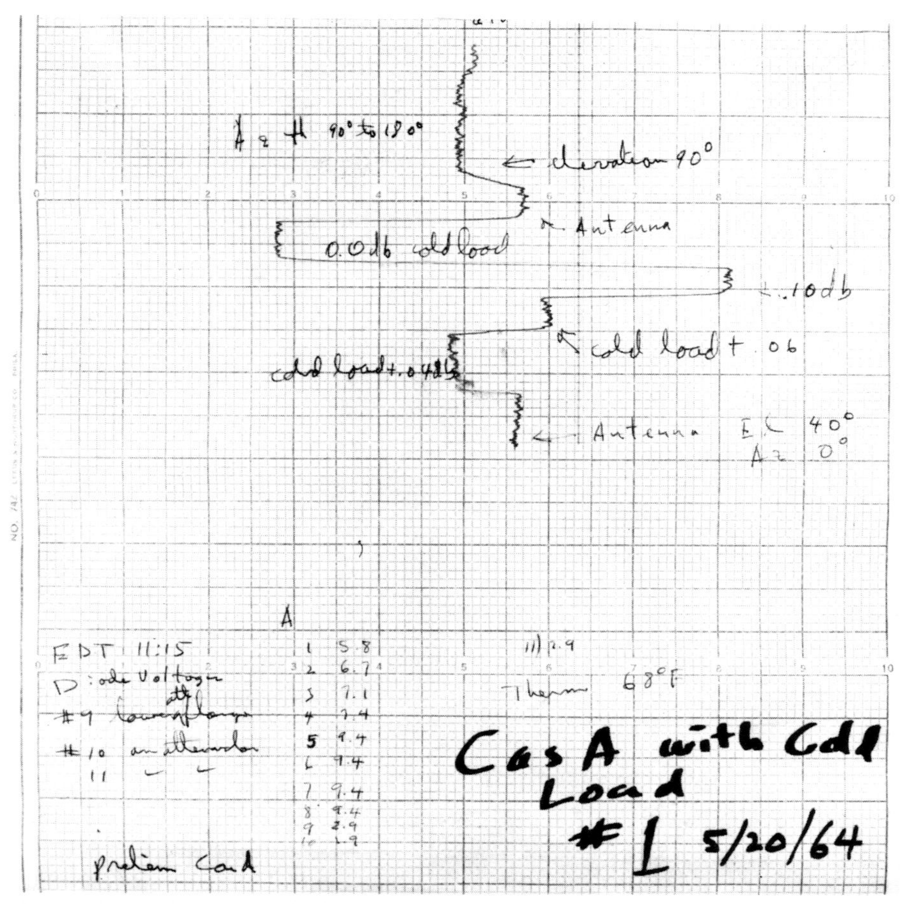

Fig. 9 The first measurement which clearly showed the presence of the microwave background. Noise temperature is plotted increasing to the right. At the top, the antenna pointed at 90° elevation is seen to have the samt noise temperature as the cold load with 0.04 db attenuation (about 7.5K). This is considerably above the expected value of 3.3K.

the antenna and has about the same temperature as the cold load plus .04 dB, corresponding to a total of about 7.5 K. This was a troublesome result. The antenna temperature should have been only the sum of the atmospheric contribution (2.3 K) and the radiation from the walls of the antenna and ground (1 K). The excess system temperature found in the previous experiments had, contrary to our expectations, all been in the antenna or beyond. We now had a direct comparison of the antenna with the cold load and had to assign our excess temperature to the antenna whereas in the previous cases only the total system temperature was measured. If we had missed some loss, the cold load might have been warmer than calculated, but it could not be colder than 4.2 K – the temperature of the liquid helium. The antenna was at least 2 K hotter than that. Unless we could understand our "antenna problem" our 21 cm galactic halo experiment would not be possible. We considered a number of possible reasons for this excess and, where warranted, tested for them. These were:

a. At that time some radio astronomers thought that the microwave absorption of the earth's atmosphere was about twice the value we were using – in other words the "sky temperature" of Figs. 6 and 8 was about 5 K instead of 2.5 K. We knew from our measurement of sky temperature such as shown in Fig. 7 that this could not be the case.
b. We considered the possibility of man-made noise being picked up by our antenna. However, when we pointed our antenna to New York City, or to any other direction on the horizon, the antenna temperature never went significantly above the thermal temperature of the earth.
c. We considered radiation from our galaxy. Our measurements of the emission from the plane of the Milky Way were a reasonable fit to the intensities expected from extrapolations of low-frequency measurements. Similar extrapolations for the coldest part of the sky (away from the Milky Way) predicted about .02 K at our wavelength. Furthermore, any galactic contribution should also vary with position and we saw changes only near the Milky Way, consistent with the measurements at lower frequencies.
d. We ruled out discrete extraterrestrial radio sources as the source of our radiation as they have spectra similar to that of the Galaxy. The same extrapolation from low frequency measurements applies to them. The strongest discrete source in the sky had a maximum antenna temperature of 7 K.

Thus we seemed to be left with the antenna as the source of our extra noise. We calculated a contribution of 0.9 K from its resistive loss using standard waveguide theory. The most lossy part of the antenna was its small diameter throat, which was made of electroformed copper. We had measured similar waveguides in the lab and corrected the loss calculations for the imperfect surface conditions we had found in those waveguides. The remainder of the antenna was made of riveted aluminum sheets, and although we did not expect any trouble there, we had no way to evaluate the loss in the riveted joints. A pair of pigeons was roosting up in the small part of the horn where it enters the warm cab. They had covered the inside with a white material familiar to all city dwellers. We evicted the pigeons and cleaned up their mess, but obtained only a small reduction in antenna temperature.

For some time we lived with the antenna temperature problem and concentrated on measurements in which it was not critical. Dave Hogg and I had made a very accurate measurement of the antenna's gain[10], and Arno and I wanted to complete our absolue flux measurements before disturbing the antenna further.

In the spring of 1965 with our flux measurements finished[5], we thoroughly cleaned out the 20-foot horn-reflector and put aluminum tape over the riveted joints. This resulted in only a minor reduction in antenna temperature. We also took apart the throat section of the antenna, and checked it, but found it to be in order.

By this time almost a year had passed. Since the excess antenna temperature had not changed during this time, we could rule out two additional sources: 1) Any source in the solar system should have gone through a large change in angle and we should have seen a change in antenna temperature. 2) In 1962, a high-altitude nuclear explosion had filled up the Van Allen belts with ionized particles. Since they were at a large distance from the surface of the earth, any radiation from them would not show the same elevation-angle dependence as the atmosphere and we might not have identified it. But after a year, any radiation from this source should have reduced considerably.

VII. IDENTIFICATION

The sequence of events which led to the unravelling of our mystery began one day when Arno was talking to Bernard Burke of M.I.T. about other matters and mentioned our unexplained noise. Bernie recalled hearing about theoretical work of P. J. E. Peebles in R. H. Dicke's group in Princeton on radiation in the universe. Arno called Dicke who sent a copy of Peebles' preprint. The Princeton group was investigating the implications of an oscillating universe with an extremely hot condensed phase. This hot bounce was necessary to destroy the heavy elements from the previous cycle so each cycle could start fresh. Although this was not a new idea[11], Dicke had the important idea that if the radiation from this hot phase were large enough, it would be observable. In the preprint, Peebles, following Dicke's suggestion calculated that the universe should be filled with a relic blackbody radiation at a minimum temperature of 10 K. Peebles was aware of Hogg and Semplak's (1961)[12] measurement of atmospheric radiation at 6 cm using the system of DeGrasse et al., and concluded that the present radiation temperature of the universe must be less than their system temperature of 15 K. He also said that Dicke, Roll, and Wilkinson were setting up an experiment to measure it.

Shortly after sending the preprint, Dicke and his coworkers visited us in order to discuss our measurements and see our equipment. They were quickly convinced of the accuracy of our measurements. We agreed to a side-by-side publication of two letters in the *Astrophysical Journal*—a letter on the theory from Princeton[13] and one on our measurement of excess antenna temperature from Bell Laboratories[14]. Arno and I were careful to exclude any discussion of the cosmological theory of the origin of background radiation from our letter because we had not been involved in any of that work. We thought, furthermore, that our measurement was independent of the theory and might outlive it. We were pleased that the mysterious noise appearing in our antenna had an explanation of any kind, especially one with such significant cosmological implications. Our mood, however, remained one of cautious optimism for some time.

VIII. RESULTS

While preparing our letter for publication we made one final check on the antenna to make sure we were not picking up a uniform 3 K from earth. We measured its response to radiation from the earth by using a transmitter located in various places on the ground. The transmitter artificially increased the ground's brightness at the wavelength of our receiver to a level high enough for the backlobe response of the antenna to be measurable. Although not a perfect measure of the structure of the backlobes of an antenna, it was a good enough method of determining their average level. The backlobe level we found in this test was as low as we had expected and indicated a negligible contribution to the antenna temperatur from the earth.

The right-hand column of Fig. 10 shows the final results of our measurement. The numbers on the left were obtained later in 1965 with a new throat on the 20-foot horn-reflector. From the total antenna temperature we subtracted the known sources with a result of 3.4± 1 K. Since the errors in this measurement are not statistical, we have summed the maximum error from each source. The maximum measurement error of 1 K was considerably smaller than the measured value, giving us confidence in the reality of the result. We stated in the original paper that "This excess temperature is, within the limits of our observations, isotropic, unpolarized, and free of seasonal variations". Although not stated explicitly, our limits on an isotropy and polarization were not affected by most of the errors listed in Fig. 10 and were about 10 percent or 0.3 K.

	New Throat		Old Throat
He Temp.	4.22	4.22	
Calculated Contribution from Cold Load Waveguide	.38	.70 ± 0.2	
Attenuator Setting for Balance	2.73	2.40 ± 0.1	
Total C.L.	7.33	7.32 ± 0.3	6.7 ± 0.3
Atmosphere	2.3 ± 0.3		2.3 ± 0.3
Waveguide and Antenna loss	1.8 ± 0.3		.9 ± 0.3
Back lobes	.1 ± 0.1		.1 ± 0.1
Total Ant.	4.2 ± 0.7		3.3 ± 0.7
Background	3.1 ± 1		3.4 ± 1

Fig. 10 Results of our 1965 measurements of the microwave background. "Old Throat" and "New Throat" refer to the original and a replacement throat section for the 20 foot horn-reflector.

At that time the limit we could place on the shape of the spectrum of the background radiation was obtained by comparing our value of 3.5 K with a 74 cm survey of the northern sky done at Cambridge by Pauliny-Toth and Shakeshaft, 1962[15]. The minimum temperature on their map was 16 K. Thus the spectrum was no steeper than $\lambda^{0.7}$ over a range of wavelengths that varied by a factor of 10. This clearly ruled out any type of radio source known at that time, as they all had spectra with variation in the range $\lambda^{2.0}$ to $\lambda^{3.0}$. The previous Bell Laboratories measurement at 6 cm ruled out a spectrum which rose rapidly toward shorter wavelengths.

IX. CONFIRMATION

After our meeting, the Princeton experimental group returned to complete their apparatus and make their measurement with the expectation that the background temperature would be about 3 K.

The first confirmation of the microwave cosmic background that we knew of, however, came from a totally different, indirect measurement. This measurement had, in fact, been made thirty years earlier by Adams and Dunhan[16–21]. Adams and Dunhan had discovered several faint optical interstellar absorption lines which were later identified with the molecules CH, CH^{+}, and CN. In the case of CN, in addition to the ground state, absorption was seen from the first rotationally excited state. McKellar[22] using Adams' data on the populations of these two states calculated that the excitation temperature of CN was 2.3 K. This rotational transition occurs at 2.64 mm wavelength, near the peak of a 3 K black body spectrum. Shortly after the discovery of the background radiation, G. B. Field[23], I. S. Shklovsky[24], and P. Thaddeus[25] (following a suggestion by N. J. Woolf), independently realized that the CN is in equilibrium with the background radiation. (There is no other significant source of excitation where these molecules are located). In addition to confirming that the background was not zero, this idea immediately confirmed that the spectrum of the background radiation was close to that of a blackbody source for wavelengths larger than the peak. It also gave a hint that at short wavelengths the intensity was departing from the $1/\lambda^2$ dependence expected in the long wavelength (Raleigh-Jeans) region of the spectrum and following the true blackbody (Plank) distribution. In 1966, Field and Hitchcock[23] reported new measurements using Herbig's plates of ζ Oph and ζ Per obtaining 3.22 ± 0.15 K and 3.0 ± 0.6 K for the excitation temperature. Thaddeus and Clauser[25] also obtained new plates and measured 3.75 ± 0.5 K in ç Oph. Both groups argued that the main source of excitation in CN is the background radiation. This type of observation, taken alone, is most convincing as an upper limit, since it is easier to imagine additional sources of excitation than refrigeration.

In December 1965 Roll and Wilkinson[26] completed their measurement of 3.0 ± 0.5 K at 3.2 cm, the first confirming microwave measurement. This was followed shortly by Howell and Shakeshaft's[27] value of 2.8 ± 0.6

K at 20.7 cm[22] and then by our measurement of 3.2 K ± 1 K at 21.1 cm[28]. (Half of the difference between these two results comes from a difference in the corrections used for the galactic halo and integrated discrete sources.) By mid 1966 the intensity of the microwave background radiation had been shown to be close to 3 K between 21 cm and 2.6 mm, almost two orders of magnitude in wavelength.

X. EARLIER THEORY

I have mentioned that the first experimental evidence for cosmic microwave background radiation was obtained (but unrecognized) long before 1965. We soon learned that the theoretical prediction of it had been made at least sixteen years before our detection. George Gamow had made calculations of the conditions in the early universe in an attempt to understand Galaxy formation[29]. Although these calculations were not strictly correct, he understood that the early stages of the universe had to be very hot in order to avoid combining all of the hydrogen into heavier elements. Furthermore, Gamow and his collaborators calculated that the density of radiation in the hot early universe was much higher than the density of matter. In this early work the present remnants of this radiation were not considered. However in 1949, Alpher and Herman[30] followed the evolution of the temperature of the hot radiation in the early universe up to the present epoch and predicted a value of 5 K. They noted that the present density of radiation was not well known experimentally. In 1953 Alpher, Follin, and Herman[31] reported what has been called the first thoroughly modern analysis of the early history of the universe, but failed to recalculate or mention the present radiation temperature of the universe.

In 1964, Doroshkevich and Novikov[32, 33] had also calculated the relic radiation and realized that it would have a blackbody spectrum. They quoted E. A. Ohm's article on the Echo receiver, but misunderstood it and concluded that the present radiation temperature of the universe is near zero.

A more complete discussion of these early calculations is given in Arno's lecture.[34]

XI. ISOTROPY

In assigning a single temperature to the radiation in space, these theories assume that it will be the same in all directions. According to contemporary theory, the last scattering of the cosmic microwave background radiation occurred when the universe was a million years old, just before the electrons and nucleii combined to form neutral atoms ("recombination").The isotropy of the background radiation thus measures the isotropy of the universe at that time and the isotropy of its expansion since then. Prior to recombination, radiation dominated the universe and the Jeans mass, or mass of the smallest gravitationally stable clumps was larger than a cluster of Galaxies. It is only in the period following recombination that Galaxies could have formed.

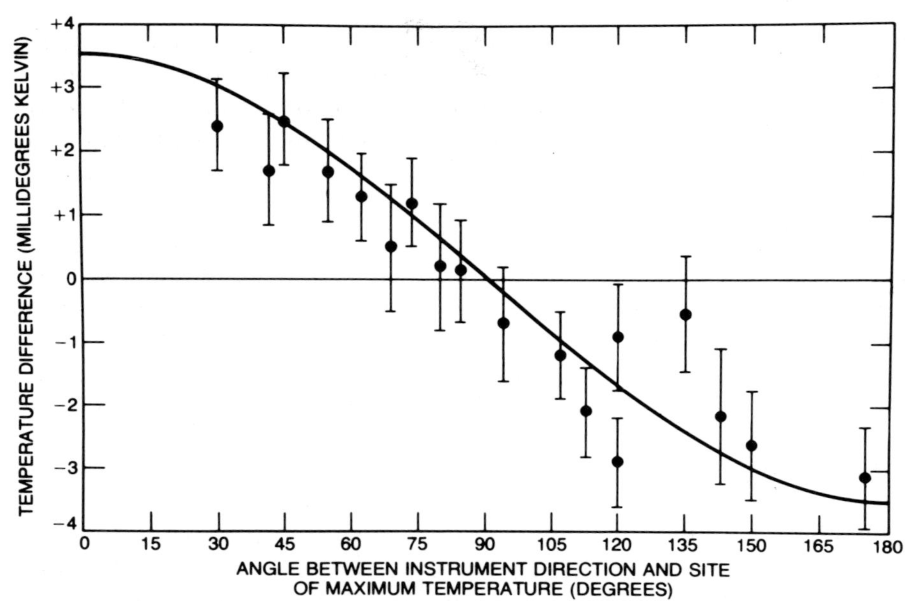

Fig. 11 Results of the large scale isotropy Experiment of Smoot, Gorenstein and Muller showing the clear cosine dependence of brightness expected from the relative velocity of the earth in the background radiation. The shaded area and arrows show the values allowed by the data of Woody and Richards. (This figure is reproduced with permission of Scientific American.)

In 1967 Rees and Sciama[35] suggested looking for large scale anisotropies in the background radiation which might have been left over from anisotropies of the universe prior to recombination.

In the same year Wilkinson and Partridge[36] completed an experiment which was specifically designed to look for anisotropy within the equatorial plane. The reported a limit of 0.1 percent for a 24 hour asymmetry and a possible 12 hour asymmetry of 0.2 percent. Meanwhile we had re-analyzed an old record covering most of the sky which was visible to us and put a limit of 0.1 K on any large scale fluctuations.[37]

Since then a series of measurements [38] [39] [40] have shown a 24-hour anisotropy due to the earth's velocity with respect to the background radiation. Data from the most sensitive measurement to date[41] are shown in Fig. 11. They show a striking cosine anisotropy with an amplitude of about .003 K, indicating that the background radiation has a maximum temperature in one direction and a minimum in the opposite direction. The generally accepted explanation of this effect is that the earth is moving toward the direction where the radiation is hottest and it is the blue shift of the radiation which increases its measured temperature in that direction. The motion of the sun with respect to the background radiation from the data of Smoot et al. is 390 ± 60 km/s in the direction 10.8^h R. A., 5° Dec. The magnitude of this velocity is not a surprise since 300 km/s is the orbital velocity of the sun around our galaxy. The direction, is different, however yielding a peculiar velocity of our galaxy of about 600 km/s. Since other nearby Galaxies, including the Virgo cluster, have a small velocity with respect to our Galaxy, they have a similar velocity with respect to the matter which last scattered the background radiation. After subtracting the 24-hour anisotropy, one can search the data for more complicated anisotropies to put observational limits on such things as rotation of the universe[41]. Within the noise of .001 K, these anisotropies are all zero.

To date, no fine-scale anisotropy has been found. Several early investigations were carried out to discredit discrete source models of the background radiation. In the most sensitive experiment to date, Boynton and Partridge[42] report a relative intensity variation of less than 3.7×10^{-3} in an 80" Arc beam. A discrete source model would require orders of magnitude more sources than the known number of Galaxies to show this degree of smoothness.

It has also been suggested by Sunyaev and Zel'dovich[43] that there will be a reduction of the intensity of the background radiation from the direction of clusters of galaxies due to inverse Compton scattering by the electrons in the intergalactic gas. This effect which has been found by Birkinshaw and Gull[44], provides a measure of the intergalactic gas density in the clusters and may give an alternate measurement of Hubble's constant.

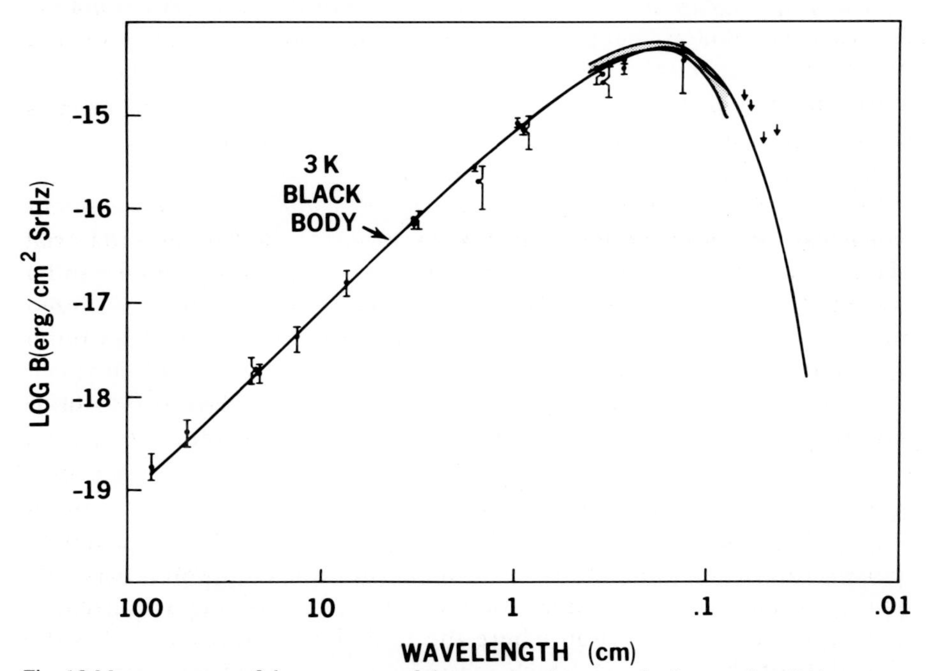

Fig. 12 Measurements of the spectrum of the cosmic microwave background radiation.

XII. SPECTRUM

Since 1966, a large number of measurements of the intensity of the background radiation have been made at wavelengths from 74 cm to 0.5 mm. Measurements have been made from the ground, mountain tops, airplanes, balloons, and rockets. In addition, the optical measurements of the interstellar molecules have been repeated and we have observed their millimeter-line radiation directly to establish the equilibrium of the excitation of their levels with the background radiation[45]. Fig. 12 is a plot of most of these measurements[46]. An early set of measurements from Princeton covered the range 3.2 to .33 cm showing tight consistency with a 2.7 K black body [47–50]. A series of rocket and balloon measurements in the millimeter and submillimeter part of the spectrum have converged on about 3 K. The data of Robson, et al. [51] and Woody and Richards[52] extend to 0.8 mm, well beyond the spectral peak. The most recent experiment, that of D. Woody and P. Richards, gives a close fit to a 3.0 K spectrum out to 0.8 mm wavelength with upper limits at atmospheric windows out to 0.4 mm. This establishes that the background radiation has a blackbody spectrum which would be quite hard to reproduce with any other type of cosmic source. The source must have been optically thick and therefore must have existed earlier than any of the other sources, which can be observed.

The spectral data are now almost accurate enough for one to test for systematic deviations from a single-temperature blackbody spectrum which could be caused by minor deviations from the simplest cosmology. Danese and DeZotti[53] report that except for the data of Woody and Richards, the spectral data of Fig. 12 do not show any statistically significant deviation of this type.

XIII. CONCLUSION

Cosmology is a science which has only a few observable facts to work with. The discovery of the cosmic microwave background radiation added one —the present radiation temperature of the universe. This, however, was a significant increase in our knowledge since it requires a cosmology with a source for the radiation at an early epoch and is a new probe of that epoch. More sensitive measurements of the background radiation in the future will allow us to discover additional facts about the universe.

XIV. ACKNOWLEDGMENTS

The work which I have described was done with Arno A. Penzias. In our fifteen years of partnership he has been a constant source of help and encouragement. I wish to thank W. D. Langer and Elizabeth Wilson for carefully reading the manuscript and suggesting changes.

REFERENCES

1. A more complete discussion of radio telescope antennas and receivers may be found in several text books. Chapters 6 and 7 of J. D. Kraus "Radio Astronomy", 1966, McGraw-Hill are good introductions to the subjects.
2. Crawford, A. B. Hogg, D. C. and Hunt, L. E. 1961, Bell System Tech. J., *40,* 1095.

3. DeGrasse, R. W. Schultz-DuBois, E. O. and Scovil, H. E. D. BSTJ *38,* 305.
4. Tabor, W. J. and Sibilia, J. T. 1963, Bell System Tech. J., *42,* 1863.
5. Penzias, A. A. 1965, Rev. Sci. Instr., *36,* 68.
6. Penzias, A. A. and Wilson, R. W. 1965 Astrophysical Journal *142,* 1149.
7. DeGrasse, R. W. Hogg, D. C. Ohm, E. A. and Scovil, H. E. D. 1959, Proceedings of the National Electronics Conference, *15,* 370.
8. Dicke, R. Beringer R. Kyhl, R. L. and Vane, A. V. Phys. Rev. *70,* 340 (1946).
9. Ohm, E. A. 1961 BSTJ, *40,* 1065.
10. Hogg, D. C. and Wilson, R. W. Bell system Tech J., *44,* 1019.
11. c.f. F. Hoyle and R. J. Taylor 1964, Nature *203,* 1108. A less explicit disussion of the same notion occurs in [29].
12. Hogg, D. C. and Semplak, R. A. 1961, Bell System Technical Journal, *40,* 1331.
13. Dicke, R. H. Peebles, P. J. E. Roll, P. G. and Wilkinson, D. T. 1965, Ap. J., *142,* 414.
14. Penzias, A. A. and Wilson R. W. 1965, Ap. J., *142,* 420.
15. Pauliny-Toth, I. I. K. and Shakeshaft, J. R. 1962, M. N. RAS., *124,* 61.
16. Adams, W. S. 1941, Ap. J. *93,* 11.
17. Adams, W. S. 1943, Ap. J., *97,* 105.
18. Dunham, T. Jr. 1937, PASP, *49,* 26.
19. Dunham, T. Jr. 1939, Proc. Am. Phil. Soc., *81,* 277.
20. Dunham, T. Jr. 1941, Publ. Am. Astron. Soc., *10,* 123.
21. Dunham, T. Jr. and W. S. Adams 1937, Publ. Am. Astron. Soc., *9,* 5.
22. McKellar, A. 1941, Publ. Dominion Astrophysical Observatory Victoria B. C. *7,* 251.
23. G. B. Field, G. H. Herbig and J. L. Hitchcock, talk at the American Astronomical Society Meeting, 22–29 December 1965, Astronomical Journal 1966, *71,* 161; G. B. Field and J. L. Hitchcock, 1966, Phys. Rev. Lett. *16,* 817.
24. Shklovsky, I. S. 1966, Astron. Circular No. *364,* Acad. Sci. USSR.
25. Thaddeus, P. and Clauser, J. F. 1966, Phys. Rev. Lett. *16,* 819.
26. Roll, P. G. and Wilkinson, D. T. 1966, Physical Review Letters, *16,* 405.
27. Howell, T. F. and Shakeshaft, J. R. 1966, Nature *210,* 138.
28. Penzias, A. A. and Wilson R. W. 1967, Astron. J. *72,* 315.
29. Gamow, G. 1948, Nature *162* 680.
30. Alpher, R. A. and Herman, R. C. 1949, Phys. Rev., *75,* 1089.
31. Alpher, R. A. Follin, J. W. and Herman, R. C. 1953, Phys. Rev., *92,* 1347.
32. Doroshkevich, A. G. and Novikov, I. D. 1964 Dokl. Akad. Navk. SSR *154,* 809.
33. Doroshkevich, A. G. and Novikov, I. D. 1964 Sov. Phys. Dokl. *9,* 111.
34. Penzias, A. A. *The Origin of the Elements,* Nobel Prize lecture 1978.
35. Rees, M. J. and Sciama, D. W. 1967, Nature, *213,* 374.
36. Partridge, R. B. and Wilkinson, D. T. 1967, Nature, *215,* 719.
37. Wilson, R. W. and Penzias, A. A. 1967, Science *156,* 1100.
38. Conklin, E. K. 1969, Nature, *222,* 971.
39. Henry, P. S. 1971, Nature, *231,* 516.
40. Corey, B. E. and Wilkinson, D. T. 1976, Bull. Astron. Astrophys. Soc., *8,* 351.
41. Smoot, G. F. Gorenstein, M. V. and Muller, R. A. 1977, Phys. Rev. Lett., *39,* 898.
42. Boynton, P. E. and Partridge, R. B. 1973, Ap. J., *181,* 243.
43. Sunyaev, R. A. and Zel'dovich, Ya B 1972, Comments Astrophys. Space Phys., *4,* 173.
44. Birkinshaw, M. and Gull, S. F. 1978, Nature *275* 40.
45. Penzias, A. A. Jefferts, K. B. and Wilson, R. W. 1972, Phys. Rev. Letters, *28,* 772.
46. The data in Fig. 11 are all referenced by Danese and Dezotti[45] except for the 13 cm measurement of T. Otoshi 1975, IEEE Trans on Instrumentation and Meas. *24* 174. I have used the millimeter measurements of Woody and Richards[48] and left off those of Robson et al[47] to avoid confusion.
47. Wilkinson, D. J. 1967, Phys. Rev. Letters, *19,* 1195.
48. Stokes, R. A. Partridge, R. B. and Wilkinson, D. J. 1967, Phys. Rev. Letters, *19,* 1199.
49. Boynton, P. E. Stokes, R. A. and Wilkinson, D. J. 1968, Phys. Rev. Letters, *21,* 462.
50. Boynton, P. E. and Stokes, R. A. 1974, Nature, *247,* 528.
51. Robson, E. I. Vickers, D. G. Huizinga, J. S. Beckman J. E. and Clegg, P. E. 1974, Nature *251,* 591.
52. Woody, D. P. and Richards, P. L. Private communication.
53. Danese, L. and DeZotti, G. 1978, Astron. and Astrophys., *68,* 157.

Physics 1979

SHELDON L GLASHOW, ABDUS SALAM and STEVEN WEINBERG

for their contributions to the theory of the unified weak and electromagnetic interaction between elementary particles, including inter alia the prediction of the weak neutral current

THE NOBEL PRIZE FOR PHYSICS

Speech by Professor BENGT NAGEL of the Royal Academy of Sciences.
Translation from the Swedish text.

Your Majesties, Your Royal Highnesses, Ladies and Gentlemen,

This year's Nobel prize in Physics is shared equally between Sheldon Glashow, Abdus Salam and Steven Weinberg "for their contributions to the theory of the unified weak and electromagnetic interaction between elementary particles, including inter alia the prediction of the weak neutral current".

Important advances in physics often consist in relating apparently unconnected phenomena to a common cause. A classical example is Newton's introduction of the gravitational force to explain the fall of the apple and the motion of the moon around the earth. – In the 19th century it was found that electricity and magnetism are really two aspects of one and the same force, the electromagnetic interaction between charges. Electromagnetism, with the electron playing the leading part and the photon–the electromagnetic quantum of light–as the swift messenger, dominates technology and our everday life: not only electrotechnics and electronics, but also atomic and molecular physics and hence chemical and biological processes are governed by this force.

When one began to study the atomic nucleus in the first decades of our century, two new forces were discovered: the strong and the weak nuclear forces. Unlike gravitation and electromagnetism these forces act only over distances of the order of nuclear diameters or less. The strong force keeps the nucleus together, whereas the weak force is responsible for the so called beta decays of the nucleus. Most radioactive substances used in medicine and technology are beta radioactive. The electron also participates in the weak interaction, but the principal part is played by the neutrino, a particle which is described as follows in a poem by the American writer John Updike:

"Cosmic Gall"

Neutrinos, they are very small.
They have no charge and have no mass
And do not interact at all.
The earth is just a silly ball
To them, through which they simply pass,
Like dustmaids down a drafty hall
Or photons through a sheet of glass.

– – –

At night, they enter at Nepal
And pierce the lover and his lass
From underneath the bed – you call
It wonderful; I call it crass.

The description is accurate, apart from the statement 'they do not interact at all'; they do interact through the weak force. The neutrinos of the poem, entering the earth at night at Nepal and exiting in the U.S. in a sort of reversed China syndrome, come to us from the centre of the sun. Solar energy, necessary for life on earth, is created when hydrogen is burnt to helium in the interior of the sun in a chain of nuclear reactions–even the advocates of "Solsverige" must ultimately rely on nuclear energy although it must be said that the fusion reactor Sun is well encapsuled and sufficiently relocated away from populated areas. The first ignating and moderating link in this chain, burning hydrogen to deuterium, is based on the weak force, which could then be called the Sunignator and Suntamer.

The theory which is awarded this year's prize, and which was developed in separate works by the prizewinners in the 60's, has extended and deepened our understanding of the weak force by displaying a close relationship to the electromagnetic force: these two forces emerge as different aspects of a unified electroweak interaction. This means e.g. that the electron and the neutrino belong to the same family of particles; the neutrino is the electron's little brother. Another consequence of the unified theory is that there should exist a new kind of weak interaction. It was formerly assumed that weak processes could occur only in connection with a change of identity of the electron to neutrino (or vice versa); such a process is said to proceed by a charged current, since the particle changes its charge. The theory implies that there should also be processes connected with a neutral current in which the neutrino–or else the electron–acts without changing identity. Experiments in the 70's have fully confirmed these predictions of the theory.

The importance of the new theory is first of all intrascientific. The theory has set a pattern for the description also of the strong nuclear force and for efforts to integrate further the interactions between elementary particles.

Let me end by giving an example of the intricate links which exist between different branches of natural science.

Our body is to a large part constructed from "stardust": the elements besides hydrogen which build our cells have been formed in the interior of stars in nuclear reactions, which form a continuation of the processes taking place in our sun. According to the astrophysicists, certain heavy elements appearing in life-important enzymes and hormones–iodine and selenium are examples of such elements–can probably only be created in connection with violent explosions of giant stars, so called supernova explosions, which occur in our Galaxy once every one or two hundred years. It is likely that neutrinos interacting via the neutral current play an important role in these explosions, in which a large part of the matter of the star is thrown out into space. Thus, for our functioning as biological beings we rely on elements formed milliards of years ago in supernova explosions, with the new kind of weak force predicted by the theory contributing in an important way; really a fascinating connection between biology, astrophysics and elementary particle physics.

Professors Sheldon Glashow, Abdus Salam, and Steven Weinberg,

In my talk I have tried to give a background to your great discoveries in the borderland between a strange but known country and the probably large unknown territory of the innermost structure of matter.

Our way of looking at this structure has changed radically in the last decade. The theory of electroweak interaction has been one of the most important forces to bring about this change of outlook.

It is a privilege and a pleasure for me to convey to you the warmest felicitations of the Royal Swedish Academy of Sciences and to invite you to receive your prizes from the hands of his Majesty the King.

Sheldon Lee Glashow

SHELDON LEE GLASHOW

My parents, Lewis Glashow and Bella née Rubin immigrated to New York City from Bobruisk in the early years of this century. Here they found the freedom and opportunity denied to Jews in Czarist Russia. After years of struggle, my father became a successful plumber, and his family could then enjoy the comforts of the middle class. While my parents never had the time or money to secure university education themselves, they were adamant that their children should. In comfort and in love, we were taught the joys of knowledge and of work well done. I only regret that neither my mother nor my father could live to see the day I would accept the Nobel Prize.

When I was born in Manhattan in 1932, my brothers Samuel and Jules were eighteen and fourteen years old. They chose careers of dentistry and medicine, to my parents' satisfaction. From an early age, I knew I would become a scientist. It may have been my brother Sam's doing. He interested me in the laws of falling bodies when I was ten, and helped my father equip a basement chemistry lab for me when I was fifteen. I became skilled in the synthesis of selenium halides. Never again would I do such dangerous research. Except for the occasional suggestion that I should become a physician and do science in my spare time, my parents always encouraged my scientific inclinations.

Among my chums at the Bronx High School of Science were Gary Feinberg and Steven Weinberg. We spurred one another to learn physics while commuting on the New York subway. Another classmate, Dan Greenberger, taught me calculus in the school lunchroom. High-school mathematics then terminated with solid geometry. At Cornell University, I again had the good fortune to join a talented class. It included the mathematician Daniel Kleitman who was to become my brother-in-law, my old classmate Steven Weinberg, and many others who were to become prominent scientists. Throughout my formal education, I would learn as much from my peers as from my teachers. So it is today among our graduate students.

I came to graduate school at Harvard University in 1954. My thesis supervisor, Julian Schwinger, had about a dozen doctoral students at a time. Getting his ear was as difficult as it was rewarding. I called my thesis "The Vector Meson in Elementary Particle Decays", and it showed an early commitment to an electroweak synthesis. When I completed my work in 1958, Schwinger and I were to write a paper summarizing our thoughts on weak-electromagnetic unification. Alas, one of us lost the first draft of the manuscript, and that was that.

I won an NSF postdoctoral fellowship, and planned to work at the Lebedev Institute in Moscow with I. Tamm, who enthusiastically supported my proposal. I spent the tenure of my fellowship in Copenhagen at the Niels Bohr Institute (and, partly, at CERN), waiting for the Russian visa that was never to come. Perhaps all was for the best, because it was in these years (1958–60) that I discovered the SU(2) × U(1) structure of the electroweak theory. Interestingly, it was also in Copenhagen that my early work on charm with Bjorken was done. This was during a brief return to Denmark in 1964.

During my stay in Europe, I was "discovered" by Murray Gell-Mann. He presented my ideas on the algebraic structure of weak interactions to the 1960 "Rochester meeting" and brought me to Caltech. Then, he invented the eightfold way, which kept Sidney Coleman and me distracted for several years. How we found various electromagnetic formulae, yet missed the discovery of the Gell-Mann-Okubo formula and of the Cabibbo current is another story.

I became an assistant professor at Stanford University and then spent several years on the faculty of the University of California at Berkeley. During this time, I continued to exploit the phenomenological successes of flavor SU(3) and attempted to understand the departures from exact symmetry as a consequence of spontaneous symmetry breakdown. I returned to Harvard University in 1966 where I have remained except for leaves to CERN, MIT, and the University of Marseilles. Today, I am Eugene Higgins Professor of Physics at Harvard.

In 1969, John Iliopoulos and Luciano Maiani came to Harvard as research fellows. Together, we found the arguments that predicted the existence of charmed hadrons. Much of my later work was done in collaboration with Alvaro de Rújula or Howard Georgi. In early 1974, we predicted that charm would be discovered in neutrino physics or in $e^+ e^-$ annihilation. So it was. With the discovery of the J/Ψ particle, we realized that many diverse strands of research were converging on a single theory of physics. I remember once saying to Howard that if QCD is so good, it should explain the $\Sigma - \Lambda$ mass splitting. The next day he showed that it did. When we spoke, in 1974, of the unification of all elementary particle forces within a simple gauge group, and of the predicted instability of the proton, we were regarded as mad. How things change!

The wild ideas of yesterday quickly become today's dogma. This year I have been honored to participate in the inauguration of the Harvard Core Curriculum Program. My students are not, and will never be, scientists. Nonetheless, in my course "From Alchemy to Quarks" they seem to be as fascinated as I am by the strange story of the search for the ultimate constituents of matter.

I was married in 1972 to the former Joan Alexander. We live in a large old house with our four children, who attend the Brookline public schools.

Education
A.B. 1954 Cornell University
A.M. 1955 Harvard University
Ph. D. 1959 Harvard University
Married 1972 Joan Shirley Alexander

Employment
NSF Post-Doctoral Fellow 1958–60
Caltech Research Fellow 1960–61
Stanford University, Assistant Professor 1961–62
University of California, Berkeley, Associate Professor 1962–66
Harvard University, Professor 1966–1982
CERN, Visiting Scientist 1968
University of Marseilles, Visiting Professor 1970
MIT, Visiting Professor 1974
Brookhaven Laboratory, Consultant 1964
Texas A&M University, Visiting Professor 1982
University of Houston, Affiliated Senior Scientist, 1982–
Boston University, Distinguished Visiting Scientist, 1984–

Member
American Physical Society
Sigma Xi
American Association for the Advancement of Science
American Academy of Arts and Sciences
National Academy of Sciences

Honors
Westinghouse Science Talent Search Finalist 1950
Alfred P. Sloan Foundation Fellowship 1962–66
Oppenheimer Memorial Medal 1977
George Ledlie Award 1978

Honorary Degrees
Yeshiva University 1978
University of Aix-Marseille 1982
Adelphi University 1985
Bar-Ilan University 1988
Gustavus Augustus University 1989

TOWARDS A UNIFIED THEORY–THREADS IN A TAPESTRY

Nobel Lecture, 8 December, 1979
by
SHELDON LEE GLASHOW
Lyman Laboratory of Physics Harvard University Cambridge, Mass., USA

INTRODUCTION

In 1956, when I began doing theoretical physics, the study of elementary particles was like a patchwork quilt. Electrodynamics, weak interactions, and strong interactions were clearly separate disciplines, separately taught and separately studied. There was no coherent theory that described them all. Developments such as the observation of parity violation, the successes of quantum electrodynamics, the discovery of hadron resonances and the appearance of strangeness were well-defined parts of the picture, but they could not be easily fitted together.

Things have changed. Today we have what has been called a "standard theory" of elementary particle physics in which strong, weak, and electromagnetic interactions all arise from a local symmetry principle. It is, in a sense, a complete and apparently correct theory, offering a qualitative description of all particle phenomena and precise quantitative predictions in many instances. There is no experimental data that contradicts the theory. In principle, if not yet in practice, all experimental data can be expressed in terms of a small number of "fundamental" masses and coupling constants. The theory we now have is an integral work of art: the patchwork quilt has become a tapestry.

Tapestries are made by many artisans working together. The contributions of separate workers cannot be discerned in the completed work, and the loose and false threads have been covered over. So it is in our picture of particle physics. Part of the picture is the unification of weak and electromagnetic interactions and the prediction of neutral currents, now being celebrated by the award of the Nobel Prize. Another part concerns the reasoned evolution of the quark hypothesis from mere whimsy to established dogma. Yet another is the development of quantum chromodynamics into a plausible, powerful, and predictive theory of strong interactions. All is woven together in the tapestry; one part makes little sense without the other. Even the development of the electroweak theory was not as simple and straightforward as it might have been. It did not arise full blown in the mind of one physicist, nor even of three. It, too, is the result of the collective endeavor of many scientists, both experimenters and theorists.

Let me stress that I do not believe that the standard theory will long survive as a correct and complete picture of physics. All interactions may be gauge interactions, but surely they must lie within a unifying group. This would imply the existence of a new and very weak interaction which mediates the decay of protons. All matter is thus inherently unstable, and can be observed to decay. Such a synthesis of weak, strong, and electromagnetic interactions has been called a "grand unified theory", but a theory is neither grand nor unified unless it includes a description of gravitational phenomena. We are still far from Einstein's truly grand design.

Physics of the past century has been characterized by frequent great but unanticipated experimental discoveries. If the standard theory is correct, this age has come to an end. Only a few important particles remain to be discovered, and many of their properties are alleged to be known in advance. Surely this is not the way things will be, for Nature must still have some surprises in store for us.

Nevertheless, the standard theory will prove useful for years to come. The confusion of the past is now replaced by a simple and elegant synthesis. The standard theory may survive as a part of the ultimate theory, or it may turn out to be fundamentally wrong. In either case, it will have been an important way-station, and the next theory will have to be better.

In this talk, I shall not attempt to describe the tapestry as a whole, nor even that portion which is the electroweak synthesis and its empirical triumph. Rather, I shall describe several old threads, mostly overwoven, which are closely related to my own researches. My purpose is not so much to explain who did what when, but to approach the more difficult question of why things went as they did. I shall also follow several new threads which may suggest the future development of the tapestry.

EARLY MODELS

In the 1920's, it was still believed that there were only two fundamental forces: gravity and electromagnetism. In attempting to unify them, Einstein might have hoped to formulate a universal theory of physics. However, the study of the atomic nucleus soon revealed the need for two additional forces: the strong force to hold the nucleus together and the weak force to enable it to decay. Yukawa asked whether there might be a deep analogy between these new forces and electromagnetism. All forces, he said, were to result from the exchange of mesons. His conjectured mesons were originally intended to mediate both the strong and the weak interactions: they were strongly coupled to nucleons and weakly coupled to leptons. This first attempt to unify strong and weak interactions was fully forty years premature. Not only this, but Yukawa could have predicted the existence of neutral currents. His neutral meson, essential to provide the charge independence of nuclear forces, was also weakly coupled to pairs of leptons.

Not only is electromagnetism mediated by photons, but it arises from the requirement of local gauge invariance. This concept was generalized in 1954 to apply to non-Abelian local symmetry groups. [1] It soon became clear that a more far-reaching analogy might exist between electromagnetism and the other forces. They, too, might emerge from a gauge principle.

A bit of a problem arises at this point. All gauge mesons must be massless, yet the photon is the only massless meson. How do the other gauge bosons get their masses? There was no good answer to this question until the work of Weinberg and Salam [2] as proven by 't Hooft [3] (for spontaneously broken gauge theories) and of Gross, Wilczek, and Politzer [4] (for unbroken gauge theories). Until this work was done, gauge meson masses had simply to be put in *ad hoc*.

Sakurai suggested in 1960 that strong interactions should arise from a gauge principle. [5] Applying the Yang–Mills construct to the isospin-hypercharge symmetry group, he predicted the existence of the vector mesons ϱ and ω. This was the first phenomenological SU(2) × U(1) gauge theory. It was extended to local SU(3) by Gell-Mann and Ne'eman in 1961. [6] Yet, these early attempts to formulate a gauge theory of strong interactions were doomed to fail. In today's jargon, they used "flavor" as the relevant dynamical variable, rather than the hidden and then unknown variable "color". Nevertheless, this work prepared the way for the emergence of quantum chromodynamics a decade later.

Early work in nuclear beta decay seemed to show that the relevant interaction was a mixture of S, T, and P. Only after the discovery of parity violation, and the undoing of several wrong experiments, did it become clear that the weak interactions were in reality V-A. The synthesis of Feynman and Gell-Mann and of Marshak and Sudarshan was a necessary precursor to the notion of a gauge theory of weak interactions. [7] Bludman formulated the first SU(2) gauge theory of weak interactions in 1958. [8] No attempt was made to include electromagnetism. The model included the conventional charged-current interactions, and in addition, a set of neutral current couplings. These are of the same strength and form as those of today's theory in the limit in which the weak mixing angle vanishes. Of course, a gauge theory of weak interactions alone cannot be made renormalizable. For this, the weak and electromagnetic interactions must be unified.

Schwinger, as early as 1956, believed that the weak and electromagnetic interactions should be combined together into a gauge theory. [9] The charged massive vector intermediary and the massless photon were to be the gauge mesons. As his student, I accepted this faith. In my 1958 Harvard thesis, I wrote: "It is of little value to have a potentially renormalizable theory of beta processes without the possibility of a renormalizable electrodynamics. We should care to suggest that a fully acceptable theory of these interactions may only be achieved if they are treated together. . ." [10] We used the original SU(2) gauge interaction of Yang and Mills. Things had to be arranged so that the charged current, but not the neutral (electromagnetic) current, would violate parity and strangeness. Such a theory is technically possible to construct, but it is both ugly and experimentally false. [11] We know now that neutral currents do exist and that the electroweak gauge group must be larger than SU(2).

Another electroweak synthesis without neutral currents was put forward by Salam and Ward in 1959. [12] Again, they failed to see how to incorporate the experimental fact of parity violation. Incidentally, in a continuation of their work in 1961, they suggested a gauge theory of strong, weak, and electromagnetic interactions based on the local symmetry group SU(2) × SU(2). [13] This was a remarkable portent of the SU(3) × SU(2) × U(1) model which is accepted today.

We come to my own work [14] done in Copenhagen in 1960, and done independently by Salam and Ward. [15] We finally saw that a gauge group larger than SU(2) was necessary to describe the electroweak interactions. Salam and Ward were motivated by the compelling beauty of gauge theory. I thought I saw a way to a renormalizable scheme. I was led to the group SU(2) × U(1) by analogy with the approximate isospin-hypercharge group which characterizes strong interactions. In this model there were two electrically neutral intermediaries: the massless photon and a massive neutral vector meson which I called B but which is now known as Z. The weak mixing angle determined to what linear combination of SU(2) × U(1) generators B would correspond. The precise form of the predicted neutral-current interaction has been verified by recent experimental data. However, the strength of the neutral current was not prescribed, and the model was not in fact renormalizable. These glaring omissions were to be rectified by the work of Salam and Weinberg and the subsequent proof of renormalizability. Furthermore, the model was a model of leptons–it could not evidently be extended to deal with hadrons.

RENORMALIZABILITY

In the late 50's, quantum electrodynamics and pseudoscalar meson theory were known to be renormalizable, thanks in part to work of Salam. Neither of the customary models of weak interactions – charged intermediate vector bosons or direct four-fermion couplings – satisfied this essential criterion. My thesis at Harvard, under the direction of Julian Schwinger, was to pursue my teacher's belief in a unified electroweak gauge theory. I had found some reason to believe that such a theory was less singular than its alternatives. Feinberg, working with charged intermediate vector mesons discovered that a certain type of divergence would cancel for a special value of the meson anomalous magnetic moment.[16] It did not correspond to a "minimal electromagnetic coupling", but to the magnetic properties demanded by a gauge theory. Tzou Kuo-Hsien examined the zero-mass limit of charged vector meson electrodynamics.[17] Again, a sensible result is obtained only for a very special choice of the magnetic dipole moment and electric quadrupole moment, just the values assumed in a gauge theory. Was it just coincidence that the electromagnetism of a charged vector meson was least pathological in a gauge theory?

Inspired by these special properties, I wrote a notorious paper.[18] I alleged that a softly-broken gauge theory, with symmetry breaking provided by explicit mass terms, was renormalizable. It was quickly shown that this is false.

Again, in 1970, Iliopoulos and I showed that a wide class of divergences that might be expected would cancel in such a gauge theory.[19] We showed that the naive divergences of order $(\alpha\Lambda^4)^n$ were reduced to "merely" $(\alpha\Lambda^2)^n$, where Λ is a cut-off momentum. This is probably the most difficult theorem that Iliopoulos or I had even proven. Yet, our labors were in vain. In the spring of 1971, Veltman informed us that his student Gerhart 't Hooft had established the renormalizability of spontaneously broken gauge theory.

In pursuit of renormalizability, I had worked diligently but I completely missed the boat. The gauge symmetry is an exact symmetry, but it is hidden. One must not put in mass terms by hand. The key to the problem is the idea of spontaneous symmetry breakdown: the work of Goldstone as extended to gauge theories by Higgs and Kibble in 1964.[20] These workers never thought to apply their work on formal field theory to a phenomenologically relevant model. I had had many conversations with Goldstone and Higgs in 1960. Did I neglect to tell them about my SU (2)×U (1) model, or did they simply forget?

Both Salam and Weinberg had had considerable experience in formal field theory, and they had both collaborated with Goldstone on spontaneous symmetry breaking. In retrospect, it is not so surprising that it was they who first used the key. Their SU (2)×U (1) gauge symmetry was spontaneously broken. The masses of the W and Z and the nature of neutral current effects depend on a single measurable parameter, not two as in my unrenormalizable model. The strength of the neutral currents was correctly predicted. The daring Weinberg-Salam conjecture of renormalizability was proven in 1971. Neutral currents were discovered in 1973[21], but not until 1978 was it clear that they had just the predicted properties.[22]

THE STRANGENESS-CHANGING NEUTRAL CURRENT

I had more or less abandoned the idea of an electroweak gauge theory during the period 1961–1970. Of the several reasons for this, one was the failure of my naive foray into renormalizability. Another was the emergence of an empirically successful description of strong interactions – the SU(3) unitary symmetry scheme of Gell-Mann and Ne'eman. This theory was originally phrased as a gauge theory, with ϱ, ω, and K* as gauge mesons. It was completely impossible to imagine how both strong and weak interactions could be gauge theories: there simply wasn't room enough for commuting structures of weak and strong currents. Who could foresee the success of the quark model, and the displacement of SU(3) from the arena of flavor to that of color? The predictions of unitary symmetry were being borne out – the predicted Ω^- was discovered in 1964. Current algebra was being successfully exploited. Strong interactions dominated the scene.

When I came upon the SU(2)×U(1) model in 1960, I had speculated on a possible extension to include hadrons. To construct a model of leptons alone seemed senseless: nuclear beta decay, after all, was the first and foremost problem. One thing seemed clear. The fact that the charged current violated strangeness would force the neutral current to violate strangeness as well. It was already well known that strangeness-changing neutral currents were either strongly suppressed or absent. I concluded that the Z^0 had to be made very much heavier than the W. This was an arbitrary but permissible act in those days: the symmetry breaking mechanism was unknown. I had "solved" the problem of strangeness-changing neutral currents by suppressing all neutral currents: the baby was lost with the bath water.

I returned briefly to the question of gauge theories of weak interactions in a collaboration with Gell-Mann in 1961.[23] From the recently developing ideas of current algebra we showed that a gauge theory of weak interactions would inevitably run into the problem of strangeness-changing neutral currents. We concluded that something essential was missing. Indeed it was. Only after quarks were invented could the idea of the fourth quark and the GIM mechanism arise.

From 1961 to 1964, Sidney Coleman and I devoted ourselves to the exploitation of the unitary symmetry scheme. In the spring of 1964, I spent a short leave of absence in Copenhagen. There, Bjorken and I suggested that the Gell-Mann–Zweig-system of three quarks should be extended to four.[24] (Other workers had the same idea at the same time).[25] We called the fourth quark the charmed quark. Part of our motivation for introducing a fourth quark was based on our mistaken notions of hadron spectroscopy. But we also wished to enforce an analogy between the weak leptonic current and the weak hadronic current. Because there were two weak doublets of leptons, we believed there had to be two weak doublets of quarks as well. The basic idea was correct, but today there seem to be three doublets of quarks and three doublets of leptons.

The weak current Bjorken and I introduced in 1964 was precisely the GIM current. The associated neutral current, as we noted, conserved strangeness. Had we inserted these currents into the earlier electroweak theory, we would have solved the problem of strangeness-changing neutral currents. We did not. I had apparently quite forgotten my earlier ideas of electroweak synthesis. The problem which was explicitly posed in 1961 was solved, in principle, in 1964. No one, least of all me, knew it. Perhaps we were all befuddled by the chimera of relativistic SU(6), which arose at about this time to cloud the minds of theorists.

Five years later, John Iliopoulos, Luciano Maiani and I returned to the question of strangeness-changing neutral currents.[26] It seems incredible that the problem was totally ignorned for so long. We argued that unobserved effects (a large K_1K_2 mass difference; decays like $K \rightarrow \pi\nu\bar{\nu}$; etc.) would be expected to arise in any of the known weak interaction models: four fermion couplings; charged vector meson models; or the electroweak gauge theory. We worked in terms of cut-offs, since no renormalizable theory was known at the time. We showed how the unwanted effects would be eliminated with the conjectured existence of a fourth quark. After languishing for a decade, the problem of the selection rules of the neutral current was finally solved. Of course, not everyone believed in the predicted existence of charmed hadrons.

This work was done fully three years after the epochal work of Weinberg and Salam, and was presented in seminars at Harvard and at M. I. T. Neither I, nor my coworkers, nor Weinberg, sensed the connection between the two endeavors. We did not refer, nor were we asked to refer, to the Weinberg–Salam work in our paper.

The relevance became evident only a year later. Due to the work of 't Hooft, Veltman, Benjamin Lee, and Zinn-Justin, it became clear that the Weinberg-Salam *ansatz* was in fact a renormalizable theory. With GIM, it was trivially extended from a model of leptons to a theory of weak interactions. The ball was now squarely in the hands of the experimenters. Within a few years, charmed hadrons and neutral currents were discovered, and both had just the properties they were predicted to have.

FROM ACCELERATORS TO MINES

Pions and strange particles were discovered by passive experiments which made use of the natural flux of cosmic rays. However, in the last three decades, most discoveries in particle physics were made in the active mode, with the artificial aid of particle accelerators. Passive experimentation stagnates from a lack of funding and lack of interest. Recent developments in theoretical particle physics and in astrophysics may mark an imminent rebirth of passive experimentation. The concentration of virtually all high-energy physics endeavors at a small number of major accelerator laboratories may be a thing of the past.

This is not to say that the large accelerator is becoming extinct; it will remain an essential if not exclusive tool of high-energy physics. Do not forget that the existence of Z^0 at ~ 100 GeV is an essential but quite untested prediction of the electroweak theory. There will be additional dramatic discoveries at accelerators, and these will not always have been predicted in advance by theorists. The construction of new machines like LEP and ISABELLE is mandatory.

Consider the successes of the electroweak synthesis, and the fact that the only plausible theory of strong interactions is also a gauge theory. We must believe in the ultimate synthesis of strong, weak, and electromagnetic interactions. It has been shown how the strong and electroweak gauge groups may be put into a larger but simple gauge group.[27] Grand unification – perhaps along the lines of the original SU (5) theory of Georgi and me – must be essentially correct. This implies that the proton, and indeed all nuclear matter, must be inherently unstable. Sensitive searches for proton decay are now being launched. If the proton lifetime is shorter than 10^{32} years, as theoretical estimates indicate, it will not be long before it is seen to decay.

Once the effect is discovered (and I am sure it will be), further experiments will have to be done to establish the precise modes of decay of nucleons. The selection rules, mixing angles, and space-time structure of a new class of effective four-fermion couplings must be established. The heroic days of the discovery of the nature of beta decay will be repeated.

The first generation of proton decay experiments is cheap, but subsequent generations will not be. Active and passive experiments will compete for the same dwindling resources.

Other new physics may show up in elaborate passive experiments. Today's theories suggest modes of proton decay which violate both baryon number and lepton number by unity. Perhaps this $\Delta B = \Delta L = 1$ law will be satisfied. Perhaps $\Delta B = -\Delta L$ transitions will be seen. Perhaps, as Pati and Salam suggest, the proton will decay into three leptons. Perhaps two nucleons will annihilate in $\Delta B = 2$ transitions. The effects of neutrino oscillations resulting from neutrino masses of a fraction of an election volt may be detectable. "Superheavy isotopes" which may be present in the Earth's crust in small concentrations could reveal themselves through their multi-GeV decays. Neutrino bursts arising from distant astronomical catastrophes may be seen. The list may be endless or empty. Large passive experiments of the sort now envisioned have never been done before. Who can say what results they may yield?

PREMATURE ORTHODOXY

The discovery of the J/Ψ in 1974 made it possible to believe in a system involving just four quarks and four leptons. Very quickly after this a third charged lepton (the tau) was discovered, and evidence appeared for a third $Q = -1/3$ quark (the b quark). Both discoveries were classic surprises. It became immediately fashionable to put the known fermions into families or generations:

$$\begin{pmatrix} u & \nu_e \\ d & e \end{pmatrix} \begin{pmatrix} c & \nu_\mu \\ s & \mu \end{pmatrix} \begin{pmatrix} t & \nu_\tau \\ b & \tau \end{pmatrix}.$$

The existence of a third $Q = 2/3$ quark (the t quark) is predicted. The Cabibbo-GIM scheme is extended to a system of six quarks. The three family system is the basis to a vast and daring theoretical endeavor. For example, a variety of papers have been written putting experimental constraints on the four parameters which replace the Cabibbo angle in a six quark system. The detailed manner of decay of particles containing a single b quark has been worked out. All that is wanting is experimental confirmation. A new orthodoxy has emerged, one for which there is little evidence, and one in which I have little faith.

The predicted t quark has not been found. While the upsilon mass is less than 10 GeV, the analogous tt particle, if it exists at all, must be heavier than 30 GeV. Perhaps it doesn't exist.

Howard Georgi and I, and other before us, have been working on models with no t quark.[28] We believe this unorthodox view is as attractive as its alternative. And, it suggests a number of exciting experimental possibilities.

We assume that b and τ share a quantum number, like baryon number, that is essentially exactly conserved. (Of course, it may be violated to the same extent that baryon number is expected to be violated.) Thus, the b,τ system is assumed to be distinct from the lighter four quarks and four leptons. There is, in particular, no mixing between b and d or s. The original GIM structure is left intact. An additional mechanism must be invoked to mediate b decay, which is not present in the $SU(3)\times SU(2)\times U(1)$ gauge theory.

One possibility is that there is an additional SU(2) gauge interaction whose effects we have not yet encountered. It could mediate such decays of b as these

$$b \rightarrow \tau^{+} + (e^{-} \text{ or } \mu^{-}) + (d \text{ or } s).$$

All decays of b would result in the production of a pair of leptons, including a τ^{+} or its neutral partner. There are other possibilities as well, which predict equally bizarre decay schemes for b-matter. How the b quark decays is not yet known, but it soon will be.

The new SU(2) gauge theory is called upon to explain CP violation as well as b decay. In order to fit experiment, three additional massive neutral vector bosons must exist, and they cannot be too heavy. One of them can be produced in $e^{+}e^{-}$ annihilation, in addition to the expected Z^{0}. Our model is rife with experimental predictions, for example: a second Z^{0}, a heavier version of b and of τ, the production of τ b in e p collisions, and the existence of heavy neutral unstable leptons which may be produced and detected in $e^{+}e^{-}$ or in νp collisions.

This is not the place to describe our views in detail. They are very speculative and probably false. The point I wish to make is simply that it is too early to convince ourselves that we know the future of particle physics. There are too many points at which the conventional picture may be wrong or incomplete. The $SU(3)\times SU(2)\times U(1)$ gauge theory with three families is certainly a good beginning, not to accept but to attack, extend, and exploit. We are far from the end.

ACKNOWLEDGEMENTS

I wish to thank the Nobel Foundation for granting me the greatest honor to which a scientist may aspire. There are many without whom my work would never have been. Let me thank my scientific collaborators, especially James Bjorken, Sidney Coleman, Alvaro De Rújula, Howard Georgi, John Iliopoulos, and Luciano Maiani; the Niels Bohr Institute and Harvard University for their hospitality while my research on the electroweak interaction was done: Julian Schwinger for teaching me how to do scientific research in the first place; the Public School System of New York City, Cornell University, and Harvard University for my formal education; my high-school friends Gary Feinberg and Steven Weinberg for making me learn too much too soon of what I might otherwise have never learned at all; my parents and my two brothers for always encouraging a child's dream to be a scientist. Finally, I wish to thank my wife and my children for the warmth of their love.

REFERENCES

1. Yang, C. N. and Mills, R., Phys. Rev. *96,* 191 (1954). Also, Shaw, R., unpublished.
2. Weinberg, S., Phys. Rev. Letters *19,* 1264 (1967); Salam, A., in *Elementary Particle Physics* (ed. Svartholm, N.; Almqvist and Wiksell; Stockholm; 1968).
3. 't Hooft, G., Nuclear Physics *B 33,* 173 and *B 35,* 167 (1971); Lee, B. W., and Zinn-Justin, J., Phys. Rev. D *5,* pp. 3121–3160 (1972); 't Hooft, G., and Veltman M., Nuclear Physics *B 44,* 189 (1972).
4. Gross, D. J. and Wilczek, F., Phys. Rev. Lett. *30,* 1343 (1973); Politzer, H. D., Phys. Rev. Lett. *30,* 1346 (1973).
5. Sakurai, J. J., Annals of Physics *11,* 1 (1960).
6. Gell-Mann, M., and Ne'eman, Y., *The Eightfold Way* (Benjamin, W. A., New York, 1964).
7. Feynman, R., and Gell-Mann, M., Phys. Rev. *109,* 193 (1958); Marshak, R., and Sudarshan, E. C. G., Phys. Rev. *109,* 1860 (1958).
8. Bludman, S., Nuovo Cimento Ser. 10 *9,* 433 (1958).
9. Schwinger, J., Annals of Physics *2,* 407 (1958).
10. Glashow, S. L., Harvard University Thesis, p. 75 (1958).
11. Georgi, H., and Glashow, S. L., Phys. Rev. Letters *28,* 1494 (1972).
12. Salam, A., and Ward, J., Nuovo Cimento *11,* 568 (1959).
13. Salam, A., and Ward, J., Nuovo Cimento *19,* 165 (1961).
14. Glashow, S. L., Nuclear Physics *22,* 579 (1961).
15. Salam, A., and Ward, J., Physics Letters *13,* 168 (1964).
16. Feinberg, G., Phys. Rev. *110,* 1482 (1958).
17. Tzou Kuo-Hsien, Comptes rendus *245,* 289 (1957).
18. Glashow, S. L., Nuclear Physics *10,* 107 (1959).
19. Glashow, S. L., and Iliopoulos J., Phys. Rev. D *3,* 1043 (1971).
20. Many authors are involved with this work: Brout, R., Englert, F., Goldstone, J., Guralnik, G., Hagen, C., Higgs, P., Jona-Lasinio, G., Kibble, T., and Nambu, Y.
21. Hasert, F. J., *et al.,* Physics Letters *46 B,* 138 (1973) and Nuclear Physics *B 73,* 1 (1974). Benvenuti, A., *et al.,* Phys. Rev. Letters *32,* 800 (1974).
22. Prescott, C. Y., *et al.,* Phys. Lett. *B 77,* 347 (1978).
23. Gell-Mann, M., and Glashow, S. L., Annals of Physics *15,* 437 (1961).
24. Bjorken, J., and Glashow, S. L., Physics Letters *11,* 84 (1964).

25. Amati, D., *et al.*, Nuovo Cimento *34,* 1732 (A 64); Hara, Y. Phys. Rev. *134,* B 701 (1964); Okun, L. B., Phys. Lett. *12,* 250 (1964); Maki, Z., and Ohnuki, Y., Progs. Theor. Phys. *32,* 144 (1964); Nauenberg, M., (unpublished); Teplitz, V., and Tarjanne, P., Phys. Rev. Lett. *11,* 447 (1963).
26. Glashow, S. L., Iliopoulos, J., and Maiani, L., Phys. Rev. D *2,* 1285 (1970).
27. Georgi, H., and Glashow, S. L., Phys. Rev. Letters *33,* 438 (1974).
28. Georgi, H., and Glashow, S. L., Harvard Preprint HUTP-79/A 053.

Abdus Salam عبدالسلام

ABDUS SALAM

Abdus Salam was born in Jhang, a small town in what is now Pakistan, in 1926. His father was an official in the Department of Education in a poor farming district. His family has a long tradition of piety and learning.

When he cycled home from Lahore, at the age of 14, after gaining the highest marks ever recorded for the Matriculation Examination at the University of the Panjab, the whole town turned out to welcome him. He won a scholarship to Government College, University of the Panjab, and took his MA in 1946. In the same year he was awarded a scholarship to St. John's College, Cambridge, where he took a BA (honours) with a double First in mathematics and physics in 1949. In 1950 he received the Smith's Prize from Cambridge University for the most outstanding pre-doctoral contribution to physics. He also obtained a PhD in theoretical physics at Cambridge; his thesis, published in 1951, contained fundamental work in quantum electrodynamics which had already gained him an international reputation.

Salam returned to Pakistan in 1951 to teach mathematics at Government College, Lahore, and in 1952 became head of the Mathematics Department of the Panjab University. He had come back with the intention of founding a school of research, but it soon became clear that this was impossible. To pursue a career of research in theoretical physics he had no alternative at that time but to leave his own country and work abroad. Many years later he succeeded in finding a way to solve the heartbreaking dilemma faced by many young and gifted theoretical physicists from developing countries. At the ICTP, Trieste, which he created, he instituted the famous "Associateships" which allowed deserving young physicists to spend their vacations there in an invigorating atmosphere, in close touch with their peers in research and with the leaders in their own field, losing their sense of isolation and returning to their own country for nine months of the academic year refreshed and recharged.

In 1954 Salam left his native country for a lectureship at Cambridge, and since then has visited Pakistan as adviser on science policy. His work for Pakistan has, however, been far-reaching and influential. He was a member of the Pakistan Atomic Energy Commission, a member of the Scientific Commission of Pakistan and was Chief Scientific Adviser to the President from 1961 to 1974.

Since 1957 he has been Professor of Theoretical Physics at Imperial College, London, and since 1964 has combined this position with that of Director of the ICTP, Trieste.

For more than forty years he has been a prolific researcher in theoretical elementary particle physics. He has either pioneered or been associated with all the important developments in this field, maintaining a constant and fertile flow of brilliant ideas. For the past thirty years he has used his academic reputation to add weight to his active and influential participation in international scientific affairs. He has served on a number of United Nations committees concerned with the advancement of science and technology in developing countries.

To accommodate the astonishing volume of activity that he undertakes, Professor Salam cuts out such inessentials as holidays, parties and entertainments. Faced with such an example, the staff of the Centre find it very difficult to complain that they are overworked.

He has a way of keeping his administrative staff at the ICTP fully alive to the real aim of the Centre – the fostering through training and research of the advancement of theoretical physics, with special regard to the needs of developing countries. Inspired by their personal regard for him and encouraged by the fact that he works harder than any of them, the staff cheerfully submit to working conditions that would be unthinkable here at the International Atomic Energy Agency in Vienna (IAEA). The money he received from the Atoms for Peace Medal and Award he spent on setting up a fund for young Pakistani physicists to visit the ICTP. He uses his share of the Nobel Prize entirely for the benefit of physicists from developing countries and does not spend a penny of it on himself or his family.

Abdus Salam is known to be a devout Muslim, whose religion does not occupy a separate compartment of his life; it is inseparable from his work and family life. He once wrote: "The Holy Quran enjoins us to reflect on the verities of Allah's created laws of nature; however, that our generation has been privileged to glimpse a part of His design is a bounty and a grace for wich I render thanks with a humble heart."

The biography was written by Miriam Lewis, now at IAEA, Vienna, who was at one time on the staff of ICTP (International Centre For Theoretical Physics, Trieste).

Date of birth:	29 January, 1926
Place of birth:	Jhang, Pakistan

Educational Career

Government College, Jhang and Lahore (1938–1946)	M.A. (Panjab University)
Foundation Scholar, St. John's College, Cambridge (1946–1949)	B.A. Honours Double first in Mathematics (Wrangler) and Physics
Cavendish Laboratory, Cambridge (1952)	Ph.D. in Theoretical Physics

Awarded Smith's Prize by the University of Cambridge for "the most outstanding pre-doctoral contribution to Physics" (1950)

D.Sc. Honoris Causa:

Panjab University, Lahore (1957)
University of Edinburgh (1971)
Panjab University, Lahore (Pakistan) (1957)
University of Edinburgh (UK) (1971)
University of Trieste (Italy) (1979)
University of Islamabad (Pakistan) (1979)
Universidad Nacional de Ingenieria, Lima (Peru) (1980)
University of San Marcos, Lima (Peru) (1980)
National University of San Antonio Abad, Cuzco (Peru) (1980)
Universidad Simon Bolivar, Caracas (Venezuela) (1980)
University of Wroclow (Poland) (1980)
Yarmouk University (Jordan) (1980)
University of Istanbul (Turkey) (1980)
Guru Nanak Dev University, Amritsar (India) (1981)
Muslim University, Aligarh (India) (1981)
Hindu University, Banaras (India) (1981)
University of Chittagong (Bangladesh) (1981)
University of Bristol (UK) (1981)
University of Maiduguri (Nigeria) (1981)
University of the Philippines, Quezon City (Philippines) (1982)
University of Khartoum (Sudan) (1983)
Universidad Complutense de Madrid (Spain) (1983)
City College, City University of New York (USA) (1984)
University of Nairobi (Kenya) (1984)
Universidad Nacional de Cuyo (Argentina) (1985)
Universidad Nacional de la Plata (Argentina) (1985)
University of Cambridge (UK) (1985)
University of Goteborg (Sweden) (1985)
Kliment Ohridski University of Sofia (Bulgaria) (1986)
University of Glasgow (UK) (1986)
University of Science and Technology, Hefei (China) (1986)
The City University, London (UK) (1986)
Panjab University, Chandigarh (India) (1987)
Medicina Alternativa, Colombo (Sri Lanka) (1987)
National University of Benin, Cotonou (Benin) (1987)
University of Exeter (UK) (1987)
University of Gent (Belgium) (1988)
"Creation" International Association of Scientists and Intelligentsia (USSR) (1989)
Bendel State University, Ekpoma (Nigeria) (1990)
University of Ghana (Ghana) (1990)
University of Warwick (UK) (1991)
University of Dakar (Senegal) (1991)

University of Tucuman (Argentina) (1991)
University of Lagos (Nigeria) (1992)

Awards

Hopkins Prize (Cambridge University) for "the most outstanding contribution to Physics during 1957–1958"
Adams Prize (Cambridge University) (1958)
First recipient of Maxwell Medal and Award (Physical Society, London) (1961)
Hughes Medal (Royal Society, London) (1964)
Atoms for Peace Medal and Award (Atoms for Peace Foundation) (1968)
J. Robert Oppenheimer Memorial Medal and Prize (University of Miami) (1971)
Guthrie Medal and Prize (1976)
Matteuci Medal (Accademia Nazionale dei Lincei, Rome) (1978)
John Torrence Tate Medal (American Institute of Physics) (1978)
Royal Medal (Royal Society, London) (1978)
Einstein Medal (UNESCO, Paris) (1979)
Shri R.D. Birla Award (India Physics Association) (1979)
Josef Stefan Medal (Josef Stefan Institue, Ljublijana) (1980)
Gold Medal for Outstanding Contributions to Physics (Czechoslovak Academy of Sciences, Prague) (1981)
Lomonosov Gold Medal (USSR Academy of Sciences) (1983)
Copley Medal (Royal Society, London) (1990)

Appointments

Professor, Government College and Panjab University, Lahore (1951– 1954)
Elected Fellow St. John's College, Cambridge (1951–1956)
Member, Institute of Advanced Study, Princeton (1951)
Lecturer, Cambridge University (1954–1956)
Professor of Theoretical Physics, London University, Imperial College, London, since 1957
Director, International Centre for Theoretical Physics, Trieste, since 1964
Elected (First) Fellow of the Royal Society, London, from Pakistan (1959)
Elected, Foreign Member of the Royal Swedish Academy of Sciences (1970)
Elected, Foreign Member of the American Academy of Arts and Sciences (1971)
Elected, Foreign Member, USSR Academy of Sciences (1971)
Elected, Honorary Fellow St. John's College, Cambridge (1971)
Elected, Foreign Associate, USA National Academy of Sciences (Washington) (1979)
Elected, Foreign Member, Accademia Nazionale dei Lincei (Rome) (1979)
Elected, Foreign Member, Accademia Tiberina (Rome) (1979)
Elected, Foreign Member, Iraqi Academy (Baghdad) (1979)
Elected, Honorary Fellow, Tata Institute of Fundamental Research (Bombay) (1979)

Elected, Honorary Member, Korean Physics Society (Seoul) (1979)
Elected, Foreign Member, Academy of the Kingdom of Morocco (Rabat) (1980)
Elected, Foreign Member, Accademia Nazionale delle Scienze dei XL (Rome) (1980)
Elected, Member, European Academy of Science, Arts and Humanities (Paris) (1980)
Elected, Associate Member, Josef Stefan Institute (Ljublijana) (1980)
Elected, Foreign Fellow, Indian National Science Academy (New Delhi) (1980)
Elected, Fellow, Bangladesh Academy of Sciences (Dhaka) (1980)
Elected, Member, Pontifical Academy of Sciences (Vatican City) (1981)
Elected, Corresponding Member, Portuguese Academy of Sciences (Lisbon) (1981)
Founding Member, Third World Academy of Sciences (1983)
Elected, Corresponding Member, Yugoslav Academy of Sciences and Arts (Zagreb) (1983)
Elected, Honorary Fellow, Ghana Academy of Arts and Sciences (1984)
Elected, Honorary Member, Polish Academy of Sciences (1985)
Elected, Corresponding Member, Academia de Ciencias Medicas, Fisicas y Naturales de Guatemala (1986)
Elected, Fellow, Pakistan Academy of Medical Sciences (1987)
Elected, Honorary Fellow, Indian Academy of Sciences (Bangalore) (1988)
Elected, Distinguished International Fellow of Sigma Xi (1988)
Elected, Honorary Member, Brazilian Mathematical Society (1989)
Elected, Honorary Member, National Academy of Exact, Physical and Natural Sciences, Argentina (1989)
Elected, Honorary Member, Hungarian Academy of Sciences (1990)
Elected, Member, Academia Europaea (1990)

Orders and other Distinctions
Order of Andres Bello (Venezuela) (1980)
Order of Istiqlal (Jordan) (1980)
Cavaliere di Gran Croce dell'Ordine al Merito della Repubblica Italiana (1980)
Honorary Knight Commander of the Order of the British Empire (1989)

Awards for contributions towards peace and promotion of international scientific collaboration
Atoms for Peace Medal and Award (Atoms for Peace Foundation) (1968)
Peace Medal (Charles University, Prague) (1981)
Premio Umberto Biancomano (Italy) (1986)
Dayemi International Peace Award (Bangladesh) (1986)
First Edinburgh Medal and Prize (Scotland) (1988)
"Genoa" International Development of Peoples Prize (Italy) (1988)
Catalunya International Prize (Spain) (1990)

United Nations Assignments

Scientific Secretary, Geneva Conferences on Peaceful Uses of Atomic Energy (1955 and 1958)

Member, United Nations Advisory Committee on Science and Technology (1964–1975)

Member, United Nations Panel and Foundation Committee for the United Nations University (1970–1973)

Chairman, United Nations Advisory Committee on Science and Technology (1971–1972)

Member, Scientific Council, SIPRI, Stockholm International Peace Research Institute (1970)

Vice President, International Union of Pure and Applied Physics (1972–1978)

Pakistan Assignments

Member, Atomic Energy Commission, Pakistan (1958–1974)

Adviser, Education Commission, Pakistan (1959)

Member, Scientific Commission, Pakistan (1959)

Chief Scientific Adviser, President of Pakistan (1961–1974)

President, Pakistan Association for Advancement of Science (1961–1962)

Chairman, Pakistan Space and Upper Atmosphere Committee (1961–1964)

Governor from Pakistan to the International Atomic Energy Agency (1962–1963)

Member, National Science Council, Pakistan (1963–1975)

Member, Board of Pakistan Science Foundation (1973–1977)

Pakistani Awards

Sitara-i-Pakistan (S.Pk.)

Pride of Performance Medal and Award (1959)

GAUGE UNIFICATION OF FUNDAMENTAL FORCES

Nobel lecture, 8 December, 1979
by
ABDUS SALAM
Imperial College of Science and Technology, London, England
and International Centre for Theoretical Physics, Trieste, Italy

Introduction: In June 1938, Sir George Thomson, then Professor of Physics at Imperial College, London, delivered his 1937 Nobel Lecture. Speaking of Alfred Nobel, he said: "The idealism which permeated his character led him to ... (being) as much concerned with helping science as a whole, as individual scientists. ... The Swedish people under the leadership of the Royal Family and through the medium of the Royal Academy of Sciences have made Nobel Prizes one of the chief causes of the growth of the prestige of science in the eyes of the world ... As a recipient of Nobel's generosity, I owe sincerest thanks to them as well as to him."

I am sure I am echoing my colleagues' feelings as well as my own, in reinforcing what Sir George Thomson said—in respect of Nobel's generosity and its influence on the growth of the prestige of science. Nowhere is this more true than in the developing world. And it is in this context that I have been encouraged by the Permanent Secretary of the Academy – Professor Carl Gustaf Bernhard—to say a few words before I turn to the scientific part of my lecture.

Scientific thought and its creation is the common and shared heritage of mankind. In this respect, the history of science, like the history of all civilization, has gone through cycles. Perhaps I can illustrate this with an actual example.

Seven hundred and sixty years ago, a young Scotsman left his native glens to travel south to Toledo in Spain. His name was Michael, his goal to live and work at the Arab Universities of Toledo and Cordova, where the greatest of Jewish scholars, Moses bin Maimoun, had taught a generation before.

Michael reached Toledo in 1217 AD. Once in Toledo, Michael formed the ambitious project of introducing Aristotle to Latin Europe, translating not from the original Greek, which he did not know, but from the Arabic translation then taught in Spain. From Toledo, Michael travelled to Sicily, to the Court of Emperor Frederick II.

Visiting the medical school at Salerno, chartered by Frederick in 1231, Michael met the Danish physician, Henrik Harpestraeng – later to become Court Physician of King Erik Plovpenning. Henrik had come to Salerno to compose his treatise on blood-letting and surgery. Henrik's sources were the medical canons of the great clinicians of Islam, Al-Razi and Avicenna, which only Michael the Scot could translate for him.

Toledo's and Salerno's schools, representing as they did the finest synthesis of Arabic, Greek, Latin and Hebrew scholarship, were some of the most memorable of international assays in scientific collaboration. To Toledo and Salerno came scholars not only from the rich countries of the East and the South, like Syria, Egypt, Iran and Afghanistan, but also from developing lands of the West and the North like Scotland and Scandinavia. Then, as now, there were obstacles to this international scientific concourse, with an economic and intellectual disparity between different parts of the world. Men like Michael the Scot or Henrik Harpestraeng were singularities. They did not represent any flourishing schools of research in their own countries. With all the best will in the world their teachers at Toledo and Salerno doubted the wisdom and value of training them for advanced scientific research. At least one of his masters counselled young Michael the Scot to go back to clipping sheep and to the weaving of woollen cloth.

In respect of this cycle of scientific disparity, perhaps I can be more quantitative. George Sarton, in his monumental five-volume History of Science chose to divide his story of achievement in sciences into ages, each age lasting half a century. With each half century he associated one central figure. Thus 450 BC–400 BC Sarton calls the Age of Plato; this is followed by half centuries of Aristotle, of Euclid, of Archimedes and so on. From 600 AD to 650 AD is the Chinese half century of Hsiian Tsang, from 650 to 700 AD that of I-Ching, and then from 750 AD to 1100 AD–350 years continuously–it is the unbroken succession of the Ages of Jabir, Khwarizmi, Razi, Masudi, Wafa, Biruni and Avicenna, and then Omar Khayam–Arabs, Turks, Afghans and Persians–men belonging to the culture of Islam. After 1100 appear the first Western names; Gerard of Cremona, Roger Bacon–but the honours are still shared with the names of Ibn-Rushd (Averroes), Moses Bin Maimoun, Tusi and Ibn-Nafis–the man who anticipated Harvey's theory of circulation of blood. No Sarton has yet chronicled the history of scientific creativity among the pre-Spanish Mayas and Aztecs, with their invention of the zero, of the calendars of the moon and Venus and of their diverse pharmacological discoveries, including quinine, but the outline of the story is the same–one of undoubted superiority to the Western contemporary correlates.

After 1350, however, the developing world loses out except for the occasional flash of scientific work, like that of Ulugh Beg–the grandson of Timurlane, in Samarkand in 1400 AD; or of Maharaja Jai Singh of Jaipur in 1720–who corrected the serious errors of the then Western tables of eclipses of the sun and the moon by as much as six minutes of arc. As it was, Jai Singh's techniques were surpassed soon after with the development of the telescope in Europe. As a contemporary Indian chronicler wrote: "With him on the funeral pyre, expired also all science in the East." And this brings us to this century when the cycle begun by Michael the Scot

turns full circle, and it is we in the developing world who turn to the Westwards for science. As Al-Kindi wrote 1100 years ago: "It is fitting then for us not to be ashamed to acknowledge and to assimilate it from whatever source it comes to us. For him who scales the truth there is nothing of higher value than truth itself; it never cheapens nor abases him."
Ladies and Gentlemen,

It is in the spirit of Al-Kindi that I start my lecture with a sincere expression of gratitude to the modern equivalents of the Universities of Toledo and Cordova, which I have been privileged to be associated with – Cambridge, Imperial College, and the Centre at Trieste.

I. FUNDAMENTAL PARTICLES, FUNDAMENTAL FORCES AND GAUGE UNIFICATION

The Nobel lectures this year are concerned with a set of ideas relevant to the gauge unification of the electromagnetic force with the weak nuclear force. These lectures coincide nearly with the 100th death-anniversary of Maxwell, with whom the first unification of forces (electric with the magnetic) matured and with whom gauge theories originated. They also nearly coincide with the 100th anniversary of the birth of Einstein—the man who gave us the vision of an ultimate unification of *all* forces.

The ideas of today started more than twenty years ago, as gleams in several theoretical eyes. They were brought to predictive maturity over a decade back. And they started to receive experimental confirmation some six years ago.

In some senses then, our story has a fairly long background in the past. In this lecture I wish to examine some of the theoretical gleams of today and ask the question if these may be the ideas to watch for maturity twenty years from now.

From time immemorial, man has desired to comprehend the complexity of nature in terms of as few elementary concepts as possible. Among his quests—in Feynman's words—has been the one for "wheels within wheels"—the task of natural philosophy being to discover the innermost wheels if any such exist. A second quest has concerned itself with the fundamental forces which make the wheels go round and enmesh with one another. The greatness of gauge ideas—of gauge field theories—is that they reduce these two quests to just one; elementary particles (described by relativistic quantum fields) are representations of certain charge operators, corresponding to gravitational mass, spin, flavour, colour, electric charge and the like, while the fundamental forces are the forces of attraction or repulsion between these same *charges*. A third quest seeks for a *unification* between the charges (and thus of the forces) by searching for a single entity, of which the various charges are components in the sense that they can be transformed one into the other.

But are all fundamental forces gauge forces? Can they be understood as such, in terms of charges—and their corresponding currents—only? And if they are, how many charges? What unified entity are the charges components of?

What is the nature of charge? Just as Einstein comprehended the nature of gravitational charge in terms of space-time curvature, can we comprehend the nature of the other charges—the nature of the entire unified set, *as a set,* in terms of something equally profound? This briefly is the dream, much reinforced by the verification of gauge theory predictions. But before I examine the new theoretical ideas on offer for the future in this particular context, I would like your indulgence to range over a one-man, purely subjective, perspective in respect of the developments of the last twenty years themselves. The point I wish to emphasize during this part of my talk was well made by G. P. Thomson in his 1937 Nobel Lecture. G. P. said ". . . The goddess of learning is fabled to have sprung full grown from the brain of Zeus, but it is seldom that a scientific conception is born in its final form, or owns a single parent. More often it is the product of a series of minds, each in turn modifying the ideas of those that came before, and providing material for those that come after."

II. THE EMERGENCE OF SPONTANEOUSLY BROKEN SU(2)×U(1) GAUGE THEORY

I started physics research thirty years ago as an experimental physicist in the Cavendish, experimenting with tritium-deuterium scattering. Soon I knew the craft of experimental physics was beyond me—it was the sublime quality of patience—patience in accumulating data, patience with recalcitrant equipment—which I sadly lacked. Reluctantly I turned my papers in, and started instead on quantum field theory with Nicholas Kemmer in the exciting department of P. A. M. Dirac.

The year 1949 was the culminating year of the Tomonaga-Schwinger-Feynman-Dyson reformulation of renormalized Maxwell-Dirac gauge theory, and its triumphant experimental vindication. A field theory must be renormalizable and be capable of being made free of infinities—first discussed by Waller—if perturbative calculations with it are to make any sense. More—a renormalizable theory, with no dimensional parameter in its interaction term, connotes *somehow* that the fields represent "structureless" elementary entities. With Paul Matthews, we started on an exploration of renormalizability of meson theories. Finding that renormalizability held only for spin-zero mesons and that these were the only mesons that empirically existed then, (pseudoscalar pions, invented by Kemmer, following Yukawa) one felt thrillingly euphoric that with the triplet of pions (considered as the carriers of the strong nuclear force between the proton-neutron doublet) one might resolve the dilemma of the origin of this particular force which is responsible for fusion and fission. By the same token, the so-called weak nuclear force—the force responsible for β-radioactivity (and described then by Fermi's non-renormalizable theory) had to be mediated by some unknown spin-zero mesons if it was to be renormalizable, If massive charged spin-one mesons were to mediate this interaction, the theory would be non-renormalizable, according to the ideas then.

Now this agreeably renormalizable spin-zero theory for the pion was a field theory, but not a gauge field theory. There was no conserved charge which determined the pionic interaction. As is well known, shortly after the theory was elaborated, it was found wanting. The $(\frac{3}{2}, \frac{3}{2})$ resonance Δ effectively killed it off as a fundamental theory; we were dealing with a complex dynamical system, not "structureless" in the field-theoretic sense.

For me, personally, the trek to gauge theories as candidates for fundamental physical theories started in earnest in September 1956—the year I heard at the Seattle Conference Professor Yang expound his and Professor Lee's ideas[1] on the possibility of the hitherto sacred principle of left-right symmetry, being violated in the realm of the *weak nuclear force*. Lee and Yang had been led to consider abandoning left-right symmetry for weak nuclear interactions as a possible resolution of the (τ, θ) puzzle. I remember travelling back to London on an American Air Force (MATS) transport flight. Although I had been granted, for that night, the status of a Brigadier or a Field Marshal—I don't quite remember which—the plane was very uncomfortable, full of crying service-men's children—that is, the children were crying, not the servicemen. I could not sleep. I kept reflecting on why Nature should violate left-right symmetry in weak interactions. Now the hallmark of most weak interactions was the involvement in radioactivity phenomena of Pauli's neutrino. While crossing over the Atlantic, came back to me a deeply perceptive question about the neutrino which Professor Rudolf Peierls had asked when he was examining me for a Ph. D. a few years before. Peierls' question was: "The photon mass is zero because of Maxwell's principle of a gauge symmetry for electromagnetism; tell me, why is the neutrino mass zero?" I had then felt somewhat uncomfortable at Peierls, asking for a Ph. D. viva, a question of which he himself said he did not know the answer. But during that comfortless night the answer came. The analogue for the neutrino, of the gauge symmetry for the photon existed; it had to do with the masslessness of the neutrino, with symmetry under the γ_5 transformation[2] (later christened "chiral symmetry"). The existence of this symmetry for the massless neutrino must imply a combination $(1+\gamma_5)$ or $(1-\gamma_5)$ for the neutrino interactions. Nature had the choice of an aesthetically satisfying but a left-right symmetry violating theory, with a neutrino which travels exactly with the velocity of light; or alternatively a theory where left-right symmetry is preserved, but the neutrino has a tiny mass—some ten thousand times smaller than the mass of the electron.

It appeared at that time clear to me what choice Nature must have made. Surely, left-right symmetry must be sacrificed in all neutrino interactions. I got off the plane the next morning, naturally very elated. I rushed to the Cavendish, worked out the Michel parameter and a few other consequences of γ_5 symmetry, rushed out again, got into a train to Birmingham where Peierls lived. To Peierls I presented my idea; he had asked the original question; could he approve of the answer? Peierls' reply was kind but firm. He said "I do not believe left-right symmetry is violated in weak nuclear forces at all. I would not touch such ideas with a pair of tongs." Thus rebuffed in Birmingham, like Zuleika Dobson, I wondered where I could go next and the obvious place was CERN in Geneva, with Pauli—the father of the neutrino—nearby in Zurich.

At that time CERN lived in a wooden hut just outside Geneva airport. Besides my friends, Prentki and d'Espagnat, the hut contained a gas ring on which was cooked the staple diet of CERN—Entrecôte à la creme. The hut also contained Professor Villars of MIT, who was visiting Pauli the same day in Zurich. I gave him my paper. He returned the next day with a message from the Oracle; "Give my regards to my friend Salam and tell him to think of something better". This was discouraging, but I was compensated by Pauli's excessive kindness a few months later, when Mrs. Wu's[3], Lederman's[4] and Telegdi's[5] experiments were announced showing that left-right symmetry was indeed violated and ideas similar to mine about chiral symmetry were expressed independently by Landau[6] and Lee and Yang[7]. I received Pauli's first somewhat apologetic leatter on 24 January 1957. Thinking that Pauli's spirit should by now be suitably crushed, I sent him two short notes[8] I had written in the meantime. These contained suggestions to extend chiral symmetry to electrons and muons, assuming that their masses were a consequence of what has come to be known as dynamical spontaneous symmetry breaking. With chiral symmetry for electrons, muons and neutrinos, the only mesons that could mediate weak decays of the muons would have to carry spin one. Reviving thus the notion of charged intermediate *spin-one* bosons, one could then postulate for these a type of gauge invariance which I called the "neutrino gauge". Pauli's reaction was swift and terrible. He wrote on 30th January 1957, then on 18 February and later on 11, 12 and 13 March: "I am reading (along the shores of Lake Zurich) in bright sunshine quietly your paper . . ." "I am very much startled on the title of your paper 'Universal Fermi interaction' . . . For quite a while I have for myself the rule if a theoretician says *universal* it just means pure nonsense. This holds particularly in connection with the Fermi interaction, but otherwise too, and now you too, Brutus, my son, come with this word. . . ." Earlier, on 30 January, he had written "There is a similarity between this type of gauge invariance and that which was published by Yang and Mills . . . In the latter, of course, no γ_5 was used in the exponent." and he gave me the full reference of Yang and Mills' paper; (Phys. Rev. *96*, 191 (1954)). I quote from his letter: "However, there are dark points in your paper regarding the vector field B_μ. If the rest mass is infinite (or very large), how can this be compatible with the gauge transformation $B_\mu \rightarrow B_\mu - \partial_\mu \Lambda$?" and he concludes his letter with the remark: "Every reader will realize that you deliberately conceal here something and will ask you the same questions". Although he signed himself "With friendly regards", Pauli had forgotten his earlier penitence. He was clearly and rightly on the warpath.

Now the fact that I was using gauge ideas similar to the Yang—Mills (non-Abelian SU(2)-invariant) gauge theory was no news to me. This was because the Yang—Mills theory[9] (which married gauge ideas of Maxwell with the internal symmetry SU(2) of which the proton-neutron system constituted a doublet) had been independently invented by a Ph. D. pupil of mine, Ronald Shaw,[10] at Cambridge at the same time as Yang and Mills had written. Shaw's work is relatively unknown; it remains buried in his Cambridge thesis. I must admit I was taken aback by Pauli's fierce prejudice against universalism—against what we would today call unification of basic forces—but I did not take this too seriously. I felt this was a legacy of the exasperation which Pauli had always felt at Einstein's somewhat formalistic attempts at unifying gravity with electromagnetism—forces which in Pauli's phrase "cannot be joined—for God hath rent them asunder". But Pauli was absolutely right in accusing me of darkness about the problem of the masses of the Yang—Mills fields; one could not obtain a mass without wantonly destroying the gauge symmetry one had started with. And this was particularly serious in this context, because Yang and Mills had conjectured the desirable renormalizability of their theory with a proof which relied heavily and exceptionally on the masslessness of their spin-one intermediate mesons. The problem was to be solved only seven years later with the understanding of what is now known as the Higgs mechanism, but I will come back to this later.

Be that as it may, the point I wish to make from this exchange with Pauli is that already in early 1957, just after the first set of parity experiments, many ideas coming to fruition now, had started to become clear. These are:

1. First was the idea of chiral symmetry leading to a V-A theory. In those early days my humble suggestion[2], [8] of this was limited to neutrinos, electrons and muons only, while shortly after, that year, Sudarshan and Marshak,[11] Feynman and Gell-Mann,[12] and Sakurai[13] had the courage to postulate γ_5 symmetry for baryons as well as leptons, making this into a universal principle of physics.[1]

Concomitant with the (V-A) theory was the result that if weak interactions are mediated by intermediate mesons, these must carry spin one.

2. Second, was the idea of spontaneous breaking of chiral symmetry to generate electron and muon masses: though the price which those latter-day Shylocks, Nambu and Jona-Lasinio[14] and Goldstone[15] exacted for this (i.e. the appearance of massless scalars), was not yet appreciated.
3. And finally, though the use of a Yang—Mills—Shaw (non-Abelian) gauge theory for describing spin-one intermediate charged mesons was suggested already in 1957, the giving of masses to the intermediate bosons through spontaneous symmetry breaking, in a manner to preserve the renormalizability of the theory, was to be accomplished only during a long period of theoretical development between 1963 and 1971.

Once the Yang—Mills—Shaw ideas were accepted as relevant to the charged weak currents—to which the charged intermediate mesons were coupled in this theory—during 1957 and 1958 was raised the question of what was the third component of the SU(2) triplet, of which the charged weak currents were the two members. There were the two alternatives: the electroweak unification suggestion, where the electromagnetic current was assumed to be this third component; and the rival suggestion that the third component was a neutral current unconnected with electroweak unification. With hindsight, I shall

[1] Today we believe protons and neutrons are composites of quarks, so that γ_5 symmetry is now postulated for the elementary entities of today—the quarks.

call these the Klein[16] (1938) and the Kemmer[17] (1937) alternatives. The Klein suggestion, made in the context of a Kaluza—Klein five-dimensional space-time, is a real tour-de-force; it combined two hypothetical spin-one charged mesons with the photon in one multiplet, deducing from the compactification of the fifth dimension, a theory which looks like Yang—Mills—Shaw's. Klein intended his charged mesons for *strong* interactions, but if we read charged *weak* mesons for Klein's *strong* ones, one obtains the theory independently suggested by Schwinger[18] (1957), though Schwinger, unlike Klein, did not build in any non-Abelian gauge aspects. With just these non-Abelian Yang—Mills gauge aspects very much to the fore, the idea of uniting weak interactions with electromagnetism was developed by Glashow[19] and Ward and myself[20] in late 1958. The rival Kemmer suggestion of a global SU(2)-invariant triplet of weak charged and neutral currents was independently suggested by Bludman[21] (1958) in a gauge context and this is how matters stood till 1960.

To give you the flavour of, for example, the year 1960, there is a paper written that year of Ward and myself[22] with the statement: "Our basic postulate is that it should be possible to generate strong, weak and electromagnetic interaction terms with all their correct symmetry properties (as well as with clues regarding their relative strengths) by making local gauge transformations on the kinetic energy terms in the free Lagrangian for all particles. This is the statement of an ideal which, in this paper at least, is only very partially realized". I am not laying a claim that we were the only ones who were saying this, but I just wish to convey to you the temper of the physics of twenty years ago—qualitatively no different today from then. But what a quantitative difference the next twenty years made, first with new and far-reaching developments in theory—and then, thanks to CERN, Fermilab, Brookhaven, Argonne, Serpukhov and SLAC in testing it!

So far as theory itself is concerned, it was the next seven years between 1961—67 which were the crucial years of quantitative comprehension of the phenomenon of spontaneous symmetry breaking and the emergence of the $SU(2)\times U(1)$ theory in a form capable of being tested. The story is well known and Steve Weinberg has already spoken about it. So I will give the barest outline. First there was the realization that the two alternatives mentioned above a pure electromagnetic current versus a pure neutral current—Klein—Schwinger versus Kemmer—Bludman—were not alternatives; they were complementary. As was noted by Glashow[23] and independently by Ward and myself[24], both types of currents and the corresponding gauge particles ($W^{\pm}$, Z^0 and γ) were needed in order to build a theory that could simultaneously accommodate parity violation for weak and parity conservation for the electromagnetic phenomena. Second, there was the influential paper of Goldstone[25] in 1961 which, utilizing a non-gauge self-interaction between scalar particles, showed that the price of spontaneous breaking of a continuous internal symmetry was the appearance of zero mass scalars—a result foreshadowed earlier by Nambu. In giving a proof of this theorem[26] with Goldstone I collaborated with Steve Weinberg, who spent a year at Imperial College in London.

I would like to pay here a most sincerely felt tribute to him and to Sheldon Glashow for their warm and personal friendship.

I shall not dwell on the now well-known contributions of Anderson[27], Higgs[28], Brout & Englert[29], Guralnik, Hagen and Kibble[30] starting from 1963, which showed the way how spontaneous symmetry breaking using spin-zero fields could generate vector-meson masses, defeating Goldstone at the same time. This is the so-called Higgs mechanism.

The final steps towards the electroweak theory were taken by Weinberg[31] and myself[32] (with Kibble at Imperial College tutoring me about the Higgs phenomena). We were able to complete the present formulation of the spontaneously broken SU(2)×U(1) theory so far as leptonic weak interactions were concerned—with one parameter $\sin^2\theta$ describing all weak and electromagnetic phenomena and with one isodoublet Higgs multiplet. An account of this development was given during the contribution[32] to the Nobel Symposium (organized by Nils Svartholm and chaired by Lamek Hulthén held at Gothenburg after some postponements, in early 1968). As is well known, we did not then, and still do not, have a prediction for the scalar Higgs mass.

Both Weinberg and I suspected that this theory was likely to be renormalizable.[2] Regarding spontaneously broken Yang—Mills—Shaw theories in general this had earlier been suggested by Englert, Brout and Thiry[29]. But this subject was not pursued seriously except at Veltman's school at Utrecht, where the proof of renormalizability was given by 't Hooft[33] in 1971. This was elaborated further by that remarkable physicist the late Benjamin Lee[34], working with Zinn Justin, and by 't Hooft and Veltman[35]. This followed on the earlier basic advances in Yang—Mills calculational technology by Feynman[36], DeWitt[37], Faddeev and Popov[38], Mandelstam[39], Fradkin and Tyutin[40], Boulware[41], Taylor[42], Slavnov[43], Strathdee[44] and Salam. In Coleman's eloquent phrase "'t Hooft's work turned the Weinberg—Salam frog into an enchanted prince". Just before had come the GIM (Glashow, Iliopoulos and Maiani) mechanism[45], emphasising that the existence of the fourth charmed quark (postulated earlier by several authors) was essential to the natural resolution of the dilemma posed by the absence of strangeness--violating currents. This tied in naturally with the understanding of the Steinberger—Schwinger—Rosenberg—Bell—Jackiw—Adler anomaly[46] and its removal for SU(2)×U(1) by the parallelism of four quarks and four leptons, pointed out by Bouchiat, Iliopoulos and Meyer and independently by Gross and Jackiw.[47]

[2] When I was discussing the final version of the SU(2)×U(1) theory and its possible renormalizability in Autumn 1967 during a post-doctoral course of lectures at Imperial College, Nino Zichichi from CERN happened to be present. I was delighted because Zichichi had been badgering me since 1958 with persistent questioning of what theoretical avail his precise measurements on (g-2) for the muon as well as those of the muon lifetime were, when not only the magnitude of the electromagnetic corrections to weak decays was uncertain, but also conversely the effect of non-renormalizable weak interactions on "renormalized" electromagnetism was so unclear.

If one has kept a count, I have so far mentioned around fifty theoreticians. As a failed experimenter, I have always felt envious of the ambience of large experimental teams and it gives me the greatest pleasure to acknowledge the direct or the indirect contributions of the "series of minds" to the spontaneously broken SU(2)×U(1) gauge theory. My profoundest personal appreciation goes to my collaborators at Imperial College, Cambridge, and the Trieste Centre, John Ward, Paul Matthews, Jogesh Pati, John Strathdee, Tom Kibble and to Nicholas Kemmer.

In retrospect, what strikes me most about the early part of this story is how uninformed all of us were, not only of each other's work, but also of work done earlier. For example, only in 1972 did I learn of Kemmer's paper written at Imperial College in 1937.

Kemmer's argument essentially was that Fermi's weak theory was not globally SU(2) invariant and should be made so—though not for its own sake but as a prototype for strong interactions. Then this year I learnt that earlier, in 1936, Kemmer's Ph. D. supervisor, Gregor Wentzel[48], had introduced (the yet undiscovered) analogues of lepto-quarks, whose mediation could give rise to neutral currents after a Fierz reshuffle. And only this summer, Cecilia Jarlskog at Bergen rescued Oscar Klein's paper from the anonymity of the Proceedings of the International Institute of Intellectual Cooperation of Paris, and we learnt of his anticipation of a theory similar to Yang—Mills—Shaw's long before these authors. As I indicated before, the interesting point is that Klein was using his triplet, of two charged mesons plus the photon, not to describe weak interaction but for strong nuclear force unification with the electromagnetic—something our generation started on only in 1972—and not yet experimentally verified. Even in this recitation I am sure I have inadvertantly left off some names of those who have in some way contributed to SU(2)×U(1). Perhaps the moral is that not unless there is the prospect of quantitative verification, does a qualitative idea make its impress in physics.

And this brings me to experiment, and the year of the Gargamelle[49]. I still remember Paul Matthews and I getting off the train at Aix-en-Provence for the 1973 European Conference and foolishly deciding to walk with our rather heavy luggage to the student hostel where we were billeted. A car drove from behind us, stopped, and the driver leaned out. This was Musset whom I did not know well personally then. He peered out of the window and said: "Are you Salam?" I said "Yes". He said: "Get into the car. I have news for you. We have found neutral currents." I will not say whether I was more relieved for being given a lift because of our heavy luggage or for the discovery of neutral currents. At the Aix-en-Provence meeting that great and modest man, Lagarrigue, was also present and the atmosphere was that of a carnival—at least this is how it appeared to me. Steve Weinberg gave the rapporteur's talk with T. D. Lee as the chairman. T. D. was kind enough to ask me to comment after Weinberg finished. That summer Jogesh Pati and I had predicted proton decay within the context of what is now called grand unification and in the flush of this excitement I am afraid I ignored weak neutral currents as a subject which had already come to a successful conclusion, and concentrated on speaking of the possible decays of the proton. I understand now that proton decay experiments are being planned in the United States by the Brookhaven, Irvine and Michigan and the Wisconsin—Harvard groups and also by a European collaboration to be mounted in the Mont Blanc Tunnel Garage No. 17. The later quantitative work on neutral currents at CERN, Fermilab., Brookhaven, Argonne and Serpukhov is, of course, history, but a special tribute is warranted to the beautiful SLAC-Yale-CERN experiment[50] of 1978 which exhibited the effective Z^0-photon interference in accordance with the predictions of the theory. This was foreshadowed by Barkov *et al*'s experiments[51] at Novosibirsk in the USSR in their exploration of parity violation in the atomic potential for bismuth. There is the apocryphal story about Einstein, who was asked what he would have thought if experiment had not confirmed the light deflection predicted by him. Einstein is supposed to have said, "Madam, I would have thought the Lord has missed a most marvellous opportunity." I believe, however, that the following quote from Einstein's Herbert Spencer lecture of 1933 expresses his, my colleagues' and my own views more accurately. "Pure logical thinking cannot yield us any knowledge of the empirical world; all knowledge of reality starts from experience and ends in it." This is exactly how I feel about the Gargamelle-SLAC experience.

III. THE PRESENT AND ITS PROBLEMS

Thus far we have reviewed the last twenty years and the emergence of $SU(2)\times U(1)$, with the twin developments of a gauge theory of basic interactions, linked with internal symmetries, and of the spontaneous breaking of these symmetries. I shall first summarize the situation as we believe it to exist now and the immediate problems. Then we turn to the future.

1. To the level of energies explored, we believe that the following sets of particles are "structureless" (in a field-theoretic sense) and, at least to the level of energies explored hitherto, constitute the elementary entities of which all other objects are made.

		$SU_c(3)$ triplets		
Family I	quarks	$\begin{Bmatrix} u_R, u_Y, u_B \\ d_R, d_Y, d_B \end{Bmatrix}$	leptons $\begin{pmatrix} \nu_e \\ e \end{pmatrix}$	SU(2) doublets
Family II	quarks	$\begin{Bmatrix} c_R, c_Y, c_B \\ s_R, s_Y, s_B \end{Bmatrix}$	leptons $\begin{pmatrix} \nu_\mu \\ \mu \end{pmatrix}$	,,
Family III	quarks	$\begin{Bmatrix} t_R, t_Y, t_B \\ b_R, b_Y, b_B \end{Bmatrix}$	leptons $\begin{pmatrix} \nu_\tau \\ \tau \end{pmatrix}$	,,

Together with their antiparticles each family consists of 15 or 16 two-component fermions (15 or 16 depending on whether the neutrino is massless or not). The third family is still conjectural, since the top quark (t_R, t_Y, t_B) has not yet been discovered. Does this family really follow the pattern of the other two? Are there more families? Does the fact that the families are replicas of each other imply that Nature has discovered a dynamical stability about a system of 15 (or 16) objects, and that by this token there is a more basic layer of structure underneath?[52]

2. Note that quarks come in three colours; Red (R), Yellow (Y) and Blue (B). Parallel with the electroweak $SU(2)\times U(1)$, a *gauge* field[3] theory ($SU_c(3)$) of strong (quark) interactions (quantum chromodynamics, QCD)[53] has emerged which gauges the three colours. The indirect discovery of the (eight) gauge bosons associated with QCD (gluons), has already been surmised by the groups at DESY.[54]
3. All known baryons and mesons are singlets of colour $SU_c(3)$. This has led to a hypothesis that colour is always confined. One of the major unsolved problems of field theory is to determine if QCD—treated non-perturbatively—is capable of confining quarks and gluons.
4. In respect of the electroweak $SU(2)\times U(1)$, all known experiments on weak and electromagnetic phenomena below 100 GeV carried out to date agree with the theory which contains one theoretically undetermined parameter $\sin^2\theta = 0.230\pm0.009$.[55] The predicted values of the associated gauge boson ($W^\pm$ and Z^0) masses are: $m_W \approx 77$—84 GeV, $m_Z \approx 89$—95 GeV, for $0.25 \geqslant \sin^2\theta \geqslant 0.21$.
5. Perhaps the most remarkable measurement in electroweak physics is that of the parameter $\rho = \left(\frac{m_W}{m_Z \cos\theta}\right)^2$. Currently this has been determined from the ratio of neutral to charged current cross-sections. The predicted value $\rho = 1$ for weak *iso-doublet Higgs* is to be compared with the experimental[4] $\rho = 1.00\pm0.02$.
6. Why does Nature favour the simplest suggestion in $SU(2)\times U(1)$ theory of the Higgs scalars being iso-doublet?[5] Is there just one physical Higgs?

[3] "To my mind the most striking feature of theoretical physics in the last thirty-six years is the fact that not a single new theoretical idea of a fundamental nature has been successful. The notions of relativistic quantum theory . . . have in every instance proved stronger than the revolutionary ideas . . . of a great number of talented physicists. We live in a dilapidated house and we seem to be unable to move out. The difference between this house and a prison is hardly noticeable"—Res Jost (1963) in Praise of Quantum Field Theory (Siena European Conference).

[4] The one-loop radiative corrections to ρ suggest that the maximum mass of leptons contributing to ρ is less than 100 GeV.[56]

[5] To reduce the arbitrariness of the Higgs couplings and to motivate their iso-doublet character, one suggestion is to use supersymmetry. Supersymmetry is a Fermi-Bose symmetry, so that iso-doublet leptons like (ν_e, e) or (ν_μ, μ) in a supersymmetric theory must be accompanied in the same multiplet by iso-doublet Higgs.

Alternatively, one may identify the Higgs as composite fields associated with bound states of a yet new level of elementary particles and new (so-called techni-colour) forces (Dimopoulos & Susskind[57], Weinberg[58] and 't Hooft) of which, at present low energies, we have no cognisance and which may manifest themselves in the 1—100 TeV range. Unfortunately, both these ideas at first sight appear to introduce complexities, though in the context of a wider theory, which spans energy scales upto much higher masses, a satisfactory theory of the Higgs phenomena, incorporating these, may well emerge.

Of what mass? At present the Higgs interactions with leptons, quarks as well as their self-interactions are non-gauge interactions. For a three-family (6-quark) model, 21 out of the 26 parameters needed, are attributable to the Higgs interactions. Is there a basic principle, as compelling and as economical as the gauge principle, which embraces the Higgs sector? Alternatively, could the Higgs phenomenon itself be a manifestation of a dynamical breakdown of the gauge symmetry.[5]

7. Finally there is the problem of the families; is there a distinct SU(2) for the first, another for the second as well as a third SU(2), with spontaneous symmetry breaking such that the SU(2) apprehended by present experiment is a diagonal sum of these "family" SU(2)'s? To state this in another way, how far in energy does the e-μ universality (for example) extend? Are there more[59] Z^0's than just one, effectively differentially coupled to the e and the μ systems? (If there are, this will constitute mini-modifications of the theory, but not a drastic revolution of its basic ideas.)

In the next section I turn to a direct extrapolation of the ideas which went into the electroweak unification, so as to include strong interactions as well. Later I shall consider the more drastic alternatives which may be needed for the unification of all forces (including gravity)—ideas which have the promise of providing a deeper understanding of the charge concept. Regretfully, by the same token, I must also become more technical and obscure for the non-specialist. I apologize for this. The non-specialist may sample the flavour of the arguments in the next section (Sec. IV), ignoring the Appendices and then go on to Sec. V which is perhaps less technical.

IV. DIRECT EXTRAPOLATION FROM THE ELECTROWEAK TO THE ELECTRONUCLEAR

4.1 *The three ideas*

The three main ideas which have gone into the electronuclear—also called grand—unification of the electroweak with the *strong* nuclear force (and which date back to the period 1972—1974), are the following:

1. First: the psychological break (for us) of grouping quarks and leptons in the *same* multiplet of a unifying group G, suggested by Pati and myself in 1972[60]. The group G must contain $SU(2)\times U(1)\times SU_c(3)$; must be simple, if all quantum numbers (flavour, colour, lepton, quark and family numbers) are to be automatically quantized and the resulting gauge theory asymptotically free.
2. Second: an extension, proposed by Georgi and Glashow (1974)[61] which places not only (left-handed) quarks and leptons but also their antiparticles in the same multiplet of the unifying group.

Appendix I displays some examples of the unifying groups presently considered.

Now a gauge theory based on a "simple" (or with discrete symmetrics, a "semi-simple") group G contains one basic gauge constant. This constant would manifest itself physically above the "grand unification mass" M, exceeding all particle masses in the theory—these themselves being generated (if possible) hierarchially through a suitable spontaneous symmetry-breaking mechanism.

3. The third crucial development was by Georgi, Quinn and Weinberg (1974)[62] who showed how, using renormalization group ideas, one could relate the observed low-energy couplings $\alpha(\mu)$, and $\alpha_s(\mu)$ ($\mu \sim 100$ GeV) to the magnitude of the grand unifying mass M and the observed value of $\sin^2\theta(\mu)$; ($\tan\theta$ is the ratio of the U(1) to the SU(2) couplings).
4. If one extrapolates with Jowett[6], that nothing essentially new can possibly be discovered—i.e. one assumes that there are no new features, no new forces, or no new "types" of particles to be discovered, till we go beyond the grand unifying energy M—then the Georgi, Quinn, Weinberg method leads to a startling result: this featureless "plateau" with no "new physics" heights to be scaled stretches to fantastically high energies. More precisely, if $\sin^2\theta(\mu)$ is as large as 0.23, then the grand unifying mass M cannot be smaller than 1.3×10^{13} GeV.[63] (Compare with Planck mass $m_P \approx 1.2\times10^{19}$ GeV related to Newton's constant where gravity must come in.)[7] The result follows from the formula[63], [64]

$$\frac{11\alpha}{3\pi}\,\ell n\,\frac{M}{\mu} = \frac{\sin^2\theta(M) - \sin^2\theta(\mu)}{\cos^2\theta(M)}, \qquad \text{(I)}$$

if it is assumed that $\sin^2\theta(M)$—the magnitude of $\sin^2\theta$ for energies of the order of the unifying mass M—equals 3/8 (see Appendix II).

This startling result will be examined more closely in Appendix II. I show there that it is very much a consequence of the assumption that the SU(2)×U(1) symmetry survives intact from the low regime energies μ right upto the grand unifying mass M. I will also show that there already is some experimental indication that this assumption is too strong, and that there may be likely peaks of new physics at energies of 10 TeV upwards (Appendix II).

[6] The universal urge to extrapolate from what we know to-day and to believe that nothing new can possibly be discovered, is well expressed in the following:

"I come first, My name is Jowett
I am the Master of this College,
Everything that is, I know it
If I don't, it isn't knowledge"

—The Balliol Masque.

[7] On account of the relative proximity of $M \approx 10^{13}$ GeV to m_P (and the hope of eventual unification with gravity), Planck mass m_P is now the accepted "natural" mass scale in Particle Physics. With this large mass as the input, the great unsolved problem of Grand Unification is the "natural" emergence of mass hierarchies (m_P, αm_P, $\alpha^2 m_P$, . . .) or $m_P \exp(-c_n/\alpha)$, where c_n's are constants.

$$\left(\frac{m_e}{m_P} \sim 10^{-22}.\right)$$

4.2 *Tests of electronuclear grand unification*

The most characteristic prediction from the existence of the ELECTRONUCLEAR force is proton decay, first discussed in the context of grand unification at the Aix-en-Provence Conference (1973)[65]. For "semi-simple" unifying groups with multiplets containing quarks and leptons only, (but no antiquarks nor antileptons) the lepto-quark composites have masses (determined by renormalization group arguments), of the order of $\approx 10^5-10^6$ GeV[66]. For such theories the characteristic proton decays (proceeding through exchanges of *three* lepto-quarks) conserve quark number+lepton number, i.e. $P = qqq \rightarrow \ell\ell\ell$, $\tau_P \sim 10^{29}-10^{34}$ years. On the contrary, for the "simple" unifying family groups like SU(5)[61] or SO(10)[67] (with multiplets containing antiquarks and antileptons) proton decay proceeds through an exchange of *one* lepto-quark into an antilepton (plus pions etc.) ($P \rightarrow \bar{\ell}$).

An intriguing possibility in this context is that investigated recently for the maximal unifying group SU(16)—the largest group to contain a 16-fold fermionic (q, ℓ, $\bar{q}$, $\bar{\ell}$). This can permit four types of decay modes: $P \rightarrow 3\bar{\ell}$ *as well as* $P \rightarrow \bar{\ell}$, $P \rightarrow \ell$ (e.g. $P \rightarrow \ell^- + \pi^+ + \pi^+$) and $P \rightarrow 3\ell$ (e.g. $N \rightarrow 3\nu + \pi^0$, $P \rightarrow 2\nu + e^+ + \pi^0$), the relative magnitudes of these alternative decays being model-dependent on how precisely SU(16) breaks down to SU(3)×SU(2)×U(1). Quite clearly, it is the central fact of the existence of the proton decay for which the present generation of experiments must be designed, rather than for any specific type of decay modes.

Finally, grand unifying theories predict mass relations like:[68]

$$\frac{m_d}{m_e} = \frac{m_s}{m_\mu} = \frac{m_b}{m_\tau} \approx 2.8$$

for 6 (or at most 8) flavours *below the unification mass*. The important remark for proton decay and for mass relations of the above type as well as for an understanding of baryon excess[69] in the Universe[8], is that for the present *these are essentially characteristic of the fact of grand unification—rather than of specific models*.

"Yet each man kills the thing he loves" sang Oscar Wilde in his famous Ballad of Reading Gaol. Like generations of physicists before us, some in our generation also (through a direct extrapolation of the electroweak gauge methodology to the electronuclear)—and with faith in the assumption of

[8] The calculation of baryon excess in the Universe—arising from a combination of CP and baryon number violations—has recently been claimed to provide teleological arguments for grand unification. For example, Nanopoulos[70] has suggested that the "existence of human beings to measure the ratio n_B/n_γ (where n_B is the numbers of baryons and n_γ the numbers of photons in the Universe) necessarily imposes severe bounds on this quantity: i.e. $10^{-11} \approx (m_e/m_p)^{1/2} \lesssim n_B/n_\gamma \lesssim 10^{-4}$ ($\approx 0(\alpha^2)$)". Of importance in deriving these constraints are the upper (and lower) bound on the numbers of flavours (≈ 6) deduced (1) from mass relations above, (2) from cosmological arguments which seek to limit the numbers of massless neutrinos, (3) from asymptotic freedom and (4) from numerous (one-loop) radiative calculations. It is clear that lack of accelerators as we move up in energy scale will force particle physics to reliance on teleology and cosmology (which in Landau's famous phrase is "often wrong, but never in doubt").

no "new physics", which leads to a grand unifying mass $\sim 10^{13}$ GeV—are beginning to believe that the end of the problems of elementarity as well as of fundamental forces is nigh. They may be right, but before we are carried away by this prospect, it is worth stressing that even for the simplest grand unifying model (Georgi and Glashow's SU(5) with just two Higgs (a 5 and a 24)), the number of presently *ad hoc* parameters needed by the model is still unwholesomely large—22, to compare with 26 of the six-quark model based on the humble $SU(2)\times U(1)\times SU_c(3)$. We cannot feel proud.

V. ELEMENTARITY: UNIFICATION WITH GRAVITY AND NATURE OF CHARGE

In some of the remaining parts of this lecture I shall be questioning two of the notions which have gone into the direct extrapolation of Sec. IV—first, do quarks and leptons represent the correct elementary[9] fields, which should appear in the matter Lagrangian, and which are structureless for renormalizaibility; second, could some of the presently considered gauge fields themselves be composite? This part of the lecture relies heavily on an address I was privileged to give at the European Physical Society meeting in Geneva in July this year.[64]

5.1 *The quest for elementarity, prequarks (preons and pre-preons)*

If quarks and leptons are elementary, we are dealing with $3\times 15 = 45$ elementary entities. The "natural" group of which these constitute the fundamental representation is SU(45) with 2024 elementary gauge bosons. It is possible to reduce the size of this group to SU(11) for example (see Appendix I), with only 120 gauge bosons, but then the number of elementary fermions increases to 561, (of which presumably $3\times 15 = 45$ objects are of low and the rest of Planckian mass). Is there any basic reason for one's instinctive revulsion when faced with these vast numbers of elementary fields.

The numbers by themselves would perhaps not matter so much. After all, Einstein in his description of gravity,[71] chose to work with *10* fields ($g_{\mu\nu}(x)$) rather than with just one (scalar field) as Nordström[72] had done before him. Einstein was not perturbed by the multiplicity he chose to introduce, since he relied on the sheet-anchor of a fundamental principle—(the equivalence principle)—which permitted him to relate the 10 fields for gravity $g_{\mu\nu}$ with the 10 components of the physically relevant quantity, the tensor $T_{\mu\nu}$ of energy and momentum. *Einstein knew that nature was not economical of structures:* only of principles of fundamental applicability. The question we must ask ourselves is this: have we yet discovered such principles in our quest for elementarity, to justify having fields with such large numbers of components as elementary.

[9] I would like to quote Feynman in a recent interview to the "Omni" magazine: "As long as it looks like the way things are built with wheels within wheels, then you are looking for the innermost wheel—but it might not be that way, in which case you are looking for whatever the hell it is you find!". In the same interview he remarks "a few years ago I was very sceptical about the gauge theories . . . I was expecting mist, and now it looks like ridges and valleys after all."

Recall that quarks carry at least three charges (colour, flavour and a family number). Should one not, by now, entertain the notions of quarks (and possibly of leptons) as being composites of some more basic entities[10] (PRE-QUARKS or PREONS), which each carry but *one* basic charge[52]. These ideas have been expressed before but they have become more compulsive now, with the growing multiplicity of quarks and leptons. Recall that it was similar ideas which led from the eight-fold of baryons to a triplet of (Sakatons and) quarks in the first place.

The preon notion is not new. In 1975, among others, Pati, Salam and Strathdee[52] introduced 4 chromons (the fourth colour corresponding to the lepton number) and 4 flavons, the basic group being SU(8)—of which the family group $SU_F(4)\times SU_C(4)$ was but a subgroup. As an extension of these ideas, we now believe these preons carry magnetic charges and are bound together by very strong short-range forces, with quarks and leptons as their magnetically neutral composites[73]. The important remark in this context is that in a theory containing *both* electric and magnetic generalized charges, the analogues of the well-known Dirac quantization condition[74] gives relations like $\frac{eg}{4\pi} = \frac{n}{2}$ for the strength of the two types of charges. Clearly, magnetic monopoles[11] of strength $\pm g$ and mass $\approx m_w/d \approx 10^4$ GeV, are likely to bind much more tightly than electric charges, yielding composites whose non-elementary nature will reveal itself only for very high energies. This appears to be the situation at least for leptons if they are composites.

In another form the preon idea has been revived this year by Curtwright and Freund[52], who motivated by ideas of extended supergravity (to be discussed in the next subsection), reintroduce an SU(8) of 3 chromons (R, Y, B), 2 flavons and 3 familons (horrible names). The family group SU(5) could be a subgroup of this SU(8). In the Curtwright-Freund scheme, the $3\times 15 = 45$ fermions of SU(5)[61] can be found among the 8+28+56 of SU(8) (or alternatively the $3\times 16 = 48$ of SO(10) among the vectorial 56 fermions of SU(8)). (The next succession after the preon level may be the pre-preon level. It was suggested at the Geneva Conference[64] that with certain developments in field theory of composite fields it could be that just two-preons may suffice. But at this stage this is pure speculation.)

Before I conclude this section, I would like to make a prediction regarding the course of physics in the next decade, extrapolating from our past experience of the decades gone by:

[10] One must emphasise however that zero mass neutrinos are the hardest objects to conceive of as composites.

[11] According to 't Hooft's theorem, a monopole corresponding to the $SU_L(2)$ gauge symmetry is expected to possess a mass with the lower limit $\frac{m_W}{\alpha}$.[75], [76] Even if such monopoles are confined, their indirect effects must manifest themselves, if they exist. (Note that $\frac{m_W}{\alpha}$ is very much a lower limit. For a grand unified theory like SU(5) for which the monopole mass is α^{-1} times the heavy lepto-quark mass.) The monopole force may be the techni-colour force of Footnote 5.

DECADE	1950—1960	1960—1970	1970—1980	1980→
Discovery in early part of the decade	The strange particles	The 8-fold way, Ω^-	Confirmation of neutral currents	W, Z, Proton decay
Expectation for the rest of the decade		SU(3) resonances		Grand Unification, Tribal Groups
Actual discovery		Hit the next level of elementarity with quarks		May hit the preon level, and composite structure of quarks

5.2 *Post-Planck physics, supergravity and Einstein's dreams*

I now turn to the problem of a deeper comprehension of the charge concept (the basis of gauging)—*which, in my humble view, is the real quest of particle physics*. Einstein, in the last thirty-five years of his life lived with two dreams: one was to unite gravity with matter (the photon)—he wished to see the "base wood" (as he put it) which makes up the stress tensor $T_{\mu\nu}$ on the right-hand side of his equation $R_{\mu\nu}-\frac{1}{2}g_{\mu\nu}R = -T_{\mu\nu}$ transmuted through this union, into the "marble" of gravity on the left-hand side. The second (and the complementary) dream was to use this unification to comprehend the nature of electric charge in terms of space-time geometry in the same manner as he had successfully comprehended the nature of gravitational charge in terms of space-time curvature.

In case some one imagines[12] that such deeper comprehension is irrelevant to quantitative physics, let me adduce the tests of Einstein's theory versus the proposed modifications to it (Brans-Dicke[77] for example). Recently (1976), the *strong* equivalence principle (i.e. the proposition that gravitational forces contribute equally to the inertial and the gravitational masses) was tested[13] to one part in 10^{12} (i.e. to the same accuracy as achieved in particle physics for $(g\text{-}2)_e$) through lunar-laser ranging measurements[78]. These measurements determined departures from Kepler equilibrium distances, of the moon, the earth and the sun to better than ±30 cms. and triumphantly vindicated Einstein.

There have been four major developments in realizing Einstein's dreams:

1. The Kaluza-Klein[79] miracle: An Einstein Lagrangian (scalar curvature) in five-dimensional space-time (where the fifth dimension is compactified in

[12] The following quotation from Einstein is relevant here. "We now realize, with special clarity, how much in error are those theorists who believe theory comes inductively from experience. Even the great Newton could not free himself from this error (Hypotheses non fingo)." This quote is complementary to the quotation from Einstein at the end of Sec. II.

[13] The *weak* equivalence principle (the proposition that all but the gravitational force contribute equally to the inertial and the gravitational masses) was verified by Eötvös to $1:10^8$ and by Dicke and Braginsky and Panov to $1:10^{12}$.

the sense of all fields being explicitly independent of the fifth co-ordinate) precisely reproduces the *Einstein-Maxwell* theory in four dimensions, the $g_{\mu 5}$ ($\mu = 0, 1, 2, 3$) components of the metric in five dimensions being identified with the Maxwell field A_μ. From this point of view, Maxwell's field is associated with the extra components of curvature implied by the (conceptual) existence of the fifth dimension.

2. The second development is the recent realization by Cremmer, Scherk, Englert, Brout, Minkowski and others that the compactification of the extra dimensions[80]—(their curling up to sizes perhaps smaller than Planck length $\lesssim 10^{-33}$ cms. and the very high curvature associated with them)—might arise through a spontaneous symmetry breaking (in the first 10^{-43} seconds) which reduced the higher dimensional space-time effectively to the four-dimensional that we apprehend directly.
3. So far we have considered Einstein's second dream, i.e. the unification of of electromagnetism (and presumably of other gauge forces) with gravity, giving a space-time significance to gauge charges as corresponding to extended curvature in extra bosonic dimensions. A full realization of the first dream (unification of spinor matter with gravity and with other gauge fields) had to await the development of supergravity[81], [82]—and an extension to extra fermionic dimensions of superspace[83] (with extended torsion being brought into play in addition to curvature). I discuss this development later.
4. And finally there was the alternative suggestion by Wheeler[84] and Schemberg that electric charge may be associated with space-time topology—with worm-holes, with space-time Gruyère-cheesiness. This idea has recently been developed by Hawking[14] and his collaborators[85].

5.3 *Extended supergravity, SU(8) preons and composite gauge fields*

Thus far I have reviewed the developments in respect of Einstein's dreams as reported at the Stockholm Conference held in 1978 in this hall and organized by the Swedish Academy of Sciences.

A remarkable new development was reported during 1979 by Julia and Cremmer[87] which started with an attempt to use the ideas of Kaluza and Klein to formulate extended supergravity theory in a higher (compactified) spacetime—more precisely in eleven dimensions. This development links up, as we shall see, with preons and composite Fermi fields—and even more important—possibly with the notion of composite gauge fields.

Recall that simple supergravity[81] is the gauge theory of supersymmetry[88]—the gauge particles being the (helicity ± 2) gravitons and

[14] The Einstein Langrangian allows large fluctuations of metric and topology on Planck-length scale. Hawking has surmised that the dominant contributions to the path integral of quantum gravity come from metrics which carry one unit of topology per Planck volume. On account of the intimate connection (de Rham, Atiyah-Singer)[86] of curvature with the measures of space-time topology (Euler number, Pontryagin number) the extended Kaluza-Klein and Wheeler-Hawking points of view may find consonance after all.

(helicity $\pm\frac{3}{2}$) gravitinos[15]. *Extended supergravity* gauges supersymmetry combined with SO(N) internal symmetry. For N = 8, the (tribal) supergravity multiplet consists of the following SO(8) families:[81], [87]

Helicity	
± 2	1
$\pm\frac{3}{2}$	8
± 1	28
$\pm\frac{1}{2}$	56
0	70

As is well known, SO(8) is too small to contain $SU(2)\times U(1)\times SU_c(3)$. Thus this tribe has no place for $W^{\pm}$ (though Z^0 and γ are contained) and no places for μ or τ or the t quark.

This was the situation last year. This year, Cremmer and Julia[87] attempted to write down the N = 8 supergravity Langrangian explicitly, using an extension of the Kaluza-Klein ansatz which states that *extended supergravity* (with SO(8) internal symmetry) has the same Lagrangian in four space-time dimensions as *simple supergravity* in (compactified) eleven dimensions. This formal—and rather formidable ansatz—when carried through yielded a most agreeable bonus. *The supergravity Lagrangian possesses an unsuspected SU(8) "local" internal symmetry* although one started with an internal SO(8) only.

The tantalizing questions which now arise are the following.

1. Could this internal SU(8) be the symmetry group of the 8 preons (3 chromons, 2 flavons, 3 familons) introduced earlier?
2. When SU(8) is gauged, there should be 63 spin-one fields. The supergravity tribe contains only 28 spin-one fundamental objects which are not minimally coupled. Are the 63 fields of SU(8) to be identified with composite gauge fields made up of the 70 spin-zero objects of the form $V^{-1}\,\partial_\mu\, V$; Do these composites propagate, in analogy with the well-known recent result in CP^{n-1} theories[89], where a composite gauge field of this form propagates as a consequence of quantum effects (quantum completion)?

The entire development I have described—the unsuspected extension of SO(8) to SU(8) when extra compactified space-time dimensions are used—and the possible existence and quantum propagation of composite gauge fields—is of such crucial importance for the future prospects of gauge theories that one begins to wonder how much of the extrapolation which took $SU(2)\times U(1)\times$

[15] Supersymmetry algebra extends Poincaré group algebra by adjoining to it supersymmetric charges Q_α which transform bosons to fermions, $\{Q_\alpha, Q_\beta\} = (\gamma_\mu P_\mu)_{\alpha\beta}$. The currents which correspond to these charges (Q_α and P_μ) are $J_{\mu\alpha}$ and $T_{\mu\nu}$—these are essentially the currents which in gauged supersymmetry (i.e. supergravity) couple to the gravitino and the graviton, respectively.

$SU_c(3)$ into the electronuclear grand unified theories is likely to remain unaffected by these new ideas now unfolding.

But where in all this is the possibility to appeal directly to experiment? For grand unified theories, it was the proton decay. What is the analogue for supergravity? Perhaps the spin $\frac{3}{2}$ massive gravitino, picking its mass from a super-Higgs effect[90] provides the answer. Fayet[91] has shown that for a spontaneously broken globally supersymmetric weak theory the introduction of a local gravitational interaction leads to a super-Higgs effect. Assuming that supersymmetry breakdown is at mass scale m_W, the gravitino acquires a mass and an effective interaction, but of conventional weak rather than of the gravitational strength—an enhancement by a factor of 10^{34}. One may thus search for the gravitino among the neutral decay modes of J/ψ—the predicted rate being 10^{-3}—10^{-5} times smaller than the observed rate for $J/\psi \rightarrow e^+e^-$. This will surely tax all the ingenuity of the great men (and women) at SLAC and DESY. Another effect suggested by Scherk[92] is antigravity—a cancellation of the attractive gravitational force with the force produced by spin-one graviphotons which exist in all extended supergravity theories, Scherk shows that the Compton wave length of the gravi-photon is either smaller than 5 cms. or comprised between 10 and 850 metres in order that there is no conflict with what is presently known about the strenght of the gravitational force.

Let me summarize: it is conceivable of course, that there is indeed a grand plateau—extending even to Planck energies. If so, the only eventual laboratory for particle physics will be the Early Universe, where we shall have to seek for the answers to the questions on the nature of charge. There may, however, be indications of a next level of structure around 10 TeV; there are also beautiful ideas (like, for example, of electric and magnetic monopole duality) which may manifest at energies of the order of $\alpha^{-1} m_W$ (= 10 TeV). Whether even this level of structure will give us the final clues to the nature of charge, one cannot predict. All I can say is that I am for ever and continually being amazed at the depth revealed at each successive level we explore. I would like to conclude, as I did at the 1978 Stockholm Conference, with a prediction which J. R. Oppenheimer made more than twenty-five years ago and which has been fulfilled to-day in a manner he did not live to see. More than anything else, it expresses the faith for the future with which this greatest of decades in particle physics ends: "Physics will change even more . . . If it is radical and unfamiliar . . . we think that the future will be only more radical and not less, only more strange and not more familiar, and that it will have its own new insights for the inquiring human spirit."

J. R. Oppenheimer
Reith Lectures BBC 1953

APPENDIX I. EXAMPLES OF GRAND UNIFYING GROUPS

*Semi-simple groups**	*Multiplet*	*Exotic gauge particles*	*Proton decay*
(with left-right symmetry)	$G_L \to \begin{pmatrix} q \\ \ell \end{pmatrix}_L$, $G_R \to \begin{pmatrix} q \\ \ell \end{pmatrix}_R$	Lepto-quarks→$(\bar{q}\ell)$	Lepto-quarks→W +(Higgs) or
Example $[SU(6)_F \times SU(6)_c]_{L\to R}$	$G = G_L \times G_R$	Unifying mass $\approx 10^6$ GeV	Proton = qqq→$\ell\ell\ell$
Simple groups		Diquarks→(qq)	qq→$\bar{q}\ell$ i.e.
Examples		Dileptons→$(\ell\ell)$	Proton P = qqq→$\bar{\ell}$
Family groups→ SU(5) ↓	or SO(10) ↓	Lepto-quarks→$(\bar{q}\ell)$, $(q\ell)$	Also possible, P→ℓ, P→$3\bar{\ell}$,
Tribal groups→ SU(11)	SO(22) $\quad G \to \begin{pmatrix} q \\ \ell \\ \bar{q} \\ \bar{\ell} \end{pmatrix}_L$	Unifying mass $\approx 10^{13}-10^{15}$ GeV	P→3ℓ

APPENDIX II

The following assumptions went into the derivation of the formula (I) in the text.

a) $SU_L(2)\times U_{L,R}(1)$ survives intact as the electroweak symmetry group from energies $\approx \mu$ right upto M. This intact survival implies that one eschews, for example, all suggestions that i) low-energy $SU_L(2)$ may be the diagonal sum of $SU_L^{I}(2)$, $SU_L^{II}(2)$, $SU_L^{III}(2)$, where I, II, III refer to the (three?) known families; ii) or that the $U_{L,R}(1)$ is a sum of pieces, where $U_R(1)$ may have differentially descended from a (V+A)-symmetric $SU_R(2)$ contained in G, or iii) that U(1) contains a piece from a four-colour symmetry $SU_c(4)$ (with lepton number as the fourth colour) and with $SU_c(4)$ breaking at an intermediate mass scale to $SU_c(3)\times U_c(1)$.

b) The second assumption which goes into the derivation of the formula above is that there are no unexpected heavy fundamental fermions, which might make $\sin^2\theta(M)$ differ from $\frac{3}{8}$—its value for the low mass fermions presently known to exist.$^+$

* Grouping quarks (q) and leptons (ℓ) together, implies treating lepton number as the fourth colour, i.e. $SU_c(3)$ extends to $SU_c(4)$ (Pati and Salam)[93]. A Tribal group, by definition, contains all known families in its basic representation. Favoured representations of Tribal SU(11) (Georgi)[94] and Tribal SO(22) (Gell-Mann[95] *et al.*) contain 561 and 2048 fermions!

$^+$ If one does not know G, one way to infer the parameter $\sin^2\theta(M)$ is from the formula:

$$\sin^2\theta(M) = \frac{\Sigma T^2_{3L}}{\Sigma Q^2}\left(= \frac{9\,N_q + 3\,N_\ell}{20\,N_q + 12\,N_\ell}\right).$$

Here N_q and N_ℓ are the numbers of the fundamental quark and lepton SU(2) doublets (assuming these are the only multiplets that exist). If we make the further *assumption* that $N_q = N_\ell$ (from the requirement of anomaly cancellation between quarks and leptons) we obtain $\sin^2\theta(M) = \frac{3}{8}$. This assumption however is not compulsive; for example anomalies cancel also if (heavy) mirror fermions exist[98]. This is the case for $[SU(6)]^4$ for which $\sin^2\theta(M) = \frac{9}{28}$.

c) If these assumptions are relaxed, for example, for the three family group $G = [SU_F(6) \times SU_c(6)]_{L \to R}$, where $\sin^2\theta(M) = \frac{9}{28}$, we find the grand unifying mass M tumbles down to 10^6 GeV.

d) The introduction of intermediate mass scales (for example, those connoting the breakdown of family universality, or of left-right symmetry, or of a breakdown of 4-colour $SU_c(4)$ down to $SU_c(3) \times U_c(1)$) will as a rule push the magnitude of the grand unifying mass M upwards[96]. In order to secure a proton decay life, consonant with present empirical lower limits ($\sim 10^{30}$ years)[97] this is desirable anyway. (τ_{proton} for $M \sim 10^{13}$ GeV is unacceptably low $\sim 6 \times 10^{23}$ years unless there are 15 Higgs.) There is, from this point of view, an indication of there being in Particle Physics one or several intermediate mass scales which can be shown to start from around 10^4 GeV upwards. *This is the end result which I wished this Appendix to lead up to.*

REFERENCES

1. Lee, T. D. and Yang, C. N., Phys. Rev. *104,* 254 (1956).
2. Abdus Salam, Nuovo Cimento *5,* 299 (1957).
3. Wu, C. S., *et al.,* Phys. Rev. *105,* 1413 (1957).
4. Garwin, R., Lederman, L. and Weinrich, M., Phys. Rev. *105,* 1415 (1957).
5. Friedman, J. I. and Telegdi, V. L., Phys. Rev. *105,* 1681 (1957).
6. Landau, L., Nucl. Phys. *3,* 127 (1957).
7. Lee, T. D. and Yang, C. N., Phys. Rev. *105,* 1671 (1957).
8. Abdus Salam, Imperial College, London, preprint (1957). For reference, see Footnote 7, p. 89, of *Theory of Weak Interactions in Particle Physics,* by Marshak, R. E., Riazuddin and Ryan, C. P., (Wiley-Interscience, New York 1969), and W. Pauli's letters (CERN Archives).
9. Yang, C. N. and Mills, R. L., Phys. Rev. *96,* 191 (1954).
10. Shaw, R., "The problem of particle types and other contributions to the theory of elementary particles", Cambridge Ph. D. Thesis (1955), unpublished.
11. Marshak, R. E. and Sudarshan, E. C. G., Proc. Padua-Venice Conference on Mesons and Recently Discovered Particles (1957), and Phys. Rev. *109,* 1860 (1958). The idea of a universal Fermi interaction for (P,N), (υ_e,e) and (Υ_μ,μ) doublets goes back to Tiomno, J. and Wheeler, J. A., Rev. Mod. Phys. *21,* 144 (1949); *21,* 153 (1949) and by Yang, C. N. and Tiomno, J., Phys. Rev. *75,* 495 (1950). Tiomno, J., considered γ_5 transformations of Fermi fields linked with mass reversal in Il Nuovo Cimento *1,* 226 (1956).
12. Feynman, R. P. and Gell-Mann, M., Phys. Rev. *109,* 193 (1958).
13. Sakurai, J. J., Nuovo Cimento *7,* 1306 (1958).
14. Nambu, Y. and Jona-Lasinio, G., Phys. Rev. *122,* 345 (1961).
15. Nambu, Y., Phys. Rev. Letters *4,* 380 (1960); Goldstone, J., Nuovo Cimento *19,* 154 (1961).
16. Klein, O., "On the theory of charged fields", Proceedings of the Conference organized by International Institute of Intellectual Cooperation, Paris (1939).
17. Kemmer, N., Phys. Rev. *52,* 906 (1937).
18. Schwinger, J., Ann. Phys. (N. Y.) *2,* 407 (1957).
19. Glashow, S. L., Nucl. Phys. *10,* 107 (1959).
20. Abdus Salam and Ward, J. C., Nuovo Cimento *11,* 568 (1959).
21. Bludman, S., Nuovo Cimento *9,* 433 (1958).

22. Abdus Salam and Ward, J. C., Nuovo Cimento *19,* 165 (1961).
23. Glashow, S. L., Nucl. Phys. *22,* 579 (1961).
24. Abdus Salam and Ward, J. C., Phys. Letters *13,* 168 (1964).
25. Goldstone, J., see Ref. 15.
26. Goldstone, J., Abdus Salam and Weinberg, S., Phys. Rev. *127,* 965 (1962).
27. Anderson, P. W., Phys. Rev. *130,* 439 (1963).
28. Higgs, P. W., Phys. Letters *12,* 132 (1964); Phys. Rev. Letters *13,* 508 (1964); Phys. Rev. *145,* 1156 (1966).
29. Englert, F. and Brout, R., Phys. Rev. Letters *13,* 321 (1964);
 Englert, F., Brout, R. and Thiry, M. F., Nuovo Cimento *48,* 244 (1966).
30. Guralnik, G. S., Hagen, C. R. and Kibble, T. W. B., Phys. Rev. Letters *13,* 585 (1964); Kibble, T. W. B., Phys. Rev. *155,* 1554 (1967).
31. Weinberg, S., Phys. Rev. Letters *27,* 1264 (1967).
32. Abdus Salam, Proceedings of the 8th Nobel Symposium, Ed. Svartholm, N., (Almqvist and Wiksell, Stockholm 1968).
33. 't Hooft, G., Nucl. Phys. *B33,* 173 (1971); *ibid. B35,* 167 (1971).
34. Lee, B. W., Phys. Rev. *D5,* 823 (1972); Lee, B. W. and Zinn-Justin, J., Phys. Rev. *D5,* 3137 (1972); *ibid. D7,* 1049 (1973).
35. 't Hooft, G. and Veltman, M., Nucl. Phys. *B44,* 189 (1972); *ibid. B50,* 318 (1972). An important development in this context was the invention of the dimensional regularization technique by Bollini, C. and Giambiagi, J., Nuovo Cimento *B12,* 20 (1972);
 Ashmore, J., Nuovo Cimento Letters *4,* 289 (1972) and by 't Hooft, G. and Veltman, M.
36. Feynman, R. P., Acta Phys. Polonica *24,* 297 (1963).
37. DeWitt, B. S., Phys. Rev. *162,* 1195 and 1239 (1967).
38. Faddeev, L. D. and Popov, V. N., Phys. Letters *25B,* 29 (1967).
39. Mandelstam, S., Phys. Rev. *175,* 1588 and 1604 (1968).
40. Fradkin, E. S. and Tyutin, I. V., Phys. Rev. *D2,* 2841 (1970).
41. Boulware, D. G., Ann. Phys. (N.Y.) *56,* 140 (1970).
42. Taylor, J. C., Nucl. Phys. *B33,* 436 (1971).
43. Slavnov, A., Theor. Math. Phys. *10,* 99 (1972).
44. Abdus Salam and Strathdee, J., Phys. Rev. *D2,* 2869 (1970).
45. Glashow, S., Iliopoulos, J. and Maiani, L., Phys. Rev. *D2,* 1285 (1970).
46. For a review, see Jackiw, R., in *Lectures on Current Algebra and its Applications,* by Treiman, S. B., Jackiw, R. and Gross, D. J., (Princeton Univ. Press, 1972).
47. Bouchiat, C., Iliopoulos, J. and Meyer, Ph., Phys. Letters *38B,* 519 (1972); Gross, D. J. and Jackiw, R., Phys. Rev. *D6,* 477 (1972).
48. Wentzel, G., Helv. Phys. Acta *10,* 108 (1937).
49. Hasert, F. J., *et al.,* Phys. Letters *46B,* 138 (1973).
50. Taylor, R. E., Proceedings of the 19th International Conference on High Energy Physics, Tokyo, Physical Society of Japan, 1979, p. 422.
51. Barkov, L. M., Proceedings of the 19th International Conference on High Energy Physics, Tokyo, Physical Society of Japan, 1979, p. 425.
52. Pati, J. C. and Abdus Salam, ICTP, Trieste, IC/75/106, Palermo Conference, June 1975; Pati, J. C., Abdus Salam and Strathdee, J., Phys. Letters *59B,* 265 (1975); Harari, H., Phys. Letters *86B,* 83 (1979); Schupe, M., *ibid. 86B,* 87 (1979); Curtwright, T. L. and Freund, P. G.O., Enrico Fermi Inst. preprint EFI 79/25, University of Chicago, April 1979.
53. Pati, J. C. and Abdus Salam, see the review by Bjorken, J. D., Proceedings of the 16th International Conference on High Energy Physics, Chicago-Batavia, 1972, Vol. 2, p. 304; Fritsch, H. and Gell-Mann, M., *ibid.* p. 135; Fritzsch, H., Gell-Mann, M. and Leutwyler, H., Phys. Letters *47B,* 365 (1973);
 Weinberg, S., Phys. Rev. Letters *31,* 494 (1973); Phys. Rev. *D8,* 4482 (1973); Gross, D. J. and Wilczek, F., Phys. Rev. *D8,* 3633 (1973); For a review see Marciano, W. and Pagels, H., Phys. Rep. *36C,* 137 (1978).
54. Tasso Collaboration, Brandelik *et al.,* Phys. Letters *86B,* 243 (1979); Mark-J. Collaboration, Barber *et al.,* Phys. Rev. Letters *43,* 830 (1979); See also reports of the Jade, Mark-J,

Pluto and Tasso Collaborations to the International Symposium on Lepton and Photon Interactions at High Energies, Fermilab, August 1979.
55. Winter, K., Proceedings of the International Symposium on Lepton and Photon Interactions at High Energies, Fermilab, August 1979.
56. Ellis, J., Proceedings of the "Neutrino-79" International Conference on Neutrinos, Weak Interactions and Cosmology, Bergen, June 1979.
57. Dimopoulos, S. and Susskind, L., Nucl. Phys. *B155,* 237 (1979).
58. Weinberg, S., Phys. Rev. *D19,* 1277 (1979).
59. Pati, J. C. and Abdus Salam, Phys. Rev. *D10,* 275 (1974);
Mohapatra, R. N. and Pati, J. C., Phys. Rev. *D11,* 566, 2558 (1975);
Elias, V., Pati, J. C. and Abdus Salam, Phys. Letters *73B,* 450 (1978);
Pati, J. C. and Rajpoot, S., Phys. Letters *79B,* 65 (1978).
60. See Pati, J. C. and Abdus Salam, Ref. 53 above and Pati, J. C. and Abdus Salam, Phys. Rev. *D8,* 1240 (1973).
61. Georgi, H. and Glashow, S. L., Phys. Rev. Letters *32,* 438 (1974).
62. Georgi, H., Quinn, H. R. and Weinberg, S., Phys. Rev. Letters *33,* 451 (1974).
63. Marciano, W. J., Phys. Rev. *D20,* 274 (1979).
64. See Abdus Salam, Proceedings of the European Physical Society Conference, Geneva, August 1979, ICTP, Trieste, preprint IC/79/142, with references to H. Harari's work.
65. Pati, J. C. and Abdus Salam, Phys, Rev. Letters *31,* 661 (1973).
66. Elias, V., Pati, J. C. and Abdus Salam, Phys. Rev. Letters *40,* 920 (1978);
Rajpoot, S. and Elias, V., ICTP, Trieste, preprint IC/78/159.
67. Fritzsch, H. and Minkowski, P., Ann. Phys. (N. Y.) *93,* 193 (1975); Nucl. Phys. *B103,* 61 (1976);
Georgi, H., Particles and Fields (APS/OPF Williamsburg), Ed. Carlson, C. E., AIP, New York, 1975, p. 575;
Georgi, H. and Nanopoulos, D. V., Phys. Letters *82B,* 392 (1979).
68. Buras, A., Ellis, J., Gaillard, M. K. and Nanopoulos, D. V., Nucl. Phys. *B135,* 66 (1978).
69. Yoshimura, M., Phys. Rev. Letters *41,* 381 (1978);
Dimopoulos, S. and Susskind, L., Phys. Rev. *D18,* 4500 (1978);
Toussaint, B., Treiman, S. B., Wilczek, F. and Zee, A., Phys. Rev. *D19,* 1036 (1979);
Ellis J., Gaillard, M. K. and Nanopoulos, D. V., Phys. Letters *80B,* 360 (1979);
Erratum *82B,* 464 (1979);
Weinberg, S., Phys. Rev. Letters *42,* 850 (1979); Nanopoulos, D. V. and Weinberg, S., Harvard University preprint HUTP-79/A023.
70. Nanopoulos, D. V., CERN preprint TH2737, September 1979.
71. Einstein, A., Annalen der Phys. *49,* 769 (1916). For an English translation, see *The Principle of Relativity* (Methuen, 1923, reprinted by Dover Publications), p. 35.
72. Nordström, G., Phys. Z. *13,* 1126 (1912); Ann. Phys. (Leipzig) *40,* 856 (1913); *ibid. 42,* 533 (1913); *ibid. 43,* 1101 (1914); Phys. Z. *15,* 375 (1914); Ann. Acad. Sci. Fenn. *57* (1914, 1915);
See also Einstein, A., Ann. Phys. Leipzig *38,* 355, 433 (1912).
73. Pati, J. C. and Abdus Salam, in preparation.
74. Dirac, P. A. M., Proc. Roy. Soc. (London) *A133,* 60 (1931).
75. 't Hooft, G., Nucl. Phys. *B79,* 276 (1974).
76. Polyakov, A. M., JETP Letters *20,* 194 (1974).
77. Brans, C. H. and Dicke, R. H., Phys. Rev. *124,* 925 (1961).
78. Williams, J. G., *et al.,* Phys. Rev. Letters *36,* 551 (1976);
Shapiro, I. I., *et al.,* Phys. Rev. Letters *36,* 555 (1976);
For a discussion, see Abdus Salam, in *Physics and Contemporary Needs* Ed. Riazuddin (Plenum Publishing Corp., 1977), p. 301.
79. Kaluza, Th., Sitzungsber. Preuss. Akad. Wiss. p. 966 (1921);
Klein, O., Z. Phys. *37,* 895 (1926).
80. Cremmer, E. and Scherk, J., Nucl. Phys. *B103,* 399 (1976); *ibid. B108,* 409 (1976); *ibid. B118,* 61 (1976);
Minkowski, P., Univ. of Berne preprint, October 1977.

81. Freedman, D. Z., van Nieuwenhuizen, P. and Ferrara, S., Phys. Rev. *D13,* 3214 (1976); Deser, S. and Zumino, B., Phys. Letters *62B,* 335 (1976);
For a review and comprehensive list of references, see D. Z. Freedman's presentation to the 19th International Conference on High Energy Physics, Tokyo, Physical Society of Japan, 1979.
82. Arnowitt, R., Nath, P. and Zumino, B., Phys. Letters *56B,* 81 (1975); Zumino, B., in Proceedings of the Conference on Gauge Theories and Modern Field Theory, Northeastern University, September 1975, Eds. Arnowitt, R. and Nath, P., (MIT Press);
Wess, J. and Zumino, B., Phys. Letters *66B,* 361 (1977);
Akulov, V. P., Volkov, D. V. and Soroka, V. A., JETP Letters *22,* 187 (1975);
Brink, L., Gell-Mann, M., Ramond, P. and Schwarz, J. H., Phys. Letters *74B,* 336 (1978);
Taylor, J. G., King's College, London, preprint, 1977 (unpublished);
Siegel, W., Harvard University preprint HUTP-77/AO68, 1977 (unpublished);
Ogievetsky, V. and Sokatchev, E., Phys. Letters *79B,* 222 (1978);
Chamseddine, A. H. and West, P. C., Nucl. Phys. *B129,* 39 (1977);
MacDowell, S. W. and Mansouri, F., Phys. Rev. Letters *38,* 739 (1977).
83. Abdus Salam and Strathdee, J., Nucl. Phys. *B79,* 477 (1974).
84. Fuller, R. W. and Wheeler, J. A., Phys. Rev. *128,* 919 (1962);
Wheeler, J. A., in ***Relativity Groups and Topology,*** Proceedings of the Les Houches Summer School, 1963, Eds. DeWitt, B. S. and DeWitt, C. M., (Gordon and Breach, New York 1964).
85. Hawking, S. W., in ***General Relativity: An Einstein Centenary Survey*** (Cambridge University Press, 1979);
See also "Euclidean quantum gravity", DAMTP, Univ. of Cambridge preprint, 1979;
Gibbons, G. W., Hawking, S. W. and Perry, M. J., Nucl. Phys. *B138,* 141 (1978);
Hawking, S. W., Phys. Rev. *D18,* 1747 (1978).
86. Atiyah, M. F. and Singer, I. M., Bull. Am. Math. Soc. *69,* 422 (1963).
87. Cremmer, E., Julia, B. and Scherk, J., Phys. Letters *76B,* 409 (1978); Cremmer, E. and Julia, B., Phys. Letters *80B,* 48 (1978); Ecole Normale Supérieure preprint, LPTENS 79/6, March 1979;
See also Julia, B., in Proceedings of the Second Marcel Grossmann Meeting, Trieste, July 1979 (in preparation).
88. Gol'fand, Yu. A. and Likhtman, E. P., JETP Letters *13,* 323 (1971);
Volkov, D. V. and Akulov, V. P., JETP Letters *16,* 438 (1972);
Wess, J. and Zumino, B., Nucl. Phys. *B70,* 39 (1974);
Abdus Salam and Strathdee, J., Nucl. Phys. *B79,* 477 (1974); *ibid. B80,* 499 (1974); Phys. Letters *51B,* 353 (1974);
For a review, see Abdus Salam and Strathdee, J., Fortschr. Phys. *26,* 57 (1978).
89. D'Adda, A., Lüscher, M. and Di Vecchia, P., Nucl. Phys. *B146,* 63 (1978).
90. Cremmer, E., *et al.,* Nucl. Phys. *B147,* 105 (1979);
See also Ferrara, S., in Proceedings of the Second Marcel Grossmann Meeting, Trieste, July 1979 (in preparation), and references therein.
91. Fayet, P., Phys. Letters *70B,* 461 (1977); *ibid. 84B,* 421 (1979).
92. Scherk, J., Ecole Normale Supérieure preprint, LPTENS 79/17, September 1979.
93. Pati, J. C. and Abdus Salam, Phys. Rev. *D10,* 275 (1974).
94. Georgi, H., Harvard University Report No. HUTP-29/AO13 (1979).
95. Gell-Mann, M., (unpublished).
96. See Ref. 64 above and also Shafi, Q. and Wetterich, C., Phys. Letters *85B,* 52 (1979).
97. Learned, J., Reines, F. and Soni, A., Phys. Letters *43,* 907 (1979).
98. Pati, J. C., Abdus Salam and Strathdee, J., Nuovo Cimento *26A,* 72 (1975);
Pati, J. C. and Abdus Salam, Phys. Rev. *D11,* 1137, 1149 (1975);
Pati, J. C., invited talk, Proceedings Second Orbis Scientiae, Coral Gables, Florida, 1975, Eds. Perlmutter, A. and Widmayer, S., p. 253.

Steven Weinberg

STEVEN WEINBERG

I was born in 1933 in New York City to Frederick and Eva Weinberg. My early inclination toward science received encouragement from my father, and by the time I was 15 or 16 my interests had focused on theoretical physics.

I received my undergraduate degree from Cornell in 1954, and then went for a year of graduate study to the Institute for Theoretical Physics in Copenhagen (now the Niels Bohr Institute). There, with the help of David Frisch and Gunnar Källén. I began to do research in physics. I then returned to the U.S. to complete my graduate studies at Princeton. My Ph.D thesis, with Sam Treiman as adviser, was on the application of renormalization theory to the effects of strong interactions in weak interaction processes.

After receiving my Ph.D. in 1957, I worked at Columbia and then from 1959 to 1966 at Berkeley. My research during this period was on a wide variety of topics – high energy behavior of Feynman graphs, second-class weak interaction currents, broken symmetries, scattering theory, muon physics, etc. – topics chosen in many cases because I was trying to teach myself some area of physics. My active interest in astrophysics dates from 1961–62; I wrote some papers on the cosmic population of neutrinos and then began to write a book, *Gravitation and Cosmology,* which was eventually completed in 1971. Late in 1965 I began my work on current algebra and the application to the strong interactions of the idea of spontaneous symmetry breaking.

From 1966 to 1969, on leave from Berkeley, I was Loeb Lecturer at Harvard and then visiting professor at M.I.T. In 1969 I accepted a professorship in the Physics Department at M.I.T., then chaired by Viki Weisskopf. It was while I was a visitor to M.I.T. in 1967 that my work on broken symmetries, current algebra, and renormalization theory turned in the direction of the unification of weak and electromagnetic interactions. In 1973, when Julian Schwinger left Harvard, I was offered and accepted his chair there as Higgins Professor of Physics, together with an appointment as Senior Scientist at the Smithsonian Astrophysical Observatory.

My work during the 1970's has been mainly concerned with the implications of the unified theory of weak and electromagnetic interactions, with the development of the related theory of strong interactions known as quantum chromodynamics, and with steps toward the unification of all interactions.

In 1982 I moved to the physics and astronomy departments of the University of Texas at Austin, as Josey Regental Professor of Science.

I met my wife Louise when we were undergraduates at Cornell, and we were married in 1954. She is now a professor of law. Our daughter Elizabeth was born in Berkeley in 1963.

Awards and Honors

Honorary Doctor of Science degrees, University of Chicago, Knox College, City University of New York, University of Rochester, Yale University
American Academy of Arts and Sciences, elected 1968
National Academy of Sciences, elected 1972
J. R. Oppenheimer Prize, 1973
Richtmeyer Lecturer of Am. Ass'n. of Physics Teachers, 1974
Scott Lecturer, Cavendish Laboratory, 1975
Dannie Heineman Prize for Mathematical Physics, 1977
Silliman Lecturer, Yale University, 1977
Am. Inst. of Physics — U.S. Steel Foundation Science Writing Award, 1977, for authorship of *The First Three Minutes* (1977)
Lauritsen Lecturer, Cal. Tech., 1979
Bethe Lecturer, Cornell Univ., 1979
Elliott Cresson Medal (Franklin Institute), 1979
Nobel Prize in Physics, 1979

Awards and Honors since 1979

Honorary Doctoral degrees, Clark University, City University of New York, Dartmouth College, Weizmann Institute, Clark University, Washington College, Columbia University
Elected to American Philosophical Society, Royal Society of London (Foreign Honorary Member), Philosophical Society of Texas
Henry Lecturer, Princeton University, 1981
Cherwell-Simon Lecturer, University of Oxford, 1983
Bampton Lecturer, Columbia University, 1983
Einstein Lecturer, Israel Academy of Arts & Sciences, 1984
McDermott Lecturer, University of Dallas, 1985
Hilldale Lecturer, University of Wisconsin, 1985
Clark Lecturer, University of Texas at Dallas, 1986
Brickweede Lecturer, Johns Hopkins University, 1986
Dirac Lecturer, University of Cambridge, 1986
Klein Lecturer, University of Stockholm, 1989
James Madison Medal of Princeton University, 1991
National Medal of Science, 1991

CONCEPTUAL FOUNDATIONS OF THE UNIFIED THEORY OF WEAK AND ELECTROMAGNETIC INTERACTIONS

Nobel Lecture, December 8, 1979
by STEVEN WEINBERG
Lyman Laboratory of Physics Harvard University and Harvard-Smithsonian Center for Astrophysics Cambridge, Mass., USA.

Our job in physics is to see things simply, to understand a great many complicated phenomena in a unified way, in terms of a few simple principles. At times, our efforts are illuminated by a brilliant experiment, such as the 1973 discovery of neutral current neutrino reactions. But even in the dark times between experimental breakthroughs, there always continues a steady evolution of theoretical ideas, leading almost imperceptibly to changes in previous beliefs. In this talk, I want to discuss the development of two lines of thought in theoretical physics. One of them is the slow growth in our understanding of symmetry, and in particular, broken or hidden symmetry. The other is the old struggle to come to terms with the infinities in quantum field theories. To a remarkable degree, our present detailed theories of elementary particle interactions can be understood deductively, as consequences of symmetry principles and of a principle of renormalizability which is invoked to deal with the infinities. I will also briefly describe how the convergence of these lines of thought led to my own work on the unification of weak and electromagnetic interactions. For the most part, my talk will center on my own gradual education in these matters, because that is one subject on which I can speak with some confidence. With rather less confidence, I will also try to look ahead, and suggest what role these lines of thought may play in the physics of the future.

Symmetry principles made their appearance in twentieth century physics in 1905 with Einstein's identification of the invariance group of space and time. With this as a precedent, symmetries took on a character in physicists' minds as *a priori* principles of universal validity, expressions of the simplicity of nature at its deepest level. So it was painfully difficult in the 1930's to realize that there are internal symmetries, such as isospin conservation, [1] having nothing to do with space and time, symmetries which are far from self-evident, and that only govern what are now called the strong interactions. The 1950's saw the discovery of another internal symmetry – the conservation of strangeness [2] – which is not obeyed by the weak interactions, and even one of the supposedly sacred symmetries of space-time – parity – was also found to be violated by weak interactions. [3] Instead of moving toward unity, physicists were learning that different interactions are apparently governed by quite different symmetries. Matters became yet more confusing with the recognition in the early 1960's of a symmetry group – the "eightfold way" – which is not even an exact symmetry of the strong interactions. [4]

These are all "global" symmetries, for which the symmetry transformations do not depend on position in space and time. It had been recognized [5] in the 1920's that quantum electrodynamics has another symmetry of a far more powerful kind, a "local" symmetry under transformations in which the electron field suffers a phase change that can vary freely from point to point in space-time, and the electromagnetic vector potential undergoes a corresponding gauge transformation. Today this would be called a U(1) gauge symmetry, beacause a simple phase change can be thought of as multiplication by a 1×1 unitary matrix. The extension to more complicated groups was made by Yang and Mills [6] in 1954 in a seminal paper in which they showed how to construct an SU(2) gauge theory of strong interactions. (The name "SU(2)" means that the group of symmetry transformations consists of 2×2 unitary matrices that are "special," in that they have determinant unity). But here again it seemed that the symmetry if real at all would have to be approximate, because at least on a naive level gauge invariance requires that vector bosons like the photon would have to be massless, and it seemed obvious that the strong interactions are not mediated by massless particles. The old question remained: if symmetry principles are an expression of the simplicity of nature at its deepest level, then how can there be such a thing as an approximate symmetry? Is nature only approximately simple?

Some time in 1960 or early 1961, I learned of an idea which had originated earlier in solid state physics and had been brought into particle physics by those like Heisenberg, Nambu, and Goldstone, who had worked in both areas. It was the idea of "broken symmetry," that the Hamiltonian and commutation relations of a quantum theory could possess an exact symmetry, and that the physical states might nevertheless not provide neat representations of the symmetry. In particular, a symmetry of the Hamiltonian might turn out to be not a symmetry of the vacuum.

As theorists sometimes do, I fell in love with this idea. But as often happens with love affairs, at first I was rather confused about its implications. I thought (as turned out, wrongly) that the approximate symmetries – parity, isospin, strangeness, the eight-fold way – might really be exact *a priori* symmetry principles, and that the observed violations of these symmetries might somehow be brought about by spontaneous symmetry breaking. It was therefore rather disturbing for me to hear of a result of Goldstone, [7] that in at least one simple case the spontaneous breakdown of a continuous symmetry like isospin would necessarily entail the existence of a massless spin zero particle – what would today be called a "Goldstone boson." It seemed obvious that there could not exist any new type of massless particle of this sort which would not already have been discovered.

I had long discussions of this problems with Goldstone at Madison in the summer of 1961, and then with Salam while I was his guest at Imperial College in 1961–62. The three of us soon were able to show that Goldstone bosons must in fact occur whenever a symmetry like isospin or strangeness is spontaneously broken, and that their masses then remain zero to all orders of perturbation theory. I remember being so discouraged by these zero masses that when we wrote our joint paper on the subject, [8] I added an epigraph to the paper to underscore the futility of supposing that anything could be explained in terms of a non-invariant vacuum state: it was Lear's retort to Cordelia, "Nothing will come of nothing: speak again." Of course, The Physical Review protected the purity of the physics literature, and removed the quote. Considering the future of the non-invariant vacuum in theoretical physics, it was just as well.

There was actually an exception to this proof, pointed out soon afterwards by Higgs, Kibble, and others. [9] They showed that if the broken symmetry is a local, gauge symmetry, like electromagnetic gauge invariance, then although the Goldstone bosons exist formally, and are in some sense real, they can be eliminated by a gauge transformation, so that they do not appear as physical particles. The missing Goldstone bosons appear instead as helicity zero states of the vector particles, which thereby acquire a mass.

I think that at the time physicists who heard about this exception generally regarded it as a technicality. This may have been because of a new development in theoretical physics which suddenly seemed to change the role of Goldstone bosons from that of unwanted intruders to that of welcome friends.

In 1964 Adler and Weisberger [10] independently derived sum rules which gave the ratio g_A/g_V of axial-vector to vector coupling constants in beta decay in terms of pion-nucleon cross sections. One way of looking at their calculation, (perhaps the most common way at the time) was as an analogue to the old dipole sum rule in atomic physics: a complete set of hadronic states is inserted in the commutation relations of the axial vector currents. This is the approach memorialized in the name of "current algebra." [11] But there was another way of looking at the Adler-Weisberger sum rule. One could suppose that the strong interactions have an approximate symmetry, based on the group $SU(2) \times SU(2)$, and that this symmetry is spontaneously broken, giving rise among other things to the nucleon masses. The pion is then identified as (approximately) a Goldstone boson, with small non-zero mass, an idea that goes back to Nambu. [12] Although the $SU(2) \times SU(2)$ symmetry is spontaneously broken, it still has a great deal of predictive power, but its predictions take the form of approximate formulas, which give the matrix elements for low energy pionic reactions. In this approach, the Adler-Weisberger sum rule is obtained by using the predicted pion nucleon scattering lengths in conjunction with a well-known sum rule [13], which years earlier had been derived from the dispersion relations for pion-nucleon scattering.

In these calculations one is really using not only the fact that the strong interactions have a spontaneously broken approximate SU(2) × SU(2) symmetry, but also that the currents of this symmetry group are, up to an overall constant, to be identified with the vector and axial vector currents of beta decay. (With this assumption g_A/g_V gets into the picture through the Goldberger-Treiman relation, [14] which gives g_A/g_V in terms of the pion decay constant and the pion nucleon coupling.) Here, in this relation between the currents of the symmetries of the strong interactions and the physical currents of beta decay, there was a tantalizing hint of a deep connection between the weak interactions and the strong interactions. But this connection was not really understood for almost a decade.

I spent the years 1965–67 happily developing the implications of spontaneous symmetry breaking for the strong interactions. [15] It was this work that led to my 1967 paper on weak and electromagnetic unification. But before I come to that I have to go back in history and pick up one other line of though, having to do with the problem of infinities in quantum field theory.

I believe that it was Oppenheimer and Waller in 1930 [16] who independently first noted that quantum field theory when pushed beyond the lowest approximation yields ultraviolet divergent results for radiative self energies. Professor Waller told me last night that when he described this result to Pauli, Pauli did not believe it. It must have seemed that these infinities would be a disaster for the quantum field theory that had just been developed by Heisenberg and Pauli in 1929–30. And indeed, these infinites did lead to a sense of discouragement about quantum field theory, and many attempts were made in the 1930's and early 1940's to find alternatives. The problem was solved (at least for quantum electrodynamics) after the war, by Feynman, Schwinger, and Tomonaga [17] and Dyson [19]. It was found that all infinities disappear if one identifies the observed finite values of the electron mass and charge, not with the parameters m and e appearing in the Lagrangian, but with the electron mass and charge that are *calculated* from m and e, when one takes into account the fact that the electron and photon are always surrounded with clouds of virtual photons and electron-positron pairs [18]. Suddenly all sorts of calculations became possible, and gave results in spectacular agreement with experiment.

But even after this success, opinions differed as to the significance of the ultraviolet divergences in quantum field theory. Many thought–and some still do think–that what had been done was just to sweep the real problems under the rug. And it soon became clear that there was only a limited class of so-called "renormalizable" theories in which the infinities could be eliminated by absorbing them into a redefinition, or a "renormalization," of a finite number of physical parameters. (Roughly speaking, in renormalizable theories no coupling constants can have the dimensions of negative powers of mass. But every time we add a field or a space-time derivative to an interaction, we reduce the dimensionality of the associated coupling constant. So only a few simple types of interaction can be renormalizable.) In particular, the existing Fermi theory of weak interactions clearly was not renormalizable. (The Fermi coupling constant has the dimensions of $[\text{mass}]^{-2}$.) The sense of discouragement about quantum field theory persisted into the 1950's and 1960's.

I learned about renormalization theory as a graduate student, mostly by reading Dyson's papers. [19] From the beginning it seemed to me to be a wonderful thing that very few quantum field theories are renormalizable. Limitations of this sort are, after all, what we most *want,* not mathematical methods which can make sense of an infinite variety of physically irrelevant theories, but methods which carry constraints, because these constraints may point the way toward the one true theory. In particular, I was impressed by the fact that quantum electrodynamics could in a sense be *derived* from symmetry principles and the constraints of renormalizability; the only Lorentz invariant and gauge invariant renormalizable Lagrangian for photons and electrons is precisely the orginal Dirac Lagrangian of QED. Of course, that is not the way Dirac came to his theory. He had the benefit of the information gleaned in centuries of experimentation on electromagnetism, and in order to fix the final form of his theory he relied on ideas of simplicity (specifically, on what is sometimes called minimal electromagnetic coupling). But we have to look ahead, to try to make theories of phenomena which have not been so well studied experimentally, and we may not be able to trust purely formal ideas of simplicity. I thought that renormalizability might be the key criterion, which also in a more general context would impose a precise kind of simplicity on our theories and help us to pick out the one true physical theory out of the infinite variety of conceivable quantum field theories. As I will explain later, I would say this a bit differently today, but I am more convinced than ever that the use of renormalizability as a constraint on our theories of the observed interactions is a good strategy. Filled with enthusiasm for renormalization theory, I wrote my Ph.D. thesis under Sam Treiman in 1957 on the use of a limited version of renormalizability to set constraints on the weak interactions, [20] and a little later I worked out a rather tough little theorem [21] which completed the proof by Dyson [19] and Salam [22] that ultraviolet divergences really do cancel out to all orders in nominally renormalizable theories. But none of this seemed to help with the important problem, of how to make a renormalizable theory of weak interactions.

Now, back to 1967. I had been considering the implications of the broken $SU(2) \times SU(2)$ symmetry of the strong interactions, and I thought of trying out the idea that perhaps the $SU(2) \times SU(2)$ symmetry was a "local," not merely a "global," symmetry. That is, the strong interactions might be described by something like a Yang-Mills theory, but in addition to the vector ϱ mesons of the Yang-Mills theory, there would also be axial vector A1 mesons. To give the ϱ meson a mass, it was necessary to insert a common ϱ and A1 mass term in the Lagrangian, and the spontaneous breakdown of the SU(2) × SU(2) symmetry would then split the ϱ and A1 by something like the Higgs mechanism, but since the theory would not be gauge invariant the pions would remain as physical Goldstone bosons. This theory gave an intriguing result, that the A1/ϱ mass ratio should be $\sqrt{2}$, and in trying to understand this result without relying on perturbation theory, I discovered certain sum rules, the "spectral function sum rules," [23] which turned out to have variety of other uses. But the SU(2) × SU(2) theory was not gauge invariant, and hence it could not be renormalizable, [24] so I was not too enthusiastic about it. [25] Of course, if I did not insert the ϱ-A1 mass term in the Lagrangian, then the theory would be gauge invariant and renormalizable, and the A1 would be massive. But then there would be no pions and the ϱ mesons would be massless, in obvious contradiction (to say the least) with observation.

At some point in the fall of 1967, I think while driving to my office at M.I.T., it occurred to me that I had been applying the right ideas to the wrong problem. It is not the ϱ mesons that is massless: it is the photon. And its partner is not the A1, but the massive intermediate boson, which since the time of Yukawa had been suspected to be the mediator of the weak interactions. The weak and electromagnetic interactions could then be described [26] in a unified way in terms of an exact but spontaneously broken gauge symmetry. [Of course, not necessarily SU(2) × SU(2)]. And this theory would be renormalizable like quantum electrodynamics because it is gauge invariant like quantum electrodynamics.

It was not difficult to develop a concrete model which embodied these ideas. I had little confidence then in my understanding of strong interactions, so I decided to concentrate on leptons. There are two left-handed electron-type leptons, the ν_{eL} and e_L, and one right-handed electron-type lepton, the e_R, so I started with the group U(2) × U(1): all unitary 2 × 2 matrices acting on the left-handed e-type leptons, together with all unitary 1 × 1 matrices acting on the right-handed e-type lepton. Breaking up U(2) into unimodular transformations and phase transformations, one could say that the group was SU(2) × U(1) × U(1). But then one of the U(1)'s could be identified with ordinary lepton number, and since lepton number appears to be conserved and there is no massless vector particle coupled to it, I decided to exclude it from the group. This left the four-parameter group SU(2) × U(1). The spontaneous breakdown of SU(2) × U(1) to the U(1) of ordinary electromagnetic gauge invariance would give masses to three of the four vector gauge bosons: the charged bosons $W^{\pm}$, and a neutral boson that I called the Z^0. The fourth boson would automatically remain massless, and could be identified as the photon. Knowing the strength of the ordinary charged current weak interactions like beta decay which are mediated by $W^{\pm}$, the mass of the $W^{\pm}$ was then determined as about 40 GeV/sinΘ, where Θ is the γ-Z^0 mixing angle.

To go further, one had to make some hypothesis about the mechanism for the breakdown of SU (2) × U (1). The only kind of field in a renormalizable SU(2) × U(1) theory whose vacuum expectation values could give the electron a mass is a spin zero SU(2) doublet (Φ^+, Φ^0), so for simplicity I assumed that these were the only scalar fields in the theory. The mass of the Z^0 was then determined as about 80 GeV/sin 2Θ. This fixed the strength of the neutral current weak interactions. Indeed, just as in QED, once one decides on the menu of fields in the theory all details of the theory are completely determined by symmetry principles and renormalizability, with just a few free parameters: the lepton charge and masses, the Fermi coupling constant of beta decay, the mixing angle Θ, and the mass of the scalar particle. (It was of crucial importance to impose the constraint of renormalizability; otherwise weak interactions would receive contributions from SU(2)×U(I)−invariant four-fermion couplings as well as from vector boson exchange, and the theory would lose most of its predictive power.) The naturalness of the whole theory is well demonstrated by the fact that much the same theory was independently developed [27] by Salam in 1968.

The next question now was renormalizability. The Feynman rules for Yang−Mills theories with unbroken gauge symmetries had been worked out [28] by deWitt, Faddeev and Popov and others, and it was known that such theories are renormalizable. But in 1967 I did not know how to prove that this renormalizability was not spoiled by the spontaneous symmetry breaking. I worked on the problem on and off for several years, partly in collaboration with students, [29] but I made little progress. With hindsight, my main difficulty was that in quantizing the vector fields I adopted a gauge now known as the unitarity gauge [30]: this gauge has several wonderful advantages, it exhibits the true particle spectrum of the theory, but it has the disadvantage of making renormalizability totally obscure.

Finally, in 1971 't Hooft [31] showed in a beautiful paper how the problem could be solved. He invented a gauge, like the "Feynman gauge" in QED, in which the Feynman rules manifestly lead to only a finite number of types of ultraviolet divergence. It was also necessary to show that these infinities satisfied essentially the same constraints as the Lagrangian itself, so that they could be absorbed into a redefinition of the parameters of the theory. (This was plausible, but not easy to prove, because a gauge invariant theory can be quantized only after one has picked a specific gauge, so it is not obvious that the ultraviolet divergences satisfy the same gauge invariance constraints as the Lagrangian itself.) The proof was subsequently completed [32] by Lee and Zinn-Justin and by 't Hooft and Veltman. More recently, Becchi, Rouet and Stora [33] have invented an ingenious method for carrying out this sort of proof, by using a global supersymmetry of gauge theories which is preserved even when we choose a specific gauge.

I have to admit that when I first saw 't Hooft's paper in 1971, I was not convinced that he had found the way to prove renormalizability. The trouble was not with 't Hooft, but with me: I was simply not familiar enough with the path integral formalism on which 't Hooft's work was based, and I wanted to see a derivation of the Feynman rules in 't Hooft's gauge from canonical quantization. That was soon supplied (for a limited class of gauge theories) by a paper of Ben Lee, [34] and after Lee's paper I was ready to regard the renormalizability of the unified theory as essentially proved.

By this time, many theoretical physicists were becoming convinced of the general approach that Salam and I had adopted: that is, the weak and electromagnetic interactions are governed by some group of exact local gauge symmetries; this group is spontaneously broken to U(1), giving mass to all the vector bosons except the photon; and the theory is renormalizable. What was not so clear was that our specific simple model was the one chosen by nature. That, of course, was a matter for experiment to decide.

It was obvious even back in 1967 that the best way to test the theory would be by searching for neutral current weak interactions, mediated by the neutral intermediate vector boson, the Z^0. Of course, the possibility of neutral currents was nothing new. There had been speculations [35] about possible neutral currents as far back as 1937 by Gamow and Teller, Kemmer, and Wentzel, and again in 1958 by Bludman and Leite-Lopes. Attempts at a unified weak and electromagnetic theory had been made [36] by Glashow and Salam and Ward in the early 1960's, and these had neutral currents with many of the features that Salam and I encountered in developing the 1967–68 theory. But since one of the predictions of our theory was a value for the mass of the Z^0, it made a definite prediction of the strength of the neutral currents. More important, now we had a comprehensive quantum field theory of the weak and electromagnetic interactions that was physically and mathematically satisfactory in the same sense as was quantum electrodynamics—a theory that treated photons and intermediate vector bosons on the same footing, that was based on an exact symmetry principle, and that allowed one to carry calculations to any desired degree of accuracy. To test this theory, it had now become urgent to settle the question of the existence of the neutral currents.

Late in 1971, I carried out a study of the experimental possibilites. [37] The results were striking. Previous experiments had set upper bounds on the rates of neutral current processes which were rather low, and many people had received the impression that neutral currents were pretty well ruled out, but I found that in fact the 1967–68 theory *predicted* quite low rates, low enough in fact to have escaped clear detection up to that time. For instance, experiments [38] a few years earlier had found an upper bound of 0.12 ± 0.06 on the ratio of a neutral current process, the elastic scattering of muon neutrinos by protons, to the corresponding charged current process, in which a muon is produced. I found a predicted ratio of 0.15 to 0.25, depending on the value of the Z^0-γ mixing angle θ. So there was every reason to look a little harder.

As everyone knows, neutral currents were finally discovered [39] in 1973. There followed years of careful experimental study on the detailed properties of the neutral currents. It would take me too far from my subject to survey these experiments, [40] so I will just say that they have confirmed the 1967–68 theory with steadily improving precision for neutrino-nucleon and neutrino-electron neutral current reactions, and since the remarkable SLAC-Yale experiment [41] last year, for the electron-nucleon neutral current as well.

This is all very nice. But I must say that I would not have been too disturbed if it had turned out that the correct theory was based on some other spontaneously broken gauge group, with very different neutral currents. One possibility was a clever SU(2) theory proposed in 1972 by Georgi and Glashow, [42] which has no neutral currents at all. The important thing to me was the idea of an exact spontaneously broken gauge symmetry, which connects the weak and electromagnetic interactions, and allows these interactions to be renormalizable. Of this I was convinced, if only because it fitted my conception of the way that nature ought to be.

There were two other relevant theoretical developments in the early 1970's, before the discovery of neutral currents, that I must mention here. One is the important work of Glashow, Iliopoulos, and Maiani on the charmed quark. [43] Their work provided a solution to what otherwise would have been a serious problem, that of neutral strangeness changing currents. I leave this topic for Professor Glashow's talk. The other theoretical development has to do specifically with the strong interactions, but it will take us back to one of the themes of my talk, the theme of symmetry.

In 1973, Politzer and Gross and Wilczek discovered [44] a remarkable property of Yang–Mills theories which they called "asymptotic freedom" – the effective coupling constant [45] decreases to zero as the characteristic energy of a process goes to infinity. It seemed that this might explain the experimental fact that the nucleon behaves in high energy deep inelastic electron scattering as if it consists of essentially free quarks. [46] But there was a problem. In order to give masses to the vector bosons in a gauge theory of strong interactions one would want to include strongly interacting scalar fields, and these would generally destroy asymptotic freedom. Another difficulty, one that particularly bothered me, was that in a unified theory of weak and electromagnetic interactions the fundamental weak coupling is of the same order as the electronic charge, e, so the effects of virtual intermediate vector bosons would introduce much too large violations of parity and strangeness conservation, of order 1/137, into the strong interactions of the scalars with each other and with the quarks. [47] At some point in the spring of 1973 it occurred to me (and independently to Gross and Wilczek) that one could do away with strongly interacting scalar fields altogether, allowing the strong interaction gauge symmetry to remain unbroken so that the vector bosons, or "gluons", are massless, and relying on the increase of the strong forces with increasing distance to explain why quarks as well as the massless gluons are not seen in the laboratory. [48] Assuming no strongly interacting scalars, three "colors" of quarks (as indicated by earlier work of several authors [49]), and an SU(3) gauge group, one then had a specific theory of strong interactions, the theory now generally known as quantum chromodynamics.

Experiments since then have increasingly confirmed QCD as the correct theory of strong interactions. What concerns me here, though, is its impact on our understanding of symmetry principles. Once again, the constraints of gauge invariance and renormalizability proved enormously powerful. These constraints force the Lagrangian to be so simple, that the strong interactions in QCD must conserve strangeness, charge conjugation, and (apart from problems [50] having to do with instantons) parity. One does not have to assume these symmetries as *a priori* principles; there is simply no way that the Lagrangian can be complicated enough to violate them. With one additional assumption, that the u and d quarks have relatively small masses, the strong interactions must also satisfy the approximate $SU(2) \times SU(2)$ symmetry of current algebra, which when spontaneously broken leaves us with isospin. If the s quark mass is also not too large, then one gets the whole eight-fold way as an approximate symmetry of the strong interactions. And the breaking of the $SU(3) \times SU(3)$ symmetry by quark masses has just the $(3,\bar{3})+(\bar{3},3)$ form required to account for the pion-pion scattering lengths [15] and Gell-Mann–Okubo mass formulas. Furthermore, with weak and electromagnetic interactions also described by a gauge theory, the weak currents are necessarily just the currents associated with these strong interaction symmetries. In other words, pretty much the whole pattern of approximate symmetries of strong, weak, and electromagnetic interactions that puzzled us so much in the 1950's and 1960's now stands explained as a simple consequence of strong, weak, and electromagnetic gauge invariance, plus renormalizability. Internal symmetry is now at the point where space-time symmetry was in Einstein's day. All the approximate internal symmetries are explained dynamically. On a fundamental level, there are no approximate or partial symmetries; there are only exact symmetries which govern all interactions.

I now want to look ahead a bit, and comment on the possible future development of the ideas of symmetry and renormalizability.

We are still confronted with the question whether the scalar particles that are responsible for the spontaneous breakdown of the electroweak gauge symmetry $SU(2) \times U(1)$ are really elementary. If they are, then spin zero semi-weakly decaying "Higgs bosons" should be found at energies comparable with those needed to produce the intermediate vector bosons. On the other hand, it may be that the scalars are composites. [51] The Higgs bosons would then be indistinct broad states at very high mass, analogous to the possible s-wave enhancement in π-π scattering. There would probably also exist lighter, more slowly decaying, scalar particles of a rather different type, known as pseudo-Goldstone bosons. [52] And there would have to exist a new class of "extra strong" interactions [53] to provide the binding force, extra strong in the sense that asymptotic freedom sets in not at a few hundred MeV, as in QCD, but at a few hundred GeV. This "extra strong" force would be felt by new families of fermions, and would give these fermions masses of the order of several hundred GeV. We shall see.

Of the four (now three) types of interactions, only gravity has resisted incorporation into a renormalizable quantum field theory. This may just mean that we are not being clever enough in our mathematical treatment of general relativity. But there is another possibility that seems to me quite plausible. The constant of gravity defines a unit of energy known as the Planck energy, about 10^{19} GeV. This is the energy at which gravitation becomes effectively a strong interaction, so that at this energy one can no longer ignore its ultraviolet divergences. It may be that there is a whole world of new physics with unsuspected degrees of freedom at these enormous energies, and that general relativity does not provide an adequate framework for understanding the physics of these superhigh energy degrees of freedom. When we explore gravitation or other ordinary phenomena, with particle masses and energies no greater than a TeV or so, we may be learning only about an "effective" field theory; that is, one in which superheavy degrees of freedom do not explicitly appear, but the coupling parameters implicitly represent sums over these hidden degrees of freedom.

To see if this makes sense, let us suppose it is true, and ask what kinds of interactions we would expect on this basis to find at ordinary energy. By "integrating out" the superhigh energy degrees of freedom in a fundamental theory, we generally encounter a very complicated effective field theory – so complicated, in fact, that it contains *all* interactions allowed by symmetry principles. But where dimensional analysis tells us that a coupling constant is a certain power of some mass, that mass is likely to be a typical superheavy mass, such as 10^{19} GeV. The infinite variety of non-renormalizable interactions in the effective theory have coupling constants with the dimensionality of negative powers of mass, so their effects are suppressed at ordinary energies by powers of energy divided by superheavy masses. Thus the only interactions that we can detect at ordinary energies are those that are renormalizable in the usual sense, plus any non-renormalizable interactions that produce effects which, although tiny, are somehow exotic enough to be seen.

One way that a very weak interaction could be detected is for it to be coherent and of long range, so that it can add up and have macroscopic effects. It has been shown [54] that the only particles whose exchange could produce such forces are massless particles of spin 0, 1, or 2. And furthermore, Lorentz's invariance alone is enough to show that the long-range interactions produced by any particle of mass zero and spin 2 must be governed by general relativity. [55] Thus from this point of view we should not be too surprised that gravitation is the only interaction discovered so far that does not seem to be described by a renormalizable field theory – it is almost the only superweak interaction that *could* have been detected. And we should not be surprised to find that gravity is well described by general relativity at macroscopic scales, even if we do not think that general relativity applies at 10^{19} GeV.

Non-renormalizable effective interactions may also be detected if they violate otherwise exact conservation laws. The leading candidates for violation are baryon and lepton conservation. It is a remarkable consequence of the SU(3) and SU(2) x U(1) gauge symmetries of strong, weak, and electromagnetic interactions, that all renormalizable interactions among known particles automatically conserve baryon and lepton number. Thus, the fact that ordinary matter seems pretty stable, that proton decay has not been seen, should not lead us to the conclusion that baryon and lepton conservation are fundamental conservation laws. To the accuracy with which they have been verified, baryon and lepton conservation can be explained as dynamical consequences of other symmetries, in the same way that strangeness conservation has been explained within QCD. But superheavy particles may exist, and these particles may have unusual SU(3) or SU(2) x SU(1) transformation properties, and in this case, there is no reason why their interactions should conserve baryon or lepton number. I doubt that they would. Indeed, the fact that the universe seems to contain an excess of baryons over antibaryons should lead us to suspect that baryon nonconserving processes have actually occurred. If effects of a tiny nonconservation of baryon or lepton number such as proton decay or neutrino masses are discovered experimentally, we will then be left with gauge symmetries as the only true internal symmetries of nature, a conclusion that I would regard as most satisfactory.

The idea of a new scale of superheavy masses has arisen in another way. [56] If any sort of "grand unification" of strong and electroweak gauge couplings is to be possible, then one would expect all of the SU(3) and SU(2) x U(1) gauge coupling constants to be of comparable magnitude. (In particular, if SU(3) and SU(2) x U(1) are subgroups of a larger simple group, then the ratios of the squared couplings are fixed as rational numbers of order unity.[57]) But this appears in contradiction with the obvious fact that the strong interactions are stronger than the weak and electromagnetic interactions. In 1974 Georgi, Quinn and I suggested that the grand unification scale, at which the couplings are comparable, is at an enormous energy, and that the reason that the strong coupling is so much larger than the electroweak couplings at ordinary energies is that QCD is asymptotically free, so that its effective coupling constant rises slowly as the energy drops from the grand unification scale to ordinary values. The change of the strong couplings is very slow (like $1/\sqrt{\ln E}$) so the grand unification scale must be enormous. We found that for a fairly large class of theories the grand unification scale comes out to be in the neighborhood of 10^{16} GeV, an energy not all that different from the Planck energy of 10^{19} GeV. The nucleon lifetime is very difficult to estimate accurately, but we gave a representative value of 10^{32} years, which may be accessible experimentally in a few years. (These estimates have been improved in more detailed calculations by several authors.) [58] We also calculated a value for the mixing parameter $\sin^2\Theta$ of about 0.2, not far from the present experimental value[40] of 0.23 ± 0.01. It will be an important task for future experiments on neutral currents to improve the precision with which $\sin^2\Theta$ is known, to see if it really agrees with this prediction.

In a grand unified theory, in order for elementary scalar particles to be available to produce the spontaneous breakdown of the electroweak gauge symmetry at a few hundred GeV, it is necessary for such particles to escape getting superlarge masses from the spontaneous breakdown of the grand unified gauge group. There is nothing impossible in this, but I have not been able to think of any reason why it should happen. (The problem may be related to the old mystery of why quantum corrections do not produce an enormous cosmological constant; in both cases, one is concerned with an anomalously small "super-renormalizable" term in the effective Lagrangian which has to be adjusted to be zero. In the case of the cosmological constant, the adjustment must be precise to some fifty decimal places.) With elementary scalars of small or zero bare mass, enormous ratios of symmetry breaking scales can arise quite naturally [59]. On the other hand, if there are no elementary scalars which escape getting superlarge masses from the breakdown of the grand unified gauge group, then as I have already mentioned, there must be extra strong forces to bind the composite Goldstone and Higgs bosons that are associated with the spontaneous breakdown of SU(2) x U(1). Such forces can occur rather naturally in grand unified theories. To take one example, suppose that the grand gauge group breaks, not into SU(3) x SU(2) x U(1), but into SU(4) x SU(3) x SU(2) x U(1). Since SU(4) is a bigger group than SU(3), its coupling constant rises with decreasing energy more rapidly than the QCD coupling, so the SU(4) force becomes strong at a much higher energy than the few hundred MeV at which the QCD force becomes strong. Ordinary quarks and leptons would be neutral under SU(4), so they would not feel this force, but other fermions might carry SU(4) quantum numbers, and so get rather large masses. One can even imagine a sequence of increasingly large subgroups of the grand gauge group, which would fill in the vast energy range up to 10^{15} or 10^{19} GeV with particle masses that are produced by these successively stronger interactions.

If there are elementary scalars whose vacuum expectation values are responsible for the masses of ordinary quarks and leptons, then these masses can be affected in order α by radiative corrections involving the superheavy vector bosons of the grand gauge group, and it will probably be impossible to explain the value of quantities like m_e/m_μ without a complete grand unified theory. On the other hand, if there are no such elementary scalars, then almost all the details of the grand unified theory are forgotten by the effective field theory that describes physics at ordinary energies, and it ought to be possible to calculate quark and lepton masses purely in terms of processes at accessible energies. Unfortunately, no one so far has been able to see how in this way anything resembling the observed pattern of masses could arise. [60]

Putting aside all these uncertainties, suppose that there is a truly fundamental theory, characterized by an energy scale of order 10^{16} to 10^{19} GeV, at which strong, electroweak, and gravitational interactions are all united. It might be a conventional renormalizable quantum field theory, but at the moment, if we include gravity, we do not see how this is possible. (I leave the topic of supersymmetry and supergravity for Professor Salam's talk.) But if it is not renormalizable, what then determines the infinite set of coupling constants that are needed to absorb all the ultraviolet divergences of the theory?

I think the answer must lie in the fact that the quantum field theory, which was born just fifty years ago from the marriage of quantum mechanics with relativity, is a beautiful but not very robust child. As Landau and Källén recognized long ago, quantum field theory at superhigh energies is susceptible to all sorts of diseases—tachyons, ghosts, etc.—and it needs special medicine to survive. One way that a quantum field theory can avoid these diseases is to be renormalizable and asymptotically free, but there are other possibilities. For instance, even an infinite set of coupling constants may approach a non-zero fixed point as the energy at which they are measured goes to infinity. However, to require this behavior generally imposes so many constraints on the couplings that there are only a finite number of free parameters left[61]—just as for theories that are renormalizable in the usual sense. Thus, one way or another, I think that quantum field theory is going to go on being very stubborn, refusing to allow us to describe all but a small number of possible worlds, among which, we hope, is ours.

I suppose that I tend to be optimistic about the future of physics. And nothing makes me more optimistic than the discovery of broken symmetries. In the seventh book of the *Republic,* Plato describes prisoners who are chained in a cave and can see only shadows that things outside cast on the cave wall. When released from the cave at first their eyes hurt, and for a while they think that the shadows they saw in the cave are more real than the objects they now see. But eventually their vision clears, and they can understand how beautiful the real world is. We are in such a cave, imprisoned by the limitations on the sorts of experiments we can do. In particular, we can study matter only at relatively low temperatures, where symmetries are likely to be spontaneously broken, so that nature does not appear very simple or unified. We have not been able to get out of this cave, but by looking long and hard at the shadows on the cave wall, we can at least make out the shapes of symmetries, which though broken, are exact principles governing all phenomena, expressions of the beauty of the world outside.

***

It has only been possible here to give references to a very small part of the literature on the subjects discussed in this talk. Additional references can be found in the following reviews:

Abers, E.S. and Lee, B.W., *Gauge Theories* (Physics Reports *9C,* No. 1, 1973).

Marciano, W. and Pagels, H., *Quantum Chromodynamics* (Physics Reports *36C,* No. 3, 1978).

Taylor, J.C., *Gauge Theories of Weak Interactions* (Cambridge Univ. Press, 1976).

REFERENCES

1. Tuve, M. A., Heydenberg, N. and Hafstad, L. R. Phys. Rev. *50,* 806 (1936); Breit, G., Condon, E. V. and Present, R. D. Phys. Rev. *50,* 825 (1936); Breit, G. and Feenberg, E. Phys. Rev. *50,* 850 (1936).
2. Gell-Mann, M. Phys. Rev. *92,* 833 (1953); Nakano. T. and Nishijima, K. Prog. Theor. Phys. *10,* 581 (1955).
3. Lee, T. D. and Yang, C. N. Phys. Rev. *104,* 254 (1956); Wu. C. S. et.al. Phys. Rev. *105,* 1413 (1957); Garwin, R., Lederman, L. and Weinrich, M. Phys. Rev. *105,* 1415 (1957); Friedman, J. I. and Telegdi V. L. Phys. Rev. *105,* 1681 (1957).
4. Gell-Mann, M. Cal. Tech. Synchotron Laboratory Report CTSL-20 (1961), unpublished; Ne'eman, Y. Nucl. Phys. *26,* 222 (1961).
5. Fock, V. Z. f. Physik *39,* 226 (1927); Weyl, H. Z. f. Physik *56,* 330 (1929). The name "gauge invariance" is based on an analogy with the earlier speculations of Weyl, H. in *Raum, Zeit, Materie,* 3rd edn, (Springer, 1920). Also see London, F. Z. f. Physik *42,* 375 (1927). (This history has been reviewed by Yang, C. N. in a talk at City College, (1977).)
6. Yang, C. N. and Mills, R. L. Phys. Rev. *96,* 191 (1954).
7. Goldstone, J. Nuovo Cimento *19,* 154 (1961).
8. Goldstone, J., Salam, A. and Weinberg, S. Phys. Rev. *127,* 965 (1962).
9. Higgs, P. W. Phys. Lett. *12,* 132 (1964); *13,* 508 (1964); Phys. Rev. *145,* 1156 (1966); Kibble, T. W. B. Phys. Rev. *155,* 1554 (1967); Guralnik, G. S., Hagen, C. R. and Kibble, T. W. B. Phys. Rev. Lett. *13,* 585 (1964); Englert, F. and Brout, R. Phys. Rev. Lett. *13,* 321 (1964); Also see Anderson, P. W. Phys. Rev. *130,* 439 (1963).
10. Adler, S. L. Phys. Rev. Lett. *14,* 1051 (1965); Phys Rev. *140,* B736 (1965); Weisberger, W. I. Phys. Rev. Lett. *14,* 1047 (1965); Phys Rev. *143,* 1302 (1966).
11. Gell-Mann, M. Physics *1,* 63 (1964).
12. Nambu, Y. and Jona-Lasinio, G. Phys. Rev. *122,* 345 (1961); *124,* 246 (1961); Nambu, Y, and Lurie, D. Phys. Rev. *125,* 1429 (1962); Nambu. Y. and Shrauner, E. Phys. Rev. *128,* 862 (1962); Also see Gell-Mann, M. and Lévy, M., Nuovo Cimento *16,* 705 (1960).
13. Goldberger, M. L., Miyazawa, H. and Oehme, R. Phys Rev. *99,* 986 (1955).
14. Goldberger, M. L., and Treiman, S. B. Phys. Rev. *111,* 354 (1958).
15. Weinberg, S. Phys. Rev. Lett. *16,* 879 (1966); *17,* 336 (1966); *17,* 616 (1966); *18,* 188 (1967); Phys Rev. *166,* 1568 (1967).
16. Oppenheimer, J. R. Phys. Rev. *35,* 461 (1930); Waller, I. Z. Phys. *59,* 168 (1930); ibid., *62,* 673 (1930).
17. Feynman, R. P. Rev. Mod. Phys. *20,* 367 (1948); Phys. Rev. *74,* 939, 1430 (1948); *76,* 749, 769 (1949); *80,* 440 (1950); Schwinger, J. Phys. Rev. *73,* 146 (1948); *74,* 1439 (1948); *75,* 651 (1949); *76,* 790 (1949); *82,* 664, 914 (1951); *91,* 713 (1953); Proc. Nat. Acad. Sci. *37,* 452 (1951); Tomonaga, S. Progr. Theor. Phys. (Japan) *1,* 27 (1946); Koba, Z., Tati, T. and Tomonaga, S. ibid. *2,* 101 (1947); Kanazawa, S. and Tomonaga, S. ibid. *3,* 276 (1948); Koba, Z. and Tomonaga, S. ibid *3,* 290 (1948).
18. There had been earlier suggestions that infinities could be eliminated from quantum field theories in this way, by Weisskopf, V. F. Kong. Dansk. Vid. Sel. Mat.-Fys. Medd. *15* (6) 1936, especially p. 34 and pp. 5–6; Kramers, H. (unpublished).
19. Dyson, F. J. Phys. Rev. *75,* 486, 1736 (1949).
20. Weinberg, S. Phys. Rev. *106,* 1301 (1957).
21. Weinberg, S. Phys. Rev. *118,* 838 (1960).
22. Salam, A. Phys. Rev. *82,* 217 (1951); *84,* 426 (1951).

23. Weinberg, S. Phys. Rev. Lett. *18,* 507 (1967).
24. For the non-renormalizability of theories with intrinsically broken gauge symmetries, see Komar, A. and Salam, A. Nucl. Phys. *21,* 624 (1960); Umezawa, H. and Kamefuchi, S. Nucl. Phys. *23,* 399 (1961); Kamefuchi, S., O'Raifeartaigh, L. and Salam, A. Nucl. Phys. *28,* 529 (1961); Salam, A. Phys. Rev. *127,* 331 (1962); Veltman, M. Nucl. Phys. B7, 637 (1968); B*21,* 288 (1970); Boulware, D. Ann. Phys. (N, Y,) *56,* 140 (1970).
25. This work was briefly reported in reference 23, footnote 7.
26. Weinberg, S. Phys. Rev. Lett. *19,* 1264 (1967).
27. Salam, A. In *Elementary Particle Physics* (Nobel Symposium No. 8), ed. by Svartholm, N. (Almqvist and Wiksell, Stockholm, 1968), p. 367.
28. deWitt, B. Phys. Rev. Lett. *12,* 742 (1964); Phys. Rev. *162,* 1195 (1967); Faddeev L. D., and Popov, V. N. Phys. Lett. *B25,* 29 (1967); Also see Feynman, R. P. Acta. Phys. Pol. *24,* 697 (1963); Mandelstam, S. Phys. Rev. *175,* 1580, 1604 (1968).
29. See Stuller, L. M. I. T., Thesis, Ph. D. (1971), unpublished.
30. My work with the unitarity gauge was reported in Weinberg, S. Phys. Rev. Lett. *27,* 1688 (1971), and described in more detail in Weinberg, S. Phys. Rev. *D7,* 1068 (1973).
31. 't Hooft, G Nucl. Phys. *B35,* 167 (1971).
32. Lee, B. W. and Zinn-Justin, J. Phys. Rev. *D5,* 3121, 3137, 3155 (1972); 't Hooft, G. and Veltman, M. Nucl. Phys. *B44,* 189 (1972), *B50,* 318 (1972). There still remained the problem of possible Adler–Bell–Jackiw anomalies, but these nicely cancelled; see D. J. Gross and R. Jackiw, Phys. Rev. *D6,* 477 (1972) and C. Bouchiat, J. Iliopoulos, and Ph. Meyer, Phys. Lett. *38B,* 519 (1972).
33. Beechi, C., Rouet, A. and Stora R. Comm. Math. Phys. *42,* 127 (1975).
34. Lee, B. W. Phys. Rev. *D5,* 823 (1972).
35. Gamow, G. and Teller, E. Phys. Rev. *51,* 288 (1937); Kemmer, N. Phys. Rev. *52,* 906 (1937); Wentzel, G. Helv. Phys. Acta. *10,* 108 (1937); Bludman, S. Nuovo Cimento *9,* 433 (1958); Leite-Lopes, J. Nucl. Phys. *8,* 234 (1958).
36. Glashow, S. L. Nucl. Phys. *22,* 519 (1961); Salam, A. and Ward, J. C. Phys. Lett. *13,* 168 (1964).
37. Weinberg, S. Phys. Rev. *5,* 1412 (1972).
38. Cundy, D. C. et.al., Phys. Lett. *31B,* 478 (1970).
39. The first published discovery of neutral currents was at the Gargamelle Bubble Chamber at CERN: Hasert, F. J. et.al., Phys. Lett. *46B,* 121, 138 (1973). Also see Musset, P. Jour. de Physique 11/12 T34 (1973). Muonless events were seen at about the same time by the HPWF group at Fermilab, but when publication of their paper was delayed, they took the opportunity to rebuild their detector, and then did not at first find the same neutral current signal. The HPWF group published evidence for neutral currents in Benvenuti, A. et.al., Phys. Rev. Lett. *32,* 800 (1974).
40. For a survey of the data see Baltay, C. *Proceedings of the 19th International Conference on High Energy Physics,* Tokyo, 1978. For theoretical analyses, see Abbott, L. F. and Barnett, R. M. Phys. Rev. *D19,* 3230 (1979); Langacker, P., Kim, J. E., Levine, M., Williams, H. H. and Sidhu, D. P. Neutrino Conference '79; and earlier references cited therein.
41. Prescott, C. Y. et.al., Phys. Lett. *77B,* 347 (1978).
42. Glashow, S. L. and Georgi, H. L. Phys. Rev. Lett. *28,* 1494 (1972). Also see Schwinger, J. Annals of Physics (N. Y.) *2,* 407 (1957).
43. Glashow, S. L., Iliopoulos, J. and Maiani, L. Phys. Rev. *D2,* 1285 (1970). This paper was cited in ref. 37 as providing a possible solution to the problem of strangeness changing neutral currents. However, at that time I was skeptical about the quark model, so in the calculations of ref. 37 baryons were incorporated in the theory by taking the protons and neutrons to form an SU(2) doublet, with strange particles simply ignored.
44. Politzer, H. D. Phys. Rev. Lett. *30,* 1346 (1973); Gross, D. J. and Wilczek, F. Phys. Rev. Lett. *30,* 1343 (1973).
45. Energy dependent effective couping constants were introduced by Gell-Mann, M. and Low, F. E. Phys. Rev. *95,* 1300 (1954).
46. Bloom, E. D. et.al., Phys. Rev. Lett. *23,* 930 (1969); Breidenbach, M. et.al., Phys. Rev. Lett. *23,* 935 (1969).

47. Weinberg, S. Phys. Rev. *D8,* 605 (1973).
48. Gross, D. J. and Wilczek, F. Phys. Rev. *D8,* 3633 (1973); Weinberg, S. Phys. Rev. Lett. *31,* 494 (1973). A similar idea had been proposed before the discovery of asymptotic freedom by Fritzsch, H., Gell-Mann, M. and Leutwyler, H. Phys. Lett. *47B,* 365 (1973).
49. Greenberg, O. W. Phys. Rev. Lett. *13,* 598 (1964); Han, M. Y. and Nambu, Y. Phys. Rev. *139,* B1006 (1965); Bardeen, W. A., Fritzsch, H. and Gell-Mann, M. in *Scale and Conformal Symmetry in Hadron Physics,* ed. by Gatto, R. (Wiley, 1973), p. 139; etc.
50. 't Hooft, G. Phys. Rev. Lett. *37,* 8 (1976).
51. Such "dynamical" mechanisms for spontaneous symmetry breaking were first discussed by Nambu, Y. and Jona-Lasinio, G. Phys. Rev. *122,* 345 (1961); Schwinger, J. Phys. Rev. *125,* 397 (1962); *128,* 2425 (1962); and in the context of modern gauge theories by Jackiw, R. and Johnson, K. Phys. Rev. *D8,* 2386 (1973); Cornwall, J. M. and Norton, R. E. Phys. Rev. *D8,* 3338 (1973). The implications of dynamical symmetry breaking have been considered by Weinberg, S. Phys. Rev. *D13,* 974 (1976); *D19,* 1277 (1979); Susskind, L. Phys. Rev. *D20,* 2619 (1979).
52. Weinberg, S. ref 51. The possibility of pseudo-Goldstone bosons was originally noted in a different context by Weinberg, S. Phys. Rev. Lett. *29,* 1698 (1972).
53. Weinberg, S. ref. 51. Models involving such interactions have also been discussed by Susskind, L. ref. 51.
54. Weinberg, S. Phys. Rev. *135,* B1049 (1964).
55. Weinberg. S. Phys. Lett. *9,* 357 (1964); Phys. Rev. *B138,* 988 (1965); *Lectures in Particles and Field Theory,* ed. by Deser, S. and Ford, K. (Prentice-Hall, 1965), p. 988; and ref. 54. The program of deriving general relativity from quantum mechanics and special relativity was completed by Boulware, D. and Deser, S. Ann. Phys. *89,* 173 (1975). I understand that similar ideas were developed by Feynman, R. in unpublished lectures at Cal. Tech.
56. Georgi, H., Quinn, H. and Weinberg, S. Phys. Rev. Lett. *33,* 451 (1974).
57. An example of a simple gauge group for weak and electromagnetic interactions (for-which $\sin^2\theta=\frac{1}{4}$) was given by S. Weinberg, Phys. Rev. *D5,* 1962 (1972). There are a number of specific models of weak, electromagnetic, and strong interactions based on simple gauge groups, including those of Pati, J. C. and Salam, A. Phys. Rev. *D10,* 275 (1974); Georgi, H. and Glashow, S. L. Phys. Rev. Lett. *32,* 438 (1974); Georgi, H. in *Particles and Fields* (American Institute of Physics, 1975); Fritzsch, H. and Minkowski, P. Ann. Phys. *93,* 193 (1975); Georgi, H. and Nanopoulos, D. V. Phys. Lett. *82B,* 392 (1979); Gürsey, F. Ramond, P. and Sikivie, P. Phys. Lett. *B60,* 177 (1975); Gürsey, F. and Sikivie, P. Phys. Rev. Lett. *36,* 775 (1976); Ramond, P. Nucl. Phys, *B110,* 214 (1976); etc; all these violate baryon and lepton conservation, because they have quarks and leptons in the same multiplet; see Pati, J. C. and Salam, A. Phys. Rev. Lett. *31,* 661 (1973); Phys. Rev. D *8,* 1240 (1973).
58. Buras, A., Ellis, J., Gaillard, M. K. and Nanopoulos, D. V. Nucl. Phys. *B135,* 66 (1978); Ross, D. Nucl. Phys. *B140,* 1 (1978); Marciano, W. J. Phys. Rev. *D20,* 274 (1979); Goldman, T. and Ross, D. CALT 68-704, to be published; Jarlskog, C. and Yndurain, F. J. CERN preprint, to be published. Machacek, M. Harvard preprint HUTP-79/AO21, to be published in Nuclear Physics; Weinberg, S. paper in preparation. The phenomenonology of nucleon decay has been discussed in general terms by Weinberg, S. Phys. Rev. Lett. *43,* 1566 (1979); Wilczek, F. and Zee, A. Phys. Rev. Lett. *43,* 1571 (1979).
59. Gildener, E. and Weinberg, S. Phys. Rev. *D13,* 3333 (1976); Weinberg, S. Phys. Letters *82B,* 387 (1979). In general there should exist at least one scalar particle with physical mass of order 10 GeV. The spontaneous symmetry breaking in models with zero bare scalar mass was first considered by Coleman, S. and Weinberg, E., Phys. Rev. D 7, 1888 (1973).
60. This problem has been studied recently by Dimopoulos, S. and Susskind, L. Nucl. Phys. *B155,* 237 (1979); Eichten, E. and Lane, K. Physics Letters, to be published; Weinberg, S. unpublished.
61. Weinberg, S. in *General Relativity – An Einstein Centenary Survey,* ed. by Hawking, S. W. and Israel, W. (Cambridge Univ. Press, 1979), Chapter 16.

Physics 1980

JAMES W CRONIN and VAL L FITCH

for the discovery of violations of fundamental symmetry principles in the decay of neutral K-mesons

THE NOBEL PRIZE FOR PHYSICS

Speach by Professor GÖSTA EKSPONG of the Royal Academy of Sciences.
Translation from the Swedish text.

Your Majesty, Your Royal Highnesses, Ladies and Gentlemen,

By decision of the Royal Swedish Academy of Sciences, this year's Nobel Prize for Physics has been awarded Professor James Cronin and Professor Val Fitch for their discovery in a joint experiment of violations of fundamental symmetry principles. The experiment was carried out in 1964 at Brookhaven National Laboratory in the United States of America and was concerned with a forbidden decay of a certain type of elementary particles, named the neutral K-meson.

Suppose the TV-news suddenly reported one evening that visitors from outer space were planning to land on Earth; that the space travelers have radioed a demand for immediate information about the composition of the Earth. Does it consist of Matter or Antimatter? The answer to this question is one of life and death. The two kinds of matter are known to annihilate each other atom by atom. The space travelers claim, furthermore, that the nature of their own kind of matter was determined before leaving. What they now want to know is, whether the same tests have been made on Earth. Thanks to Cronin's and Fitch's discovery it is now possible to give them a clear-cut answer, so they can avoid a disastrous landing. Let us now leave the world of science fiction, remembering, however, what a fortunate circumstance it was that no space visits occurred before 1964.

Symmetries are science's lodestars and symmetry principles act as guiding rules to help us discover the mathematical laws of Nature. Three mirror symmetries are of immediate interest in relation to the prize-winning discovery. One of them is ordinary mirror reflection, which corresponds to switching left and right. The other two symmetries of interest concern reflection of time and of charge, which implies switching forward and backward movements and switching matter and antimatter, respectively. In the latter case it is positive and negative electric charges that are switched.

The beauty of spacial symmetries is well known in the realm of art and architecture, from the ornamental arabesques of the old Alhambra to the recent intricate woodcuts signed by Escher, from the palace of the Doges in Venice to the Town Hall in Stockholm. A master such as Johann Sebastian Bach has created music with ingenious symmetries, generated both by reflection in space of the theme and by reflection in time when the theme is played backwards. The laws of physics resemble a canon by Bach. They are symmetric in space and time. They do not distinguish between left and right, nor between forward and backward movements. For a long time everyone thought it had to be like that. A remarkable exception exists, however, in the law governing radioactive decay, which violates the left-right symmetry. Lee and Yang were awarded the Nobel prize for physics in 1957 for this revolutionary discovery.

The third mirror symmetry is not present in art. The laws for electric and magnetic phenomena contain a complete symmetry between the two kinds of electric charge. The discoveries of antimatter with plus and minus charges in exchanged roles are among the most profound of the last half-century. Nowadays microscopic amounts of antiparticles are produced with relative ease in such special laboratories as Brookhaven National Laboratory in the U.S. or CERN in Europe.

Cronin and Fitch elected to carry out tests to find out whether a certain decay of K-mesons occurred, in spite of being forbidden by symmetry. Their research team found that two out of a thousand K-mesons did in fact decay in the forbidden manner. This means that some law of Nature now must be changed or a new law invoked. In what way does this discovery concern antimatter? As early as 1955 Gell-Mann and Pais had analyzed the neutral K-mesons and found that they are strange, indeed unique in their ambivalence with respect both to matter and antimatter. If perfect symmetry were to prevail, a decaying K-meson would have to be antimatter in exactly half the cases and in the other half, matter. Lee's and Yang's Chinese revolution did not change the conclusions, but new arguments were required. Cronin and Fitch interpreted the results of their experiment as a small but clear lack of symmetry. Their conclusion has been confirmed in a long series of other experiments. The new symmetry violation constitutes the basic prerequisite for the claim that a definite answer can be relayed to our visitors from outer space.

The discovery also implied consequences for time reflection. At least one theme is played more slowly backwards than forwards by Nature.

Artists nearly always introduce symmetry breaking elements into their works. Perhaps, the laws of nature, too, are in the deepest sense works of art. Violations of perfect symmetry open roads to new insights, or in the words of a poet:

"A knot there is in th'entendrill'd arabesque
No mortal eye but mine has ever seen".

Professor Fitch, Professor Cronin,

The scientific world was shocked when you first announced your discovery. Nobody, absolutely nobody, had anticipated anything like it. You had pursued your experiment with skill and determination and found the impossible to be possible.

On behalf of the Royal Swedish Academy of Sciences I have the pleasure and the honour of extending to you our warmest congratulations. I now invite you to receive your Prizes from the hands of His Majesty the King.

James W Cronin

JAMES W. CRONIN

I was born on September 29, 1931 in Chicago, Illinois, while my father, James Farley Cronin, was a graduate student at the University of Chicago. He was a student of classical languages. My mother, Dorothy Watson, had met my father in a Greek class at Northwestern University. After a brief stay at a small school in Alabama, my father became Professor of Latin and Greek at Southern Methodist University in Dallas, Texas, in September 1939. My primary and secondary education was provided by the Highland Park Public School System. I received my undergraduate degree from Southern Methodist University with a major in physics and mathematics in 1951. In high school my natural interest in science was encouraged by an excellent physics teacher, Mr. Charles H. Marshall. He stressed analytical methods as applied to simple physical systems as well as practical experimental problems.

My real education began when I entered the University of Chicago in September 1951 as a graduate student. I was fortunate to have among my classroom teachers, Enrico Fermi, Maria Mayer, Edward Teller, Gregor Wentzel, Val Telegdi, Marvin Goldberger and Murray Gell-Mann. I did a thesis in experimental nuclear physics under the direction of Samuel K. Allison. While at Chicago my interest in the new field of particle physics was stimulated by a course given by Gell-Mann, who was developing his ideas about Strangeness at the time.

It was also at the University of Chicago that I met my future wife, Annette Martin, in the summer of 1953. It was a wonderful, happy summer; I had passed my Ph.D. qualifying exams the previous winter, and I realized that I had met my lifetime companion. We were married in September 1954. The stable point in my life became our home. On even the worst days, when nothing was working at the lab, I knew that at home I would find warmth, peace, companionship, and encouragement. As a consequence, the next day would surely be better. Annette, with great patience and good spirit, tolerated my many long absences when experiments were carried out at distant laboratories.

After receiving my Ph.D. in 1955 I had the opportunity to join the group of Rodney Cool and Oreste Piccioni who were working at the Brookhaven Cosmotron, a newly completed 3 GeV accelerator. That period was an exciting time in physics. The famous τ—θ puzzle led to the prediction of parity violation and the experimental demonstration of its violation. The long-lived K meson was discovered at Brookhaven.

When the violation of parity was discovered I began a series of electronic experiments to investigate parity violation in hyperon decays. In early 1958 the

Cosmotron suffered a severe magnet failure. As a consequence, we moved our experiment to the Berkeley Bevatron. Here I had the good fortune to meet William Wenzel and Bruce Cork. These physicists had a great influence on me. From their example I learned not to be intimidated by complex pieces of apparatus.

While at Brookhaven I met Val Fitch who was responsible for my coming to Princeton University in the fall of 1958. At Princeton all the work in particle physics was supported through a contract with the Office of Naval Research. The Director of the Laboratory, George Reynolds, was most supportive of my efforts to work independently. There followed for ten years a glorious time for research. I was much involved in the development of the spark chamber as a practical research tool. During this period, with a series of excellent students, we further studied hyperon decays. Then we joined with Val Fitch to study neutral K meson decays which led to the discovery of CP violation.

Following the discovery in the summer of 1964, I spent a year in France working at the Centre d'Etudes Nucléaires at Saclay with René Turlay. In addition to the research, I enjoyed learning French and assimilating the culture of another country. One of the greatest joys in my life was giving a lecture in French at the Collège de France.

On returning to Princeton in 1965, I began with students a series of experiments to study the neutral CP violating decay modes of the long lived neutral K meson. These experiments lasted until 1971. In 1971 I returned to the University of Chicago as Professor of Physics. The fact that the new Fermilab 400 GeV Accelerator was being built near Chicago made this move an attractive one. At Fermilab, with younger associates and students, I carried out experiments on the production of particles at high transverse momentum, and on the production of direct leptons. At present with my colleague at Chicago, Bruce Winstein, I am preparing to study with much greater accuracy some of the CP violating parameters of the neutral K meson.

I now live in Chicago near the campus with my wife Annette, and son Daniel. My oldest daughter Cathryn lives and works in New York City. My daughter Emily attends the University of Minnesota. My mother remained in Dallas, Texas, after the death of my father in 1959. For recreation we have a cabin in the woods in Wisconsin which we visit year-round. In the summer we spend some time in Aspen, Colorado. Our whole family assembles in Chicago at Christmas and usually in Aspen in the summer.

Education
B.S., Southern Methodist University, 1951
M.S., University of Chicago, 1953
Ph.D., (Physics) University of Chicago, 1955

Career
National Science Foundation Fellow, 1952—1955
Assistant Physicist, Brookhaven National Laboratory, 1955—1958

Assistant Professor of Physics, Princeton University, 1958—1962
Associate Professor of Physics, Princeton University, 1962—1964
Professor of Physics, Princeton University, 1964—1971
University Professor of Physics, University of Chicago, 1971—

Member
American Academy of Arts and Sciences
American Physical Society
National Academy of Sciences

Recipient
Research Corporation Award, 1968
John Price Wetherill Medal of the Franklin Institute, 1975
Ernest O. Lawrence Award, 1977

CP SYMMETRY VIOLATION — THE SEARCH FOR ITS ORIGIN

Nobel lecture, 8 December, 1980

by

JAMES W. CRONIN

The University of Chicago, Illinois 60637, USA

The greatest pleasure a scientist can experience is to encounter an unexpected discovery. I am always astonished when a simple apparatus, designed to ask the right question of nature, receives a clear response. Our experiment, carried out with James Christenson, Val Fitch and René Turlay, gave convincing evidence that the long-lived neutral K meson (K_L) decayed into two charged pions, a decay mode forbidden by CP symmetry. The forbidden decay mode was found to be a small fraction $(2.0 \pm 0.4) \times 10^{-3}$ of all charged decay modes. Professor Fitch has described our discovery of CP symmetry violation. He has discussed how it was preceded by brilliant theoretical insights and incisive experiments with K mesons. My lecture will review the knowledge that we have obtained about CP violation since its discovery.[1] The discovery triggered an intense international experimental effort. It also provoked many theoretical speculations which in turn stimulated a variety of experiments.

At present there is no satisfactory theoretical understanding of CP violation. Such understanding as we do have has come entirely from experimental studies. These studies have extended beyond the high energy accelerator laboratories into nuclear physics laboratories and research reactor laboratories. The experiments which have sought to elucidate the tiny effect have involved both ingenuity and painstaking attention to detail.

Upon learning of the discovery in 1964, the natural reaction of our colleagues was to ask what was wrong with the experiment. Or, if they were convinced of the correctness of the measurements, they asked how could the effect be explained while still retaining CP symmetry. I remember vividly a special session organized at the 1964 International Conference on High Energy Physics at Dubna in the Soviet Union. There, for an afternoon, I had to defend our experiment before a large group of physicists who wanted to know every detail of the experiment—more details than could have been given in the formal conference session.

As the session neared a close, one of my Soviet colleagues suggested that, perhaps, the effect was due to regeneration of short-lived K mesons (K_S) in a fly unfortunately trapped in the helium bag. We did a quick "back of the evelope" estimate of the density of the fly necessary to produce the effect. The density required was far in excess of uranium.

More serious questions were raised at this session and by many other physicists who had thought deeply about our result. While we were confident that the experiment had been correctly carried out and interpreted, many sought reassurance through confirmation of the experiment by other groups. This confirmation came quickly from experiments at the Rutherford Laboratory[2] in England, and at CERN[3] in Geneva, Switzerland.

Another important issue was raised. In the original experiment, the decay to two pions was inferred kinematically, but no proof was given that these pions were identical to the ordinary pions or that the decay was not accompanied by a third light particle emitted at a very low energy. The direct proof that the effect was indeed a violation of CP symmetry was the demonstration of interference between the decay of the long-lived and short-lived K meson to two charged pions. This interference was first demonstrated in a simple and elegant experiment by my colleague Val Fitch with Roth, Russ and Vernon.[4]

Their experiment compared the rate of decay of a K_L beam into two charged pions in vacuum and in the presence of a diffuse beryllium regenerator. The density of the regenerator was adjusted so that the regeneration amplitude A_r was equal to the CP violating amplitude η_{+-}. These amplitudes are defined by

$$\eta_{+-} = \frac{\text{amplitude } (K_L \rightarrow \pi^+\pi^-)}{\text{amplitude } (K_S \rightarrow \pi^+\pi^-)}$$

and

$$A_r = i\pi N\Lambda \left(\frac{f-\bar{f}}{k}\right) \left(i\delta+\frac{1}{2}\right)^{-1}.$$

The yield of $K_L \rightarrow \pi^+\pi^-$ in the presence of the regenerator is proportional to

$$|A_r+\eta_{+-}|^2.$$

In the expression for A_r, δ is given by $(M_S—M_L)/\Gamma_S$ where M_S and M_L are the K_S and K_L masses, and Γ_S the decay rate of the K_S meson, Λ is the mean decay length of the K_S meson, k is the wave number of the incident K_L beam and f and $\bar{f}$ are the forward scattering amplitudes for K and $\bar{K}$, respectively on the nuclei of the regenerator. The regeneration amplitude is proportional to N, the number density of the material. The quantity $(f-\bar{f})/k$ was determined in an auxiliary experiment with a dense regenerator. Then a regenerator of appropriate density was constructed using the formula for A_r.[5] The actual regenerator was constructed of 0.5 mm sheets separated by 1 cm. Such an arrangement behaves as a homogeneous regenerator of (1/20) normal density if the separation of the sheets is small compared to the quantity $\delta\Lambda$.

In the earliest experiment Fitch and his colleagues found that with $|A_r|$ chosen to be equal to $|\eta_{+-}|$ the rate of $\pi^+\pi^-$ decays was about *four times* the rate without the regenerator. This result showed not only that there was interference, but also that the interference was fully constructive. Complete analysis of this experiment reported subsequently[6] gave the $\pi^+\pi^-$ yield as a function of density as shown in Fig 1. The quantity α in the figure is the relative phase between the regeneration amplitude and the CP violating amplitude.

The result of this experiment also permits the experimental distinction

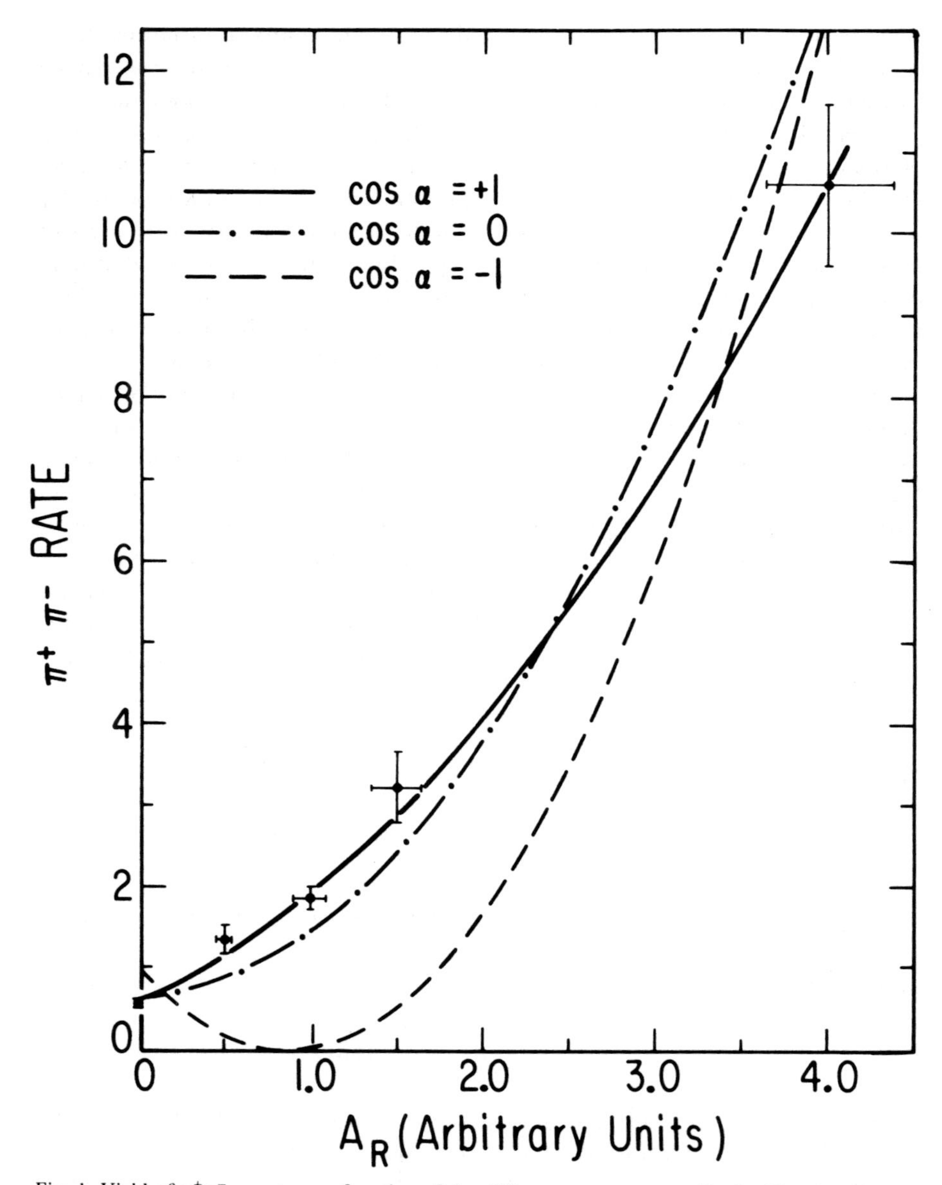

Fig. 1. Yield of $\pi^+\pi^-$ events as a function of the diffuse regenerator amplitude. The three curves correspond to the three stated values of the phase between the regeneration amplitude A_r and the CP violating amplitude η_{+-}.

between a world composed of matter and a world composed of antimatter.[7] Imagine that this experiment were performed in the antiworld. The only difference would be that the regenerator material would be antimatter. If we assume C invariance for the strong interactions, the forward scattering amplitudes for K and $\bar{K}$ would be interchanged so that A_r would have the opposite sign. Thus, in the antiworld an investigator performing the interference experiment would observe destructive interference similar to the dashed curve of Fig 1, an unmistakable difference from the result found in our world. The interference experiment of Fitch and collaborators eliminated alternate explanations of the $K_L \rightarrow \pi^+\pi^-$ decay, since the effect was of such a nature that an experiment distinguishing a world of matter and antimatter was possible.

It was also suggested that the effect might be due to a long range vector field of cosmological origin.[8] Such a source of the effect would lead to a decay rate for $K_L \rightarrow \pi^+\pi^-$ which would be proportional to the square of the K_L energy in the laboratory. Our original experiment was carried out at a mean K_L energy of 1.1 GeV. The confirming experiments at the Rutherford Laboratory and CERN were carried out at mean K_L energies of 3.1 and 10.7 GeV, respectively. Since the three experiments found the same branching ratio for $K_L \rightarrow \pi^+\pi^-$, the possibility of a long range vector field was eliminated.

Before continuing, it is necessary to state some of the phenomenology which describes the CP violation in the neutral K system. The basic notation was introduced by Wu and Yang.[9] For this discussion CPT conservation is assumed. Later we shall refer to the evidence from K-meson decays which show that all data are consistent with a corresponding T violation. Any CPT violation is consistent with zero within the present sensitivity of the measurements.

There are two basic complex parameters which are required to discuss CP violation as observed in the two pion decays of K_L mesons. The first quantity ε is a measure of the CP impurity in the eigenstates $|K_S\rangle$ and $|K_L\rangle$. These eigenstates are given by

$$|K_S\rangle = \frac{1}{\sqrt{2}\sqrt{1+|\varepsilon|^2}}[(1+\varepsilon)|K\rangle+(1-\varepsilon)|\bar{K}\rangle],$$

and

$$|K_L\rangle = \frac{1}{\sqrt{2}\sqrt{1+|\varepsilon|^2}}[(1+\varepsilon)|K\rangle-(1-\varepsilon)|\bar{K}\rangle].$$

The quantity ε can be expressed in terms of the elements of the mass and decay matrices which couple and control the time evolution of the $|K\rangle$ and $|\bar{K}\rangle$ states. It is given by

$$\varepsilon = \frac{-\mathrm{Im}M_{12}+i\mathrm{Im}\Gamma_{12}/2}{i(M_S-M_L)+(\Gamma_S-\Gamma_L)/2}$$

Limits on the size of $\mathrm{Im}\Gamma_{12}$ can be obtained from the observed decay rates of K_S and K_L to the various decay modes. If $\mathrm{Im}\Gamma_{12}$ were zero, then the phase of ε would be determined by the denominator which is just the difference in eigenvalues of the matrix which couples K and $\bar{K}$. These quantities have been experimentally measured and give $\arg \varepsilon \sim 45°$.

The second quantity ε' is defined by

$$\varepsilon' = \frac{i}{\sqrt{2}}\mathrm{Im}\left(\frac{A_2}{A_0}\right)e^{i(\delta_2-\delta_0)}.$$

Here A_0 and A_2 are respectively the amplitudes for a K meson to decay to standing wave states of two pions in the isotopic spin 0 and 2 states, respectively. Time reversal symmetry demands that A_0 and A_2 be relatively real.[10] The quantities δ_0 and δ_2 are the s-wave $\pi\pi$ scattering phase shifts for the states

I = 0 and I = 2, respectively. The parameters ε and ε' are related to observable quantities defined by

$$|\eta_{+-}|\, e^{i\phi_{+-}} = \frac{\text{amp}\,(K_L \rightarrow \pi^+\pi^-)}{\text{amp}\,(K_S \rightarrow \pi^+\pi^-)},$$

$$|\eta_{oo}|\, e^{i\phi_{oo}} = \frac{\text{amp}\,(K_L \rightarrow \pi^o\pi^o)}{\text{amp}\,(K_S \rightarrow \pi^o\pi^o)},$$

and

$$\delta_\ell = \frac{\Gamma(K_L \rightarrow \pi^-\ell^+\nu_\ell) - \Gamma(K_L \rightarrow \pi^+\ell^-\overline{\nu}_\ell)}{\Gamma(K_L \rightarrow \pi^-\ell^+\nu_\ell) + \Gamma(K_L \rightarrow \pi^+\ell^-\overline{\nu}_\ell)}$$

These experimentally measured quantities are related to ε and ε' by the following expressions:[11]

$$\eta_{+-} = \varepsilon + \varepsilon'$$
$$\eta_{oo} = \varepsilon - 2\varepsilon'$$
$$\delta_\ell = 2\,\text{Re}\varepsilon.$$

The magnitude and phase of the quantity η_{+-} have been most precisely measured by studying the time dependence of $\pi^+\pi^-$ decays from a K beam which was prepared as a mixture of K_S and K_L. This experimental technique was suggested by Whatley,[12] long before the discovery of CP violation. If we let ρ be the amplitude for K_S at t = 0, relative to the K_L amplitude, then the time dependence of $\pi^+\pi^-$ decays will be given by[13]

$$N_{+-}(t) = |\rho \exp\,[(-i\Delta M - \Gamma_s/2)t] + \eta_{+-}|^2.$$

The initial amplitude for the K_S component can be prepared by two different methods. In the first method we pass a K_L beam through a regenerator. Then ρ is the regeneration amplitude. Here the interference term is $2|\rho|\,|\eta_{+-}|\,e^{-\Gamma_s t/2}\cos(-\Delta M t + \phi_\rho - \phi_{+-})$. In the second method we produce a beam which is pure K (or $\overline{K}$) at t = 0. In practice protons of $\approx$20 GeV produce at small angles about three times as many K as K. The K dilution is a detail which need not be of concern here. In this case $\rho = +1$, and the interference term is $2|\eta_{+-}|\,e^{-\Gamma_s t/2}\cos(-\Delta M t - \phi_{+-})$.

The important CP parameters are $|\eta_{+-}|$ and ϕ_{+-}. We see, however, that a knowledge of the auxiliary parameters Γ_s and ΔM is also required. In the first method one measures $\phi_{+-} - \phi_\rho$ and one must also have a technique to independently measure ϕ_ρ. In both cases the $\pi^+\pi^-$ yield is most sensitive to the interference term when the two interfering amplitudes are of the same size. For the second method we require observation at 12 K_S lifetimes. (We want $e^{-\Gamma_s t/2} \approx |\eta_{+-}| \approx 2\times10^{-3}$.) As a consequence, a small error in ΔM can lead to a large uncertainty on ϕ_{+-}, and, more importantly, a systematic error in ΔM can lead to an incorrect value for ϕ_{+-}. A one percent error in ΔM corresponds to an error in ϕ_{+-} of about 3°. The measurement of ΔM with satisfactory precision has required an effort as formidable as the interference experiments themselves.[14]

Time and space do not permit a survey which does justice to the many groups at CERN, Brookhaven, Argonne, and SLAC who have made the meticulous measurements which have led to the following parameters:[15]

$$\eta_{+-} = [(2.27\pm0.02)\times10^{-3}]\exp[i(44.7°\pm1.2°)],$$
$$\Delta M = M_S - M_L = -(0.535\pm0.002)\times10^{10}/\text{sec},$$
$$\Gamma_S = (1.121\pm0.003)\times10^{10}/\text{sec}.$$

As an example of the quality of the measurements mentioned above, Fig 2 shows a time distribution of $\pi^+\pi^-$ decays following the passage of a K_L beam of 4 to 10 GeV/c momentum through an 81 cm thick carbon regenerator.[16]

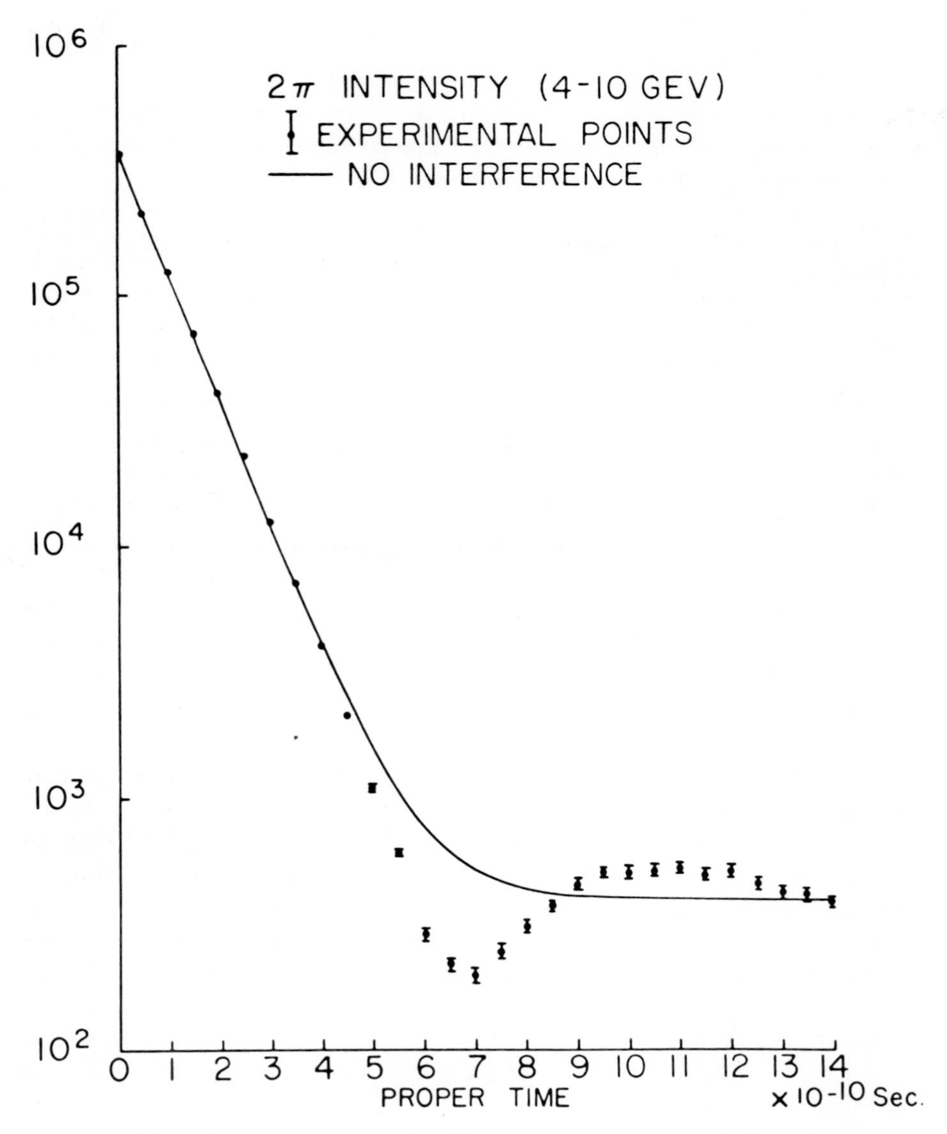

Fig. 2. Yield of $\pi^+\pi^-$ events as a function of proper time downstream from an 81 cm carbon regenerator placed in a K_L beam.

The destructive interference is clearly seen. If the experiment were carried out with a regenerator of anticarbon, then constructive interference would have been observed.

Measurements of the charge asymmetry δ_ℓ for K_L decays began in 1966. This asymmetry is found in the abundant semileptonic decay modes $K_L \to \pi^\pm \ell^\mp \nu$, where ℓ is either an electron or muon. It basically measures the difference in amplitude of K and $\overline{K}$ in the eigenstate of the K_L. It does so by virtue of the $\Delta S = \Delta Q$ rule, which states that all semileptonic decays have the change in charge of the hadron equal the change in strangeness. Thus, K mesons decay to $\pi^- \ell^+ \nu$ and $\overline{K}$ mesons decay to $\pi^+ \ell^- \overline{\nu}$. The validity of the $\Delta S = \Delta Q$ rule was in doubt for many years, but it has finally been established that the $\Delta Q = -\Delta S$ transitions are no more than about 2% of the $\Delta Q = +\Delta S$ transitions.[17] The size of the charge asymmetry expected is $\sim\sqrt{2}\,|\eta_{+-}| \approx 3\times10^{-3}$. Millions of events are required to measure δ_ℓ accurately, and excellent control of the symmetry of the apparatus and understanding of charge dependent biases are needed to reduce systematic errors.

Again, we must omit a detailed review of all asymmetry measurements. These have been carried out at CERN, Brookhaven, and SLAC. The net result of these measurements gives[15]

$$\delta_e = (3.33\pm0.14)\times10^{-3}$$

and

$$\delta_\mu = (3.19\pm0.24)\times10^{-3}.$$

We expect these two asymmetries to be equal since they both are a measure of 2 Reε. These asymmetries are measured for a pure K_L beam. For a beam which is pure K at t = 0 the charge asymmetry shows a strong oscillation term with angular frequency ΔM. Figure 3 shows the time dependence of the

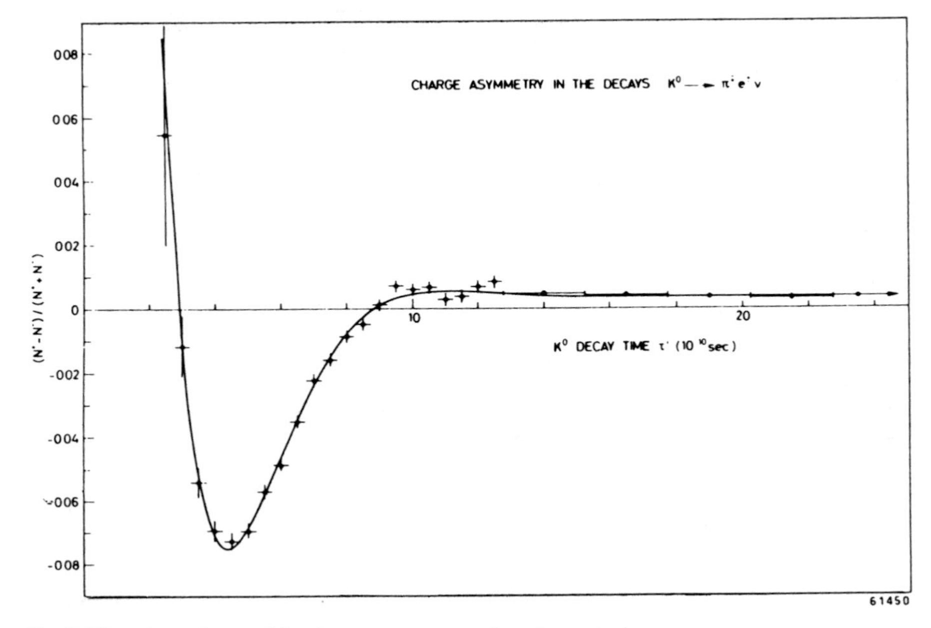

Fig. 3. Time dependence of the charge asymmetry of semileptonic decays.

charge asymmetry taken from the thesis of V. Lüth.[18] The small residual charge asymmetry of the K_L decays after the oscillations have died out is clearly resolved.

The charge asymmetry is a manifest violation of CP, and as such also permits an experimental distinction between a world and an antiworld. In our world we find that the positrons in the decay are slightly in excess. The positrons are leptons which have the same charge as our atomic nuclei. In the antiworld the experimenter will find that the excess leptons have opposite charge to his atomic nuclei; hence, he would report a different result for the same experiment.

Simple examination of the relations between the experimentally measurable parameters and the complex quantities ε and ε' show that measurements of $|\eta_{oo}|$ and ϕ_{oo} are essential to finding ε and ε'. The path to reliable results for $|\eta_{oo}|$ and ϕ_{oo} has been torturous. This statement is based on personal experience; six years of my professional life have been spent on the measurement of $|\eta_{oo}|$.

Measurement of the parameters associated with $K_L \to \pi^o\pi^o$ is complicated by the fact that each π^o decays rapidly (10^{-16} sec) into two photons. For typical K_L beams used in these experiments the photon energies are in the range of 0.25 to 5 GeV. It is difficult to measure accurately the direction and energy of such photons. In addition to that difficulty, the CP conserving decay $K_L \to 3\pi^o$ occurs at a rate which is about 200 times as frequent, and presents a severe background.

Early results suggested that $|\eta_{oo}|$ was about twice $|\eta_{+-}|$ with the consequence that ε' was a large number. By 1968 however, an improved experiment using

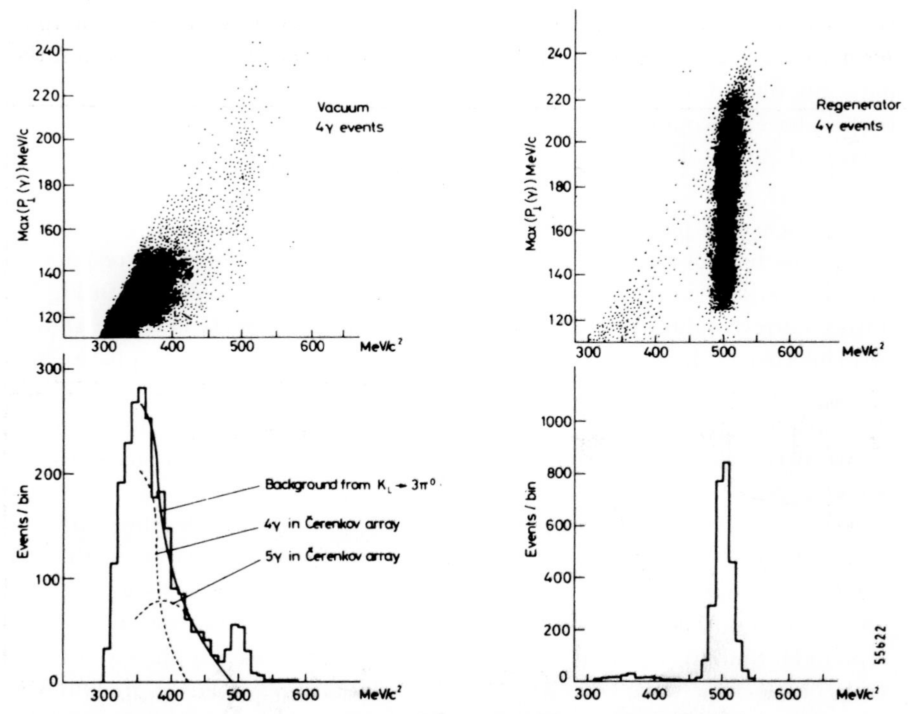

Fig. 4. Distributions of reconstructed $K_L \to \pi^o\pi^o$ events, and regenerated $K_S \to \pi^o\pi^o$ events.

spark chambers[19] and a painstaking heavy-liquid bubble chamber experiment from CERN[20] showed that $|\eta_{oo}|$ was rather close in value to $|\eta_{+-}|$. Figure 4 shows the results from the most accurate measurement of $|\eta_{oo}|/|\eta_{+-}|$.[21] Shown are reconstructed events from free K_L decays as well as a sample of $K_S \rightarrow \pi^o\pi^o$ from a regenerator used to determine the resolution of the apparatus. The serious background from the $3\pi^o$ decays is clearly seen. The result $|\eta_{oo}|/|\eta_{+-}| = 1.00\pm0.06$ is based on only 167 events. The equality of $|\eta_{oo}|$ and $|\eta_{+-}|$ means that the ratio of charged 2π decays to neutral 2π decays is the same for CP violating K_L decays as for CP conserving K_S decays. This result implies that ε' is very small providing ϕ_{oo} is close to ϕ_{+-}.

The $K_L \rightarrow \pi^o\pi^o$ events cannot be collected at the rate of the $\pi^+\pi^-$ decays, nor can they be separated so cleanly from backgrounds. As a consequence, the precision with which we know the parameters $|\eta_{oo}|$ and ϕ_{oo} is much less than the charged parameters. A weighted average of all the data presently available gives[15]

$$|\eta_{oo}|/|\eta_{+-}| = 1.02\pm0.04,$$

and

$$\phi_{oo}-\phi_{+-} = 10°\pm6°.$$

The results are quoted with reference to the charged decay mode parameters because the most accurate experiments have measured the quantity $|\eta_{oo}|/|\eta_{+-}|$ directly. The result for ϕ_{oo} is principally due to a recent experiment by J. Christenson et al.[22]

The phase of the quantity ε' is given by the angle $\pi+\delta_2-\delta_0$. Information concerning the pion-pion scattering phase shifts comes from several sources.[23] A compendium of these sources gives $\delta_2-\delta_0 = -45°\pm10°$. The phase of ε is naturally related to $\phi_n \equiv \arg\left([i(M_S-M_L)+(\Gamma_S-\Gamma_L)/2]^{-1}\right) = 43.7°\pm0.2°$. This is the phase ε would have if there were no contributions from $\mathrm{Im}\Gamma_{12}$. The measured phase of η_{+-} $(44.7°\pm1.2°)$ is within measurement precision equal to ϕ_n.

The measured parameters are plotted on the complex plane in Fig 5a. The size of the box for η_{+-} and η_{oo} and the width of the bar for δ_ℓ correspond to one standard deviation. The derived quantities ε and ε' are plotted in Fig 5b. Boxes corresponding to both one and two standard deviations are shown. Also plotted is the constraint coming from the π—π scattering phase shifts which defines the phase of ε' to be $45°\pm10°$. With this constraint we find that ε, ε', η_{oo} and η_{+-} lie nearly on a common line. There is a mild disagreement between the π—π phase shift constraint and the result of Christenson et al. for ϕ_{oo}.

A more general analysis of the neutral K system which includes the possibility of violation of CPT with T conservation as well as CP violation with CPT conservation has been given by Bell and Steinberger.[24] The analysis does depend on the assumption of unitarity which requires that the M and Γ matrices remain Hermitian. The Bell—Steinberger analysis has been applied to the data with the conclusion that while a small CPT violation is possible, the predominant effect is one of CP violation. All experiments are consistent with exact CPT

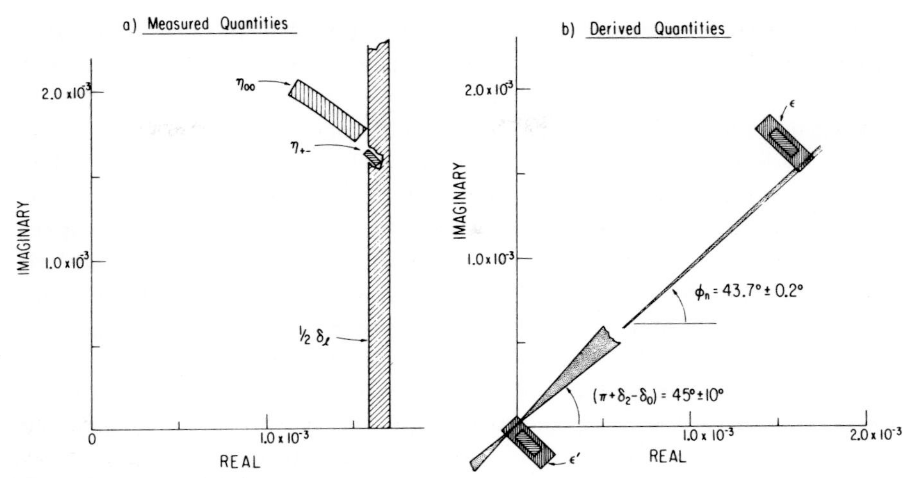

Fig. 5. Summary of CP violating parameters in the neutral K system.
(a) Measured quantities.
(b) Derived quantities.

conservation,[25] and, hence, imply a violation of time reversal symmetry. The conservation or non-conservation of CPT remains, however, a question that must continue to be addressed by experiment. A brief discussion of the unitarity analysis is given in an appendix.

The essential point of this analysis rests on the measurement of the phase of η_{+-}. Limits on the contribution of $\mathrm{Im}\Gamma_{12}$ can be estimated from measured decay rates to all modes of decay of the neutral K mesons. The absence, within present experimental limits, of CP violation in the decay modes other than the 2π modes limits the contribution of $\mathrm{Im}\Gamma_{12}$ to ε to be $\leqslant 0.3\times10^{-3}$, a value small compared to $|\eta_{+-}|$. Thus the phase of ε and hence η_{+-} is expected to be close to ϕ_n. We can examine the other extreme, namely, that CP and CPT symmetry are both violated while time reversal symmetry remains valid. Under these conditions we would find the natural phase ϕ_n to be $\sim 135°$, and would expect ϕ_{+-} to be close to 135°. The fact that this is not the case is the essence of the argument that CPT is not violated.

We note that the natural phase depends on the sign of the mass difference. We have assumed $\Delta M = (M_S - M_L) < 0$. If the sign of the mass difference were opposite, we would expect the phase of ε to be equal to 135° or $-45°$ for CP violation with CPT symmetry. The phase of ε' would remain the same, however, since it does not depend on ΔM in any way. Thus, the conclusion that the phase of ε and ε' are approximately the same is a consequence of the fact that the long lived K is heavier than the short lived K. The sign of the mass difference has been measured by several groups with complete agreement.[26]

Independent of any particular theory, we would expect results which are similar to those observed. The constraint of unitarity and $\pi\pi$ scattering phase shifts force $\phi_{oo} \approx \phi_{+-}$ for $\varepsilon' \ll \varepsilon$. Under these circumstances, a measurement of the ratio $(|\eta_{oo}|/|\eta_{+-}|)^2$ is a direct measurement of the quantity ε' by means of the relation $\varepsilon'/\varepsilon \approx [1-(|\eta_{oo}|/|\eta_{+-}|)^2]/6$. Applying this relation to the present data we have $\varepsilon'/\varepsilon = -0.007 \pm 0.013$. New experiments at the Fermilab and at Brookhaven will attempt to increase the sensitivity of the measurement by a factor 10.

As we have shown, detailed analysis of the CP violation in the neutral K meson system leads to the conclusion that time reversal is also violated. Table I gives a representative set of experiments which have searched for T violation, CP violation, and C violation (in non-weak interactions). None of these experiments has led to a positive result. Many of the experiments are approaching a sensitivity for the violation of 10^{-3}, but few have attained this value. A strength of 10^{-3} in amplitude or relative phase is what we might expect for the CP violation based on the results of K-decay. For experiments involving decays with electromagnetic interactions in the final states, an apparent T-violation effect is usually expected at the 10^{-3} level. An example of this is the result for the ^{191}Ir decay in which a significant effect is found, but it is of the size expected on the basis of the final state electromagnetic interaction.

Table I. Searches for CP, T, and C Violation

Measurement	Result	Test	Ref
$\frac{\Gamma(K^+ \to \pi^+\pi^+\pi^-) - \Gamma(K^- \to \pi^-\pi^-\pi^+)}{\text{average}}$	$(0.8 \pm 1.2) \times 10^{-3}$	CP	37
$\frac{\Gamma(K^+ \to \pi^+\pi^\circ\pi^\circ) - \Gamma(K^- \to \pi^-\pi^\circ\pi^\circ)}{\text{average}}$	$(0.8 \pm 5.8) \times 10^{-3}$	CP	38
$\frac{a_{\tau^+} - a_{\tau^-}}{\text{average}}$, where $a_{\tau^\pm}$ is the slope of the odd pion in the $K^\pm \to \pi^\pm\pi^\pm\pi^\mp$ Dalitz plot	$(-7.0 \pm 5.3) \times 10^{-3}$	CP	37
Muon polarization transverse to decay plane in $K_L \to \pi^-\mu^+\nu_\mu$	$(2.1 \pm 4.8) \times 10^{-3}$	T	39
Coefficient of T odd correlation $<\vec{J} \cdot \vec{P}_e \times \vec{P}_\nu>$ in the β-decay of polarized ^{19}Ne	$(-0.5 \pm 1.0) \times 10^{-3}$	T	40
Coefficient of T odd correlation $<\vec{\sigma}_n \cdot \vec{P}_e \times \vec{P}_\nu>$ in the β-decay of the neutron	$(-1.1 \pm 1.7) \times 10^{-3}$	T	41
Asymmetry in distribution of $(T_{\pi^+} - T_{\pi^-})$ in the decay of $\eta \to \pi^+\pi^-\pi^\circ$	$(1.2 \pm 1.7) \times 10^{-3}$	C	42
Electric dipole moment of the neutron	$(0.4 \pm 1.5) \times 10^{-24}$ e-cm $(0.4 \pm 0.75) \times 10^{-24}$ e-cm	T	43 44
Angular correlation in γ decay of polarized iridium, $^{191}Ir^* \to {}^{191}Ir + \gamma$. Measure phase angle between E_2 and M_1 decay amplitudes.	$(4.7 \pm 0.3) \times 10^{-3}$	T	45
Result expected on basis of electromagnetic interaction in final state	4.3×10^{-3}		46
Detailed balance in nuclear rections, e.g., $^{24}Mg + \alpha \rightleftarrows {}^{27}A\ell + p$ Measure: $\frac{\text{amplitude T violating}}{\text{amplitude T conserving}}$	$\leqslant 3 \times 10^{-3}$	T	47

Among the many measurements listed in Table I, we would like to single out the electric dipole moment of the neutron. The first measurement of this quantity was made in 1950 by Purcell, Ramsey and Smith[27] with the avowed purpose of testing the assumptions on which one presumed the electric dipole moment would be zero. Today, outside of the K-system, the search for an electric dipole moment of the neutron is the most promising approach to the detection of T violation. At present the upper limit is $\sim 10^{-24}$ e-cm. New experiments using ultra-cold neutrons give promise of an increase in intensity by 100-fold within the next several years. The significance of a negative result for the electric dipole moment, or for any of the measurements in Table I, is difficult to assess without a theory of CP violation.[28]

Up to now our discussion has been entirely experimental. In the analysis of the CP violation in the neutral K system general principles of quantum mechanics have been used. The manifest charge asymmetry of the K_L semi-leptonic decays requires no assumptions at all for its interpretation. The literature abounds with theoretical speculations about CP violation. One of these speculations by Wolfenstein[29] is frequently referred to. He hypothesizes a direct $\Delta S = 2$ superweak interaction which is contructed to produce a CP violation. This direct interaction interferes with the second order weak interaction to produce the CP-violating $\Delta S = 2$ coupling between K and $\bar{K}$. Since the hypothesized superweak transition is first order, it need have only $\sim 10^{-7}$ of the strength of the normal weak interaction. As such the only observable consequence is a CP violation in $K \to 2\pi$ decay characterized by a single number, the value of $\mathrm{Im}M_{12}$ in the mass matrix.

At present the data are in agreement with this hypothesis, which leads to predictions that $|\eta_{oo}| = |\eta_{+-}|$, and $\phi_{oo} = \phi_{+-} = \phi_n$. However, the relation $\phi_{oo} = \phi_{+-} = \phi_n$ to a good approximation follows from the constraints of unitarity and the π—π scattering phase shifts with no further assumptions. On the other hand, the relation $|\eta_{oo}| = |\eta_{+-}|$ has not been tested to very high accuracy, especially considering the difficulty of experiments which attempt to measure the properties of $K_L \to \pi^o\pi^o$. These experiments are more prone to systematic errors, and in truth $|\eta_{oo}|$ and $|\eta_{+-}|$ could differ considerably more than appears to be allowed by the experiments. Thus, while the superweak hypothesis is in agreement with the present data, the data by no means make a compelling case for the superweak hypothesis.

In 1973, Kobayashi and Maskawa[30] in a remarkable paper pointed out that with the (then) current understanding of weak interactions, CP violation could be accommodated only if there were three or more pairs of strongly interacting quarks. The paper was remarkable because at that time only three quarks were known to exist experimentally. Since then, strong evidence has been accumulated to support the existence of a charmed quark and a fifth bottom quark. It is presumed that the sixth quark, top, will be eventually found. With six quarks the weak hadronic current involving quarks can be characterized by three Cabibbo-angles, and a phase δ. This phase, if non-zero, would imply a CP violation in the weak interaction.

In principle, the magnitude of this phase δ which appears in the weak currents of quarks can be related to the CP violation observed in the laboratory. Unfortunately, all the experimental investigations are carried out with hadrons, which are presumed to be structures of bound quarks, while the parameter one wants to establish, δ, is expressed in terms of interactions between free quarks. The theoretical "engineering" required to relate the free quark properties to bound quark properties is difficult and, as a consequence, is not well developed. A balanced and sober view of this problem is given in a paper by Guberina and Peccei.[31] Even if the CP violation has its origin in the weak currents, it is not clear whether the experimental consequences with respect to K decay can be distinguished from the superweak hypothesis. If we are successful in establishing the fact that CP violation is the result of a phase in the weak currents between quarks, we will still have to understand why it has the particular value we find.

There are, however, on the horizon new systems which have some promise to give additional information about CP violation. These are the new neutral mesons, D°, B°, B_s°, (composed of $c\bar{u}$, $b\bar{d}$, and $b\bar{s}$ quarks), and their antiparticles $\bar{D}^{\circ}$, $\bar{B}^{\circ}$, $\bar{B}_s^{\circ}$. These mesons have the same general properties as K mesons. They are neutral particles that, with respect to strong interactions, are distinct from their own antiparticles, and yet are coupled to them by common weak decay modes. While we may not expect any stronger CP impurities on the eigenstates (the parameter analogous to ε), we might expect stronger effects in the decay amplitudes (the parameter analogous to ε'). We might expect this since the CP violation comes about through the weak interactions of the heavy quarks, c, b, t, which participate only virtually in K decay, but can be more influential in heavy neutral meson decay. At present, D mesons can be made rather copiously at the e^+e^- storage ring SPEAR at SLAC,[32] and B mesons are beginning to be produced at the e^+e^- storage ring CESR at Cornell.[33]

It is conceivable that the effect of CP violation may become stronger with energy. Soon collisions of protons with antiprotons will be observed at CERN with a total center of mass energy greater than 500 GeV. It will be most interesting to look for C violations in the spectra of particles produced in those collisions. Also, improvements in technology of detectors over the next several decades may permit sensitive searches for time reversal violating observables in high energy neutrino interactions.

Recently, much attention has been given to the role that CP violation may play in the early stages of the evolution of the universe.[34] A mechanism has been proposed with CP violation as one ingredient which leads from matter--antimatter symmetry in the early universe to the small excess of matter observed in the universe at the present time. The first published account of this mechanism, of which I am aware, was made by Sakharov[35] in 1967. He explicitly stated the three ingredients which form the foundation of the mechanism as it is presently discussed. These ingredients are: (1) baryon instability, (2) CP violation, and (3) appropriate lack of thermal equilibrium. The recent intense interest in this problem has risen because baryon instability is a natural consequence of the present ideas of unification of the strong interactions with the successfully unified electromagnetic and weak interactions. This latter unification was discussed in the 1979 Nobel lectures of Glashow, Salam, and Weinberg.[36]

A very oversimplified explanation of the process which leads to a net baryon number can be given with the aid of Fig 6a. Quarks and leptons are linked by a very heavy boson X and its antiparticle $\bar{X}$. While the total decay rates

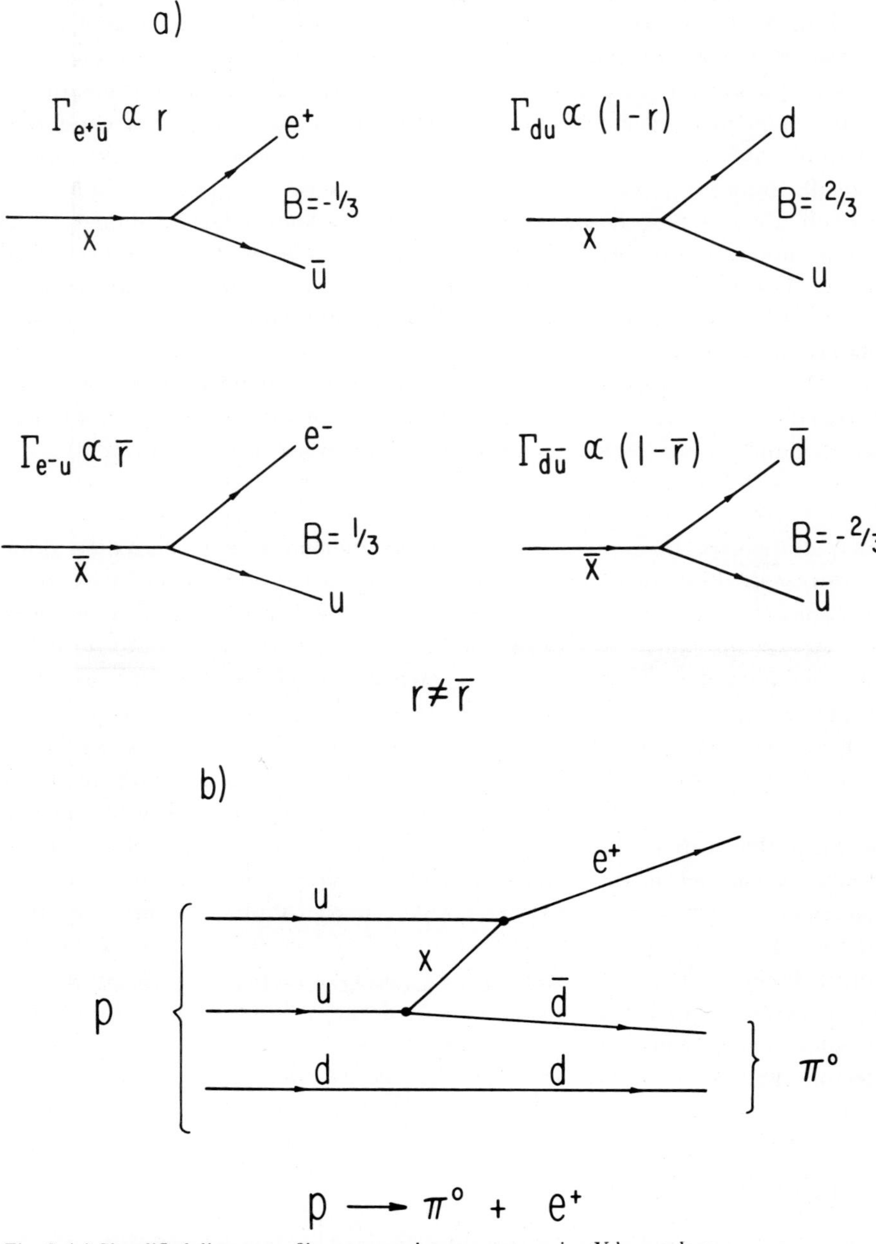

Fig. 6. (a) Simplified diagrams of baryon number non-conserving X boson decays. (b) A proton decay mediated by an X boson.

of X and X may be equal, with CP violation the fractional partial rates r and $\bar{r}$ to $B = -\frac{1}{3}$ and $B = +\frac{1}{3}$ decay channels of X and $\bar{X}$, respectively, can differ. At an early stage where the temperature is large compared to the mass of X, the density of X and $\bar{X}$ may be equal. On decay, however, the net evolution of baryon number is proportional to $(r - \bar{r})$. The excess can be quite small since the ratio of baryons to photons today is $\sim 10^{-9}$. Figure 6b shows how such an X boson can mediate the decay $p \rightarrow e^{+} + \pi^{\circ}$. If nucleon decay is discovered it will give a strong support to these present speculations.

Whether the CP violation that we observe today is a "fossil remain" of these conjectured events in the early universe is a question that cannot be answered at present. That is to say, does the CP violation we observe today provide supporting evidence for these speculations? We simply do not know enough about CP violation. Our experimental knowledge is limited to its observation in only one extraordinarily sensitive system that nature has provided us. We need to know the theoretical basis for CP violation and we need to know how to reliably extrapolate the behavior of CP violation to the very high energies involved.

At present our experimental understanding of CP violation can be summarized by the statement of a single number. If we state that the mass matrix which couples K and $\bar{K}$ has an imaginary off-diagonal term given by

$$\mathrm{Im} M_{12} = -1.16 \times 10^{-8}\,\mathrm{eV},$$

then all the experimental results related to CP violation can be accounted for. If this is all the information nature is willing to provide about CP violation it is going to be difficult to understand its origin. I have emphasized, however, that despite the enormous experimental effort, punctuated by some experiments of exceptional beauty, we have not reached a level of sensitivity for which a single parameter description should either surprise or discourage us.

We must continually remind ourselves that the CP violation, however small, is a very real effect. It has been used almost routinely as a calibration signal in several high energy physics experiments. But more importantly, the effect is telling us that there is a fundamental asymmetry between matter and antimatter, and it is also telling us that at some tiny level interactions will show an asymmetry under the reversal of time. We must continue to seek the origin of the CP symmetry violation by all means at our disposal. We know that improvements in detector technology and quality of accelerators will permit even more sensitive experiments in the coming decades. We are hopeful then, that at some epoch, perhaps distant, this cryptic message from nature will be deciphered.

APPENDIX

The evolution of a neutral K system characterized by time dependent amplitudes a and $\bar{a}$ for the $|K\rangle$ and $|\bar{K}\rangle$ components, respectively, is given by

$$-\frac{d}{dt}\begin{pmatrix} a \\ \bar{a} \end{pmatrix} = \left(iM+\frac{1}{2}\Gamma\right)\begin{pmatrix} a \\ \bar{a} \end{pmatrix},$$

where M and Γ are each Hermitian matrices, and t is the time measured in the rest system of the K meson. Expressed in terms of their elements the matrices are

$$\begin{pmatrix} M_{11} & M_{12} \\ M_{12}^* & M_{22} \end{pmatrix} \text{ and } \begin{pmatrix} \Gamma_{11} & \Gamma_{12} \\ \Gamma_{12}^* & \Gamma_{22} \end{pmatrix}.$$

The matrix $iM+\frac{1}{2}\Gamma$ has eigenvalues $\gamma_S = iM_S+\frac{1}{2}\Gamma_S$ and $\gamma_L = iM_L+\frac{1}{2}\Gamma_L$. We define small parameters $\varepsilon = (-\mathrm{Im}M_{12}+i\mathrm{Im}\Gamma_{12}/2)/(\gamma_S-\gamma_L)$ and $\Delta = [i(M_{11}-M_{22})+(\Gamma_{11}-\Gamma_{22})/2]/[2(\gamma_S-\gamma_L)]$. We can then express the eigenvectors as

$$|K_S\rangle = \frac{1}{\sqrt{2}}\frac{1}{\sqrt{1+|\varepsilon+\Delta|^2}}\left[(1+\varepsilon+\Delta)\,|K\rangle+(1-\varepsilon-\Delta)\,|\bar{K}\rangle\right]$$

and

$$|K_L\rangle = \frac{1}{\sqrt{2}}\frac{1}{\sqrt{1+|\varepsilon-\Delta|^2}}\left[(1+\varepsilon-\Delta)\,|K\rangle-(1-\varepsilon+\Delta)\,|\bar{K}\rangle\right].$$

The parameter ε represents a CP violation with T non-conservation. The parameter Δ represents a CP violation with CPT non-conservation.

If we form a state $|K(t)\rangle$ which is an arbitrary superposition of $|K_S\rangle$ and $|K_L\rangle$ with amplitudes a_S and a_L at $t = 0$, we can compute its norm $\langle K(t)|K(t)\rangle$ as a function of time. At $t = 0$ by conservation of probability we have the relation.

$$-\frac{d}{dt}\langle K(t)|K(t)\rangle\Big|_{t=0} = \sum_f |a_S\,\mathrm{amp}(K_S \to f)+a_L\,\mathrm{amp}(K_L \to f)|^2,$$

where f represents the set of final states. Explicit evaluation of the expression gives

$$[-i(M_S-M_L)+(\Gamma_S+\Gamma_L)/2]\,\langle K_S|K_L\rangle = \sum_f (\mathrm{amp}(K_S \to f))^*\,(\mathrm{amp}(K_L \to f)).$$

A number of definitions and a particular phase convention are used. We define $\tilde{\Delta} = \Delta-(A_0-\bar{A}_0)/(A_0+\bar{A}_0)$ where A_0 and $\bar{A}_0$ are the standing wave amplitudes for K and $\bar{K}$, respectively, to decay to the $I = 0$ state of two pions. A_0 and $\bar{A}_0$ are chosen real and define the phase convention used in the analysis. From the experimental parameters we define $\varepsilon_0 = \frac{2}{3}\eta_{+-}+\frac{1}{3}\eta_{oo}$ and $\varepsilon_2 = \frac{\sqrt{2}}{3}(\eta_{+-}-\eta_{oo})$, and $\alpha(f) = (1/\Gamma_S)\,(\mathrm{amp}(K_S \to f))^*\,(\mathrm{amp}(K_L \to f))$. With these definitions we find to a good approximation that

$$\left[-i\Delta M/\Gamma_S+\frac{1}{2}\right]\left[2\mathrm{Re}\varepsilon-2i\mathrm{Im}\tilde{\Delta}\right] = \varepsilon_0+\sum_f \alpha(f) \quad (1)$$

and

$$\varepsilon-\tilde{\Delta} = \varepsilon_0 \quad (2)$$

The sum over f, which now excludes the I = 0 $\pi\pi$ state, consists of the following terms:

$$\alpha(\pi\pi, I = 2) = A_2/A_0 e^{i(\delta_2-\delta_0)}\varepsilon_2^*,$$

$$\alpha(\pi^+\pi^-\pi^\circ) = (\Gamma(K_L \rightarrow \pi^-\pi^+\pi^\circ)/\Gamma_S)\eta^*_{+-o},$$

$$\alpha(\pi^\circ\pi^\circ\pi^\circ) = (\Gamma(K_L \rightarrow \pi^\circ\pi^\circ\pi^\circ)/\Gamma_S)\eta^*_{ooo},$$

$$\alpha(\pi e\nu) = (\Gamma(K_L \rightarrow \pi e\nu)/\Gamma_S)2i\,\mathrm{Im}\,x_e,$$

and
$$\alpha(\pi\mu\nu) = (\Gamma(K_L \rightarrow \pi\mu\nu)/\Gamma_S)2i\,\mathrm{Im}\,x_\mu,$$

where η_{+-o} = amp $(K_S \rightarrow \pi^+\pi^-\pi^\circ)$/amp $(K_L \rightarrow \pi^+\pi^-\pi^\circ)$, η_{ooo} = amp $(K_S \rightarrow \pi^\circ\pi^\circ\pi^\circ)$/ amp $(K_L \rightarrow \pi^\circ\pi^\circ\pi^\circ)$, and x_ℓ is the ratio, amp$(\Delta Q = -\Delta S)$/amp$(\Delta Q = \Delta S)$, for $K \rightarrow \pi\ell\nu_\ell$. The quantities η_{+-o} and η_{ooo} are CP violating ratios. (The final state $\pi^+\pi^-\pi^\circ$ can be CP even or odd. Here we refer only to the odd state.) The measurements of η_{ooo} and η_{+-o} are not at present very accurate and are consistent with zero. If we use the experimental limits which exist,[15] we find

$$\mathrm{Re}\alpha = \mathrm{Re}\sum_f \alpha(f) = (0.14 \pm 0.19) \times 10^{-3}$$

and
$$\mathrm{Im}\alpha = \mathrm{Im}\sum_f \alpha(f) = (-0.19 \pm 0.25) \times 10^{-3}.$$

The equations (1) and (2) take a very simple form if we resolve the components of ε and $\tilde{\Delta}$ parallel and perpendicular to the direction which makes an angle ϕ_n with the real axis, where

$$\phi_n = \tan^{-1}\left[-\frac{2(M_S - M_L)}{(\Gamma_S - \Gamma_L)}\right].$$

We then find
$$\varepsilon_\parallel = \varepsilon_{o_\parallel} + \cos\phi_n\,\mathrm{Re}\alpha,$$

$$\varepsilon_\perp = -\cos\phi_n\,\mathrm{Im}\alpha,$$

$$\tilde{\Delta}_\parallel = \cos\phi_n\,\mathrm{Re}\alpha,$$

and
$$\tilde{\Delta}_\perp = -\varepsilon_{o\perp} - \cos\phi_n\,\mathrm{Im}\alpha.$$

The experimental values of $\varepsilon_{o_\parallel}$ and $\varepsilon_{o\perp}$ are, respectively, $(2.27 \pm 0.03) \times 10^{-3}$ and $(0.16 \pm 0.09) \times 10^{-3}$. We then find

$$\varepsilon_\parallel = (2.37 \pm 0.19) \times 10^{-3},$$

$$\varepsilon_\perp = (0.14 \pm 0.18) \times 10^{-3},$$

$$\tilde{\Delta}_\parallel = (0.10 \pm 0.14) \times 10^{-3},$$

and
$$\tilde{\Delta}_\perp = (-0.02 \pm 0.20) \times 10^{-3}.$$

Within the present experimental limits we find that all the measurements are consistent with T violation and CPT conservation. In particular, we see the limit on $\varepsilon_\perp$ is very small so that we cannot expect ϕ_{+-} and ϕ_{oo} to differ greatly from ϕ_n. Further, if the values of η_{ooo}, η_{+-}, x_e, and x_μ were $< 10^{-2}$, then we would find $|\varepsilon_{\perp}| \leq 10^{-5}$. Such an expectation is reasonable if the strength of the CP violation is roughly the same in all modes.

ACKNOWLEDGEMENTS

I would like to thank Professors S. Chandrasekhar, R. Oehme, R. G. Sachs and B. Winstein for their advice and critical comments concerning this lecture. I would also like to thank Professors V. Telegdi, S. Treiman, and L. Wolfenstein for many valuable discussions concerning CP violation over the years.

REFERENCES

1. This lecture cannot cover all the important details, refer to all the important work concerning CP violation, or do justice to the work that is referred to in the text. For a more complete discussion, the reader is referred to the most recent review by K. Kleinknecht, *Annual Reviews of Nuclear Science*, Vol. 26, 1, 1976. Also, a good perspective of the progress in the field can be found by reading the appropriate sections of the Proceedings of the biannual International Conferences on High Energy Physics, 1964—1974.
2. Galbraith, W. et al., Phys. Rev. Letters *14*, 383 (1965).
3. de Bouard, X. et al., Phys. Letters *15*, 58 (1965).
4. Fitch, V. L. et al., Phys. Rev. Letters *15*, 73 (1965).
5. The value of A_r depends on the K_S-K_L mass difference $|\delta|$ which was, at the time, measured to be 0.5 ± 0.1 by J. H. Christenson et al., Phys. Rev. *140B*, 74 (1965).
6. Fitch, V. L. et al., Phys. Rev. *164*, 1711 (1967).
7. This argument was presented in the literature by Wattenberg, A. and Sakurai, J., Phys. Rev. *161*, 1449 (1967).
8. Bell, J. S. and Perring, J. K., Phys. Rev. Letters *13*, 348 (1964); Bernstein, J., Cabibbo, N. and Lee, T. D., Phys. Letters *12*, 146 (1964).
9. Wu, T. T. and Yang, C. N., Phys. Rev. Letters *13*, 380 (1964); the phenomenology was first discussed by Lee, Oehme and Yang, Phys. Rev. *106*, 340 (1957).
10. Wigner, E. P., Göttinger Nachrichten *31*, 546 (1932); see also paper of Lee, Oehme, and Yang in Ref. 9.
11. Some approximations have been made. The first two expressions should read: $\eta_{+-} = \varepsilon+\varepsilon'/(1+\omega)$ and $\eta_{00} = \varepsilon-2\varepsilon'/(1-2\omega)$ where $\omega = 1/\sqrt{2}\ \mathrm{Re}(A_2/A_0)e^{i(\delta_2-\delta_0)}$. The magnitude of $\omega \approx 0.05$, so that its effect is not large. The charge asymmetry δ_L should be given by $\delta_L = 2\frac{(1-|x|^2)}{|1-x|^2}\ \mathrm{Re}\varepsilon$, where x is the ratio of the $\Delta Q = -\Delta S$ amplitude to the $\Delta Q = +\Delta S$ amplitude in the semi-leptonic decay. Evidence strongly favors $x \approx 0$; see Ref. 17.
12. Whatley, M. C., Phys. Rev. Letters *9*, 317 (1962).
13. Here $\Delta M = M_S-M_L$ and we have neglected Γ_L compared to Γ_S. The interference experiments are always performed over a time scale such that $t \ll \frac{1}{\Gamma_L}$ so that the decay of the K_L amplitude is negligible.
14. Precise measurements of ΔM have been reported by: Aronson, S. H. et al., Phys. Rev. Letters *25*, 1057 (1970); Cullen, M. et al., Phys. Letters *32B*, 523 (1970); Carnegie, R. K. et al., Phys. Rev. *D4*, 1 (1971); Geweniger, C. et al., Phys. Letters *52B*, 108 (1974).
15. The data for this compilation are most readily available from the Particle Data Group, Barash-Schmidt, N. et al., Rev. Mod. Phys. *52*, S1 (1980).

16. Figure taken from thesis of Modis, T., Columbia University (1973), (unpublished); a published version of this work is given by Carithers, W. et al., Phys. Rev. Letters *34,* 1244 (1975).
17. Niebergall, F. et al., Phys. Letters *49B,* 103 (1974).
18. Figure taken from thesis of Lüth, V., Heidelberg University (1974), (unpublished); a published version of this work can be found in Gjesdal, S. et al., Phys. Letters *52B*, 113 (1974).
19. Banner, M. et al., Phys. Rev. Letters *21,* 1103 (1968).
20. Budagov, I. A. et al., Phys. Letters *28B,* 215 (1968).
21. Holder, M. et al., Phys. Letters *40B*, 141 (1972).
22. Christenson, J. H. et al., Phys. Rev. Letters *43,* 1209 (1979).
23. From extrapolation of the phase shift analysis of Ke_4 decays, one finds $\delta_2-\delta_0 = 36°\pm10°$, Rosselet, J. et al., Phys. Rev. *D15*, 574 (1977). From analysis of K^+ and K_S decays to $\pi^+\pi^-$ and $\pi^\circ\pi^\circ$, one finds $\delta_2-\delta_0 = \pm(53°\pm6°)$, Abbud, Lee, and Yang, Phys. Rev. Letters, *18,* 980 (1967); Particle Data Group, Barash-Schmidt, N. et al., Rev. Mod. Phys. *52,* S1 (1980). From analysis of pion production by pions, one finds $\delta_2-\delta_0 = 40°\pm6°$, see for example, Baton, J. et al., Phys. Letters *33B*, 528 (1970); Estabrooks, P. and Martin, A. D., Nucl. Phys. *B79,* 301 (1974).
24. Bell, J. S. and Steinberger, J., Proceedings of the Oxford International Conference on Elementary Particles, 1965, edited by Walsh, T. et al., (Rutherford Laboratory, Chilton, England 1966). An analysis with a similar purpose has also been given by Sachs, R. G., Progr. Theor. Phys. (Japan) *54,* 809 (1975). References to the literature concerning the analysis of the neutral K meson decay data into a T conserving part and T violating part are given by Sachs.
25. Schubert, K. R. et al., Phys. Letters *31B*, 662 (1970).
26. Canter, J. et al., Phys. Rev. Letters *17,* 942 (1966); Meisner, G. W. et al., Phys. Rev. Letters *17,* 492 (1966); Mehlhop, W. A. W. et al., Phys. Rev. *172,* 1613 (1968). The last reference uses a highly innovative technique, and will give pleasure to those who take the time to read it.
27. Purcell, E. M. and Ramsey, N. F., Phys. Rev. *78*, 807 (1950); also, Smith, J. H. Ph.D. thesis Harvard University (1951) (unpublished).
28. Weinberg has suggested a mechanism whereby the CP violation is due to Higgs mesons. The suggestion is attractive because the CP violation can be maximal and a neutron electric dipole moment of $\sim 10^{-24}$ might be expected, see Weinberg, S., Phys. Rev. Letters *37,* 657 (1976).
29. Wolfenstein, L., Phys. Rev. Letters *13,* 569 (1964).
30. Kobayashi, M. and Maskawa, K., Progr. Theor. Phys. *49,* 652 (1973).
31. Guberina, B. and Peccei, R. D., Nucl. Phys. *B163,* 289 (1980).
32. Lüth, V., Proceedings of the 1979 International Symposium on Lepton and Photon Interactions at High Energies, 1979, edited by Kirk, T. B. W., p. 83 (Fermilab, Batavia, IL, USA, 1980).
33. Andrews, D. et al., Phys. Rev. Letters *45,* 219 (1980); Finocchiaro, G. et al., Phys. Rev. Letters *45,* 222 (1980).
34. Yoshimura, M., Phys. Rev. Letters *41,* 381 (1978); Dimopoulos, S. and Susskind, L., Phys. Rev. *D18*, 4500 (1978); Toussaint, B. et al., Phys. Rev. *D19*, 1036 (1979); Ellis, J., Gaillard, M. K. and Nanopoulos, D. V., Phys. Letters *80B*, 360 (1979); Weinberg, S., Phys. Rev. Letters *42,* 850 (1979).
35. Sakharov, A. D., ZhETF Pis'ma *5,* 32 (1967); English translation JETP Letters *5,* 24 (1967).
36. Glashow, S. L., Rev. Mod. Phys. *52,* 539 (1980); Salam, A., ibid. *52,* 525 (1980); Weinberg, S., ibid, *52,* 515 (1980).
37. Ford, W. T. et al., Phys. Rev. Letters *25,* 1370 (1970).
38. Smith, K. M. et al., Nucl. Phys. *B60*, 411 (1970).
39. Schmidt, M. et al., Phys. Rev. Letters *43,* 556 (1979).
40. Baltrusaitis, R. M. and Calaprice, F. P., Phys. Rev. Letters *38,* 464 (1977).
41. Steinberg, R. I. et al., Phys. Rev. Letters *33,* 41 (1974).
42. Layter, J. G. et al., Phys. Rev. Letters *29,* 316 (1972); Jane, M. R. et al., Phys. Letters *48B,* 260 (1974).
43. Dress, W. B. et al., Phys. Rev. *D15,* 9 (1977).
44. Altarev, I. S. et al., "Search for an Electric Dipole Moment by Means of Ultra-cold Neutrons," p. 541, Proceedings of the Third International Symposium on Neutron Capture Gamma-Ray

Spectroscopy and Related Topics, 1978, Brookhaven National Laboratory, Upton, NY and State University of New York, Stony Brook, NY, ed. By Chrien, Robert E. and Kane, Walter R., (Plenum Press).

45. Gimlett, J. L., Phys. Rev. Letters *42,* 354 (1979).
46. Davis, B. R. et al., Phys. Rev. *C22,* 1233 (1980).
47. Weitkamp, W. G. et al., Phys. Rev. *165,* 1233 (1968); von Witsch, W. et al., Phys. Rev. *169,* 923 (1968); Thornton, S. T. et al., Phys. Rev. *C3,* 1065 (1971); Driller, M. et al., Nucl. Phys. *A317,* 300 (1979).

Val L. Fitch

VAL LOGSDON FITCH

I was born the youngest of three children, on a cattle ranch in Cherry County, Nebraska, not far from the South Dakota border, on March 10, 1923. This is a very sparsely populated part of the United States and remote from any center of population. It seems incredible by modern standards that by the age of 20 my father, Fred Fitch, had acquired a ranch of more than 4 square miles and had persuaded a local school teacher, Frances Logsdon, to marry and join him in living there. They moved to the ranch just 20 years after the battle of Wounded Knee, which occurred about 40 miles northwest. I mention this because our living close to their reservation made the Sioux Indians very much a part of our environment. My father, while not fluent, spoke their language. They recognized his friendly interest on their behalf by making him an honorary chief.

Not long after my birth my father was badly injured when a horse he was riding fell with him. He subsequently had to give up the physically strenuous activity associated with running a ranch and raising cattle. The family moved to Gordon, Nebraska, a town about 25 miles away, where my father entered the insurance business. All of my formal schooling through high school was in the public schools of Gordon. During this period my parents retained ownership of the ranch but the operation was largely left to others. E. B. White has defined farming as 10 % agriculture and 90 % fixing something that has gotten broken. My memories of ranching are primarily not the romantic ones of rounding up and branding cattle but rather of oiling windmills and fixing fences.

Probably the most significant occurrence in my education came when, as a soldier in the U.S. Army in WWII, I was sent to Los Alamos, New Mexico, to work on the Manhattan Project. The work I did there under the direction of Ernest Titterton, a member of the British Mission, was highly stimulating. The laboratory was small and even as a technician garbed in a military fatigue uniform I had the opportunity to meet and see at work many of the great figures in physics: Fermi, Bohr, Chadwick, Rabi, Tolman. I have recorded some of the experiences from those days in a chapter in *All in Our Time,* a book edited by Jane Wilson and published by the Bulletin of Atomic Scientists. I spent 3 years at Los Alamos and in that period learned well the techniques of experimental physics. I observed that the most accomplished experimentalists were also the ones who knew most about electronics and electronic techniques were the first I learned. But mainly I learned, in approaching the measurement of new phenomena, not just to consider using existing apparatus but to allow the mind to wander freely and invent new ways of doing the job.

Robert Bacher, the leader of the physics division in which I worked, offered me a graduate assistantship at Cornell after the war but I still had to finish the work for an undergraduate degree. This I did at McGill University. And then another opportunity for graduate work came from Columbia and I ended up there working with Jim Rainwater for my Ph. D. thesis. One day in his office, which he shared at the time with Aage Bohr, he handed me a preprint of a paper by John Wheeler devoted to μ-mesic atoms. This paper emphasized, in the case of the heavier nuclei, the extreme sensitivity of the ls level to the size of the nucleus. Even though the radiation from these atoms had never been observed, these atomic systems might be a good thesis topic. At this same time a convergence of technical developments took place. The Columbia Nevis cyclotron was just coming into operation. The beams of π-measons from the cyclotron contained an admixture of μ-measons which came frome the decay of the π's and which could be separated by range. Sodium iodide with thallium activation had just been shown by Hofstadter to be an excellent scintillation counter and energy spectrometer for γ rays. And there were new phototubes just being produced by RCA which were suitable matches to sodium iodide crystals to convert the scintillations to electrical signals. The other essential ingredient to make a γ-ray spectrometer was a multichannel pulse height analyzer which, utilizing my Los Alamos experience, I designed and built with the aid of a technician. The net result of all the effort for my thesis was the pioneering work on μ-mesic atoms. It is of interest to note that we came very close to missing the observation of the γ-rays completely. Wheeler had calculated the 2p-ls transition energy in Pb, using the then accepted nuclear radius 1.4 $A^{1/3}$ fermi, to be around 4.5 MeV. Correspondingly, we had set our spectrometer to look in that energy region. After several frustrating days, Rainwater suggested we broaden the range and then the peak appeared—not at 4.5 MeV but at 6 MeV! The nucleus was substantially smaller than had been deduced from other effects. Shortly afterwards Hofstadter got the same results from his electron scattering experiments. While the μ-mesic atom measurements give the rms radius of the nucleus with extreme accuracy the electron scattering results have the advantage of yielding many moments to the charge distribution. Now the best information is obtained by combining the results from both μ-mesic atoms and electron scattering.

Subsequently, in making precise γ-ray measurements to obtain a better mass value for the μ-meson, we found that substantial corrections for the vacuum polarization were required to get agreement with independent mass determinations. While the vacuum polarization is about 2% of the Lamb shift in hydrogen it is the very dominant electrodynamic correction in μ-mesic atoms.

My interest then shifted to the strange particles and K mesons but I had learned from my work at Columbia the delights of unexpected results and the challenge they present in understanding nature. I took a position at Princeton where, most often working with a few graduate students, I spent the next 20 years studying K-mesons. The ultimate in unexpected results was that which was recognized by the Nobel Foundation in 1980, the discovery of CP-violation.

At any one time there is a natural tendency among physicists to believe that we already know the essential ingredients of a comprehensive theory. But each time a new frontier of observation is broached we inevitably discover new phenomena which force us to modify substantially our previous conceptions. I believe this process to be unending, that the delights and challenges of unexpected discovery will continue always.

It is highly improbable, a priori, to begin life on a cattle ranch and then appear in Stockholm to receive the Nobel prize in physics. But it is much less improbable to me when I reflect on the good fortune I have had in the ambiance provided by my parents, my family, my teachers, colleagues and students. I have two sons from my marriage to Elise Cunningham who died in 1972. In 1976 I married Daisy Harper who brought with her three step-children into my life.

Honors and Distinctions:
I am a fellow of the American Physical Society and the American Association for the Advancement of Science, a member of the American Academy of Arts and Sciences and the National Academy of Sciences. I hold the Cyrus Fogg Brackett Professorship of Physics at Princeton University and since 1976 have served as chairman of the Physics Department. I received the E. O. Lawrence award in 1968. In 1967 Jim Cronin and I received the Research Corporation award for our work on CP violation and in 1976 the John Price Witherill medal of the Franklin Institute.

THE DISCOVERY OF CHARGE—CONJUGATION PARITY ASYMMETRY

Nobel lecture, 8 December, 1980

by

VAL L. FITCH

Princeton University, Department of Physics,
Princeton, New Jersey 08540

Physics as a science has made incredible progress because of the delicate interplay between theory and experiment. Astonishing predictions based on theories devised to account for known phenomena have been confirmed by experiment. Experiments probing previously unexplored areas often reveal physical effects which are completely unanticipated by theoretical conjecture. The incorporation of the new effects into a theoretical framework then follows.

This year Prof. Cronin and I are being honored for a purely experimental discovery, a discovery for which there were no precursive indications, either theoretical or experimental. It is a discovery for which after more than 16 years there is no satisfactory accounting. But showing as it does a lack of charge-conjugation parity symmetry and, correspondingly, a violation of time-reversal invariance, it touches on our understanding of nature at its deepest level.

The discovery of failure of CP symmetry was made in the system of K mesons. This observation is especially interesting because it was the study of these same particles that led to the overthrow of parity conservation, the notion that interactions and their mirror-reflected counterparts must be equal.

My own interest in K particles started in 1952—53 while I was at Columbia working with Jim Rainwater on $\mu-$mesonic atoms. At that time the strange behavior of the particles newly discovered in cosmic rays[1] was a major topic of conversation in the corridors and over coffee. By strange behavior I am referring to the copious production but slow decay. Protons bombarded by pions would result in the production of Λ^0's at 10^{13} times the rate of their decay back to pions and protons. Pais came to Columbia and talked of his ideas on associated production to explain this anomaly.[2] Gell Mann visited and discussed the scheme which he and independently, Nakano and Nishijima, had devised to account for associated production.[3, 4]

Their idea was implausible and daring in the face of available data. The scheme assigned the K mesons to two doublets, $K^+ K^0$, and the antiparticles K^- and $\bar{K}^0$. The natural assignment would have been the same as for pions, a triplet of particles K^+, K^0, K^-. Nishijima also assigned quantum numbers, subsequently called stangeness, which were conserved in the strong interaction but not in the weak. The $K^+ K^0$ were assigned +1, the $\bar{K}^0 K^-$ as well as the Λ^0, −1.

Standing alone among the particles with positive strangeness were the K^+ and K^0 mesons, and I idly thought that if the situation was ever to be understood these objects might be the key. Most often experiments in physics are long and difficult. It takes some special tweaking of interest to make the commitment to a new area of research. The original motivation is, in the end, apt to appear naive. However, I did in fact join the Princeton Cosmic Ray Group headed by George Reynolds, and spent the summer of 1954 on a mountain in Colorado learning about the ongoing experiments. During the same period the energy of the cosmotron at Brookhaven was being raised to 3 GeV. Associated production was clearly seen by Shutt and his group at Brookhaven[5], and K mesons produced in the cosmotron were identified in photographic emulsion.[6] By the end of the summer I reluctantly decided the future was not in studying cosmic rays in the mountains I loved, but with the accelerators.

The following fall, with Bob Motley, a graduate student, we began to design an apparatus to detect K mesons using purely counter techniques at the cosmotron. As this work progressed the cascading interest in the tau-theta puzzle[7] led us naturally to explore the lifetime of the K particles as a function of their decay mode. We were successful with our detectors and Motley and I published our results simultaneously with those from the Alvarez group at Berkeley which was using the bevatron as a source.[8, 9] These results showed the degeneracy in the lifetime of the tau and theta mesons. Independently the masses of tau and theta had been shown to be the same to within 1 %.[10] The situation then set the stage for the famous work of Lee and Yang[11] followed by the experiments with the striking results showing maximal parity violation in the weak interactions.[12] This remarkable story was told by Lee and Yang on this occasion in 1957.

At about this time there appeared a paper by Landau written before the results of the beta decay experiments were known.[13] Addressing the tau-theta problem, he observed that a simple rejection of parity conservation would create difficult problems in physics. However, with what he called "combined inversion", that is, space inversion and the simultaneous transformation of particle into antiparticle, the difficulties would be avoided. Indeed, this is a path that nature appeared to take. Subsequent experiments showed parity violation was compensated by a failure of charge conjugation. The weak interactions were therefore invariant under the combined operations of particle-antiparticle interchange and mirror reflection, charge conjugation-parity, CP.

One symmetry had been shown to be invalid but had been replaced by a still deeper one. This new symmetry was especially appealing because of the CPT theorem. This theorem, which is based on little more than special relatively and locality and which is the foundation of all quantum field theory, says that all interactions must be invariant under C, P, and T, time reversal, all combined. If CP is good so also is T, in complete accord with all experimental data. The subject was left in a highly satisfactory state. "Who would have dreamed in 1953 that studies of the decay properties of the K particles would lead to a new revolution in our understanding of invariance principles," wrote Sakurai in 1963.[14] But then in 1964 these same particles, in effect, dropped the other shoe.

It is difficult to give a better example of the mutually complementary roles of theory and experiment than in telling the story of the neutral K mesons. For a physicist the pleasures are special because there is scarcely a physical system which contains so many of the elements of modern physics. Two-state systems, of which this is an example, abound, but this one has special properties which give it a unique beauty. I hope that I can convey to you some of the reasons why this system has held such a fascination for us. The story begins with the isotopic spin, strangeness assignment of Gell Mann and Nishijima. The assignment of the K mesons to two doublets makes the K^0 and $\bar{K}^0$ distinct entities. But both particles decay to two π mesons. If the physicist sees π^+ and π^- mesons in his detector, which is the source, the K^0 or $\bar{K}^0$? The problem was solved through the remarkable insight of Gell Mann and Pais in their 1955 paper.[15] In the spirit of quantum mechanics it is necessary that the source of the $\pi^+ \pi^-$ mesons be some linear combination of K^0 and $\bar{K}^0$ states. They observed that a $\pi^+ \pi^-$ final state is even under charge conjugation. By even we mean that the wavefunction does not change its algebraic sign upon interchange of particle and antiparticle. This evenness condition is obviously met by the combination $K^0+\bar{K}^0$. This they called the K_1^0.[16] If this is the case, there must be another state equally probable, the $K^0-\bar{K}^0$, the K_2^0, which is odd under charge conjugation and, correspondingly, is forbidden to decay to $\pi^+ \pi^-$. But it can decay to many other states, three-body states such as $\pi^+ \pi^- \pi^0$. It was expected that the decay to the three-body states would be substantially inhibited compared to the two-body. The particle corresponding to the K_2^0 would have a longer lifetime than the K_1^0 by about 500. In addition, it was expected that the K_1^0 and K_2^0 would have somewhat different masses even though the masses of K^0 and $\bar{K}^0$ are strictly equal by the CPT theorem.

This long-lived neutral K meson, predicted by Gell-Mann and Pais, was then looked for and found by a Columbia group working at the Brookhaven cosmotron.[17] The theoretical model, based on the notion of charge conjugation invariance in the weak interactions, had been confirmed. Then suddenly parity was found to be violated in the weak interactions along with charge conjugation! This dark cloud was almost immediately removed with the observation that one had only to replace C with CP and the story of the neutral K mesons would remain the same.[13] With CP invariance the K_2^0 would continue to be absolutely forbidden to decay to two pions. The successfull description of the neutral system of K mesons has been characterized by Feynman as "one of the greatest achievements of theoretical physics."[18]

Additional features of the K^0 $\bar{K}^0$ system become evident if we write the wavefunction including the lifetime and energy terms for the case of production of a K^0 at t = 0.

$$\Psi(t) = \frac{1}{\sqrt{2}}\{|K_1^0> e^{-t/2\tau_1+i\omega_1 t}+|K_2^0> e^{-t/\tau_2+i\omega_2 t}\}$$

$$|K_1^0> = \frac{1}{\sqrt{2}}[|K^0> + |\bar{K}^0>]$$

$$|K_2^0> = \frac{1}{\sqrt{2}}[|K^0> - |K^0>]$$

It is seen that after a time, long compared to the K_1^0 lifetime and short compared to the K_2^0 lifetime, the state that was originally a pure K^0 will become a K_2^0 which in turn is an equal mixture of K^0 and $\overline{K}^0$. To give a measure of the magnitudes involved we should point out that the K_1^0 meson, in a typical experimental situation, travels an average of a few centimeters before it decays, whereas the K_2^0 travels tens of meters. At a distance greater than about one meter from the point of production of a K^0 a nearly pure K_2^0 beam will be present.

Another important characteristic of the system becomes apparent when we consider the interaction of K_2^0's with matter. The K^0's and $\overline{K}^0$'s, by virtue of their opposite strangeness, have quite different interaction cross sections. Passing a beam of K_2^0's through a block of material will result in a mixture of K^0 and $\overline{K}^0$'s which, because of differential absorption of the two components, is no longer 50–50, but instead a mixture equivalent to a new combination of K_1^0's and K_2^0's. The newly produced short-lived K_1^0's decaying to $\pi^+ \pi^-$ will appear behind the material. This phenomenon is called regeneration.[19] In the case of the absorbing material being completely transparent to K^0's and opaque to $\overline{K}^0$'s the intensity of the K_1^0's after the absorber will be 1/4 the initial intensity of the K_2^0 incident on the absorber.

In the late 1950's M. L. Good[20] observed that with a very small mass difference between the K_1^0 and K_2^0 the regeneration phenomena just discussed would result in a coherent process. By coherent we mean that the scattering process of K_2^0 to K_1^0 would not be from individual nuclei but from the whole block of scattering material! That is, the block of material would remain in its initial quantum mechanical state during the scattering process. In this case, as with ordinary light passing through glass, the regeneration material could be treated as having an index of refraction. The K_1^0's regenerated coherently would have precisely the same energy as the incident K_2^0's and an angular distribution identical to the incident beam but broadened by diffraction effects determined by the size of the regenerating material perpendicular to the beam. A characteristic wavelength for the K_2^0 mesons in a typical experiment is about 10^{-13} cm. The transverse dimensions are typically 10 cm. The corresponding diffraction pattern has a width of the order of 10^{-14} radians! In addition, the coherent addition of K_1^0 waves has been observed over distances greater than 10^{14} wave lengths. The unique feature of this coherently regenerated K_1^0 beam is that it can be distinguished from the original beam since it decays with a short lifetime to $\pi^+ \pi^-$. To my knowledge, it is the only instance where a forward coherently scattered beam can be distinguished from the original beam.

It has become evident to physics students in the audience that the K_1^0 K_2^0 story has an analogy in polarized light. The K_1^0 and K_2^0 correspond to the left and right circularly polarized light, and the K^0 and $\overline{K}^0$ states are equivalent to the x and y components of linear polarization. The passage of a K_2^0 beam through a block of condensed material is equivalent to the passage or left circular polarized light through a doubly refractive medium like calcite which has a different index refraction for the x and y components of polarization. The general picture of regeneration, coherent and incoherent, was confirmed in a definitive bubble chamber experiment.[21]

There are many associated phenomena still to be explored. For example, experiments coherently regenerating K_1^0's from the planes in crystals have yet to be done. At the particle momenta commonly available the Bragg angles are exceedingly small, and the extinction factor, the Debye-Waller factor, comes into play at correspondingly small angles, but the experiments could be done.

Unexpectedly, the K^0 $\bar{K}^0$ system provides us with important and highly precise information about the gravitational interaction. It relates to the question of strong universality; that is, whether different objects, in this case particle and antiparticle, with the same inertial mass behave the same in a gravitational field. As observed by M. L. Good,[22] if the K^0 and its antiparticle, the $\bar{K}^0$, had an opposite gravitational potential energy, the K^0 $\bar{K}^0$ system would mix so quickly that the long-lived particle would never be seen. By analyzing the system in more detail one can show that if the gravitational interaction of particle and antiparticle differ by a fraction, κ, then κ must be less than 10^{-10} if we're dealing with the gravitational field of the earth, 10^{-11} for the solar system, and 10^{-13} for the galaxy.

Voyages of discovery can be made in new uncharted waters but also in the familiar bays close to port provided one has observing apparatus that can see familiar objects with detail greater than that previously possible. In 1963 we had the opportunity to investigate the neutral K meson phenomena with resolution greater than that permitted before. The introduction of spark chambers as charged particle detectors permitted precise track position determination, but also the chambers could be selectively triggered on appropriate classes of events.

Using such new devices with our colleagues, Jim Christenson and Rene Turlay, Jim Cronin and I initiated a systematic study of (1) the regeneration phenomena, (2) what we called CP invariance, and (3) neutral currents. We were interested in the regeneration phenomena in particular because of an anomaly that had just been reported by a group studying the passage of K_2^0's through a liquid hydrogen bubble chamber.[23] Not many of our colleagues would have given us much credit for studying CP invariance, but we did anyway, and neutral currents, of long interest, were discussed by Professor Glashow on this occasion one year ago.

A plan view of the apparatus we used for these studies is shown in Figure 1. It is a two-armed spectrometer, each arm with spark chambers before and after a magnet for track delineation. Cerenkov and scintillation counters in both arms operated in coincidence provided the signals to trigger the spark chambers, which were recorded photographically. The apparatus was placed in a beam of neutral particles at the Brookhaven A. G. S. at a distance such that K_1^0's would have decayed away leaving K_2^0's. The angle between the spectrometer arms was chosen to optimize the detection of K^0 mesons decaying to two π mesons. In the regeneration studies blocks of various solid materials were placed in the neutral beam. In the studies of the free decay of $K_2^0 \rightarrow 2$ pions, the decay volume was filled with helium gas to minimize the interactions.

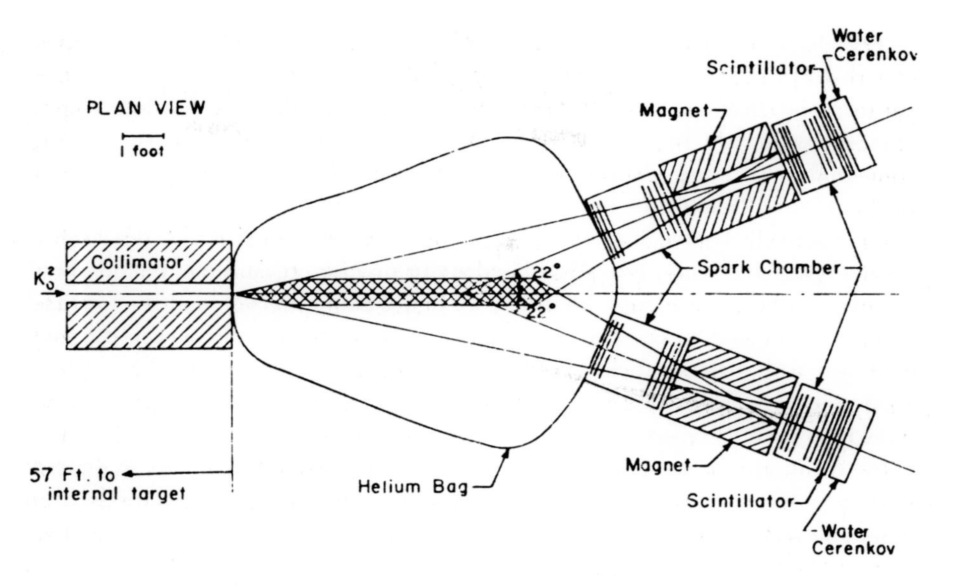

Fig. 1. Plan view of the apparatus as located at the A. G. S.

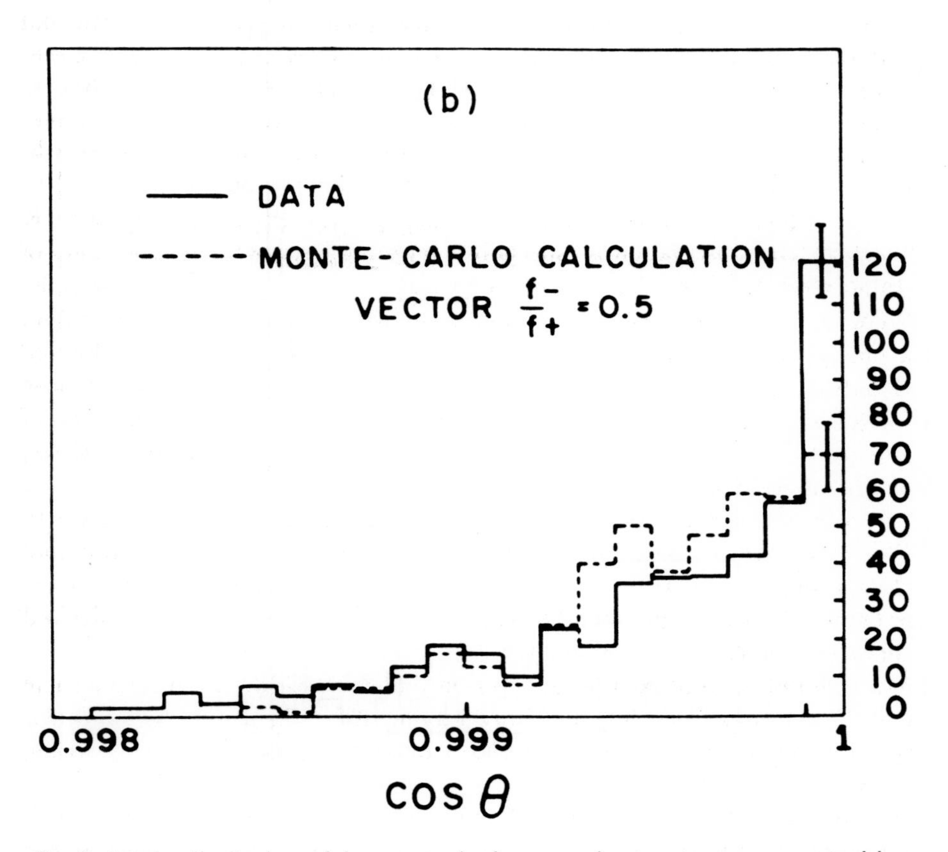

Fig. 2. Angular distributions of those events in the appropriate mass range as measured by a coarse measuring machine.

The decay to 2 pions is distinguished from the copious three-body decays in two ways. The sum of the momenta of the two detected particles must line up with the direction of the incident K_2^0's. In general this will not happen for three-body decay. In addition, the mass computed for the parent particle must match the mass of the K^0 meson. The original data are shown in Figure 2 and 3. Figure 2 shows the data after measurement of the photographic records on a relatively coarse measuring machine. The presence of the peaking of events along the beam line stimulated more precise measurements and these results are shown in Figure 3. Clearly there are about 56 events in the forward peak in the proper mass interval where the background is 11. From this data we established that the branching ratio of K_2^0 to 2 pions relative to all the charge modes decay was 2×10^{-3}. Here was the first evidence for the decay completely forbidden by CP conservation.[23] We were acutely sensitive to the importance of the result and, I must confess, did not initially believe it ourselves. We spent nearly half a year attempting to invent viable alternative explanations, but failed in every case.

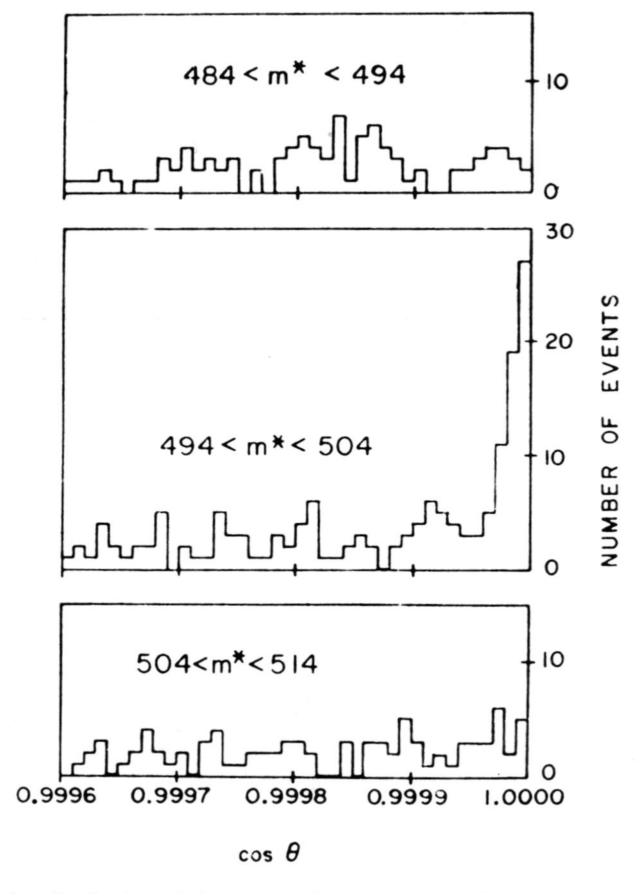

Fig. 3. Angular distribution of the events after measurement by a precise machine in three relevant mass regions.

The study of coherent regeneration was important for the CP measurement for several reasons. First, the results we found were entirely consistent with expectations; there were no anomalies. The measured coherent regeneration rates in tungsten, copper, carbon, and liquid hydrogen enabled us to show that coherent regeneration in the gaseous helium which filled the decay volume would produce a totally negligible contribution to the signal we observed. Second, the coherent regeneration of the K_1^0's, which subsequently decrayed to $\pi^+\pi^-$ mesons, provided an invaluable calibration of the apparatus.

It is appropriate now to look at the neutral K system in a somewhat more quantitative way.[24] Because of the mixing of the K^0 and $\overline{K}^0$ through the weak interaction, the time rate of change of a K^0 wave will not only depend on the K^0 amplitude, but also on the $\overline{K}^0$ amplitude, viz.,

$$-\frac{d\,K^0}{dt} = A\,K^0 + p^2\overline{K}^0$$

and

$$-\frac{d\,\overline{K}^0}{dt} = B\,\overline{K}^0 + q^2\,K^0.$$

We have let the particle symbol stand for the amplitude of the corresponding wave. With invariance under CPT, particle and antiparticle masses and lifetimes must be precisely identical. In terms of the above equations, A must be equal to B. Now, CP violation can, in fact, occur in two ways, either through terms in the set of equations above, or in the amplitudes for the decay. Subsequent experiments show that most, if not all, of the violation is in the equations above, involving the so-called mass-decay matrix. Professor Cronin will discuss the ramificatioins of the effect being present also in the decay terms. Suffice is to say here that any departure of p^2 from q^2 will result in the decay of the $K_2^0 \rightarrow 2$ pions. With CP nonconservation the short and long-lived particles are no longer the K_1^0 and K_2^0 previously defined but rather

$$K_S^0 = \frac{1}{\sqrt{p^2+q^2}}\ \{p|K^0> + \,q|\overline{K}^0>\}$$

and

$$K_L^0 = \frac{1}{\sqrt{p^2+q^2}}\ \{\,p|K^0> - \,q|\overline{K}^0>\}.$$

The fact that K_L^0 decays to 2 pions shows that the amplitude for particle to antiparticle transitions, in this case $K^0 \rightarrow \overline{K}^0$, does not quite equal the reverse, $\overline{K}^0 \rightarrow K^0$, and indeed we now know rather precisely that not only are the magnitudes somewhat different but that there is a small phase angle between the two amplitudes. See Figure 4.

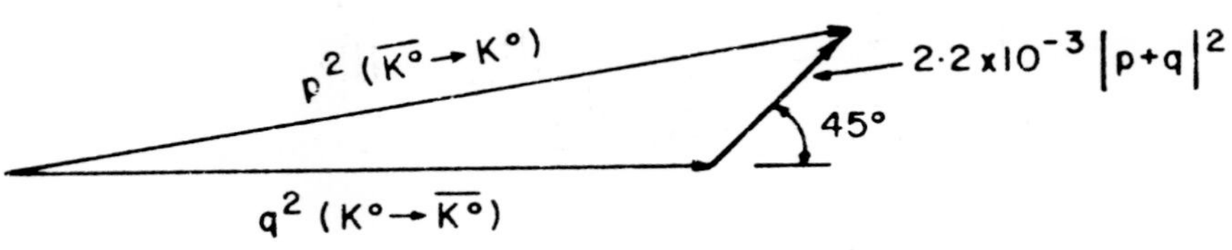

Fig. 4. Vector diagram showing schematically the difference in the amplitudes for $K^0 \rightarrow \overline{K}^0$ and $\overline{K}^0 \rightarrow K^0$.

We indicated earlier that, through the CPT theorem, a violation of CP is equivalent to a violation of time reversal invariance. As Professor Cronin will show, the CPT theorem has been shown to hold in the neutral K system independently, so in a self-contained way a violation of time reversal invariance is demonstrated.

We are all familiar with the time asymmetry associated with entropy. Entropy in a closed system increases with time. This kind of time asymmetry results from the boundary conditions. But for the first time we have in the neutral K mesons a physical system that behaves asymmetrically in time as a result of an interaction, not a boundary condition.

Since the microscopic physical laws had always been thought to be invariant under time reversal, this discovery opens up a very wide range of profound questions. Professor Cronin will go into some of these questions in greater detail. I will mention two. Can this effect be used to decrease the entropy of an isolated system? We look out from the earth and see a highly ordered universe. With entropy always increasing how can this be? Is CP violation an effect that can be used, in effect, to wind up the universe? The answers to these questions appear to be no.[(25)]

At the same time we look out from the earth and see the remains of an earlier much hotter universe. In that earlier time one expects that matter and antimatter would condense out in equal amounts and eventually annihilate to gamma radiation. However no evidence of antimatter is seen. The gauge theories described on this occasion one year ago allow for the possibility of proton (and antiproton) decay. This process, coupled with CP violation, drives the universe towards a preponderance of matter over antimatter and can account for the observed ratio of the amount of matter to radiation.[(26)]

Lewis Thomas, whose essays on science grace our literature, has written, "You measure the quality of the work by the intensity of the astonishment." After 16 years, the world of physics is still astonished by CP and T non-invariance. I suspect that the Nobel Committee was motivated by considerations similar to those of Thomas in awarding to Professor Cronin and myself this highest of honors.

REFERENCES

1. For a review ca 1953 see Rochester, G. D. and Butler, C. C., Reports Progress in Phys. *16,* 364 (1953).
2. Pais, A. Phys. Rev. *86,* 663 (1952).
3. Gell-Mann, M. Phys. Rev. *92,* 833 (1953).
4. Nakano, T. and Nishijima, K. Progr. Theoret. Phys. *10,* 581 (1953).
5. Fowler, W. B., Shutt, R. P., Thorndike, A. M. and Whittemore, W. L. Phys. Rev. *93,* 861 (1954).
6. Rochester Conference Proc. (1954).
7. Dalitz, R. H. Phil. Mag. *44,* 1068 (1953); Fabri, E. Nuovo Cimento *11,* 479 (1954).
   Among the strange particles some were seen to decay to two and some to three pions. By using the analysis of Dalitz and Fabri, it was shown, with very few examples in hand, that the parity of the three pion system was opposite to that of the two pion system. If parity is conserved in the decay interaction then there must be distinguishable parents of opposite parity, the theta that decays to two and the tau that decays to three pions. The puzzle was in the question, if the particles are distinct entities why should they have the same mass and lifetime? Now with parity violation both are recognized as K mesons, $K_{\pi 2}$ and $K_{\pi 3}$.
8. Alvarez, L. W., Crawford, F. S., Good, M. L. and Stevenson, M. L. Phys. Rev. *101,* 503 (1956); Harris, G., Orear, J. and Taylor, S. Phys. Rev. *100,* 932 (1955).
9. Fitch, V. and Motley, R. Phys. Rev. *101,* 496 (1956); Phys. Rev. *105,* 265 (1957).
10. Birge, R. W., Perkins, D. H., Peterson, J. R., Stork, D. H. and Whitehead, M. N. Nuovo Cimento *4,* 834 (1956).
11. Lee, T. D. and Yang, C. N. Phys. Rev. *104,* 254 (1956).
12. Wu, C. S., Ambler, E., Hayward, R. W., Hoppes, D. D. and Hudson, R. P. Phys. Rev. *105,* 1413 (1957); Garwin, R., Lederman, L. and Weinrich, M. Phys. Rev. *105,* 1415 (1957); Friedman, J. I. and Telegdi, V. L. Phys. Rev. *105,* 1681 (1957).
13. Landau, L. Nucl. Phys. *3,* 254 (1957).
14. Sakurai, J. J. Invariance Principles and Elementary Particles, Princeton University Press (1964), Princeton, N. J., p. 296.
15. Gell-Mann, M. and Pais, A. Phys. Rev. *97,* 1387.
16. We have changed the notation to correspond to recent custom. Gell-Mann and Pais called them Θ_1 and Θ_2.
17. Lande, K., Booth, E. T., Impeduglia, J., Lederman, L. M. and Chinowsky, W. Phys. Rev. *103,* 1901 (1956).
18. Feynman, R. P. Theory of Fundamental Processes, Benjamin, W. A. Inc. New York, p. 50.
19. Pais, A. and Piccioni, O. Phys. Rev. *100,* 1487 (1955).
20. Good, M. L. Phys. Rev. *106,* 591 (1957).
21. Good, R. H., Matsen, R. P., Muller, F., Piccioni, O., Powell, W. M., White, H. S., Fowler, W. B. and Birge, R. W. Phys. Rev. *124,* 1223 (1961).
22. Good, M. L. Phys. Rev. *121,* 311 (1961).
23. Christenson, J., Cronin, J. W., Fitch, V. L. and Turlay, R. Phys. Rev. Letters *13,* 138 (1964).
24. Lee, T. D., Cehme, K. and Yang, C. N. Phys. Rev. *106,* 340 (1957).
25. Ne'eman, Y. Erice Summer School Lectures, June 16—July 6, 1972.
26. Sakharov, A. D. JETP Letters *5,* 24 (1967). For a non-technical discussion see Wilczek, F. W. Scientific American *243,* 82 (Dec. 1980).